HADRONIC PHYSICS

HADRONIC PHYSICS

Winter School held at Folgaria, Italy
Third Course, February 15-20, 1988

Edited by:

Roberto CHERUBINI

I.N.F.N. - Laboratori Nazionali di Legnaro
Legnaro, Padova
Italy

Pietro DALPIAZ

Dipartimento di Fisica
Universita' di Ferrara
I.N.F.N. - Laboratori Nazionali di Legnaro
Legnaro, Padova
Italy

Bruno MINETTI

Dipartimento di Fisica
Politecnico di Torino
I.N.F.N. - Sezione di Torino
Torino
Italy

1989

NORTH-HOLLAND
AMSTERDAM • OXFORD • NEW YORK • TOKYO

©Elsevier Science Publishers B.V., 1989

All rights reserved. No part of this publication may be reproduced, stored in a retrieval system, or transmitted, in any form or by any means, electronic, mechanical, photocopying, recording or otherwise, without the written permission of the Publishers, North-Holland (Elsevier Science Publishers B.V.), P.O. Box 103, 1000 AC Amsterdam, The Netherlands.

Special regulations for readers in the U.S.A.: This publication has been registered with the Copyright Clearance Center Inc. (CCC), Salem, Massachusetts. Information can be obtained from the CCC about conditions under which photocopies of parts of this publication may be made in the U.S.A. All other copyright questions, including photocopying outside of the U.S.A., should be referred to the Publisher, unless otherwise specified.

No responsibility is assumed by the Publisher for any injury and/or damage to persons or property as a matter of products liability, negligence or otherwise, or from any use or operation of any methods, products, instructions or ideas contained in the material herein.

ISBN: 0 444 87486 0

Published by:

North-Holland
(Elsevier Science Publishers B.V.)
P.O. Box 103
1000 AC Amsterdam
The Netherlands

Sole distributors for the U.S.A. and Canada:

Elsevier Science Publishing Company, Inc.
655 Avenue of the Americas
New York, N.Y. 10010
U.S.A.

Library of Congress Cataloging-in-Publication Data

```
Hadronic physics : winter school held at Folgaria, Italy, third
  course, February 15-20, 1988 / edited by Roberto Cherubini, Pietro
  Dalpiaz, Bruno Minetti.
       p.   cm.
  Includes indexes.
  ISBN 0-444-87486-0 (U.S.)
   1. Hadrons--Congresses.  2. Hadron interactions--Congresses.
 I. Cherubini, Roberto.  II. Dalpiaz, P.  III. Minetti, Bruno.
 QC793.5.H32H343  1989
 539.7'216--dc20                                          89-9473
                                                             CIP
```

PRINTED IN THE NETHERLANDS

PREFACE

The Third Winter School on "Hadronic Physics" was held on Folgaria (Trento), Italy, on Febraury 15-20, 1988.
The Course, organized by the Laboratori Nazionali di Legnaro - INFN with the generous support of the Istituto Nazionale di Fisica Nucleare (I.N.F.N.), was attended by 86 participants from many institutions, belonging to different disciplines.

The Course with the same aim as of the previous two Schools was focussed on several subjects, including theoretical and experimental nuclear and elementary particle physics, and on related topics concerning detectors, data analysis, particle accelerators and technology underlining the interdisciplinary aspects.

One full day was devoted to underground physics, cosmology and related subjects. Particular emphasis was placed on the experiments planned at the Laboratori Nazionali del Gran Sasso (LNGS), Italy.

The spirit of the School was to guide the freshly graduated physicists how to carry out research and to propose present open problems.
So, the lectures were given at a tutorial level, also for very advanced subjects, as it is reflected in this book.

Thanks are due to the Folgaria Tourist Agency and in particular to Drs. G. Dorigati and Dr. M. Struffi for their assistance in the organization.
Sincere acknowledgements are due to Mrs. C. Zecchin for precious support she gave in organizing the School; to Mr. P. Schiavon for his valuable technical assistance; to Mrs. M. Stefani for her patience and efficient work in respect of manuscript preparation and typing.

R. CHERUBINI
P. DALPIAZ
B. MINETTI

TABLE OF CONTENTS

Preface   v

I: THEORETICAL BASIS OF THE HADRONIC INTERACTIONS

Ideas on Hadronic Physics at Short Distances
   G. Preparata   3

Charm Decay and a New Hadronization Scheme
   E. Predazzi   27

Hadronic Structure and Deep Inelastic Proton-Proton Collisions
   N. Paver   39

Perspectives in Superstring Theory
   M. Bregola   61

II: NUCLEAR PHYSICS

Nuclear Collective Excitations Carrying Isospin
   R. Leonardi   79

Large Shape Changes in Nuclei
   F. Barranco, E. Vigezzi and R.A. Broglia   103

Temperature Dependence of the Lifetime of Giant Dipole Resonances
   R.A. Broglia and A. Bracco   111

The Darmstadt Effect
   E. Remiddi   117

The Optical Model in the Nucleon (Antinucleon)-Nucleus Scattering
   F. Iazzi, B. Minetti and G. Puddu   123

Charm, Beauty and Nuclei
   T. Bressani and F. Iazzi   151

III: SUBNUCLEAR PHYSICS

Measurements of the $\bar{p}$ Annihilation Cross-Sections at Very Low Energies
    E. Lodi Rizzini   163

Probing the Standard Model at the SPS Collider. Experimental Results from UA1
    A. Bettini   169

Hadronic Total Cross Section Measurements at High Energies
    R. Castaldi   211

Windsurfing on the Colourless Sea
    F.L. Navarria   229

Experimental Problems in the Measurement of Structure Functions from Muon Deep-Inelastic Scattering
    U. Dosselli   257

Nuclear Effects in Deep Inelastic Muon Scattering
    C. Peroni   283

IV: UNDERGROUND AND COSMOLOGICAL PHYSICS

Elementary Physics of the Gravitational Collapse
    G. Auriemma   295

On the Standard Cosmological Model
    P. Fortini   303

SN 1987a and Extra Solar Neutrino Astrophysics
    P. Galeotti   319

Supernovae: Theory versus Observations.
(The Case of SN 1987a and the Case of Type 1b Supernovae)
    A. Tornambe'   327

A Very Large Telescope for Neutrino and Gamma Astronomy
    P. Pistilli   339

X-Ray Astronomical Research and the Italian Satellite SAX
    F. Frontera   361

GALLEX: An Experiment to Measure the Solar Neutrino Flux with Radiochemical Techniques
    R. Santonico   377

The LVD Experiment
  G. Bari, M. Basile, G. Bruni, G. Cara Romeo, A. Castelvetri, L. Cifarelli,
  A. Contin, C. Del Papa, P. Giusti, G. Iacobucci, G. Maccarrone, T. Massam,
  R. Nania, V. O'Shea, F. Palmonari, E. Perotto, G. Sartorelli, M. Willutzky,
  M. Aryal, K. De, A.M. Shapiro, M. Widgoff, J.A. Chincellato,
  C. Dobrigkeit Chincellato, A.C. Fauth, A. Turtelli, F. Rohrbach, A. Zichichi,
  L. Caputi, G. Susinno, G. Barbagli, G. Conforto, G. Landi, P. Pelfer,
  G. Anzivino, S. Bianco, R. Casaccia, F. Cindolo, M. Defelice, Y. Dong,
  M. Enorini, F.L. Fabbri, C. Jing, I. Laakso, S. Qian, A. Rindi, Z. Shi, A. Spallone,
  Y. Sun, L. Votano, A. Zallo, K. Lau, F. Lipps, B. Mayes, G.H. Mo, L. Pinsky,
  J. Pyrlik, D. Sanders, W.R. Sheldon, R. Weinstein, Y. Dai, L. Din, G. Jing, Z. Lu,
  P. Shen, Q. Zhu, D. Alyea, T. Kitamura, Y. Minorikawa, G. Di Sciascio,
  R. Scrimaglio, P. Rotelli, G.E. Kocharov, V. Vasileyev, M. Deutsch, E.S. Hafen,
  P. Haridas, B. Jeckelmann, G. Ji, H.H. Huang, C.S. Mao, A. Pitas, I.A. Pless,
  S.W. Wang, Y.R. Wu, Y.R. Yuan, C.Z. Zhao, V.S. Berezinsky, V.L. Dadykin,
  F.F. Khaichukov, E.V. Korolkova, P.V. Kortchaguin, V.B. Kortchaguin,
  V.A. Kudryavtsev, A.S. Markov, V.G. Ryassny, O.G. Ryazhskaya,
  V.P. Talochkin, V.F. Yakushev, G.T. Zatsepin, J. Moromisato, E. Saletan,
  D. Shambroom, E. von Goeler, N. Takahashi, I. Yamamoto, T. Wada, G. D'Ali,
  S. De Pasquale, B. Alpat, F. Artemi, C. Cappelletti, P. Diodati, P. Salvadori,
  A. Misaki, N. Inoue, T. Hara, C. Aglietta, G. Badino, L. Bergamasco,
  C. Castagnoli, A. Castellina, G. Cini, M. Dardo, W. Fulgione, P. Galeotti,
  P. Ghia, C. Morello, G. Navarra, L. Periale, P. Picchi, O. Saavedra,
  G.C. Trinchero, P. Vallania, S. Vernetto, F. Grianti, F. Vetrano   389

The MACRO Detector at the Gran Sasso Laboratory
  S. Petrera   409

Ultra-High Sensitivity Mass Spectometry: Applications to Rare Nuclear and
Cosmological Processes Archived in Geological Samples
  C. Tuniz and J. Klein   425

V: EXPERIMENTAL TECHNIQUES AND PARTICLE ACCELERATORS

Introduction to Accelerator Physics
  M. Pusterla   447

The Physics Case for EHF
  F. Bradamante   465

Hadron Calorimetry and the Mechanism of Compensation
  L. Piemontese   485

Semiconductor Detectors for Colliders Physics
  G. Tonelli   505

List of Participants	527
Author Index	533
Subject Index	535

I

THEORETICAL BASIS OF THE HADRONIC INTERACTIONS

IDEAS ON HADRONIC PHYSICS AT SHORT DISTANCES

Giuliano PREPARATA

Dipartimento di Fisica - Universita' di Milano and
Istituto Nazionale di Fisica Nucleare - Sezione di Milano - Italy

1. INTRODUCTION - A BRIEF HISTORY

Hadronic physics at short (Light-Cone) distances has received in the last twenty years an enormous amount of attention, both experimental and theoretical. The reason, as we shall have ample opportunity to appreciate, is that in processes involving high energies and momentum transfers very simple dynamical behaviours have been experimentally ascertained, the best known of which being Bjorken scaling in deep inelastic lepton nucleon scattering.

The puzzle that complicated physical objects, as the hadrons are known to be, could "read the free field theory book" as Fritzsch and Gell-Mann put it[1], appeared to almost everybody solved when the discovery was announced, back in 1973 [2], that QCD is asymptotically free, i.e. when the energy-momentum scale Q^2 increase the "running" coupling constant $\alpha_s(Q^2)$ vanishes. Even though the sceptics had a very relevant point in the fact that $\alpha_s(Q^2) \to 0$ implies a perturbative dynamics <u>only if the perturbative</u> (zero classical background field) <u>ground state is stable</u>, and the quark elusiveness (confinement) clearly does not speak in its favour, this warning has been basically ignored.

Thus the testing ground of the hadronic part of the Standard Model, QCD, in the last decade has been basically confined to short-distance physics, while leaving the rest to the hopeful development of the computing power needed to achieve results in Lattice Gauge Theories (LGT).

However, though unneccessarily limited, I believe that such testing of short distance dynamics can achieve full potentiality provided the predictions of Asymptotic Freedom (AF) and its hypothetical consequence Perturbative QCD(PQCD) can be contrasted with those of some rival framework, which draws its main ideas from a different approach to QCD dynamics both at short and long distances. This I will do throughout these Lectures, where I shall try to first describe the ideas underlying both PQCD and its rival theory ACD/QGD [3], that I have been developing with a number of collaborators for the better part of the last fifteen years. In these Lectures, after a brief theoretical comparison, I shall pass to confront them upon an anthology of experimental data in e^+e^- annihilation, deep inelastic scattering and high p_T physics.

But, before doing this, I believe it will be interesting to give a bird-eye, and obviously personal, view of the history of this facinating chapter of contemporary physics.

It all begins about twenty years ago (1967) when the Stanford Linear Accelerator Center (SLAC), a two-mile long linear electron accelerator believed by many important scientists of the time to be the wrong accelerator, begun to produce a large number of data on the double-differential cross-section of deep inelastic scattering of electrons (Fig. 1) as a function of Q^2, the "mass" of the virtual photon, and ν, the energy loss.

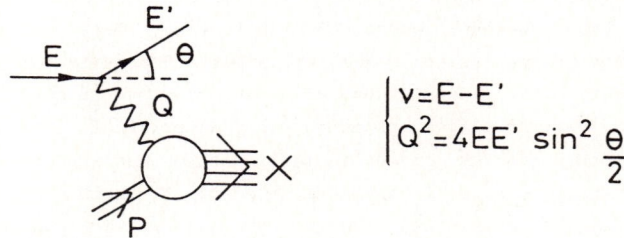

Fig. 1. The kinematics of electron proton deep inelastic scattering

In the SLAC collisions the typical Q^2 values were of the order of 10GeV^2, which allowed, according to the Heisenberg principle, to probe the target proton down to distances

$$|\Delta x| \leq \frac{1}{\sqrt{Q^2}} \simeq 10^{-14} \div 10^{-15} \text{ cm}, \tag{1.1}$$

between ten and hundred times smaller than the size of the proton.

It was J.D. Bjorken, then in the theory group of SLAC, the first to unravel the great discovery that was being made at SLAC. Following his investigations [4] on the Algebra of weak and electromagnetic currents he was lead to hypotesize[5] that in DIS very simple scaling laws would hold for the two structure functions, $W_{1,2}(Q^2,\nu)$, that fully describe the dynamics of this process. Thus according to Bjorken, in the Bjorken limit($\nu \to \infty, Q^2 \to \infty, \frac{Q^2}{2M\nu}=$ x fixed)

$$W_1(\nu,Q^2) \underset{Bj}{\to} F_1(x),$$
$$\nu W_2(\nu,Q^2) \underset{Bj}{\to} F_2(x); \tag{1.2}$$

and it was soon realized that this is what the SLAC data show even for modest values of Q^2 ($Q^2 \geq 1 \div 2 \text{GeV}^2$) and ($\nu \geq 2 \div 4 \text{GeV}$). This remarkable early achievement of asymptopia was later christened "precocity" (*).

How can we reconcile these simple laws, totally consistent with the free-field behaviour, that are at the basis of the parton model of Feynman [6], with the well known complications of hadron dynamics? As already mentioned, Gell-Mann's Current Algebra (CA) (1963) [7] was showing the way in that it postulated for the equal-time commutators of elctroweak currents in the fully interacting theory the same structure as in free field theory. The successful applications of CA that were made in the sixties strengthened the conviction of some theorists, foremost among them Bjorken, that these ideas could be further generalized. After the Bjorken hypothesis of scaling behaviours in DIS, a relevant step was made in 1969 when R. Jackiw and myself showed that a simple generalization of CA leads to the prediction:

$$R = \frac{\sigma(e^+e^- \to \text{hadrons})}{\sigma(e^+e^- \to \mu^+\mu^-)} \xrightarrow[s \to \infty]{} \Sigma_q Q_q^2 \qquad (1.3)$$

i.e. in e^+e^--annihilation at high CM energy s, the ratio between the cross-section for annihilation into hadrons is proportional to the "point-like" $\mu^+\mu^-$-annihilation by the factor $\Sigma_q Q_q^2$, the sum of the squares of the charges of the quarks involved. Eq. (1.3) is rightly regarded as central in our understanding of QCD.

In 1969 a paper appeared, by K. Wilson, which would rightly become a milestone [9]. It contained the ideas necessary to generalize CA to more involved dynamical situations. Wilson's short distance expansion for the product of local operators introduced for the first time the all-important notion of dimensionality of a local operator, and its relationship with the singularities of the product of two local operators when their distance $x_\mu \to 0$.
In 1970 R. Brandt and myself discovered the way to generalize Wilson's expansion to the product of local operators when their distance is very close to light-like, i.e. when $x^2 \to 0$ [10]. In DIS, as we shall see, the product of the current operators receives its dominant contribution from the region $x^2 \to 0$, and not $x_\mu \to 0$. Having done that, we were able to show that Bjorken scaling was a simple consequence of the fact

(*) Precocity, and precocious scaling represent one of the very few instances where my notational suggestions have been accepted by the high energy community.

that the dimensions of the operators of the Light-Cone Operator Product Expansion (LCOPE) remained very close to the free-field ones. This observation not only clarified the type of dynamics underlying the parton model, but it held very firmly that the partons ought to coincide with the fundamental fields of the hadronic theory, i.e. in QCD quarks and gluons.

Thus the LCOPE indicated very clearly the way to make good sense of the rapidly increasing number of DIS data which were being collected at the biginning of the Seventies. One had simply to find a theory for which at short (light cone) distances the effective ("running") coupling constant vanished, thus yielding local operators whose dimensions were, in the limit, canonical (free field). This discovery came, as already recalled, in 1973 and it was a matter of a very short time until all the high energy community gave up all scepticism and adhered with enthusiasm to the new PQCD ghospel. As well known, the enthusiasm that follows important revelations tends to dilute the critical attitude, that is one the most distinguishing features of scientific thought, and PQCD was soon generalized to a large number of short-distance phenomena, well beyond the areas where, through the LCOPE, one could (somewhat) avoid coming to grip with the possible effects of confinement (*). Thus, the last ten years have witnessed a very large effort to compute a great number of physical processes at short distances, most of which could only be tested at the very high energies of the supercolliders.

This course has been seriously and consistently challenged only by one theoretical approach which by a continuous metamorphosis through the Massive Quark Model (MQM 1972) [12], Quark Geometro Dynamics (QGD, 1975) [13] and Anisotropic Chromo Dynamics (ACD, 1980) [14], aimed at building a bridge between data and the fundamental QCD theory, whose relevance for hadronic physics was never questioned (**).

To conclude this brief historical survey, I think it appropriate to mention that in the last few years the link between this line of research and the fundamental QCD theory has finally come to light, and since 1986 it has been exposed to a theoretical challenge(***), that so far has not materialized.

(*) The Altarelli-Parisi [11] equation, and its simple procedures have substantially contributed to this "relaxed" attitude.
(**) See on this point my 1986 Folgaria Lectures [3].
(***) The basic ideas are described in the already mentioned 1986 Folgaria Lectures [3].

2. THE THEORETICAL BASES OF TWO RIVAL THEORIES: PQCD AND QGD/ACD.

As I have mentioned, short-distance physics has been analyzed in a systematic way only within two different -and rival- theoretical approaches: perturbative QCD and QGD/ACD. It is the purpose of this Section to briefly describe and discuss the theoretical bases upon which these approaches stand.

PQCD

Asymptotic Freedom (AF) is, as we have seen, the basis of the contention that PQCD should give a very accurate account of the short distance dynamics of hadronic matter.

According to AF, the effective coupling constant $\alpha_s(\mu)$ governing the quark-gluon dynamical correlations at light-cone distances:

$$\frac{1}{\mu^2} = (\vec{\Delta x})^2 - (\Delta t)^2$$

with increasing μ behaves as

$$\alpha_s(\mu) = \frac{g_s^2(\mu^2)}{4\pi} \to \frac{1}{\log\left(\frac{\mu^2}{\Lambda_{QCD}^2}\right)}, \qquad (2.1)$$

experimental data on scaling violation implying $\alpha_s(2\text{GeV}) \simeq .27$, a rather small number already at $Q^2 = 4$ GeV^2, as "precocity" would have let us suspect.

Fig. 2 illustrates how the correlations between two local quark operators become closer and closer to their free-field (partonic) description by increasing μ.

Is AF a true property of QCD? There is some non-perturbative evidence from LGT's that the answer is positive. Thus, one would hastily conclude, PQCD is fully justified. That this conclusion is totally unwarranted can be appreciated by the fact that in order for perturbation theory to be a good representation of QCD not only must the effective coupling constant be small, but also <u>the perturbative ground state must be stable</u>.

Here I shall only mention that I have carried out a detailed study of this very important question and concluded that the perturbative ground state is <u>essentially</u> unstable, thus preventing <u>at all scales μ</u> PQCD from being a faithful representation of QCD [15]. It is somewhat surprising that in spite of the fact that my conclusions have been left totally unchallenged, PQCD is still considered <u>the</u> correct approach to short-distance dynamics. We shall see later how also its experimental support is beginning to falter.

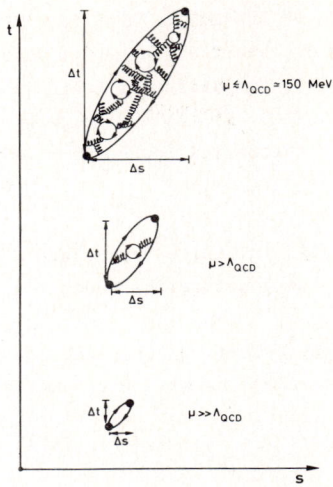

Fig. 2. As the momentum scale μ increases the corrections to free field behaviour become smaller.

QGD/ACD

As explained, for instance, in my 1986 Folgaria Lectures the following are the essential points of QGD, a phenomenological theory of confined quarks, of which ACD –an effective gauge theory of the colour SU(3) group on a peculiar magnetic ground state– is a realization at a deeper, more fundamental level:

(i) hadronic states consist of small number ($q\bar{q}$, qqq, ...) of quarks "quasi" free, grouped in colour singlets, confined in space-time domains whose extension increases linearly with the mass of such states;

(ii) there is a basic <u>perturbative structure</u> in the number of quarks that take part in a given process;

(iii) multihadron final states are the result of a (perturbative) evolution of the primitive hadronic states ($q\bar{q}$, qqq, ...).

It turns out that (i) yields very easily an approximate meson ($q\bar{q}$) spectrum in terms of the quark masses m_q, and a parameter R, carrying the dimension of a length, which controls the linear increase of the radius R_b of the hadronic "bag" with its mass ($R_b \sim R^2 M$). Neglecting spin one obtains the spectrum:

$$M^2_{nl} \simeq \frac{\pi}{R^2} (2n+l) \qquad (2.2)$$

where l is the $q\bar{q}$-angular momentum and n is a "radial" quantum number. Eq.(2.2), when plotted in a Chew-Frautschi plot (see Fig. 3) leads to a set of parallel linear Regge trajectories without odd daughters (note the factor 2n in Eq. (2.2)), a result that since a long time is known to be an important feature of the meson spectrum.

Hadronic Physics at Short Distances

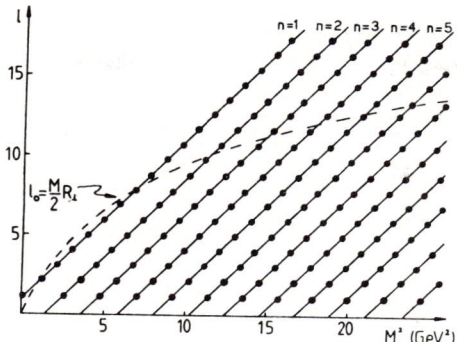

Fig. 3. The Chew-Frautschi plot of the meson spectrum (2.2).

One can also derive the asymptotic structure of the relativistic wave function (Fig. 4).

$$\Psi_{nlm}{}^m(p,k) \to F(p_1^2, p_2^2) \, Y_l^m(\Omega_{\vec{k}}) \tag{2.3}$$

where $\Omega_{\vec{k}}$ is the direction of the vector $\vec{k}$ in the meson rest-frame. The dependence on the quark invariants p_1^2, p_2^2 $F(p_1^2, p_2^2)$ — is approximately factorizable

$$F(p_1^2, p_2^2) \sim G(p_1^2) G(p_2^2), \tag{2.4}$$

and the "quark propagator" $G(p^2)$ behaves as:

$$G(p^2) \underset{p^2 \to m_q^2}{\longrightarrow} \frac{\sin R^2(p^2 - m_q^2)}{(p^2 - m_q^2)}, \tag{2.5}$$

$$G(p^2) \underset{p^2 \gg m_q^2}{\longrightarrow} \frac{1}{p^2}. \tag{2.6}$$

In accordance with confinement (2.5) has no singularity when $p^2 \to m_q^2$, and in accordance with freedom at short distances (2.6) is just the asymptotic form of the free-field propagator.

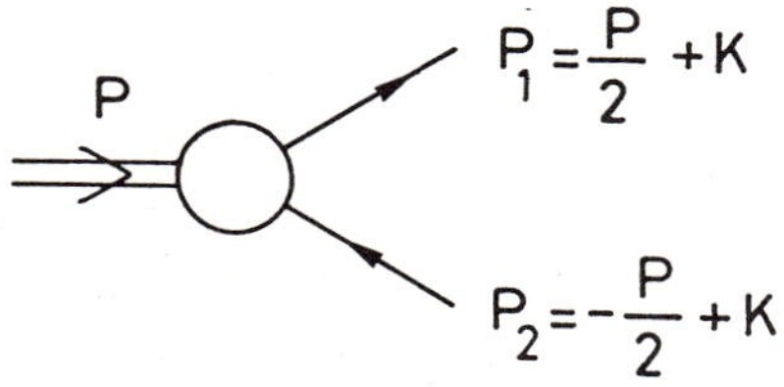

$$P_1 = \frac{P}{2} + K$$
$$P_2 = -\frac{P}{2} + K$$

Fig. 4. The relativistic meson wave function.

A notable feature of (2.2) is the mass-degeneracy of states with different angular momenta (spin) that have the same value of (2n+1). Due to this degeneracy we can construct a new set of hadronic states of a given mass M, by taking the linear combinations ($l_o = \frac{M}{2} R_1$, R_1 is a constant tranverse radius)

$$\Psi(p,k,\Omega_o) = \sum_{l=0}^{l_o} \sum_{m=-1}^{1} \Psi_{nlm}(p,k) Y_1^m(\Omega_o) \qquad (2.7)$$

whose spatial structure is reported in Fig. 5.

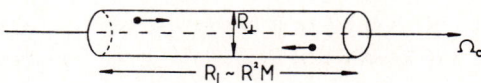

Fig. 5. The space-structure of the Fire-String (2.7).

These states have been named "Fire Strings" (FS) for their string-like character oriented in the direction Ω_o. The theoretical relevance of the "coherent states" (2.7) consists in the fact that these states have the "maximum" overlap with the highly excited states that one produces in high energy collisions. We shall see (*) that the Fire-Strings are the "parents" of hadronic jets.

It is remarkable that ACD [14] turns out to realize faithfully in the framework of an effective, more fundamental, gauge theory all the above points. Indeed:

(i) the interaction among quarks increases linearly with distance, thus implying the linear relation between radius and mass (Energy) assumed in QGD. Furthermore the interaction at short distances is very weak, thus justifying free field behaviour for quarks well inside the "hadronic bag";

(ii) that part of the interaction, unrelated with confinement, responsible for quark pair creation is rather weak and can be expanded perturbatively;

(iii) the perturbative dynamics of multihadron production turns out to be a simple consequence of (i) and (ii).

A point of very important and relevant difference with PQCD is that the interaction at short distances of light quarks is given (see Fig. 6) by the exchange of an infinite number of $q\bar{q}$-states, whose spin-properties differ

(*) See also the lectures by L. Nitti in last year's Folgaria School [16].

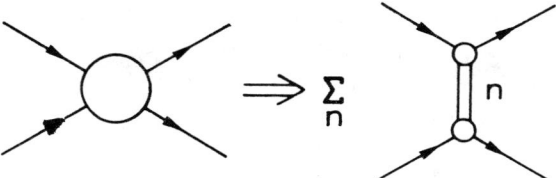

Fig. 6. The QGD amplitude for large angle qq-scattering.

drastically from one gluon-exchange, the former conserves quark-spin, while the latter helicity.

It seems useful to conclude this Section with the following synopsis, where the features of the rival theories are assessed with respect to different categories of problems.

PQCD vs QGD/ACD, A SYNOPSIS

PROBLEM	PQCD	QGD/ACD
Hadronic Spectrum	No prediction	Colour singlets built from quarks and gluons lying on Regge trajectories
low p_T physics	No prediction	Perturbative structure governed by geometrical distributions of hadronic matter
deep-inelastic interactions	perturbative amplitudes convoluted with phenomenological structure functions (parton model)	Elementary quark amplitudes built from the exchange of an infinite number of hadronic states
Jets	perturbative amplitudes superimposed to a phenomenological model largely similar to QGD (Lund)	sequential decays of Firestrings
high p_T physics	gluon-exchange; helicity conservation	quark elementary amplitudes
Theoretical bases	(Doubtful) conjecture: AF→PQCD	Effective theory of QCD on a magnetic vacuum

3. COMPARISON WITH EXPERIMENT

In this section, as anticipated in the Introduction, I shall try to carry out a comparison between the two rival approaches to QCD for a selection of

significant experimental results in
   a) e^+e^- annihilation;
   b) deep inelastic lepton nucleon scattering;
   c) large p_T scattering.

a) $\underline{e^+e^- \text{ ANNIHILATION}}$

The kinematics of e^+e^- annihilation is depicted in Fig. 7

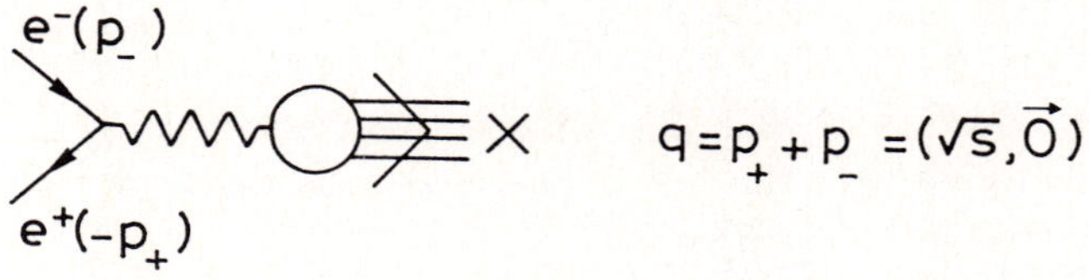

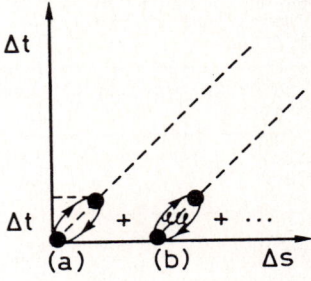

Fig. 7. The kinematics of e^+e^- annihilation to hadrons.

By standard techniques it is easy to write the total cross section as:

$$\sigma(e^+e^- \to \text{hadrons}) \propto l^\mu l^{\nu *} \int dx e^{iqx} \langle 0| (J_\mu(x) J_\nu(0)) |0\rangle \qquad (3.1)$$

with $l_\mu = \bar{v}(p_-)\gamma_\mu u(p_+)$ and $J_\mu(x)$ the local electromagnetic current operator. For large $\sqrt{s}$ the Riemann-Lebesgue theorem of Fourier transforms stipulates that the x-integration is dominated by the region where $x_\mu \to 0$ (the tip of the light cone), or better one has

$$|\Delta x_\mu| \lesssim \frac{1}{\sqrt{s}} \ ;$$

we are thus at short distance and the PQCD dictum implies that the diagrams that dominate the current correlation in (3.1) are those in Fig. 8,

Fig. 8. The space-time diagrams that dominate e^+e^--annihilation at high s:
a) the leading term, b) the one-gluon correction.

accordingly one derives:

$$R(s) = \frac{\sigma(e^+e^- \to \text{hadrons})}{\sigma(e^+e^- \to \mu^+\mu^-)} \xrightarrow{\text{large } s} 3 \left(\sum_{\substack{\text{quark}\\\text{flavours}}} \right) Q_k^2 \left(1 + \frac{\alpha_s(s)}{\pi} + \ldots \right) \quad (3.2)$$

which experimentally is rather good.

As for the structure of final states, PQCD is unable to make any well defined statement, for the $q\bar{q}$ and $q\bar{q}$-gluon final states that build up the annihilation cross section must obviously undergo a massive restructuring due to the long distance confining forces, which ought to be responsible for the emergence in the final states of a large number of hadrons, as experimentally observed.

It is however remarkable that in Nature such a restructuring is realized in a most "gentle" way. Jets get formed along the directions of the quarks and gluons, thus keeping a surprising memory of the fundamental fields. It goes without saying that PQCD cannot say anything about this fascinating experimental finding, and it is left to the future theory of long distance physics to solve the puzzle of such a "gentle confinement". We shall see in a moment that these facts can all be easily understood in QGD/ACD, to which we now turn.

Let me first recall how R is computed in ACD [17].

The diagram in Fig.7 gets now changed into one reported in Fig. 9, where we must sum over an infinite set of vector meson resonances ($J^{PC}=1^{--}$) that are

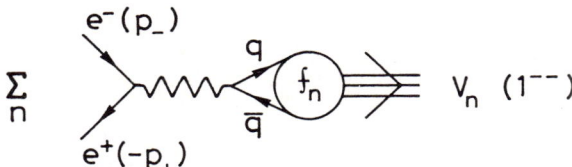

Fig. 9. The e^+e^--annihilation as a production of the vector meson states predicted by ACD.

contained in the $q\bar{q}$-spectrum. Calling M_n, Γ_n and f_n their masses, widths and coupling constants to the virtual timelike photon respectively, we can write

$$R = 12\pi \sum_n \frac{1}{\sqrt{s}} \frac{1}{M_n} f_n^2 \frac{M_n \Gamma_n}{(s-M_n^2)^2 + M_n^2 \Gamma_n^2}$$

$$\xrightarrow{s \to \infty} 3 \sum_k Q_k^2 \, (1 + \text{small kinematical corrections}) \quad (3.3)$$

a result essentially equivalent to (3.2) and in good agreement with experiment. The most remarkable aspect of (3.3) is that a typical AF behaviour is being simulated by a peculiar organization of the infinite number of confined $q\bar{q}$-states that constitute the hadronic spectrum in the channel $J^{PC}=1^{--}$. One may say that we have here a bizarre physical realization of the "1984" slogan "freedom is slavery, and slavery is freedom"[18]!

Differently from PQCD, the understanding of e^+e^- annihilation afforded by ACD/QGD does not stop with (3.3) but extends further into the detailed structure of the final states emerging from the decay of the basic 1^{--} resonances. As explained, for instance, by L. Nitti in last year's Folgaria lectures [16] one can predict the structure of hadronic final states by expanding the 1^{--} resonances in a FS-basis (the inverse of the operation which led us to (2.7)) and then letting the FS decay according to their two FS decay reported in Fig. 10.

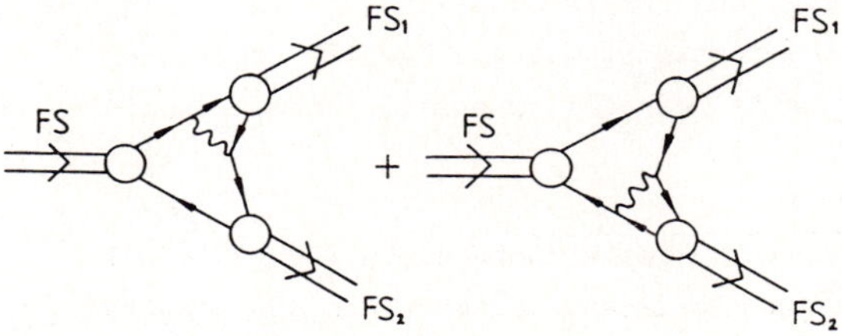

Fig. 10 The diagram describing the mesonic FS-decay.

The angular distribution of the FS-direction just coincides with the $q\bar{q}$-jet distribution in e^+e^--annihilation, i.e.

$$\frac{dN}{d\cos\theta} \sim 1 + \cos^2\theta. \qquad (3.4)$$

The FS-decay dynamics is such that hadrons are predominantly produced with low p_T with respect to the FS direction, thus yielding a nice dynamical explanation of the remarkable jet phenomena experimentally observed.

While addressing the interested readers to the more detailed literature on this object and to the experimental comparisons there reported [19], I would like to end this subsection by emphasizing that ACD/QGD affords a description of e^+e^--physics which, not only goes much beyond PQCD, but does so in an admirably realistic fashion.

b) DEEP INELASTIC SCATTERING

PQCD is able to derive the basic results of the Parton Model [6] with small scaling violations by use of the Light Cone Operator Product Expansion [10].

Let's briefly see how.

It is well known that the inclusive cross-section for lepton-nucleon deep inelastic scattering is completely determined by the tensor

$$W_{\mu\nu}(p,G) = \int d^4x \, e^{iqx} \langle p | J_\mu^+(x) J_\nu(0) | p \rangle , \qquad (3.5)$$

the 4-dim Fourier transform of the product of the electroweak current $J_\mu(x)$ with its hermitian conjugate. In the Bjorken-limit [$-q^2=Q^2 \to \infty$, $pq=M\nu \to \infty$ with $x=Q^2/2M\nu$ fixed] the important region of integration in (3.5) is all concentrated around $|x^2| \lesssim 1/Q^2 \to 0$, i.e. around the Light Cone (LC). The relevance of AF in the Bjorken limit can be appreciated by looking at Fig. 11, where one can see that when the distance between the currents $J_\mu^+(x)$ and $J_\nu(0)$

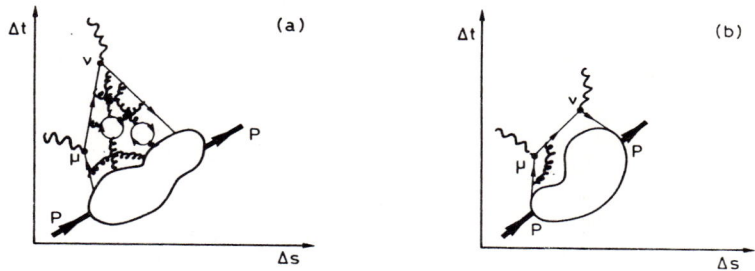

Fig. 11. The PQCD space-time diagrams for: a) away from LC, b) near the LC.

gets close to the LC their correlations are well described by the (quasi-) free quark propagator. The emergence of LCOPE can be then understood from the diagrams in Fig. 12, which in formulae read:

$$J_\mu^+(x) J_\nu(0) \underset{x^2 \to 0}{\longrightarrow} \sum_{rn} C_{\mu\nu}^{rn}(x^2) \, O_{\alpha_1 \ldots \alpha_n}^{rn}(0) \, x^{\alpha_1} \ldots x^{\alpha_n} + \ldots \qquad (3.6)$$

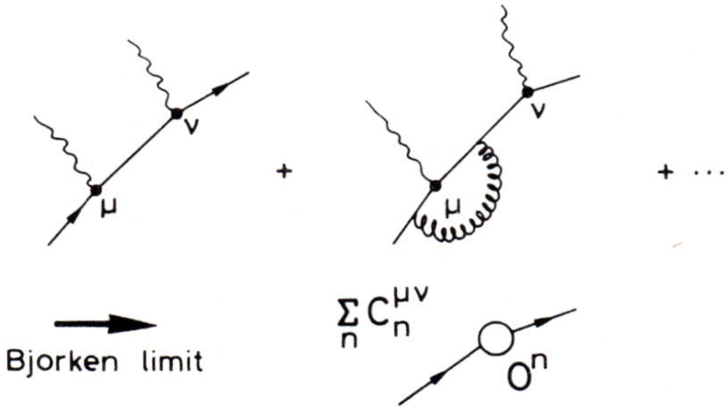

Fig. 12. At short LC distances the current-correlation function is dominated by the lowest order perturbative diagrams.

where the coefficients $C^{rn}_{\mu\nu}(x^2)$, singular when $x^2 \to 0$, have according to AF the following perturbative structure:

$$C^{rn}_{\mu\nu}(x^2) = C^{r}_{\mu\nu}(x^2)_0 + \frac{\alpha_s}{\pi} C^{rn}_{\mu\nu}(x^2)_1 + \cdots , \qquad (3.7)$$

and the operators $0^{rn}_{\alpha_1\ldots\alpha_n}(0)$ have the same structure, in terms of the unrenormalized field, as in free field theory, i.e. $[\overset{\leftrightarrow}{D}_\mu = (\overset{\leftarrow}{D}_\mu - \vec{D}_\mu)$, and D_μ is the QCD covariant derivative]

$$0^{nr}_{\alpha_1\ldots\alpha_n}(0) = \bar{q}(0) \, \Gamma_r \, \overset{\leftrightarrow}{D}_{\alpha_1} \cdots \overset{\leftrightarrow}{D}_{\alpha_n} q(0). \qquad (3.8)$$

By inserting the LCOPE in the Fourier-transform (3.5) we readily obtain relationship between moments of structure functions $F(x,Q^2)$ and matrix elements of the local operators (3.8). Typically one gets:

$$P_{\alpha_1}\cdots P_{\alpha_n} \int_0^1 dx\, x^n F(x, Q^2) = f_{nr}(\alpha_s) \langle p| 0^{nr}_{\alpha_1\ldots\alpha_n}(0) |p\rangle , \qquad (3.9)$$

where the coefficient

$$f_{nr}(\alpha_s) = f^{(0)}_{nr} + \frac{\alpha_s}{\pi} f^{(1)}_{nr} + \cdots , \qquad (3.10)$$

is given fully by its perturbative expansion.

Thus, modulo small $-O(\alpha_s/\pi)-$ corrections, one obtains the Parton Model $[f_{nr}(\alpha_s) = f^{(0)}_{nr}]$ predictions, and in particular the celebrated Bjorken scaling relations (1.2). Note that all the non-perturbative dynamics of the DIS structure functions, according to (3.9), is contained (factorized) in the matrix elements of the local operators $0^{nr}_{\alpha_1\ldots\alpha_n}(0)$.

For special values of n and r it turns out that the matrix element of $0^{nr}_{\alpha_1\ldots\alpha_n}(0)$ is known from some other place. In these cases (i.e. for electroweak currents) one easly obtains sum rules such as:

(i) The Gross-Llewellyn Smith sum rule [20]:

$$\int_0^1 dx\, F_3^{\nu N}(x, Q^2) = (1 - \frac{\alpha_s}{\pi}) \langle N| v_0^0(0) |N\rangle \qquad (3.11)$$

$$= 3 (1 - \frac{\alpha_s}{\pi}) \quad [\text{Exp. } 2.8 \pm .16] ,$$

where $F_3^{\nu N}$ is the parity violating structure-function in νN scattering and v_0^0 is the quark number charge. Experimentally, as we can see, it is rather well obeyed.

(ii) The Adler sum rule [21]:

$$\frac{1}{2} \int_0^1 \frac{dx}{x} [F_2^{\nu n} - F_2^{\nu p}] = 2 \langle p| v_0^3(0) |p\rangle = 1 \quad [\text{Exp. } 1.07 \pm .20] , \qquad (3.12)$$

note that V_o^3 is the isospin charge.

(iii) The Gottfried sum rule [22]:

$$\int_0^1 \frac{dx}{x} [F_2^{\mu p} - F_2^{\mu n}] = \frac{1}{2} (\langle p| V_o^Q(0)|p\rangle - \langle n| V_o^Q(0)|n\rangle)$$

$$= \frac{1}{6} \quad [Exp. \quad .24 \pm .11], \tag{3.13}$$

where in terms of quark fields:

$$V_o^Q = \frac{4}{9} \bar{u}\gamma_o u + \frac{1}{9} \bar{d}\gamma_o d. \tag{3.14}$$

Note that so far, all the sum rules considered have on the right hand side the matrix elements of <u>conserved vector currents</u>.

(iv) The Bjorken sum-rule [23]:

If both the lepton and the nucleon of electromagnetic DIS are polarized, the asymmetry:

$$A(x,Q^2) = \frac{d\sigma^{\rightleftarrows} - d\sigma^{\rightrightarrows}}{d\sigma^{\rightleftarrows} + d\sigma^{\rightrightarrows}} \tag{3.15}$$

allows us to define a new structure function:

$$g_1(x,Q^2) = \frac{A(x,Q^2) F_2(x,Q^2)}{2x(1+R)}, \tag{3.16}$$

where $R = \frac{\sigma_L}{\sigma_T} \underset{Bj}{\to} 0$ is the ratio between the longitudinal and transverse cross sections. The Bjorken sum rule involves g_1 for both proton and neutron and reads:

$$\int_0^1 dx [g_1^p(x,Q^2) - g_1^n(x,Q^2)] = \left(1 - \frac{\alpha_s}{\pi}\right) \langle \vec{p}| \frac{1}{6} (\bar{u}\gamma_3\gamma_5 u - \bar{d}\gamma_3\gamma_5 d) |\vec{p}\rangle$$

$$= \frac{1}{6} \left(-\frac{G_A}{G_N}\right) \left(1 - \frac{\alpha_s}{\pi}\right) = 0.191 \pm 0.002 \quad [\alpha_s = .27] \tag{3.17}$$

Note that on the right-hand side we have the matrix element on a polarized proton of the well-known axial current of β-decay. Unfortunately the Bjorken sum rule cannot be compared with experiment for, while the proton asymmetry has been recently measured [24], we still lack information on the asymmetry of the neutron.

(v) The Ellis-Jaffe sum rule [25]

If we concentrate upon g_1^p only we can write:

$$\int_0^1 dx\, g_1^p(x,Q^2) = \frac{1}{18} \langle \vec{p}| \left\{ 4 \left(1 - \frac{\alpha_s}{2\pi}\right) \bar{u}\gamma_3\gamma_5 u + \left(1 + \frac{\alpha_s}{\pi}\right) \bar{d}\gamma_3\gamma_5 d + $$

$$\left(1 + \frac{\alpha_s}{\pi}\right) \bar{s}\gamma_3\gamma_5 s \right\} |\vec{p}\rangle + O(\alpha_s^2) \quad [\text{Exp } .112 \pm .021] \quad . \quad (3.18)$$

The experimental result [24] is extremely surprising for, neglecting the strange quarks as suggested by the Quark Model, one would calculate from the right-hand side of (3.13) the much larger value $.200 \pm .001$. Actually it turns out that the contribution from strange quarks can be bound by the tiny number $.017 \; {}^{-\;.017}_{+\;.007}$ [30]. Thus if PQCD is correct one must have

$$\int_0^1 dx \; g_1^p(x,Q^2) \geq .183 \; {}^{+\;.017}_{-\;.007} \quad , \quad (3.19)$$

which, unless the experimental numbers are grossly mistaken, implies a flagrant falsification of PQCD.

In view of this it seems crucial that the experimental result (3.18) be checked with great care and that also the neutron asymmetry be measured. As they stand, for the first time things look rather desperate for PQCD. The contrary is true for ACD/QGD, as I am going to show.

Back in 1972, in the framework of MQM [12], I was able to show that all scaling phenomena were a simple consequence of the fact that at high energy the $q\bar{q}(q\bar{q})$ elementary cross section is dominated by the same diffractive singularity (the Pomeron) that controls high energy hadron-hadron scattering. In QGD the DIS mechanism are analyzed in greater detail, according to the scheme reported in Fig. 13.

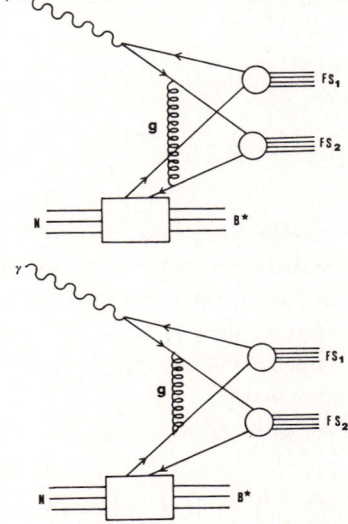

Fig. 13. The QGD diagrams contributing to DIS: a) the single FS production, b) the double FS production.

Three are the key elements of the QGD analysis of DIS [26]:

(i) The leading effect, which originates from the fact that in order to scatter tha target hadron must break its colour flux by creating a (virtual) $q\bar{q}$ in addition to a (slightly) excited hadron which carries a good memory of the target [Fig. 13 (a),(b)].

(ii) One FS-production, in which the electroweak current excites one of the quarks of the initial virtual pair to a single FS [Fig. 13(a)]. One can show [26] that this mechanism produces a scaling structure function $F_2(x)$ which dominates at large x and vanishes for $x \to 0$ [Fig. 14]. The hadronic final state, comprising the FS debris together with those of the leading hadron, has a rather low multiplicity [27].

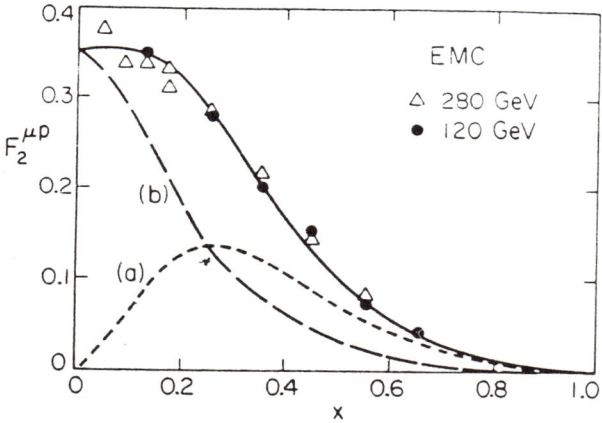

Fig. 14. The calculated structure function, compared with experiments: a) denotes the single FS-mechanism while b) describes the double FS-production.

(iii) Two FS-production, which stems from the scattering of the virtual pair produced by the electroweak current with the initial pair [Fig. 13(b)]. The $F_2(x)$ calculated from this mechanism [Fig. 14(b)] vanishes rapidly when $x \to 1$, but it tends (within logs) to a constant when $x \to 0$. The hadronic final state has a large multiplicity [28].

Figs. 15,16,17 report, for different values of x and Q^2, the QGD result for the inclusive DIS cross section. Particularly noteworthy are the results on the proton asymmetry (Fig. 17) which were obtained long before the recent measurements that have created such a sensation in the

PQCD camp. As one can see QGD has <u>predicted</u> with good accuracy the experimental measurements.

Let me conclude this subsection by mentioning that a large effort is being carried out on the analysis of the structure of the hadronic final states, with results which are very encouraging.

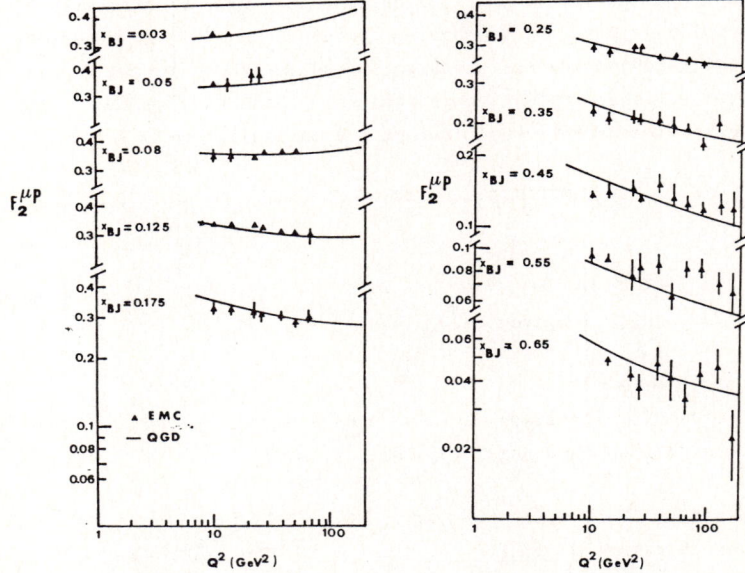

Fig. 15. The DIS scaling function calculated in Ref. 26, for different values of x and Q^2.

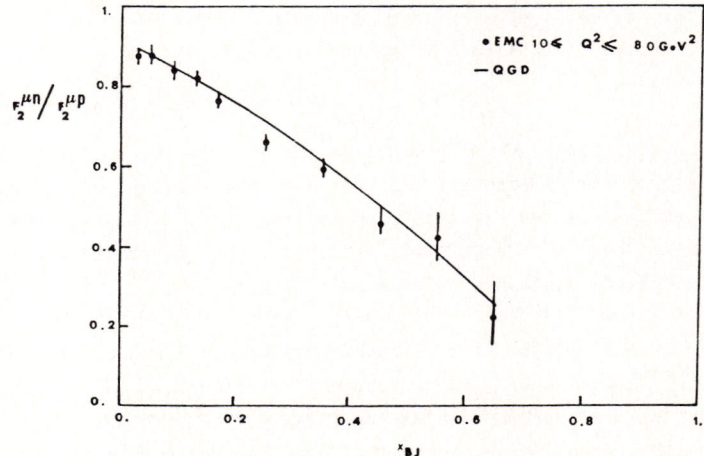

Fig. 16. The predicted ratio of the structure functions $F_2^{\mu n}(x)/F_2^{\mu p}(x)$.

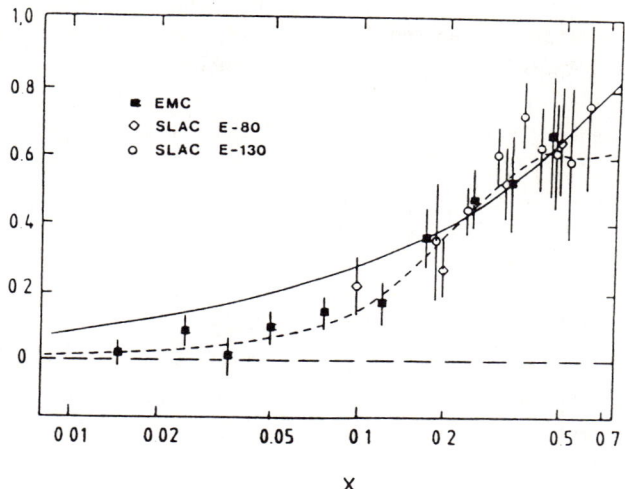

Fig. 17. The predicted asymmetry in polarized proton DIS, compared with data and a PQCD inspired model (Ref. 29).

c) HIGH p_T PHYSICS

Notwithstanding its likely decease resulting from the data just discussed, let's see first how PQCD deals with high p_T physics.

The basic PQCD picture of high p_T particle production in hadron-hadron scattering is reported in Fig. 18, where one describes the high p_T event in terms of a covolution of partonic quark and gluon distribution functions with

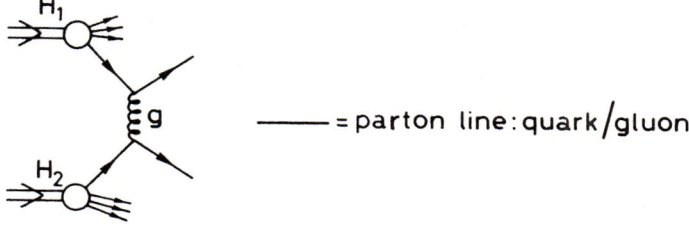

Fig. 18. The PQCD picture of high p_T physics.

the elementary large angle scattering amplitudes which, according to PQCD, are dominated by single gluon exchange. The most notable physical consequences of this picture are

(i) The parton model scaling laws (dimensional analysis) modified by the

log-factors entering in $\alpha_s(P_T^2)$;
(ii) quark-helicity conservation.

According to (i) one predicts for the jet-jet production cross-section at large p_T, a behaviour like:

$$\frac{d\sigma^{jj}}{dp_1^2} \sim \frac{1}{p_T^4} f\left(\frac{2p_T}{\sqrt{s}}\right) \text{ (modulo logs)} , \qquad (3.20)$$

where the scaling function $f(2p_T/\sqrt{s})$ depends on the various phenomenological distribution functions that one can obtain with varying ambiguities (especially for gluons), from the structure function of DIS. Both UA1 and UA2 collaborations have given good support to such predictions, within the mentioned limitations.

As for esclusive scattering, an eaxmple of a leading large angle diagram is

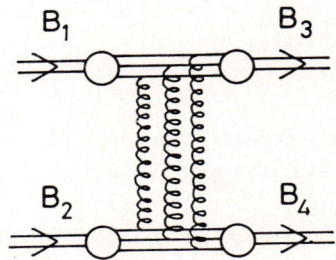

Fig. 19. A typical PQCD diagram in large angle Baryon-Baryon scattering.

reported in Fig. 19, yielding predictions for the high-t differential cross-sections of the type:

$$\frac{d\sigma}{dt} = \frac{1}{s^p} F(\theta) \text{ (modulo logs)} \qquad (3.21)$$

where $p=2n_q - 2$, n_q being the number of quarks initially present in the hadron wave-functions. Thus $p=10$ for baryon-baryon scattering ($n_q=3+3=6$) and $p=8$ for meson-baryon scattering ($n_q=3+2=5$).

These predictions fare comfortably well experimentally. As for (ii) above, it implies the absence of any polarization phenomenon at short distances. I have discussed elsewhere [29] the experimental situation of this interesting problematics, and its conflicts with the PQCD expectations. The interested reader is referred to this review and to the references therein.

As for ACD/QGD it is immune from these PQCD failures, for, as already

noticed, the fundamental qq-interaction:
a) conserves spin and not helicity;
b) for large momentum transfer t behaves like $1/t^{3/2}$ logs instead of $1/t$ (one-gluon exchange), thus obtaining pratically the same predictions [Eq. (3.21)] as PQCD for large angle scattering.

As for jet-production at high p_T it is presently being worked out quantitatively. Qualitatively one has that:

α) The inclusive cross-section behaves as:

$$\frac{d\sigma^{jj}}{dp_T^2} \sim \frac{1}{p_T^6} \log^2 p_T^2 \; f\left(\frac{2p_T}{\sqrt{s}}\right), \qquad (3.22)$$

to be compared with (3.15).

β) The behaviour of $f(2p_T/\sqrt{s})$ can be computed explicitely in the model.

Finally in a recent paper of the Bari group [31] a calculation has been carried out of the π° inclusive cross section over the full p_T span and at energies up to ISR ($\sqrt{s}=60$ GeV). As it can be seen from Fig. 20 the agreement with CERN data [32] is remarkable.

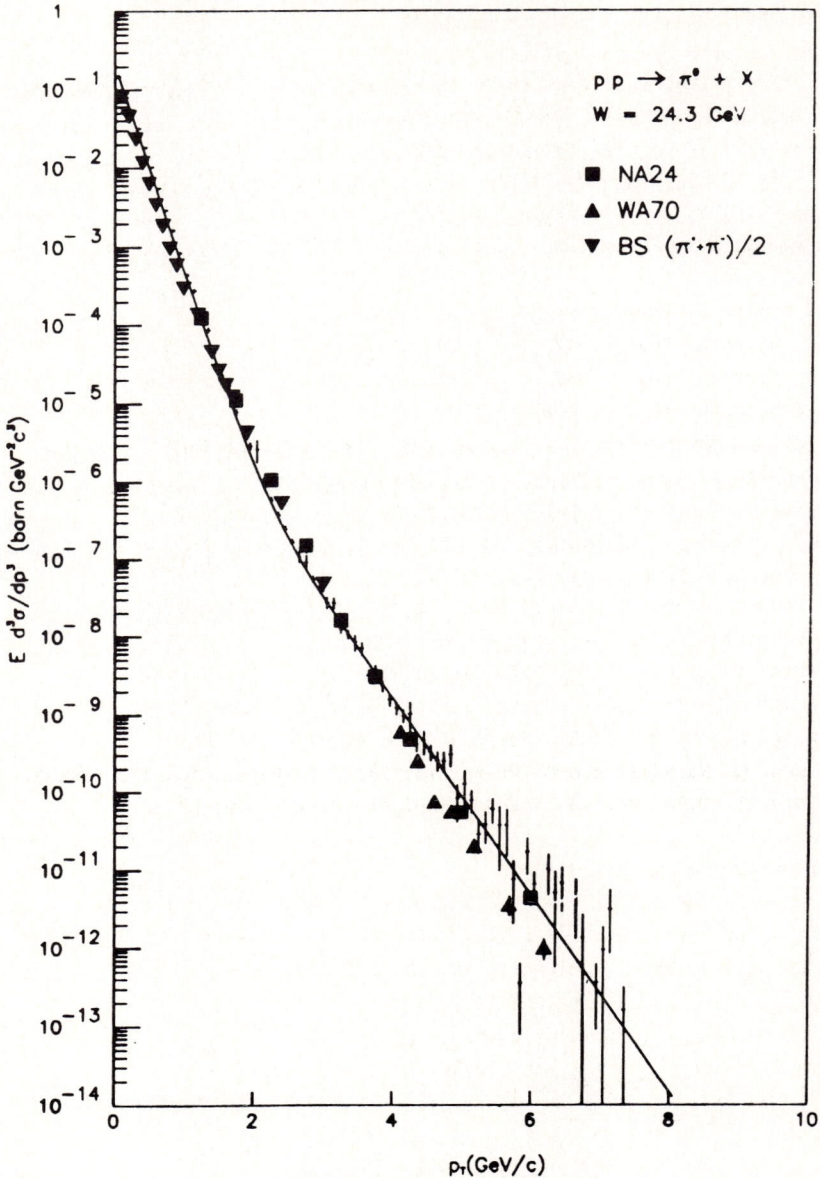

Fig. 20. The invariant cross-section $pp \to \pi^0 + X$ as a function of p_T at $\sqrt{s}=24$ GeV.

REFERENCES

1. H. Fritzsch and M. Gell-Mann, in Broken Scale invariance and the Light Cone, p.1, Gordon and Breach, New York (1971).
2. D.J. Gross and F. Wilczek, Phys. Rev. Lett. $\underline{30}$, 1342 (1973); H.D. Politzer, Phys. Rev. Lett., $\underline{30}$, 1376 (1973).
3. See for instance G. Preparata, Parton vs Hadrons; A physical theory of fragmentation, Proceedings of the 1984 SLAC Summar Institute - Stanford (1985).
4. J.D. Bjorken, Phys. Rev. $\underline{148}$, 1467 (1966).
5. J.D. Bjorken, Phys. Rev. $\underline{179}$, 1547 (1969).
6. R.P. Feynman, Photon-Hadron Interactions, Benjamin, Reading Mass. (1972).
7. M. Gell-Mann, Physics $\underline{1}$,63 (1964).
8. R. Jackiw and G. Preparata, Phys. Rev. Lett. $\underline{22}$, 975 (1969).
9. K.G. Wilson, Phys. Rev. $\underline{179}$, 1499 (1969).
10. R. Brandt and G. Preparata, Nucl. Phys. $\underline{B27}$, 541 (1971).
11. G. Altarelli and G. Parisi, Nucl. Phys. $\underline{B126}$, 298 (1977).
12. G. Preparata, Phys. Rev. $\underline{D7}$, 2973 (1973).
13. G. Preparata and N. Craige, Nucl. Phys. $\underline{B102}$, 478 (1976).
14. G. Preparata, Phys. Lett. $\underline{102B}$, 327 (1981).
15. G. Preparata, Nuovo Cim. $\underline{96A}$, 366 (1986).
16. L. Nitti, in Hadronic Physics at intermediate energy II, T. Bressani, B. Minetti and G. Pauli eds. North-Holland (1987).
17. P. Cea, G. Nardulli and G. Preparata, Zeit. f. Phys. $\underline{C16}$, 135 (1983).
18. G. Orwell, 1984, Hartcourt and Brace, New York (1949).
19. L.Angelini, L. Nitti, M. Pellicoro, G. Preparata and G. Valenti, Rivista del Nuovo Cim. $\underline{6}$, 1 (1983).
20. D.J. Gross and C.H. Llewellyn-Smith, Nucl. Phys. $\underline{B14}$, 337 (1969).
21. S.L. Adler, Phys. Rev.$\underline{143}$, 1144 (1966).
22. K. Gottfried, Phys. Rev. Lett. $\underline{18}$, 1154 (1967).
23. J.D. Bjorken, Phys. Rev. $\underline{D1}$, 1976 (1970).
24. J. Ashman et al., Phys. Lett. $\underline{B206}$, 364 (1988).
25. J. Ellis and R.L. Jaffe, Phys. Rev. $\underline{D9}$, 1444 (1974).
26. A. Giannelli, L. Nitti, G. Preparata and P. Sforza, Phys. Lett. $\underline{150B}$, 214 (1985).
27. E. Ferrari et al., Zeit. fur Phys. $\underline{C}$ (to be published).
28. R. Carlitz and J. Kaur, Phys. Rev. Lett. $\underline{38}$, 673 (1977).
29. G. Preparata, Spin Physics: a challenge to the generally accepted picture of QCD - Talk at the Adriatico Conference on Spin and Polarization Dynamics, Trieste, Jan. 1988.

30. G. Preparata and J. Soffer, Phys. Rev. Lett. 61, 1167 (1988).
31. L. Angelini, L. Nitti, M. Pellicoro, G. Preparata, High P_T physics in pp and $\bar{p}p$ scattering, to be published in Phys. Lett. B.
32. British-Scandinavian, Nucl. Phy. B180, 281 (1986) (ISR)
    NA24 Coll, Phys. Rev. D36, 16 (1987);
    WA70 Coll, CERN-EP/87/222 (1987).

CHARM DECAY AND A NEW HADRONIZATION SCHEME

Enrico PREDAZZI

Dipartimento di Fisica Teorica, Università di Torino, Italy
and
Istituto Nazionale di Fisica Nucleare, Sezione di Torino.

Abstract: The comparison between theory and data on charm decay suggests that the discrepancy one encounters within the naive decay model of charmed quarks can be explained in terms of hadronization . A simple first quantization scheme formulated in the c.m. of the decaying meson is discussed within which most of the above discrepancy is removed and various predictions are made in substantial agreement with the experimental findings. The covariants reformulation of the model through curved space geometry is given and the extension of the model to other reactions which are presently being analyzed is briefly discussed.

INTRODUCTION

The data of hadronic and semileptonic decays of pseudoscalar charmed mesons (D^+, D^o and D_s^+) have traditionally represented a most serious challenge to theory. On the one hand, in fact, given that the mass of the charmed quark is fairly large, one could invoke asymptotic freedom and treat these processes as if the produced quarks were like free particles; in this case one expects[1] the relevant diagrams to be those of Fig. 1 in which the charmed quark decays with the other quark (the light one) acting as a spectator.

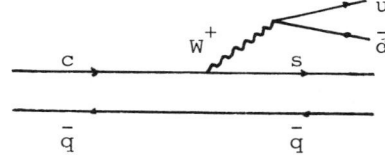

FIGURE 1

Spectator diagram for the decay of charmed pseudoscalar mesons.

On the other hand, this simple picture is bluntly contradicted by the data[2] showing that the lifetime of D^+ is almost twice that of D^0 and D_s^+. The same ratio[3] holds for $BR(D^+ \to \ell^+X)/BR(D^0 \to \ell^+X)$. Furthermore[3], the two-body decays $D^+ \to M_0^0 \pi^+$ (where $M_0 = K^0$ or $\bar{K}^0$) are systematically suppressed as compared to the predictions of the naive model whereas the opposite is true for the decays $D^0 \to M^- \pi^+$ (where $M^- = K^-, K^{-*}$) and $D^0 \to M^0 \pi^0$. In addition to all this, decay modes have experimentally found[4] such as $D^0 \to \bar{K}^0 \phi$ which are totally forbidden in the naive scheme of Ref. 1.

The moral to be learned from the above considerations is that it can not be true that the decays of charmed mesons go entirely through the spectator diagrams of Fig. 1. If, in fact, the spectator diagram of Fig. 1 is the only contribution to the decay of D^+, this does not hold neither for D^0 nor for D_s^+ for which the non-spectator diagrams of Fig. 2 could, in principle, contribute and better the agreement with the data.

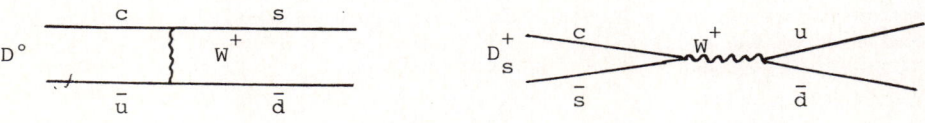

FIGURE 2

Non spectator diagrams for charmed mesons decay.

To the extent, however, that the quarks can be considered as free particles, the contribution of the diagrams of Fig. 2 are negligible being proportional to $m_s - m_d$ and to $m_u + m_d$ respectively; differently stated, they are suppressed by the conservations of the vector and axial meson currents respectively

$$\partial_\mu \bar{\psi}' \gamma^\mu \psi = i(m'-m) \bar{\psi}' \psi$$
$$\partial_\mu \bar{\psi}' \gamma^\mu \gamma_5 \psi = i(m'+m) \bar{\psi}' \gamma_5 \psi. \tag{1}$$

The crucial point, however, is that all the previous considerations hinge on a totally non trivial assumption, namely that the quarks can be considered as free. Now, the popular wisdom is that the quarks behave indeed as free particles when they are produced and their separation distance is zero but as soon as they tend to separate they experience a color force which grows with the separation distance (linearly if we are to believe lattice and potential model calculations[5] and if we ignore possible screening effects). So little do quarks behave as free particles, that they are "known" to fragment i.e. to disappear giving rise to color singlet particles (hadrons) in the final state.

Whereas, however, most everybody believes this to be the main mechanism at work within QCD, no one has really so far been able to prove that this is indeed so, i.e. that QCD has built in quark confinement and hadronization. For this reason a number of authors have proposed various schemes in which these two basic ingredients are somehow put in by hand. Alternative ingenious schemes have also been suggested[6] which, however, share with QCD the property that only hadrons appear in the final state, not quarks.

What we plan to do here is to construct an extremely simple (and thus over-simplified) scheme within which we show that the cure to the previously mentioned difficulties of the naive model[1] in analyzing the data on charm decay lies in fact in having neglected hadronization i.e. in having treated the quarks as free particles. Within these schemes, as it will turn out, the contributions of the non spectator diagrams of Fig. 2 will not at all be negligible in the charm sector and most of the previous discrepancies between theory and the data will be removed. Moreover, in spite of its simplicity (and partly because of it) we are able to apply this scheme to other reactions and to come to terms with a number of their properties which are not obvious at all. The latter development is just presently under analysis but the results are very encouraging.

The novel feature of our scheme is that we do not a priori commit ourselves to "how" hadronization occurs; we shall only insist that indeed hadronization occurs i.e. that quarks disappear from the final state. For this we shall use a first quantization approach formulated in the c.m. of the decaying charmed meson: an a priori not relativistically invariant scheme. As we shall see later on, the latter restriction will be rather easily removed by assuming the quarks to be coupled to a curved space geometry. This will make the scheme a formally covariant one but we have not been able so far to remove the limitation of inherently working in a first quantization scheme (although we have not given up hopes that we may be able to do so some time).

THE MODEL[7].

The model consists in replacing every where the free plane wave solution for the quarks produced by the weak interaction in the c.m. of the decaying meson by the confining wave function

$$\psi(\vec{x},t) = \omega(p) \, e^{ip_\mu x^\mu} \, e^{-\vec{x}^2/2x_0^2} \qquad (2)$$

where x_0 is a parameter with dimensions of a length.

Notice that at small distances $|\vec{x}| \ll x_0$, eq. (2) is, basically, a free wave which it is damped like a gaussian at large separations $|\vec{x}| \gg x_0$. As it will turn out when trying to extend the model to processes for which the available data cover a wider span in energy (such as $e^+e^- \to hX$), x_0 will turn out to be mildly energy dependent but, for the time being, we will consider it as a fixed free parameter (in fact the only free parameter) which we will determine requesting that it should reproduce the charm decay data.

The Dirac equation which eq. (2) satisfies is

$$(i\gamma^\mu \partial_\mu + i\vec{\gamma}\cdot\vec{x}/x_0^2 - m)\psi(\vec{x},t) = 0 \tag{3}$$

corresponding to the Hamiltonian

$$H = -\bar{\psi}(i\gamma\cdot\partial + i\vec{\gamma}\cdot\vec{x}/x_0^2 - m)\psi \tag{4}$$

where the non-hermitian "potential" $i\vec{\gamma}\cdot\vec{x}/x_0^2$, is irrelevant at small distances but grows with distance and disappears in the limit $x_0 \to \infty$ (which we shall "the free particle limit") when the particle is correspondingly represented by a free plane wave like a lepton.

MATHEMATICAL PROPERTIES.

If we start from the free Hamiltonian

$$H = \bar{\psi}(\vec{\gamma}\cdot\vec{p}+m)\psi \tag{5}$$

and make the replacement

$$\vec{p} \to \vec{p} = -i\vec{\partial} - i\vec{x}/x_0^2 \tag{6}$$

eq. (6) can be viewed as a sort of a "minimal coupling" with a vector potential $\vec{A} = \nabla\lambda(x)$ where

$$\lambda(x) = i\vec{x}/2x_0^2 \tag{7}$$

corresponding to a pure imaginary abelian gauge ($\vec{B}=0$) which can be gauged away putting

$$\psi(\vec{x},t) = e^{i\lambda(\vec{x})}\varphi(\vec{x},t) \tag{8}$$

The fact that $\lambda(\vec{x})$ is purely imaginary means that probability is not conserved in time. $\varphi(\vec{x},t)$ obeys the usual free wave Dirac equation.

To see more clearly in the above problem, if $\varphi(\vec{x},t)$ is a wave packet peaked at $\vec{x}=0$ at $t=0$ (so that $\langle\vec{x}\rangle = 0$) and has a group velocity $\vec{v}_0$, then $\psi(\vec{x},t)$ (eq. (8)) is the wave packet $\varphi(\vec{x},t)$ travelling with the same group velocity v_0 damped by the factor $e^{-x^2/2x_0^2}$ or, equivalently, the peak of $|\varphi(\vec{x},t)|$ is damped by a factor $\exp(-v_0^2 t^2)$. As a consequence, the total probability decreases with time. This, we take it to correspond to a quark produced free with probability 1 at $\vec{x}=0$ which disappears (i.e. hadronizes) as $\vec{x}^2 \gg x_0^2$. Using Ehrenfest theorem one can show that

$$\frac{d\,P(t)}{dt} = -\frac{d}{dt}(\vec{x}^2/x_0^2)$$

i.e., the probability decreases as the distance from the center increases.

COVARIANT REFORMULATION IN CURVED SPACE.

As already noticed, the scheme is not formally covariant being formulated in the c.m. of the decaying meson.

An elegant remedy to such a disease has been suggested[8] and shown to be a viable one. The idea in Ref. 8 was to consider eq. (3) as the small distance ($|\vec{x}| \ll x_0$) limit of a Dirac equation coupled to an anti-de Sitter curved space vacuum. In curved spacetime the global Lorentz symmetry is broken which explains the non-covariance of eq. (3). When, however, the interaction with the geometry is included to all orders, the local Lorentz invariance is restored in the complete equation.

A different, and perhaps more appropriate philosophy is taken in Ref. 9 whereby one inquires which curved line element leads exactly to eq. (3), i.e. which is the manifold in which the generalized form (3) of the Dirac equation acquires a covariant geometric interpretation. In Ref. 9 it is shown that, neglecting the mass term (which is quite correct so long as only light quarks are produced) eq. (3) can be interpreted as the general covariant Dirac equation, locally Lorentz invariant, in the space time described by the conformally flat metric

$$g_{\mu\nu} = \exp\frac{2|\vec{x}|^2}{3x_0^2}\,\eta_{\mu\nu} \tag{9}$$

where $\eta_{\mu\nu}$ denotes the usual Minkowski metric.

A further step is taken in Ref. 10 where arguments are given of why quarks but not leptons are confined so that particles with different interactions are embedded in different geometries.

In a nutshell, the argument is the following. As shown by a recent geometric realization of the commutation rules for the one-particle quantum mechanics[11], the relativistic invariant line element of four-dimensional spacetime should be replaced, in the quantum regime, by a line element defined in eight dimensional phase space[12,13] (which reduces simply to the usual four-dimensional interval in the classical limit $\hbar \to 0$). One may thus formulate, in this context, an eight-dimensional generalization of special relativity[13] and of general relativity[14] but the important point is that, in the quantum regime, different trajectories in momentum space are associated to different spacetime intervals. Given a classical background metric $g_{\mu\nu}$, the four-dimensional effective geometry in which a particle is embedded with respect to a geodesic observer depends on the acceleration of the particle relative to that observer[15].

If we have a particle of mass m and a velocity distribution inside the world type with acceleration $\ddot{x}^\mu(x) \neq 0$, the effective four dimensional metric $g'_{\mu\nu}$ is given by $g'_{\mu\nu} = g_{\mu\nu} (1 - |\ddot{x}|^2/m^2)^{-1}$ where $|\ddot{x}|^2 = |\ddot{x}^\mu \ddot{x}_\mu|$.

At the microscopic level we have thus effective quantum corrections to the classical four dimensional geometry that are particle-dependent: particles with different accelerations are seen to interact with different metric tensors.

A particle accelerated in flat space, in particular, is embedded, with respect to an inertial observer, in a generally curved manifold described by the conformally flat metric

$$g_{\mu\nu} = \eta_{\mu\nu} (1 - |\ddot{x}|^2/m^2)^{-1}. \tag{10}$$

Comparing eq. (9) with eq. (10) the geometric hadronization model under discussion is seen to correspond to a particle acceleration

$$|\ddot{x}|^2 = m^2 [1 - e^{-2|\vec{x}|^2/3x_0^2}]. \tag{11}$$

In the small distance limit $|\vec{x}|^2 \ll x_0^2$, eq. (11) gives $|\ddot{x}|^2 = (2m^2/3x_0^2) \cdot |\vec{x}|^2$ so that the effective force grows linearly with the distance.

In the above kinematic interpretation of the metric (9) one easily justifies why, inside hadrons, the spacetime should be curved only for the quarks: the leptons, which are not affected by the strong interactions, are not accelerated and their effective geometry coincides with that of an inertial observer, $g_{\mu\nu} = \eta_{\mu\nu}$; the quarks, on the contrary being accelerated by colour fields, live in a curved manifold.

PHYSICAL PROPERTIES OF THE MODEL.

The quarks produced are not in the asymptotic state; as a consequence, strict three-momentum conservation at the quark's level is relaxed; the usual Dirac δ function is replaced by a gaussian smearing. Exact momentum conservation is recovered in the "free limit" $x_0 \to \infty$.

As a consequence, in turn, new current violating terms arise so that eq. (1) is replaced by

$$\partial_\mu \bar{\psi}' \gamma^\mu \psi = i (m'-m) \bar{\psi}'\psi - 2\bar{\psi}'(\vec{\gamma}\cdot\vec{x}/x_0^2) \psi$$

$$\partial_\mu \bar{\psi}' \gamma^\mu \gamma_5 \mu = i (m'+m) \bar{\psi}'\gamma_5\psi - 2\bar{\psi}'(\vec{\gamma}\cdot\vec{x}/x_0^2)\gamma_5 \psi$$

(12)

Again, these new terms vanish as $x_0 \to \infty$. A (major) consequence of eq. (12) is that we expect a violation of the $\Delta I=1$ rule i.e. decays of the type $F^+ \to \pi^+\pi^0$ ought to be sizeable.

Ultimately, of course, the non spectator diagrams (Fig. 2) will not be zero anymore.

APPLICATIONS OF THE MODEL.

The recipe that we shall follow will be the following one: for every process, write down the matrix element as if all particles were free; next insert the above matrix element in the appropriate expression for the rate (or for the cross section) and carry on the integrations using the "new" phase space where the three-momentum conservation is replaced by the gaussian distribution which reduces to the usual Dirac δ function as $x_0 \to \infty$.

A) INCLUSIVE CHARM DECAY.

The non spectator diagrams give the following contribution to the decay rate

$$\Gamma^{NS} = 1/2 \, G^2 \, f_P^2 \, M_P^3 \, \{(2\pi \, M_P^2 \, x_0^2)^{-1} \, \text{erf}(x_0 M_P/\sqrt{2}) - [(\pi^{3/2}\sqrt{2} \, M_P \, x_0)^{-1} + \sqrt{2} \, x_0 \, M_P/6\pi^{3/2}] \, \exp(-x_0^2 M_P^2/2)\}$$

(13)

where erf(z) is the error function.

Notice that Γ^{NS} vanishes both when $x_0 \to 0$ as well as when $x_0 \to \infty$ ("free limit").

The procedure is now the following: we take the usual spectator diagram contribution to coincide with $\Gamma(D^+)$ (as mentioned earlier, D^+ has only the spectator contribution)

$$\Gamma^S(D°) = \Gamma_{tot}(D^+) \simeq 10^{12} \text{ sec}^{-1} \tag{14}$$

Taking next

$$\begin{aligned} f_{D°} &\simeq 200 \text{ MeV} \\ f_{D_s^+} &\simeq 180 \text{ MeV} \end{aligned} \tag{15}$$

and

$$x_0 \simeq 0.3 \text{ Fm} \simeq 1.5 \text{ GeV}^{-1} \tag{16}$$

we get

$$\Gamma^{NS}(D_s^+) \simeq \Gamma^{NS}(D°) \simeq 10^{12} \text{ sec}^{-1} \tag{17}$$

so that the correct lifetimes[2]

$$\begin{aligned} \Gamma(D^+) &\simeq 10^{-12} \text{ sec} \\ \tau(D°) &\simeq \tau(D_s^+) \simeq 4.8 \cdot 10^{-13} \text{ sec} \end{aligned} \tag{18}$$

are obtained.

Also the semileptonic branching rarios are in the correct ball park

$$1/2 \text{ BR}(D^+ \to \ell^+ X) = \text{BR}(D° \to^+ X) \sim 9.4\%$$

with $\text{BR}(D° \to \ell^+ X)|_{exp} \simeq 7.1 \pm 1.1 \%$.

B) EXCLUSIVE CHARM DECAY.

The prediction for the decays of pseudoscalar charmed mesons into two mesons are also quite well reproduced as shown below (in units of 10^{10} sec^{-1}).

Reaction	Exp.	Theor.
$\Gamma(D^+ \to \bar{K}^\circ \pi^+)$	4.5±1	input
$\Gamma(D^\circ \to K^- \pi^+)$	14.0±2.0	13.9
$\Gamma(D^\circ \to \bar{K}^\circ \pi^\circ)$	5.5±1.2	4.1
$\Gamma(D^\circ \to \bar{K}^\circ \eta)$	4.6±2.2	5.8
$\Gamma(D^\circ \to \pi^+ \pi^-)$	0.5±0.2	0.86
$\Gamma(D^+ \to \pi^+ \pi^\circ)$	<0.8±0.2	0.81
$\Gamma(D^\circ \to K^- K^+)$	1.7±0.4	1.1
$\Gamma(D^+ \to \bar{K}^\circ K^+)$	1.4±0.5	1.4
$\Gamma(D_s^+ \to \bar{K}^\circ K^+)$	∼ 9	7.2
$BR(D^\circ \to \bar{K}^\circ \phi)$	∼ 1%	1%
$\Gamma(D_s^+ \to \pi^+ \pi^\circ)$		8.0
$\Gamma(D_s^+ \to \eta \pi^+)$		13.6
$BR(D_s^+ \to \pi \rho)$		4-8%

C) PREDICTIONS ABOUT THE B MESON LIFETIMES.[16]

The main difference, as compared with the charmed mesons is now that one cannot neglect the charm quark mass in the various integrations which have thus to be estimated numerically.

Assuming[3] that $\Gamma_{tot}(Bu) \sim 8.3 \; 10^{11} \; sec^{-1}$ and assuming

$$f_B \sim f_D \sim 200 \; MeV$$

(which is somewhat controversial[17]) one finds $\tau(Bu)$ to be some 5% longer than $\tau(Bd)$ and $\tau(Bu)$ to be some 10-20% longer than $\tau(Bs)$.

These predictions await confirmation but the effect is anyway expected to be much less dramatic than in the charm sector.

D) THE RATIO $R = \sigma(e^+e^- \to HADRONS)/\sigma(e^+e^- \to \mu^+\mu^-)$.

Applying the model to this ratio, one finds[18], with the same value of $x_0 \sim 1.5 \; GeV^{-1}$ that R grows smoothly from zero to the experimental value in much better agreement with the data than in the naive parton model where R=2 from $\sqrt{s} = 0$ to the charm threshold.

E) INCLUSIVE $e^+e^- \to \ell X$.

This is a much more stringent test of the model since the data (cosθ dependence, Z dependence, p_T^2 dependence) are much more abundant and extend over a large energy span.

The analysis will be completed soon[19] but the preliminary results are in quite good agreement with the data provided a slight variation of x_0 with energy is allowed (i.e. a slow decrease of x_0 with increasing s). We also plan to apply this scheme to deep inelastic scattering.

CONCLUSIONS.

A general approach to discuss charm decay has been presented which promises to be useful to mimick effects due to hadronization. The agreement with the data with which the model has been compared so far is quite encouraging towards a wider application of the model to other reactions.

REFERENCES

1) N. Cabibbo and L. Maiani: Phys. Lett. 73B (1978) 418;
   D. Fakirov and B. Stech: Nucl. Phys. 133B (1978) 315.

2) J. Anjos et al. (exp. E691): Phys. Rev. Lett. 58 (1987) 311, ibid 58 (1987) 1318 and references therein.

3) Particle data group: Phys. Lett. 170B (1986) 1.

4) ARGUS Coll. H. Albrecht et al.: Phys. Lett. B158 (1985) 525.

5) See, for instance, D.B. Lichtenberg: Int. J. Mod. Phys. A (to be published) for a complete review and references on this subject.

6) G. Preparata: these proceedings and references therein.

7) J.L. Basdevant, I. Bediaga and E. Predazzi: Nuc. Phys. B294 (1987) 1054;
   J.L. Basdevant, I. Bediaga, E. Predazzi and J. Tiomno: Nucl. Phys. B294 (1987) 1071.

8) M. Gasperini: Phys. Lett. 195B (1987) 453.

9) I. Bediaga, M. Gasperini, M. Novello and E. Predazzi: "Geometric Description of Hadronization in curved Spacetime". Torino preprint DFTT 6/88 (1988).

10) E.R. Caianiello, M. Gasperini, E. Predazzi and G. Scarpetta: "On the Confining Aspects of a Conformally Flat Geometry". Torino preprint 22/88 (1988).

11) E.R. Caianiello: Il Nuovo Cimento **59B** (1981) 350.

12) E.R. Caianiello and G. Vilasi: Lett. Nuovo Cimento **30** (1981) 469;
    E.R. Caianiello: Lett. Nuovo Cimento **32** (1981) 65;
    E.R. Caianiello, S. De Filippo, G. Marmo and G. Vilasi: Lett. Nuovo Cimento **34** (1982) 112.

13) G. Scarpetta: Lett. Nuovo Cimento **41** (1984) 51.

14) H.E. Brandt in Proc. XIIIth Int. Colloquium on Group Theoretical Methods in Physics, ed. by W.W. Zachary (World Scientific, Singapore, 1984) p. 519.

15) E.R. Caianiello, M. Gasperini and G. Scarpetta: to be published.

16) I. Bediaga, A. Correa and E. Predazzi: Predictions about the decays of the B-mesons; Univ. of Torino, preprint DFTT 19/88 (1988).

17) A. Soni: Phys. Rev. Lett. **53** (1984) 1407;
    I. Bediaga et al.: Nuovo Cimento **81A** (1984) 485;
    C.A. Dominguez and N. Paver: Phys. Lett. **B197** (1987) 423.

18) I. Bediaga and E. Predazzi: Phys. Lett. **195B** (1987) 272.

19) I. Bediaga, E. Predazzi and A. Santoro: to be published.

HADRONIC STRUCTURE AND DEEP INELASTIC PROTON-PROTON COLLISIONS

Nello Paver

Dipartimento di Fisica Teorica, University of Trieste, Italy and
Istituto Nazionale di Fisica Nucleare - Sezione di Trieste, Italy.

We review some significant aspects of large p_T jet production from high energy pp collisions, and their interest as test of the QCD parton model.

1. INTRODUCTION

Inelastic events from high energy particle interactions are in general quite complex, because the multiplicities of hadrons produced in the final state are enormous. The topology of these events is however remarkably simple if one looks at kinematical configurations where large virtualities are involved (compared to M_{proton} ~ 1 GeV), and which are thus sensitive to particles interactions at short distances (compared to r_{proton} ~ 1 fm). This is the case of the so-called deeply inelastic reactions, such as the deep-inelastic lepton-nucleon scattering, the e^+e^- annihilation into hadrons at high C.M. energy, and the large p_T hadron production from high energy pp and $p\bar{p}$ collisions. In these cases the produced hadronic particles flow out of the primary interaction region in well-separated bunches, called "jets". This indicates that, while the detailed multiparticle structure of the hadronic final state is so unmanageably complicated, important aspects of the underlying dynamics can be phrased rather simply in terms of jets of particles.

Indeed conceptually the notion of jet is a really intuitive one, as it is characterized simply by an axis and an aperture angle in space (roughly speaking the cone determined by the vectorial sum and by the average spread of the particles momenta belonging to the bunch), and by the jet energy, identified to the hadronic energy flowing in the cone. In practice however the experimental identification of jets out of complicated final multihadron states is quite often not obvious, and necessitates the implementation of nontrivial reconstruction algorithms.

The formation of hadronic jets is a spectacular confirmation of the parton model description of deep inelastic phenomena, based on pointlike constituents of hadronic matter and on their elementary "hard interactions[1].

The simplest example, where jets have been first observed unambiguously, is e^+e^- annihilation into hadrons at large E_{CM}[2]. This should proceed through the elementary, short distance subprocess $e^+e^- \to q\bar{q}$, followed by q and $\bar{q}$ "hadronization" into a pair of back-to-back jets, as depicted in Fig. 1.

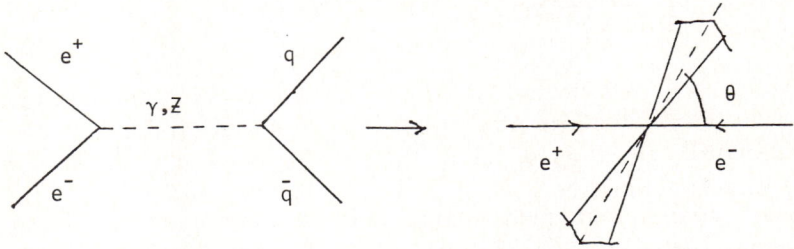

Figure 1

The hadron formation should be a "soft" process, governed by long distance and time scales, of the order of 1 fm, or equivalently by low virtualities of the order of M_{proton}. Consequently the elementary constituents "hard" processes should not be obscured by hadronization, so that the observed hadronic jets configurations should reflect the kinematics as well as the dynamics (and the quantum numbers) of the q and of the $\bar{q}$ they originate from. Thus, spin one-half quarks would imply in the present example the angular distribution $d\sigma^{jet} \sim 1 + \cos^2\theta$, with θ the angle between the jet axis and the e^+e^- beam direction. The observed e^+e^- two-jet events, a typical example being shown in Fig.2, nicely verify this prediction (Fig.3).

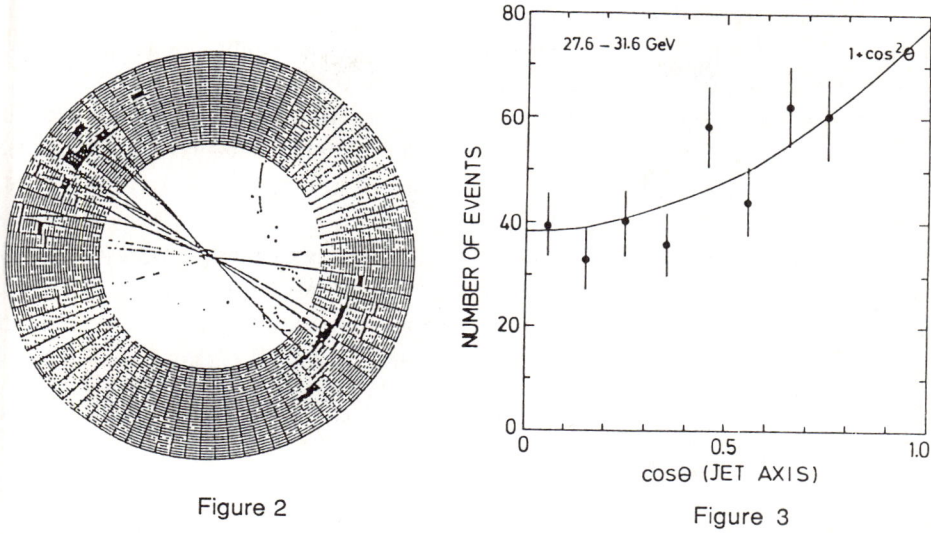

Figure 2

Figure 3

Moreover, with all intrinsic masses in the game negligible with respect to the driving "large" mass scale (represented in this case of e^+e^- by E_{CM}), cross sections should have purely dimensional behaviour, as implied by scale invariance. Thus:

$$R(E_{CM}) = \frac{\sigma(e^+e^- \to \text{hadrons})}{\sigma(e^+e^- \to \mu^+\mu^-)} \sim 3 \sum_i Q_i^2 , \qquad (1)$$

where Q_i are the quark electric charges in units of $|e|$, and the factor 3 results from quark colours. Also this kind of expectation is verified, to some approximation, by the experimental data (Fig. 4).

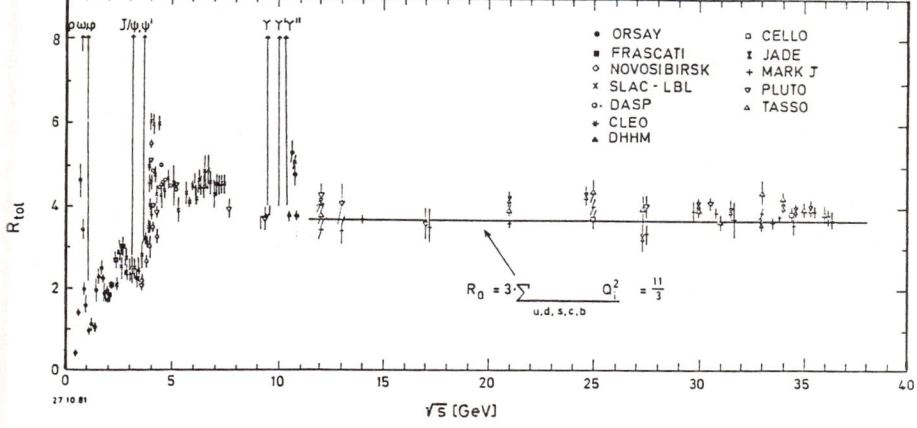

Figure 4

As well known these general parton model ideas have evolved into the notion of partons (quarks and gluons) as the elementary quantum fields describing hadron dynamics at the constituent level, and into the construction of the fundamental QCD quark and gluon interaction Lagrangian based on colour as the "gauge" symmetry of strong interactions[3]. The asymptotic freedom of QCD[4], essential in order to give the parton model a satisfactory field-theoretical basis from the phenomenological point of view, assures the logarithmic decrease of the quark-gluon and gluon-gluon coupling $\alpha_s(Q^2)$ for large virtualities Q^2. Very importantly, this allows perturbative calculations of scale breaking effects as well as of the constituents "hard" scattering cross sections. The other fundamental aspect is that "hard" quantum emission can occur, so

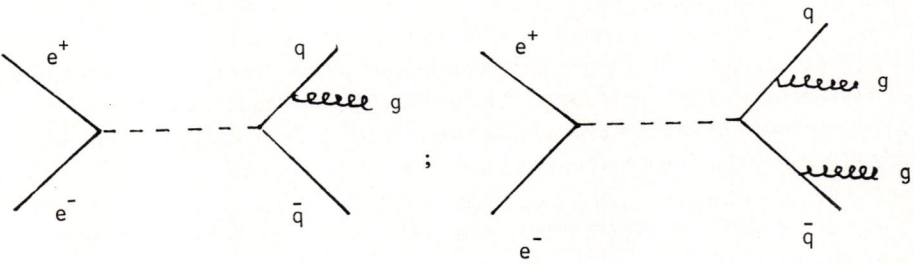

Figure 5

that this underlying theory can manifest itself in three or more jets events, as sketched in Fig.5 . Since each emission is proportional to the (small) coupling α_s, we thus expect a hyerarchy of multijet events, with rates ordered according to powers of α_s. Moreover, we introduce the notion of "gluon jets", as compared to "quark jets", with in principle different characteristics. These considerations, directly following from QCD, are well supported by the observed three-and four-jet events[2].

Turning now to large p_T hadron production from pp (and $p\bar{p}$) collisions at collider energies (which are the highest attainable), the leading QCD-parton model mechanism is represented in Fig.6 . Basically, the

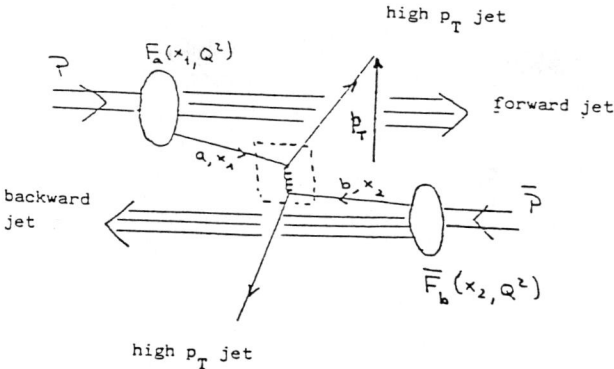

Figure 6

colliding hadrons are regarded as two "wide band" beams of collinear partons (quarks and gluons), and two partons have a hard scattering to a final partonic state, which ultimately hadronizes as a pair of high p_T jets. The remaining p and $\bar{p}$ debris continue along their initial beam line direction, and give rise to a pair of (low p_T) "spectator" jets. Such a predicted jet structure, already observed at the ISR energy ($E_{CM} \sim 60$ GeV)[5], shows up fully unambiguously and with copious rates at the Sp$\bar{p}$S collider ($E_{CM} \sim 600$ GeV)[6,7]. A spectacular example of collider two-jet event is shown in Fig. 7, representing the solid angle transverse energy flow distribution.

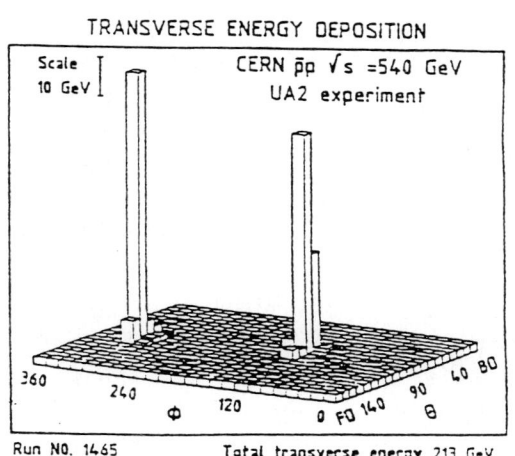

Figure 7

Multijets events have also been observed at the Collider, with relative rates down by α_s as naively predicted by QCD[7]. Particularly impressive examples of Collider three-and four-jet events are shown in Fig. 8 and in Fig.9 respectively.

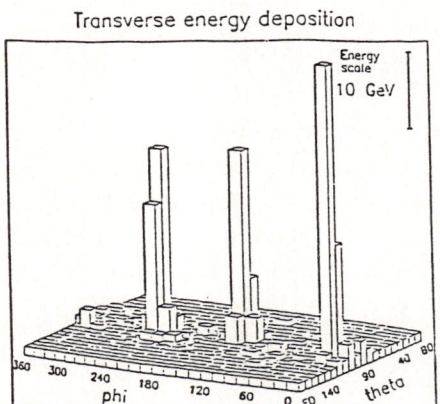
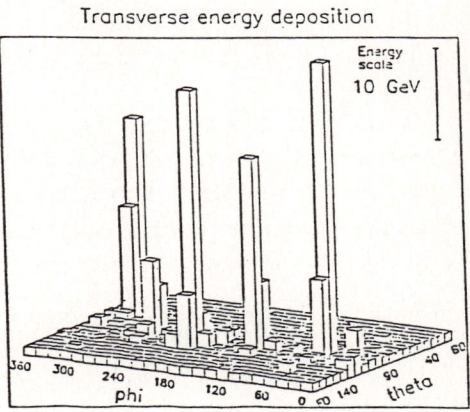

Figure 8   Figure 9

The case of large p_T hadronic jet production from high energy hadron-hadron collisions involves, in addition to the perturbative QCD parton-parton amplitudes (explicitly evidenced in Fig.6), the hadronic structure functions, of nonperturbative character, describing the way partons are distributed within the hadrons. This indicates that hadronic jets represent the ideal tool both to test the fundamental quark and gluon dynamics and to explore the structure of hadrons in terms of these elementary constituents. In practice there are some limitations, due to theoretical as well as to experimental limitations. The former are due to the partons hadronization into the observable colorless hadrons, necessary in practice to compare the theoretical prediction with the experimental data. We do not have yet a full understanding of this process from first principles, so that some non perturbative modelling has to be used[8]. Moreover, in most cases perturbative QCD calculations are limited to just the lowest orders in α_s. There are finally the experimental jet reconstruction ambiguities already mentioned above.

Indeed in this respect e^+e^- and $p\bar{p}$ colliders seem to present advantages and disadvantages which are somehow complementary. The advantage of e^+e^- (Fig.1) is that the initial particles are truly elementary, and do not "fragment" into hadrons, so that the QCD strong interactions only operate in the final state. This allows better identification of jets and of jets energies and thus easier reconstruction of the elementary parton subprocesses (all occurring at an initial energy identical to that of the e^+e^- beams). Also, complete higher order QCD calculations are available to some extent. The disadvantage in this case is that only limited E_{CM} can be explored, because the cross section decreases and the jet rate becomes too small.

Conversely, the pp (and pp) are composite objects (Fig.6), and strong interactions of the constituents (e.g. gluon radiation) operate both in the initial and in the final state. Therefore a fraction of the observed final hadronic transverse energy originates from the initial partons, and actually even the "non-partonic" spectator jets can in principle contribute some amount. Since these effects are not fully under control, the identification of high p_T QCD jets is more ambiguous, and consequently the elementary "hard" partonic subprocesses can be unfolded with larger uncertainty. In fact they can be assessed only on a statistical basis, through the hadronic structure functions, which are not uniquely determined. Also, complete higher order (in α_s) calculations are not yet available for all QCD subprocesses. The great advantage of pp (and p$\bar{\text{p}}$), on the other hand, is that cross sections are large, leading to a great number of jet events, and that many GeV/ parton are available, which allows the exploration of larger p_T.

Although keeping the above limitations in mind, we should say that the amount of information on hadronic jets, collected so far both at the e^+e^- and at the pp colliding rings, is enormous, so that it has been possible to test the physics underlying the QCD-parton model in detail, and in a great number of situations.

An exhaustive presentation of this subject being impossible, we will limit in the sequel to briefly review some significant aspects of high p_T jet physics at the p$\bar{\text{p}}$ collider. This represents the application of the parton model in the toughest situation of both projectile and target composites, and of all QCD parton-parton processes being effective. Among other things, it has allowed to probe the proton structure down to the shortest distance (the presently attained resolution being of the order of 10^{-16} cm, corresponding to measured jets p_T as high as 150 GeV).

2. JET CROSS SECTIONS AND ANGULAR DISTRIBUTIONS

In the approximation of neglecting partons masses and "intrinsic" transverse momenta, Fig.6 results into the following convolution formula for the inclusive jet yield[9]

$$E\frac{d\sigma^{jet}}{d^3p} = \sum_{a,b} \int dx_1 dx_2 F_a(x_1,Q^2) \bar{F}_b(x_2,Q^2) \frac{\hat{s}}{\pi} \delta(\hat{s}+\hat{t}+\hat{u}) \frac{d\hat{\sigma}_{ab}}{d\hat{t}} , \qquad (2)$$

where the summation is over all independent parton-parton subprocesses. In eq. (2) $d\hat{\sigma}_{ab}$ are the elementary, short distance, parton-parton 2→2 cross sections (with Mandelstam variables $\hat{s}$, $\hat{t}$ and $\hat{u}$), which have to be computed using perturbative QCD[10]. It the leading order $(\alpha_s(Q^2)^2)$ the most important subprocesses are gg→gg (which largely dominates), gq→gq and qq→qq (Fig.10). The variable Q denotes a characteristic "large" mass scale (or virtuality) involved in the process. of order $Q \sim p_T$ much larger than M_{proton} , such that the proton structure is probed to a short distance of order 1/Q.

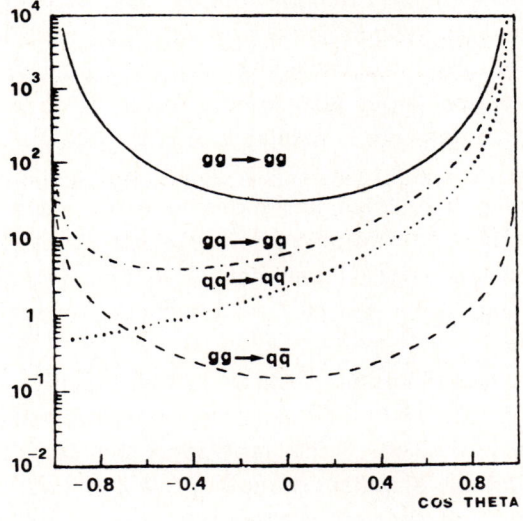

Figure 10

In eq.(2) the $F(x,Q^2)$ are the partons longitudinal fractional momentum distributions in the p and in the $\bar{p}$ (0<x<1) , and characterize the incoming quark and gluon fluxes for the various elementary subprocesses. They are "universal", in the sense that they should only characterize the colliding hadrons structure, but should not depend on the particular inelastic hadronic process. Being genuinely nonperturbative objects, they cannot be computed at the moment directly from L_{QCD}, but have to be measured independently in other reactions. The dependence of the F's on the variable Q^2, in addition to x, represents the scale breaking effects briefly alluded to in the Introduction. They are governed by perturbative QCD, hence are calculable, through the evolution equation[11]

$$\frac{d F(x,Q^2)}{d \ln Q^2} = \frac{a_s(Q^2)}{2\pi} \int_x^1 \frac{dz}{z} K(z) \otimes F(\frac{x}{z}, Q^2), \tag{3}$$

where K is a known kernel.

Thus, in the application of eq.(2) the input F's are those measured at some scale Q_0 (usually of the order of a few GeV) from deeply inelastic lepton-nucleon scattering, evolved to the larger Q ~ p_T relevant to high p_T jet physics by numerically solving eq. (3). In this way different possible parametrizations of the quarks and gluon densities in the proton have been derived[12], reflecting different sets of data. In particular the gluon distribution is somewhat uncertain, because it can be determined only indirectly via eq.(3) (to leading order leptons do not interact with the gluon component of the proton). The various solutions have however a very important common trend, namely the partons distributions sharply increase for decreasing x, and moreover at sufficiently small x gluons largely dominate this large parton flux. An indicative example is shown in Fig.11, borrowed from Ref.13, which shows the parton distributions at Q ≅ 80GeV.

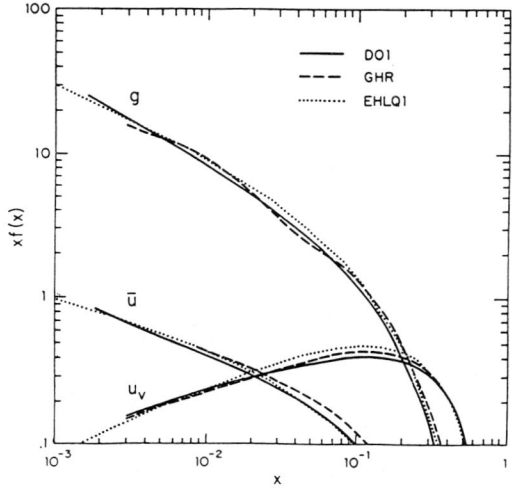

Figure 11

This has a consequence of really fundamental importance. In fact it can be easily seen, by simple kinematics, that the fractional momenta of initial partons relevant to jets of a given p_T are of order $x_{1,2} \sim 2p_T/E_{CM}$, with E_{CM} the Collider total center of mass energy. Consequently one expects the jets cross section to sharply increase for increasing E_{CM}, as the consequence of parton distributions being probed at smaller and smaller values of x. Moreover by measuring at decreasing p_T with fixed E_{CM} (so that the x-depending parton flux composition changes from "quark" to "gluon" dominated), one reaches the region of gluon-gluon dynamics and of "gluon" jets.

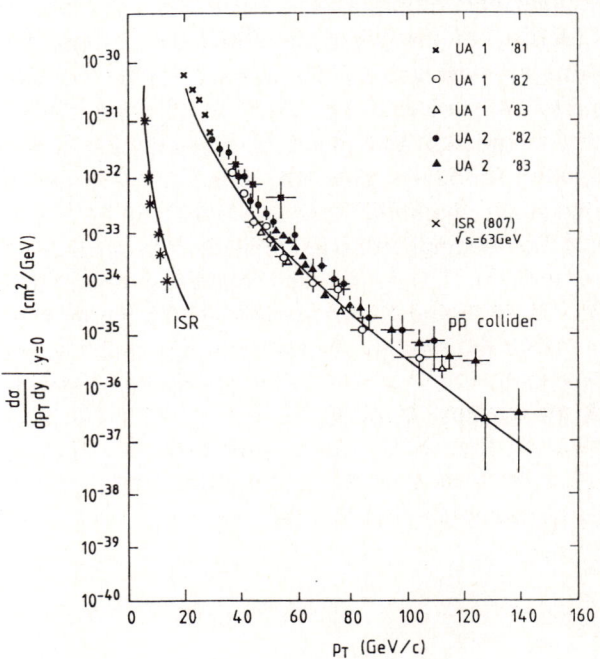

Figure 12

In Fig.12 we show the inclusive jet cross section as a function of the p_T. Actually this figure is rather old, and the most recent results are reported in Ref.7, where also one can appreciate the extensive range in p_T explored (from 5-10 GeV to 150 GeV), and the quality of the data. Fig. 12 has however the virtue of showing both the ISR and the Collider data, and thus of explicitly displaying the sharp increase of the jet cross section with the Collider energy, as expected from the considerations above.

As one can see, the QCD prediction of eq.(2), represented by the solid lines, is in good agreement with the experimental results. In this regard, if one wanted to be really conservative, one should keep in mind that eq.(2) has an intrinsic theoretical uncertainty, due in part to the inaccurate knowledge of the gluon distribution and in part to the fact that parton subprocesses are only known to the leading order QCD approximation. At this stage of approximation the definition of the scale Q in $\alpha_s(Q^2)$ and in $F(x,Q^2)$ is rather ambiguous: while $Q=p_T$ is certainly plausible, other choices are a priori equally possible, and the corresponding numerical estimates are different. Such a scale ambiguity would be significantly removed by the complete next order calculation for all parton-parton processes, which will be accomplished soon[14,15]. These higher order corrections are usually lumped into a so - called K-factor as $\sigma_{corr} = K\sigma_{lead.ord.}$. Preliminary indications are that the K-factor might be large in some cases, of the order of 1.5 or 2. All in all, the normalization of eq.(2) is thus uncertain by a factor between one and two, which anyway is within the uncertaintly of the present experimental data.

On the other hand, the agreement between the experimental data and QCD, as shown in Fig.12, is over many orders of magnitude of variation of the cross section. This is true for the dependence on p_T as well as on E_{CM}. As there are in practice no free parameters to be fixed, such an agreement is not the result of any fit procedure. We therefore conclude that this is certainly a significant success of QCD.

A stringent test of the QCD matrix elements is offered by the jets angular distributions. To a very good approximation the (dominant) gg→gg, gq→gq and qq→qq subprocesses have the "universal", Rutherford like angular dependence[16]

$$\frac{d\sigma}{d\cos\theta} \sim \frac{\alpha_s^2}{\hat{s}} \frac{1}{(1-\cos\theta)^2} , \qquad (4)$$

where θ is the scattering angle in the parton center of mass frame (i.e. the angle between jets and p$\bar{p}$ beam direction in the two-jet center of mass frame). Such a dependence on θ reflects the spin one, gluon exchange, QCD elementary dynamics. The three processes above thus differ by only a colour factor, so that we can define an "effective" structure function:

$$F(x,Q^2) = g(x,Q^2) + \frac{4}{9}[q(x,Q^2) + \bar{q}(x,Q^2)] , \qquad (5)$$

in terms of which the two-jet cross section has the factorized form:

Finally, one can anticipate distinctive multijet topologies and angular dependences, reflecting the "hard" gluon-bremsstrahlung mechanism, characteristic of QCD. Dalitz plots and angular distributions of observed multijet samples confirm these general expectations, and thus represent ever more detailed tests of the underlying dynamics. For a more extensive account we refer to Ref. 7.

3. RISING JET CROSS SECTION

Recently jet searches have been extended down to transverse energies as low as $E_T > E_T^{min} = 5 - 10$ GeV, and up to E_{CM} as high as $E_{CM} \sim 1$ TeV. The interesting result[21] is that jet events represent a sizable fraction of the totality of inelastic events, fastly rising with E_{CM} to as much as 20% at $E_{CM} = 900$ GeV. Fig. 15 shows the jet cross section, integrated over

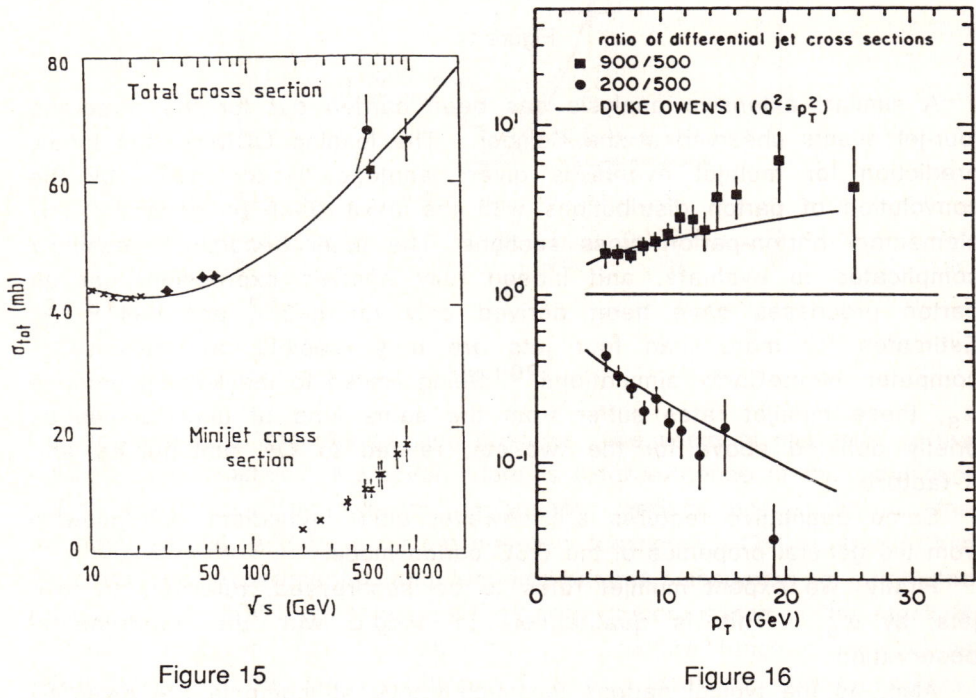

Figure 15 Figure 16

$E_T > E_T^{min}$, and also the $p\bar{p}$ total cross section for comparison. Such an E_T-cut, although considerably smaller than the typical values of the high-p_T jets discusses above, is still much greater than M_{proton}, so that perturba-

As one can see, the QCD prediction of eq.(2), represented by the solid lines, is in good agreement with the experimental results. In this regard, if one wanted to be really conservative, one should keep in mind that eq.(2) has an intrinsic theoretical uncertainty, due in part to the inaccurate knowledge of the gluon distribution and in part to the fact that parton subprocesses are only known to the leading order QCD approximation. At this stage of approximation the definition of the scale Q in $\alpha_s(Q^2)$ and in $F(x,Q^2)$ is rather ambiguous: while $Q=p_T$ is certainly plausible, other choices are a priori equally possible, and the corresponding numerical estimates are different. Such a scale ambiguity would be significantly removed by the complete next order calculation for all parton-parton processes, which will be accomplished soon[14,15]. These higher order corrections are usually lumped into a so-called K-factor as $\sigma_{corr} = K\sigma_{lead.ord.}$. Preliminary indications are that the K-factor might be large in some cases, of the order of 1.5 or 2. All in all, the normalization of eq.(2) is thus uncertain by a factor between one and two, which anyway is within the uncertaintly of the present experimental data.

On the other hand, the agreement between the experimental data and QCD, as shown in Fig.12, is over many orders of magnitude of variation of the cross section. This is true for the dependence on p_T as well as on E_{CM}. As there are in practice no free parameters to be fixed, such an agreement is not the result of any fit procedure. We therefore conclude that this is certainly a significant success of QCD.

A stringent test of the QCD matrix elements is offered by the jets angular distributions. To a very good approximation the (dominant) gg→gg, gq→gq and qq→qq subprocesses have the "universal", Rutherford like angular dependence[16]

$$\frac{d\sigma}{d\cos\theta} \sim \frac{\alpha_s^2}{\hat{s}} \frac{1}{(1-\cos\theta)^2} , \qquad (4)$$

where θ is the scattering angle in the parton center of mass frame (i.e. the angle between jets and p$\bar{p}$ beam direction in the two-jet center of mass frame). Such a dependence on θ reflects the spin one, gluon exchange, QCD elementary dynamics. The three processes above thus differ by only a colour factor, so that we can define an "effective" structure function:

$$F(x,Q^2) = g(x,Q^2) + \frac{4}{9}[q(x,Q^2) + \bar{q}(x,Q^2)] , \qquad (5)$$

in terms of which the two-jet cross section has the factorized form:

$$\frac{d\sigma^{2\text{jet}}}{dx_1 dx_2 d\cos\theta} = \frac{F(x_1)}{x_1} \frac{F(x_2)}{x_2} \left(\frac{d\hat{\sigma}}{d\cos\theta}\right)_{gg \to gg} \tag{6}$$

In Fig.13 eq.(6) is compared to the experimental jet angular distribution.

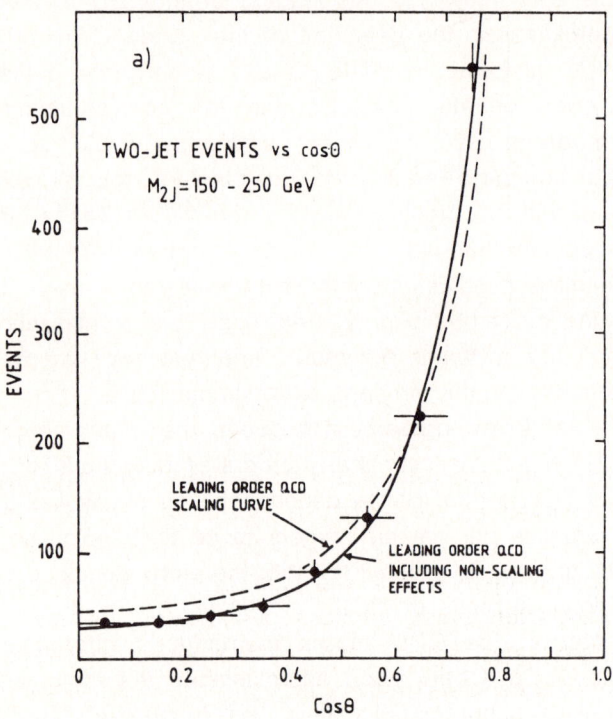

Figure 13

The agreement is spectacular, and confirms the vectorial character of the basic QCD interaction. Moreover, there is clear evidence of the significant role of the QCD scale breaking corrections. Indeed one might turn the argument around, and try to fit the structure function $F(x,Q^2)$ in eqs. (5) and (6) from the experimental angular distributions, assuming a priori the validity of the QCD elementary parton-parton subprocesses. The result is shown in Fig. 14, and compared to the structure function $F(x,Q^2)$ as determined from the lepton- nucleon deep - inelastic scattering data, evolved via eq. (3) . This simple example shows the role of jets as a tool to explore the hadronic structure.

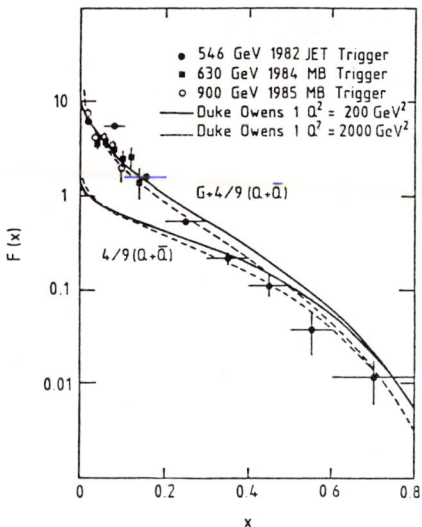

Figure 14

A similar, extensive analysis has been carried out for the three-and four-jet events observed at the Collider. The leading QCD- parton model prediction for multijet events is given, analogously to eq.(2), by the convolution of parton distributions with the 2→3, 2→4 (in general 2→n) elementary parton-parton cross sections. The latter become increasingly complicated to evaluate, and indeed fully analytic expressions for all parton processes have been derived only for n=3[17] and n=4[18,19]. Estimates for more than four jets are only feasible, at present, by computer MonteCarlo simulations[20]. Being limited to the leading order in α_s, these multijet rates suffer from the same kind of uncertainties as briefly outlined above for the two-jets, related to Q^2 ambiguities and K-factors.

Some qualitative features are however quite immediate, as following from the general properties of the QCD parton model.

Firstly, we expect multijet rates to be suppressed, relatively to two-jets, by α_s. This is qualitatively in accord with the experimental observation.

Also, as the typical partons incoming fractional momenta are given by the ratio $x_{1,2} \sim 2E_T/E_{CM}$, with E_T the total transverse energy carried by the jets, multijet rates should sharply increase with E_{CM} in a way analogous to two-jet events.

Finally, one can anticipate distinctive multijet topologies and angular dependences, reflecting the "hard" gluon-bremsstrahlung mechanism, characteristic of QCD. Dalitz plots and angular distributions of observed multijet samples confirm these general expectations, and thus represent ever more detailed tests of the underlying dynamics. For a more extensive account we refer to Ref. 7.

3. RISING JET CROSS SECTION

Recently jet searches have been extended down to transverse energies as low as $E_T > E_T^{min} = 5 - 10$ GeV, and up to E_{CM} as high as $E_{CM} \sim 1$ TeV. The interesting result[21] is that jet events represent a sizable fraction of the totality of inelastic events, fastly rising with E_{CM} to as much as 20% at $E_{CM} = 900$ GeV. Fig. 15 shows the jet cross section, integrated over

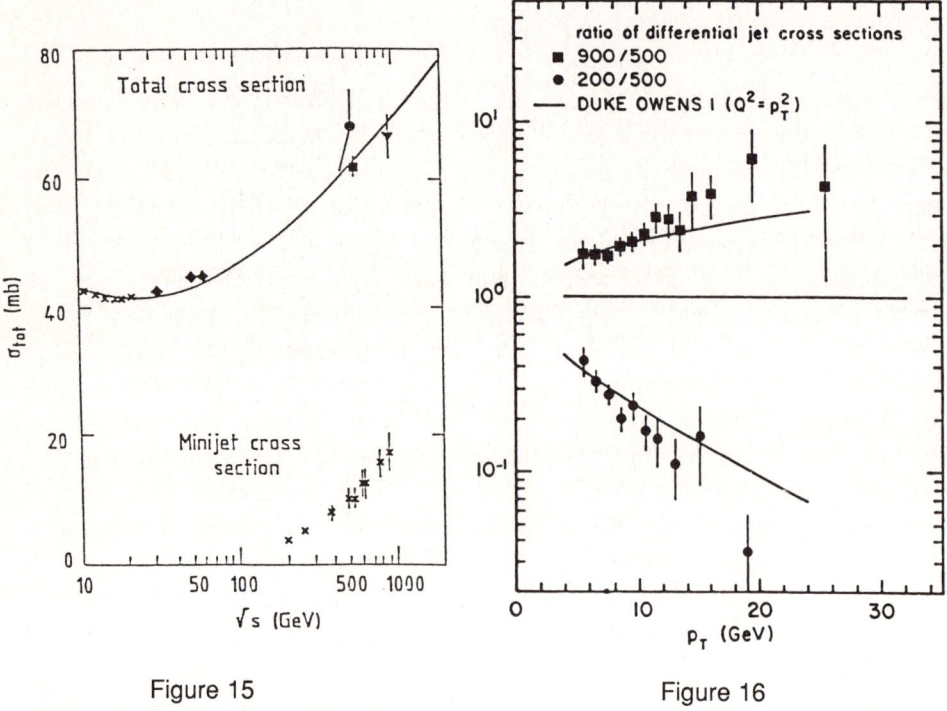

Figure 15   Figure 16

$E_T > E_T^{min}$, and also the $p\bar{p}$ total cross section for comparison. Such an E_T-cut, although considerably smaller than the typical values of the high-p_T jets discusses above, is still much greater than M_{proton}, so that perturbative QCD should presumably apply. For this reason the nomenclature "minijets" is sometimes used. The observed clusters of particles have characteristics similar to the higher p_T jets, and could thus in principle be accounted for by the mechanism of hard parton collisions.

Indeed the fast rise of the integrated jet cross section with the Collider energy E_{CM} is a well-expected phenomenon in the QCD parton model, reflecting the small-x behaviour of the parton densities, as discussed previously. In Fig.16 we show the ratios of jet cross sections at different energies, compared to the QCD leading order prediction of eq.(2). The agreement looks very reasonable, although, as already stated several times, one should keep in mind that the lower the E_T the larger the systematic uncertainty affecting the jets. Indeed, a detailed analysis seems to suggest that with increasing E_{CM} the experimental data tend to be higher than the leading QCD prediction, so that there is room for extra, non-leading dynamical effects[21].

In this regard the great interest of these "minijets" is that they are approaching the borderline between "hard" and "soft" hadronic physics, so that they should give important indications on the limits of validity of the leading QCD parton model eq.(2)[22,23].

Indeed the relevant parton fractional momenta range from $x \sim 10^{-2}$ at the present Collider down into the region $x \sim 10^{-3}\text{-}10^{-4}$ at future multiTev colliders. This is the so-called small-x region, where the transition from partonic degrees of freedom to Regge-dominated physics should gradually take place. If we insist in speaking the parton model language as far as possible in that region also, then we should expect the emergence of important modifications to the basic leading QCD parton model description, giving rise to a new class of "semihard" phenomena, and signalling the onset of the above mentioned transition[24,25].

The first theoretical problem one encounters in the region of very small x is that the standard perturbative methods can be spoiled by terms of order $\alpha_s(p_T)$ ln x, which become large in spite of $\alpha_s(p_T)$ being small. Such terms must be resummed in order that the perturbative expansion is reset[24]. Phenomenologically these ln x corrections, if properly summed, seem to be relatively unimportant in the range of x and p_T relevant to minijets[26].

The other important effect is that at small x partons can "overcrowd"[24], as the consequence of the parton density increasing faster than 1/x for $x \rightarrow 0$[27]. Basically, the relative parton density within the nucleon (at small x only gluons matter):

$$W(x,Q^2) = \alpha_s(Q^2)\frac{x\,g(x,Q^2)}{R^2 Q^2} \tag{7}$$

becomes of order unity for sufficiently small x (R is a typical hadronic radius). At such a value of x partons start to spatially overlap, and eq.(2), based on the incoherent sum of parton cross sections, starts loosing its full validity.

As the consequence of this parton overcrowding one expects the insurgence of coherence effects among partons. For example, in such a dense system partons should recombine and "shadow" the x→0 indefinite increase of $g(x,Q^2)$[28]. Notice that the behaviour of $g(x,Q^2)$ more singular than 1/x as x→0 must saturate at some value of x, otherwise unitarity would be violated. Moreover, multiple hard parton interactions per proton-proton interaction can occur, with considerable rates, and represent a competitive mechanism for multijet production with respect to the leading QCD (LQCD). The simplest example is the double-disconnected parton scattering[29] depicted in Fig. 17: two partons of one hadron have

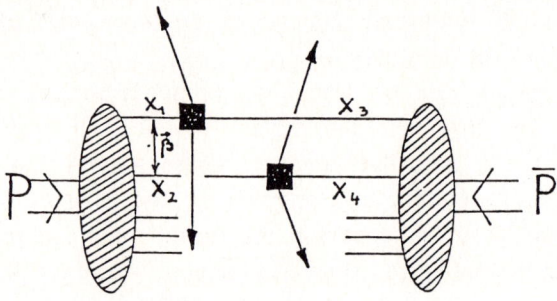

Figure 17

independent hard collisions with two different partons of the other hadron, giving origin to four minijets. Notice that the 4 →4 partonic amplitude is disconnected, in the sense that it is factorized into the product of two 2→2 amplitudes, separated (within the hadrons) by a distance of the order of the hadronic radius, therefore much larger than $1/E_T$.

The four-jet cross section corresponding to the mechanism of Fig.17 can be written as

$$d\sigma_{4jet}^{DDS} \sim \sum_{\substack{a,b\\c,d}} \int d^2\beta\, F_{ab}(x_1,x_2;\beta) F_{cd}(x_3,x_4;\beta)\, d\hat{\sigma}_1 d\hat{\sigma}_2, \tag{8}$$

where $d\hat{\sigma}_1$ and $d\hat{\sigma}_2$ are the hard cross sections for the two independent scatterings, given by perturbative QCD, and $F(x_1, x_2;\beta)$ represent two-parton distributions, with fractional momenta x_1, x_2 and with transverse distance β within the proton . Close inspection of the kinematics shows that double distributions, governing the incoming flux of parton pairs, is fully specified by this variable, in addition to x_1, x_2 (and to the "large" mass scale Q^2, not explicitly indicated in eq.(8)).

The double (in general the multi-) parton distribution[30] represent a new fundamental quantity, which contains more detailed knowledge of the hadronic structure , compared to the single distributions $F(x,Q^2)$. Like the F's they depend on non-perturbative physics, and are thus a priori not known. However, to make some phenomenological estimates, a reasonable ansatz for $x<<1$ could be the product of the single distributions:

$$F(x_1,x_2; Q^2) = F(x_1, Q^2) F(x_2, Q^2) \frac{1}{\pi R^2} , \qquad (9)$$

where R has been introduced before. The $1/R^2$ scale factor is simply related to the dimensionality of the double distribution, as compared to the dimensionless single parton distributions.

Replacing eq.(9) into eq.(8) , one finds the following simple expression for the DDS jet cross section:

$$\frac{d\sigma^{4jet}_{DDS}}{dE_T} = \int\int_{E_T^{min}}^{E_T-E_T^{min}} dE_1 dE_2 \frac{d\sigma^{2jet}}{dE_1} \frac{d\sigma^{2jet}}{dE_2} \frac{\delta(E_1+E_2-E_T)}{\sigma_{eff}} \qquad (10)$$

where $\sigma_{eff} = \pi R^2$.

The question at this point is how to evidence signatures of double parton processes, signalling the onset of parton overcrowding. Clearly, one has to look at four-minijet events, where the values of x are the smallest and the effect should be enhanced by the highest fluxes of gluon pairs. The problem is to distinguish the DDS from the (same order α_s^4) leading QCD mechanism for producing four-jets, as in general

$$d\sigma_{4jet} = d\sigma^{LQCD}_{4jet} + d\sigma^{DDS}_{4jet} . \qquad (11)$$

Referring to Refs. 31 and 32 for a detailed discussion, we briefly sketch here the most important aspects.

The most obvious observation is that, being a "dimensional" process, the DDS can be considered as a "power correction" with respect to the LQCD term, namely DDS/LQCD $\sim 1/R^2 E^2_T$. Accordingly it should exhibit a sharper dependence on the E_T, in particular on the E_T^{min} used to define the jets.

Conversely, increasing E_{CM} at fixed E_T, the DDS should rise much faster than the LQCD.

The other relevant feature, very important in practice, is that the LQCD and the DDS mechanisms should result into quite distinct topologies of the four-jet events. The DDS four jets should be predominantly pairwise balanced in p_T, as there are two separate conservations in p_T, one for each parton-parton scattering in Fig.17. The LQCD, being dominated by "double gluon emission", should give non-pairwise balanced four jets events, with jets correlations typical of the breamsstrahlung. Consequently it is possible to define jet correlation variables[32], such that the DDS and the LQCD tend to differently populate the range of values kinematically allowed. A convenient strategy to pick up DDS jets in thus to look for an excess of events in those regions of phase space where LQCD jets are less likely. Clearly in this kind of search the role of experimental cuts is crucial. Once such an excess of events is eventually found, important tests can be provided by the E_T- and the E_{CM} behaviours discussed above.

The present experimental results concerning double parton scattering seem somewhat contradictory.

Evidence of DDS four-jet events has been claimed recently at the ISR. The data there can be fit by the addition, to the LQCD, of a substantial contribution of DDS, modeled as in eq.(10), and obtainable by an appropriate choice of the hadronic scale parameter R[33]. Interestingly, such a value of R is supported by the analyses of double Drell-Yan lepton pair production[34], and of proton proton scattering at large momentum transfer, where it is expected to be dominated by triple- disconnected gluon exchange among the valence quarks[35].

On the contrary, there appears to be no evidence of DDS in the four-jet sample observed at the Collider. This can be translated in principle into a lower bound on R[36].

The results of Refs. 33 and 36 are however not necessarily in contradiction, because the jet samples are obtained in quite different ranges of p_T and of E_{CM}, and with different cuts, so that a direct comparison is not so obvious.

Clearly the subject of multiple parton scattering has to be investigated further, in order to clarify the present situation. This would lead to a great progress in our understanding of the role of the partonic hadron structure in inelastic reactions and of perturbative QCD. Also, it would allow considerable developments concerning other important aspects, not discussed here, such as the particle multiplicities[37], and the "shadowing" of the hard QCD jet cross section[38].

4. CONCLUSIONS

We can just summarize very briefly the preceding sections by saying that high p_T jet studies have allowed to test the parton model of deeply inelastic hadron reactions in its various aspects and over a wide range of x and Q^2. In particular, quarks and gluons presently appear as "pointlike".

As mentioned several times in the course of the discussion, significant improvements are foreseeable in the future, both theoretical and experimental, which should greatly reduce the present systematic uncertainties and bring jet analyses to the fully quantitative level.

One point of great interest, specially for hadronic physics at multiTeV supercolliders, is represented by minijets (in particular by multiparton processes), which are expected to give information on the interface between "hard" and "soft" physics.

In this regard, as ever more detailed analyses will be required, the description of jets fragmentation into the observed hadrons becomes a crucial ingredient. This involves nonperturbative physics at some stage, and leads us back to the problem of relating the "parton" quarks and gluons to the "constituents" quarks, i.e. to confinement.

ACKNOWLEDGEMENTS

I wish to thank Prof. P.Dalpiaz for a pleasant and interesting meeting.

I also want to express my gratitude to Mrs. E.Talocchi for the hard work of preparing the manuscript.

REFERENCES

1) R.P.Feynman, Photon Hadron Interactions, Benjamin (1972).

2) For reviews on e^+e^- physics see for instance:
   B.Naroska, DESY 86-113 (1986);
   S.L.Wu, Phys.Rep. 107(1984) 89.

3) For reviews of the QCD-parton model see e.g.:
   G.Altarelli, Phys.Rep. 81 (1982) 1;
   A.J.Buras, Rev.Mod.Phys. 52(1980) 189.

4) D.J.Gross and F.Wilczek, Phys.Rev.Lett. 30(1973) 1343;
   H.Politzer, Phys.Rev.Lett. 30(1973) 1346.

5) T.Akesson et al., Phys.Lett. 118B(1982)185; 123B(1983)133.

6) M.Banner et al., Phys.Lett, 118B(1982)203;
G.Arnison et al.,Phys.Lett. 123B(1983)115.

7) For recent reviews see:
F.Ceradini, in Proceeding of the 23rd International Conference on High Energy Physics, Ed. S.C.Loken World Scientific (1987);
P.Bagnaia, in Proceedings of the 6th Topical Workshop on Proton-Antiproton Collider Physics, Eds.K.Eggert , H.Faissner and E.Radermacker, World Scientific 1987.

8) W.Hoffmann, in Proceedings of the 1987 International Symposium on Lepton and Photon Interactions at High Energies , Eds. W.Bartel and R.Rückl, North Holland 1988.

9) M.Jacob and P.V.Landshoff, Phys.Rep. 48(1978)285;
R.Horgan and M.Jacob, Physics at Collider Energy, CERN-DESY School of Physics (1980), CERN 81-04.

10) B.Combridge and J.Kripfganz, Phys.Lett. 70B(1977)234;
R.Cutler and D.Sivers, Phys.Rev. D16(1977)248;
M.Glück, J.F.Owens and E.Reya, Phys.Rev. D18(1978)1501.

11) G.Altarelli and G.Parisi, Nucl.Phys. B126(1977)248.

12) M.Glück, E.Hoffman and E.Reya. Zeit.Phys. C13(1982)119;
D.W.Duke and J.F.Owens, Phys.Rev.D30(1984)461;
E.Eichten, I.Hinchliffe, K.Lane and C.Quigg, Rev.Mod.Phys. 56(1984)579;
M.Diemoz, F.Ferroni and E.Longo, Phys.Rep. 130(1986)293.

13) G.Altarelli , R.K.Ellis and G.Martinelli, Zeit.Phys. C27(1985)617.

14) R.K.Ellis and J.C. Sexton, Nucl.Phys.B269(1986)445;
W.Slominski and W.Furmanski, Krakow preprint TPJU-11/81(1981);
R.K.Ellis, M.A.Furman, H.E.Haber and I.Hinchliffe, Nucl.Phys. B173(1980) 397.

15) F.Aversa, P.Chiappetta, M.Greco and J.Ph.Guillet, Marseille preprint CPT-88/P 2089 (1987).

16) F.Halzen and P.Hoyer, Phys.Lett.130B(1983)326;
B.Combridge and C.Maxwell, Nucl.Phys. B239(1984)429.

17) T.Gottshalk and D.Sivers, Phys.Rev. D21(1980)102;
Z.Kunszt and E.Pietarinnen, Nucl.Phys.B164(1980)45.

18) J.F.Gunion and Z.Kunszt, Phys.Lett. 159B(1985)167; 176B (1986)163,477;
    S.Parke and T.Taylor, Fermilab-Pub.-85/118-5; Fermilab-Pub-85/162-T;
    Z.Kunszt, Nucl.Phys.B271(1986)333.

19) Z.Kunszt and W.J.Stirling, Phys.Lett. 171B(1986)307.

20) For a review see for instance:
    T.D.Gottshalk, in Physics Simulations at High Energy , Eds.V.Barger,
    T.Gottshalk and F.Halzen, World Scientific 1987.

21) C.Albajar et al., CERN-EP/88-29, and references there.

22) M.Jacob and P.V.Landshoff, Mod.Phys.Lett. A1(1986)657.

23) P.V.Landshoff, Cambridge report DAMPT/87-22, review talk at the
    Workshop for the Supercollider, Berkeley 1987.

24) An extensive account is given in:
    L.V.Gribov, E.M.Levin and M.G.Ryskin, Phys.Rep. 100(1983)1

25) J.C.Collins,in Supercollider Physics, Ed.D.E.Soper,World Scientific 1986.

26) J.Kwiecinski, Krakow report 1328/PH (1986); Zeit.Phys. C29 (1985) 561.

27) Wu-Ki Tung, Fermilab-Conf. 87/122-T (1987).

28) A.H.Muller and J.Qiu, Nucl.Phys. B268(1986)427.

29) N.Paver and D.Treleani, Nuovo Cimento 70A(1982)215; 73A(1983)392;
    Phys.Lett.146B(1984)252; Zeit.Phys. C28(1985)187;
    B.Humpert and R.Odorico, Phys.Lett. 154B(1985)211;
    M.Jacob, CERNTH-3639 (1983);
    P.V.Landshoff and J.C.Polkinghorne, Phys.Rev.D19 (1978)3344.

30) H.D.Politzer, Nucl.Phys. B172(1980)349;
    H.R.Gerhold, Nuovo Cimento 59A(1980)373;
    R.K.Ellis, R.Petronzio and W.Furmanski, Nucl.Phys. B207(1982)1;
    M.Mekhfi , Phys.Rev. D32(1985)2371,2380;
    M.Mekhfi and X.Artru, Orsay preprint LPTHE 87/63 (1987).

31) N.Paver, in Proceedings of the Workshop on New Aspects of High Energy
    Proton-Proton Collisions, Erice 1987, Ed. A.Ali, World Scientific .

32) Ll.Ametller, in Proceedings of the 7th International Symposium on Multiparticle Dynamics, Eds. M.Markytan, W.Majerotto and J.MacNaughton, World Scientific 1987.

33) T.Akesson et al. Zeit.Phys. C34(1987)163.

34) F.Halzen, P.Hoyer and J.W. Stirling, Phys.Lett. 188B(1987)375.

35) A.Donnachie and P.V.Landshoff, Zeit.Phys. C2(1979)55.

36) F.Pastore, in Proceedings of the EPS Conference on High Energy Physics, Ed. O.Botner, Uppsala 1987.

37) T.Sjöstrand and M. van Zijil, Phys.Rev. D36 (1987)2019.

38) Ll.Ametller and D.Treleani, Int.Journ.Mod.Phys. A3(1988)521;
L.Durand and P.Hong,Phys.Rev. Lett. 58(1987)303;
T.Sjöstrand and M. van Zijil, Phys.Lett. 188B(1987)149;
J.Kwiecinski, P*hys.Lett. 184B(1987)386.

PERSPECTIVES IN SUPERSTRING THEORY

Mauro Bregola

Dipartimento di Fisica, Università di Ferrara
and INFN, Sezione di Ferrara
via Paradiso 12, I 44100, Ferrara, Italy

A brief and elementary review of the structure of (super)string theory is presented together with the discussion of possible connections with low-energy physics. At the end some new developments regarding the effective theory and the conformal invariance are briefly analized.

1. INTRODUCTION

Strings are one-dimensional curves moving in space-time. In the last three-four years a new point of view is exploded in the theoretician's community, considering strings, instead of point particles, as the real elementary objects of nature and string theory as the most serious candidate to unify all fundamental interactions, gravity included. This modern attitude is essentially based on the following advantages that string theory seems to have in comparison with a unified point particle quantum field theory:

Advantage 1

Gravity is contained in a quite natural way, i.e there is in the spectrum a massless spin-two excitation (graviton), which interacts at low energies according to the requirement of general covariance.

Advantage 2

Due to the apparent finiteness of string theory (really proved only at the one loop level[1]), the chronic unrenormalizability of gravity[2] seems to be overcome.

Advantage 3

The anomaly desease, characteristic of many quantum field theories, can be cured and the cure has the pleasant consequence of strongly restricting the set of acceptable gauge symmetries[3].

As an important result of the above three aspects, only a few (super)string theories exist consistent with the requirements of quantum mechanics, relativity, causality, etc. . They seem in addition to be able to explain some fundamental questions left open by the standard model, such as the chiral asymmetry and the small number of chiral fermion families[4]. Nevertheless, there has been an impressive scale shift between around 1970, when strings were considered for the first time as an approach to hadronic physics[5], and nowadays that strings are proposed for unification: at that time their typical sizes were supposed to be about one fermi. Now, due to the presence of gravity, the natural scales are:

(Planck length) $\quad L_P = (\hbar G / c^3)^{1/2} \sim 1.6 \times 10^{33}$ cm, $\quad$ (1.1)

(Planck mass) $\quad M_P = (\hbar c / G)^{1/2} \sim 1.2 \times 10^{19}$ GeV/c^2, $\quad$ (1.2)

where G is Newton's constant.

So, string theory is far from really being experimentally testable. However, for much larger distances or lower energies, an ordinary quantum field theory emerges, carrying some consequences from the original string theory, which could make contact with experiments.

2. STRINGS AND SUPERSTRINGS

Let us start with a discussion of bosonic strings, which though not supersymmetric, contain already the most important aspects of any string theory.

To construct the theory and in particular to quantize it, we need an action, from which obtaining by variation the equations of motion for the string. A right action is

$$S = - \frac{1}{4\pi\alpha'} \int_0^\pi d\sigma \int d\tau \sqrt{-g} \, g^{\alpha\beta} \, \partial_\alpha X^\mu \, \partial_\beta X^\nu \, \eta_{\mu\nu} , \qquad (2.1)$$

where $X^\mu(\sigma,\tau)$ ($\mu = 0,1,\ldots,D-1$) is the vector field describing, at any given time τ, the coordinates of the string in a D-dimensional space-time in terms of the curvilinear parameter σ ($0 \leq \sigma \leq \pi$), $\eta_{\mu\nu}$ is the space-time Lorentz metric, $g^{\alpha\beta}(\sigma,\tau)$ is the inverse of the riemannian metric $g_{\alpha\beta}(\sigma,\tau)$ of the (σ,τ)-space (world-sheet). g is the determinant of $g_{\alpha\beta}$, which is chosen with one positive and one negative eigenvalue. It is important to remark that while S, due to the presence of g and

$g^{\alpha\beta}$, is invariant with respect to world-sheet local reparametrizations, satisfying an obvious physical requirement, the precise choice of $\eta_{\mu\nu}$ makes it invariant only with respect to global Poincaré transformations in D-dimensional space-time (target-space). But the set of invariances of S contemplates another one: the Weyl rescaling (or conformal) invariance under transformations of the form

$$g_{\alpha\beta} \to \exp[\lambda(\sigma,\tau)] \cdot g_{\alpha\beta} \qquad (2.2)$$

with X^μ inert.

Thus, at the classical level, the theory contained in S is a conformally invariant, general covariant 2-dimensional field theory, with the main field X^μ having values in the D-dimensional Minkowski space-time.

The two dimensional enery-momentum tensor is given by the variational derivative of S with respect to $g^{\alpha\beta}$, so that

$$T_{\alpha\beta} = - \frac{4\pi\alpha'}{\sqrt{-g}} \frac{\delta S}{\delta g^{\alpha\beta}} , \qquad (2.3)$$

$$T_{\alpha\beta} = \partial_\alpha X^\mu \partial_\beta X_\mu - \frac{1}{2} g_{\alpha\beta} g^{\gamma\lambda} \partial_\gamma X^\mu \partial_\lambda X_\mu . \qquad (2.4)$$

Due to the Weyl invariance, (2.4) is automatically traceless, i.e $g^{\alpha\beta} T_{\alpha\beta} = 0$, and the field equation $\delta S/\delta g^{\alpha\beta} = 0$ is equivalent to the requirement $T_{\alpha\beta} = 0$. This allows us to express $g_{\alpha\beta}$ in terms of derivatives of X^μ and one easily recognizes that S becomes

$$S_1 = - \frac{1}{2\pi\alpha'} \int_0^\pi d\sigma \int d\tau \, [- \det(\partial_\alpha X^\mu \partial_\beta X_\mu)]^{1/2}. \qquad (2.5)$$

Apart from the constant $-1/\pi\alpha'$, the new form of the action is nothing but the area of the world-sheet. An interesting analogy is now clear: while for the point particle the classical trajectories are those of minimal length, so for the string are those of minimal area.

However, one of the main informations we wish to extract from string theory is the structure of space-time, which means we should put in (2.1), in place of $\eta_{\mu\nu}$, a riemannian background metric $G_{\mu\nu}(X^\mu)$, to be determined by the consistent evolution of the string. We will come back later on this important point; for the moment we avoid this complication and even we symplify further the theory. Indeed, by using reparametrization invariance, we can go to the conformal gauge in which

$$g_{\alpha\beta} = \exp[\phi(\sigma,\tau)] \times \eta_{\alpha\beta} \ , \tag{2.6}$$

where $\eta_{\alpha\beta} = \begin{pmatrix} -1 & 0 \\ 0 & 1 \end{pmatrix}$ is the 2-dimensional Minkowski metric. Making this choice, the action simplifies to

$$S_2 = -\frac{1}{4\pi\alpha'} \int_0^\pi d\sigma \int d\tau \ \eta^{\alpha\beta} \ \partial_\alpha X^\mu \ \partial_\beta X_\mu \ . \tag{2.7}$$

In fact, the gauge choice (2.6) leaves a residual conformal invariance. The field ϕ does not appear in S_2 because the original action S is invariant under Weyl rescalings, but this decoupling is only true at the classical level[6]. In the quantized version the decoupling of the field ϕ will persist only if the space-time dimension has the critical value D=26.

The equation of motion derived from (2.7) is simply the free 2-dimensional wave equation

$$\Box X^\mu \equiv (\partial^2/\partial\sigma^2 - \partial^2/\partial\tau^2) X^\mu = 0 \ . \tag{2.8}$$

Of course, strings show two possible topologies, closed and open strings, and the boundary conditions for stationary solutions of (2.8) are different in the two cases. When we derive (2.8) by variation of (2.7) we need to satisfy the condition of vanishing of the surface term:

$$\int d\tau \ [\ X'\delta^\mu X|_{\sigma=\pi} - X'\delta^\mu X|_{\sigma=0} \ ] = 0 \ . \tag{2.9}$$

Eq.(2.9) gives for open strings the boundary conditions $X'_\mu|_{\sigma=\pi} = X'_\mu|_{\sigma=0} = 0$, while for closed strings the periodicity of X^μ is necessary and sufficent.

As usual in two dimensions, the general solution of (2.8) can be written in terms of arbitrary "right moving" modes X^μ_R and "left moving" modes X^μ_L

$$X^\mu(\sigma) = X^\mu_R(\sigma^+) + X^\mu_L(\sigma^-) \ , \tag{2.10}$$

where $\sigma^- = \tau - \sigma$, $\sigma^+ = \tau + \sigma$ and the related derivatives are $\partial_\pm = 1/2 \ (\partial_\tau \pm \partial_\sigma)$. Now, if we write the world-sheet energy-momentum-tensor in the new coordinate system, we find

$$T_{++} = 1/2 \ (T_{00} + T_{01}) = \partial_+ X^\mu \ \partial_+ X_\mu \ , \tag{2.11}$$

$$T_{--} = 1/2 \ (T_{00} - T_{01}) = \partial_- X^\mu \ \partial_- X_\mu \ , \tag{2.12}$$

while the tracelessness condition becomes

$$T_{+-} = T_{-+} = 0 \ . \tag{2.13}$$

Eq. (2.13) has the consequence that the energy-momentum conservation law, which in 2-dimensional quantum field theory is

$\partial_- T_{++} + \partial_+ T_{-+} = 0$, takes the form

$$\partial_- T_{++} = 0 \ . \qquad (2.14)$$

Now, we know that, because of the equations of motion for $g_{\alpha\beta}$, the constraints $T_{\alpha\beta} = 0$ have to be considered. They give of course the additional conditions

$$T_{++} = T_{--} = 0 \ . \qquad (2.15)$$

In fact, (2.14) states that the above constraints are conserved quantities; it is because they are conserved that we can set them to zero; if they vanish at one time, they will vanish at all later times. Moreover, these conserved quantities are really infinitely many. This unusual fact is due to a residual symmetry group left over by the gauge fixing procedure: the group of conformal transformations of 2-dimensional Minkowski space (the world sheet in our case). It is worth noting that only in two dimensions the conformal group is infinite dimensional[7].

Now that we have explored enough the classical structure, we can briefly outline the quantization procedure.

There are esssentially two equivalent approaches to quantization: the Lorentz-covariant approach, with an action (as in (2.7)) containing the full vector of coordinates X^μ, and the light-cone gauge approach, where the residual conformal invariance is exploited to eliminate the time-like and longitudinal degrees of freedom, ending up with only the physical transversal ones.

The covariant procedure can be performed, in its own turn, in two different equivalent approaches: the more modern one involves the introduction of Faddeev-Popov ghosts and the identification of BRST symmetries and currents; the old covariant approach has the problem of ghost states with negative probabilities, due to the presence of unphysical degrees of freedom and in particular of the time-like component X^o, with $\eta_{oo} = -1$. The ghost states have to be removed by imposing the infinite set of constraints on the physical states

$$L_n \ |\text{physical state}\rangle = 0 \ , \quad n \geq 1 \ , \qquad (2.16)$$

supplemented by

$$(L_o - \alpha) \ |\text{physical state}\rangle = 0 \ , \qquad (2.17)$$

where L_n of (2.16) are nothing but the operators corresponding to the positive frequency Fourier components of the classical energy-momentum tensor. In other words, (2.16) reflects at the quantum level the classical constraint $T_{++}= 0$. The equation (2.17) arises from the normal ordering problem and it is easily shown that a necessary condition for the absence of ghosts is $a \leq 1$.

The whole set of the Fourier component operators L_n satisfies the so called Virasoro algebra[8] with a central extension arising from normal ordering

$$[L_n, L_m] = (n - m) L_{n+m} + D/12 (n^3 - n) \delta_{n,-m}, \qquad (2.18)$$

where D is the space-time dimension. Unfortunately, but not so surprisingly, even the procedure for eliminating ghosts is possible only in the same critical dimension D = 26, which allows to get rid of the field ϕ at the quantum level.

However, we can make the choice of loosing the explicit Lorentz invariance, by working with the light-cone action containing only physical degrees of freedom; it is

$$S_{LC} = - \frac{1}{4 \pi \alpha'} \int_0^\pi d\sigma \int d\tau \, \eta^{\alpha\beta} \partial_\alpha X^i \partial_\beta X^i, \qquad (2.19)$$

with $i = 1, 2, ..., D-2$. The corresponding equation of motion is obviously the same as (2.8) with X^i in place of X^μ. Here the critical dimension arises from the fact that Lorentz invariance of the quantized theory holds, again, only for D=26. On the other hand, in the light-cone gauge the theory exhibits directly the complete physical spectrum of the strings. This spectrum, corresponding to vibration modes, will of course be different for open and closed strings, due to the different boundary conditions.

For the open string the solution of the equation of motion is

$$X^i(\sigma,\tau) = x^i + 2\alpha' p^i \tau + i \sum_{n \neq 0} (n)^{-1} \alpha_n^i \cos(\sigma n) e^{-in\tau}. \qquad (2.20)$$

Now, if we quantize in the canonical way, the usual commutation relations between field variables and conjugate momenta lead to the following algebra for the operators α_n^i:

$$[\alpha_n^i, \alpha_m^j] = n \delta^{ij} \delta_{n,-m}, \qquad (2.21)$$

This means that X^i contains an infinite set of harmonic oscillators with creation operators

$$(n)^{-1/2} \alpha_{-n}^i, \quad n = 1, 2, ..., \qquad (2.22)$$

and destruction operators

$$(n)^{-1/2} \alpha_n^i, \quad n = 1, 2, ... \qquad (2.23)$$

Taking into account zero point energies for the quantized oscillators, at the end we obtain the mass relation

$$\alpha' M^2 = N - 1 = \sum_{n=1}^{\infty} a^i_{-n} a^i_n - 1 . \qquad (2.24)$$

Thus, the ground state $|0\rangle$ defined by

$$a^i_n |0\rangle = 0 , \quad n=1, 2,\ldots \qquad (2.25)$$

forms the N=0 level

$$N=0 : \quad |0\rangle , \quad M^2 = - (\alpha')^{-1} . \qquad (2.26)$$

It is a scalar and has the unpleasant property of a negative squared mass: so, it is called scalar tachion. Starting from $|0\rangle$, by acting with the creation operators (2.22), we can build the whole spectrum of string states. For the massless sector (N=1) in particular we have the vector state

$$N=1 : \quad a^i_- |0\rangle , \quad M^2 = 0 . \qquad (2.27)$$

For closed strings the different boundary conditions lead to the following mode expansion

$$X^i(\sigma,t) = x^i + 2\alpha' p^i + \qquad (2.28)$$
$$+ i/2 \sum_{n \neq 0} (n)^{-1} (a^i_n e^{-2in(\sigma-\tau)} + b^i_n e^{-2in(\sigma+\tau)}) .$$

There are now two independent set of oscillators, corresponding to left- and right moving harmonic waves. In addition to the algebra (2.21), holding for both sets, we have

$$[a^i_n , b^j_m] = 0 . \qquad (2.29)$$

The mass formula for the closed bosonic string states is

$$1/4 \ \alpha' M^2 = N_a - 1 = N_b - 1 , \qquad (2.30)$$

where

$$N_a = \sum_{n=1}^{\infty} a^i_{-n} a^i_n , \quad N_b = \sum_{n=1}^{\infty} b^i_{-n} b^i_n . \qquad (2.31)$$

Thus, we have again a scalar tachion ($N_a=N_b=0$). The massless sector consists of the states

$$a^i_{-1} b^j_{-1} |0\rangle , \qquad (2.32)$$

the symmetric part (under SO(24)) of which contains a scalar and a symmetric tensor to be interpreted as the graviton. The other massless states form an antisymmetric SO(24) tensor.

Of course, to get a sensible physical theory, we need to implement a consistent interaction between strings. This is not an easy matter, due to the difficulties arising from causality requirement. However, at least in the light-cone gauge, interactions have been consistently constructed in terms of vertices for the emission of a "light" final string.

A fact which is immediately clear is that all theories containing open strings necessarily contain closed strings as well. The reason is simply that the same basic interaction, allowing two open strings to join ends and give one open string, in the same way allows one open string to join its ends for giving a closed string. On the contrary, it is possible to have theories containing only closed strings.

Summarizing, free bosonic strings have a spectrum, which in the massless sector can be conveniently interpreted in terms of known elementary particles: depending on the topology of the strings considered, there is a vector state possibly generalizable to a Yang-Mills vector potential and a symmetric tensor to be identified with the graviton. Unfortunately, already at the free level and for any topology, bosonic strings suffer two serious diseases, which spoil their consistency as quantum theories: they live only in 26 space-time dimensions and the ground state is a tachion.

In fact, a way to lower the critical dimension is possible: it consists in the introduction of supersymmetry. To supersymmetrize the theory we add a fermionic term to the action. The number of bosonic and fermionic degrees of freedom has to be the same; so, it is natural to add variables ψ_α^μ transforming at the same time like world-sheet spinor and space-time vector components. However, after some suitable observations, all the added degrees of freedom can be made fermionic, obtaining at the end, in the light-cone gauge,

$$S = \int_0^\pi d\sigma \int d\tau \, [-(4\pi\alpha')^{-1} \partial^\alpha X^i \partial^\beta X^i \, \eta_{\alpha\beta} + \qquad (2.33)$$
$$+ (4\pi)^{-1} \overline{\psi}^a \rho^\alpha \partial_\alpha \psi_a] \, ,$$

where ψ_α^a, with $\alpha = 1,2$ and $a = 1,...,8$, is both a world-sheet and a space-time spinor. Of course, ρ^α are two-dimensional γ matrices.

Nevertheless, the incorporation of fermionic variables, which hopefully should be suitable to have quarks and leptons into the theory, can be made in several different ways leading us, in general, to different superstring theories: we have however the attractive feature that only few of these are acceptable superstring theories. They can be conveniently classified by the local supersymmetry and topology of their world sheets and are called: i) type I superstring theory, ii) type II superstring theory, iii) heterotic string theory[9].

In all these theories, due to the presence of chiral fermions, gauge anomalies could emerge, thus spoiling the consistency of the whole scheme. However, it has been shown[3] (and it was a fundamental step in the string theory story) that this problem can be avoided if the gauge group is chosen to be SO(32), $E_8 \times E_8$ or SO(16)×SO(16).

Coming to the question of the critical dimension, it is customary to say that these theories have ten space-time dimensions, some (six) of which have to be consistently compactified. However, coherently with a useful distinction between theories and solutions, it is probably better to consider the space-time dimension as a property characterizing a particular solution rather than the theory itself.

Anyway, we prefer to pick up from the increasing amount of proposed solutions (the validity of which is far from being proved) the 10-dimensional, $E_8 \times E_8$ heterotic one[9], which seems the most promising for a realistic phenomenology and which we will refer to (unless otherwise stated) from now on.

3. Compactification and low-energy phenomenology.

To analyze the problem of relating a fundamental superstring theory with the standard model, we need to go to the low-energy limit $E \ll (\alpha')^{-1/2}$. In this limit, string states with masses of the order of $(\alpha')^{-1/2}$ are "frozen" and we are left with a field theory of massless modes only: its spectrum, in the case of the heterotic string[9] with D=10, $E_8 \times E_8$, coincides with the spectrum of the N=1, D=10 supergravity coupled to super-Yang-Mills (SUGRA-SYM) theory[10] with gauge group $E_8 \times E_8$. It is:

SUGRA		SYM	
$g_{\mu\nu}$	: graviton,	A^a_μ	: gauge boson,
$B_{\mu\nu} = -B_{\nu\mu}$	: graviphoton,	χ^a	: gaugino,
ϕ	: dilaton,		
ψ_μ	: gravitino,		
λ	: M-W spinor,		

(3.1)

where M-W means Majorana-Weyl, $\mu,\nu = 0,1,\ldots,9$ are space-time tensor indices and upper indices in the SYM sector are related to the adjoint representation of $E_8 \times E_8$; spinor indices have not been manifested.

The lagrangian of the effective field theory is not yet completely known (on this point we will return later for a short comment). However, a phenomenon which should occur is the compactification of six dimensions, to recover in the low-energy limit the physics of the 4-dimensional space-time. The most attractive situation would certainly be

$$M_c \sim (\alpha')^{-1/2} \sim M_p, \qquad (3.2)$$

where, as usual, M_p is the Planck mass of (1.2) and M_c is the compactification scale, i.e. essentially the inverse of the radius of the 6-dimensional compact space K_6.

At energies much lower than the Planck or compactification scale, we should check whether the field content of the effective theory is able to achieve a symmetry breaking path, ending with the residual group $SU(3) \times SU(2) \times U(1)$ in the 10^2 GeV energy range. We now briefly outline a possible scenario of compactification, but it should be clear that, nowadays, we can only discuss whether this scenario can possibly occur, not whether it is really given by the dynamics.

For phenomenological reasons we assume that the compactification leaves one unbroken supersymmetry in 4-dimensions: this should be useful to explain the hierarchy problem, i.e. the smallness of the weak interaction scale M_w with respect to M_p. Moreover, the holonomy group (the group acting on the connections) of K_6 is restricted by the above assumption to be in general $SU(3)$ [11].

To generate the compactification mechanism, the Riemann curvature tensor and gauge field strengths assume non zero background values, according to the following relation [12]:

$$\langle R^{ab}_{[\mu\nu} R^{ab}_{\rho\sigma]}\rangle = \langle F^{\alpha}_{[\mu\nu} F^{\alpha}_{\rho\sigma]}\rangle , \qquad (3.3)$$

where a,b are SO(1,9) indices, α runs over the adjoint representation of $E_8 \times E_8$, $\langle \ \rangle$ means background value and a sum is assumed over repeated indices; the symbol $[\mu\nu...\rho]$ means antisymmetrization w.r.t. $\mu,\nu,...,\rho$.

The preservation of one unbroken supersymmetry in the compactification process has the consequence that the right-hand side of (3.3) selects an SU(3) subgroup according to $E_8 \supset SU(3) \times E_6$, leaving $E_6 \times E_8$ as unbroken gauge group[11].

The non zero SU(3) field strenghts in (3.3) generate massless modes in the sector, non trivial and complex under SU(3), and we get at the end a set of chiral multiplets, which transform under E_6 according to

$$N\ 27 + Q\ (\ 27 + \overline{27}\ ) , \qquad (3.4)$$

where N and Q (which are of course non negative natural numbers) depend on the geometry of K_6.

Now, it is known that in each multiplet 27 of E_6 takes place one generation of quarks and leptons. Thus, if K_6 allows the number of 27's to be larger than the number of $\overline{27}$'s, we can interpret the number of exceeding 27's as the number of chiral fermion families. It is important to observe that one of two factors in $E_8 \times E_8$ is a hidden spectator, i.e. chiral multiplets never transform under it.

The problem of compactification is then reduced to find a space K_6, with SU(3) holonomy, N = 3 (or 4 ?) and Q $\geq$ 1 (for reasons related to the symmetry breaking pattern): of course, this space should also be an acceptable background for string dynamics.

There is by now a vast literature of solutions to this problem, but none of these seems completely satisfactory. Either they have problems with the consistency conditions coming from conformal invariance requirements, or do not allow to recover the standard symmetry group SU(3)×SU(2)×U(1) at energies smaller than the Planck scale. Typical examples are respectively Calabi-Yau spaces (Ricci flat, Kähler manifolds with SU(3) holonomy) and orbifolds[13] (manifolds divided by some discrete groups).

Of course, before having adequately understood the vacuum (i.e. the compactified solution) of the theory, it will be difficult to make a significative phenomenology: in particular, it will be hard to use successfully the breaking of the residual supersymmetry for solving the hierarchy problem.

4. Progress on the D = 10 effective theory.

Even if the massless spectrum is the same, the lagrangian of the low energy (effective) field theory, coming as point-like limit of the heterotic string, is different, due to string effects, from the lagrangian of the corresponding SUGRA-SYM model.

We have already stated that the lagrangian of this effective theory is not yet completely known; anyway, important progresses have been made, taking into account the consistency conditions imposed by the supersymmetric nature of the heterotic string.

Already at one loop level in string effects, the effective field theory is modified, with respect to the anomalous SUGRA-SYM model, by the presence of higher (higher than two) derivative terms, which are protagonists of the anomaly cancellation mechanism. In particular, a different coupling between SUGRA and SYM sectors is implemented by a different expression for the graviphoton two-form B, which, in place of defining condition

$$H = dB + (1/30) W_{3YM} \tag{4.1}$$

of the usual N = 1, D = 10 SUGRA-SYM model[10], satisfies

$$H = dB + (1/30) W_{3YM} - W_{3L}, \tag{4.2}$$

where W_{3YM} and W_{3L} are respectively the YM- and Lorentz-Chern Simons three-forms. Their expressions in space-time components are:

$$(W_{3YM})_{\mu\nu\rho} = \text{Tr}\left(A_{[\mu} F_{\nu\rho]} - (2/3) A_{[\mu} A_{\nu} A_{\rho]} \right), \tag{4.3}$$

$$(W_{3L})_{\mu\nu\rho} = \text{tr}\left(W_{[\mu} R_{\nu\mu]} - (2/3) W_{[\mu} W_{\nu} W_{\rho]} \right), \tag{4.4}$$

where A_μ is the gauge potential (which has values in the Lie algebra of $E_8 \times E_8$), $F_{\mu\nu}$ its associate field strength and the trace Tr of eq. (4.3) is taken over the adjoint representation of $E_8 \times E_8$. Regarding eq. (4.4), W_μ is the Lorentz connection (having values on the Lie algebra of SO(1,9)), $R_{\mu\nu}$ is the Riemann tensor and the trace tr is taken over the vector representation of SO(1,9).

Unfortunately, the Lorentz-Chern Simmons term in (4.2) (which cures the anomaly problem[3]) breaks supersymmetry. However, it has been possible to show[14], working with superspace techniques and a simple set of torsion constraints, that the theory related to (4.2) can be completely supersymmetrized, provided that an infinite series of terms is present into the effective lagrangian: these terms arise from the low-energy limit of the whole string theory (not only from few perturbative orders).

To conclude this section, we want to mention an interesting tentative[15] to make clearer the structure of the effective field theory for the massless modes. It is based on the immersion of the heterotic string in Yang-Mills and curved superspace background, hence on the study of the so-called k-symmetry both at classical and quantum level. Consistency conditions for the anomalies have been found, which seem to indicate a possible unicity of the theory.

Then, there are several signals that a better understanding of the effective field theory can be reached: this fact should be of great relevance for discussing compactification and phenomenology.

5. Progress in the conformal symmetry approach.

We have seen in section 2. that local conformal symmetry and the necessity of cancelling its anomaly play a fundamental role in string theory: so, any progress in this subject is of great importance for a deeper comprehension of the whole theory.

Conformal symmetry is probably the best tool for studying the structure of space-time background, in which (super)strings can be considered immersed and interacting with background fields as the metric $G_{\mu\nu}$ or the antisymmetric field $B_{\mu\nu}$, which are nothing but the massless modes of the (super)string itself. Then, a so-called nonlinear σ-model action can be considered of the form

$$S(X^\mu) = \frac{1}{2} \int d^2x \, [\sqrt{-g} \, g_{\alpha\beta} \, G_{\mu\nu}(X^\rho) \, \partial^\alpha X^\mu \, \partial^\beta X^\nu + \qquad (5.1)$$
$$+ \varepsilon_{\alpha\beta} \, B_{\mu\nu}(X^\rho) \, \partial^\alpha X^\mu \, \partial^\beta X^\nu \,] \, ,$$

where $\varepsilon_{\alpha\beta}=-\varepsilon_{\beta\alpha}(\varepsilon_{10}=1)$ and d^2x is the usual volume element $d\sigma d\tau$.

The space-time coordinates of the string x^μ are the only quantum variables of the σ-model (5.1) and the equations resulting from imposing the conformal invariance can be interpreted as low-energy equations for the massless modes of the string.

If a Wick rotation $\tau \longrightarrow i\tau$ is performed, the world-sheet action becomes Euclidean and the surface swept out by the string can be regarded as a Riemann surface described (in coordinate patches) by a holomorphic variable z ($\sigma+i\tau$) and its antiholomorphic conjugate $\bar{z}$. The Riemann surfaces are topologically classified by their number of handles (genus) : for instance, sphere and torus have respectively genus zero and one.

Physically, the genus corresponds to the number of loops, i.e. to the perturbation order at which is considered the string propagation.

To verify whether conformal invariance is preserved at the quantum level, it is sufficient to prove the vanishing of the σ-model β-function[16]. This approach leads to specific conditions on the space-time geometry, which restrict the acceptable background manifolds. Anyway, these techniques become more and more difficult to be controlled as the loop number increases.

Now, another method of dealing with conformal symmetry is in progress. It is based on the generalization that Krichever and Novikov[17] (KN) have done of the Virasoro algebra (whose quantum version is (2.18)) to a Riemann surface of arbitrary genus. Of course, we cannot go here into the details of this work, but we limit ourselves to give the formal expression of the KN-algebra with central extension (which is unique). It is

$$[e_i, e_j] = \sum_{s=-g_0}^{g_0} c_{i,j}^s e_{i+j-s} + t\,\chi(e_i, e_j)\ ,\quad [t, e_i]=0\ ,\qquad (5.2)$$

where $g_0 = 3g/2$ (g is the genus), $c_{i,j}^s$ are the structure constants and the e_i's are (in the differential-geometric representation) meromorphic vector fields, generating the algebra. In the case g=0 we have $c_{i,j}^0 = j-i$ and the usual Virasoro algebra for the sphere (the only known in global terms before the KN-generalization) is recovered.

The most interesting aspect of (5.2) is that it has a universal aspect and contains at the same time the information coming from the genus (topology), through a graduation determined by the genus itself. It has been shown[18] that, using the KN-algebra, it is possible to define a BRST charge on arbitrary genus Riemann surface, i.e. to consider the quantum problem of the propagating bosonic string at an arbitrary perturbation order: the critical dimension turns out to be the universal D=26.

Moreover, we can consider the string interacting with background fields as in (5.1) and try to investigate the conformal invariance, at any perturbation order (any genus) in the string propagation, by the quantum implementation of the KN-algebra.

The consequences of this approach on the space-time structure are under investigation: they seem very interesting to find confirmations of knowns results and, eventually, to discover new possibilities for the low-energy theory.

6. Conclusions.

A significative progress has been made in string theory in the last three years. So, in spite of increasing mathematical difficulties, we can still be confident that the string approach can show us the solution to the long-standing problem of quantum gravity and, in consequence, to the problem of fundamental forces unification. Of course, we have touched only few aspects and possibilities of the theory: in particular, those related to heterotic strings seem promising for future phenomenology.

However, only the knowledge of a general "string gauge principle"[19] could give a definite structure to the theory in terms of "field theory of interacting strings".

Lacking till now this knowledge, reliable signals that the way followed is right should come from the computation of low-energy field theory limit. We expect from this computation a better understanding of aspects as: particle masses, coupling constants, neutrimo mass spectrum, decay rates (nucleon decay), etc..

Finally, even if string theory should not be expected to be a TOE (theory of everything), it will certainly have a great influence on the modern physics and in particular on the unification perspective.

References.

1) M.B. Green and J.H. Schwarz, Phys. Lett. 151B (1985) 21.

2) For a review, see for instance: M. Veltman, in "Methods in Field Theory", Les Houches 1975, North Holland Publishing Co. Amsterdam, 1976.

3) M.B. Green and J.H. Schwarz, Phys. Lett. 149B (1984) 117; Nucl. Phys. B255 (1985) 93.

4) For a beautiful discussion, see: M.B. Green, J.H. Schwarz and E. Witten, "Superstring theory" Cambridge Monographs in Math. Phys. (1987).

5) J.H. Schwarz, "Dual resonance theory", Phys. Reports 8 269 (1973).

6) A.M. Polyakov, Phys. Lett. 103B (1981) 207.

7) A.A. Belavin, A.M. Polyakov and A.B. Zamolodchikov, Nucl. Phys. B241 (1984) 333.

8) M. Virasoro, Phys.Rev. D1 (1970) 2933.

9) D.J. Gross, J.A. Harvey, and R. Rohn, Phys. Rev. Lett. 54 (1985) 502; Nucl. Phys. B256 (1985) 253 and B267 (1986) 75.

10) G.F. Chapline and N.S. Manton, Phys. Lett. 120B (1983) 105.

11) P. Candelas, G.T. Horowitz, A. Strominger and E. Witten, Nucl. Phys. B258 (1985) 46.

12) E. Witten, Phys. Lett. 149B (1984) 351.

13) L. Dixon, J.A. Harvey, C. Vafa and E. Witten, Nucl. Phys. B261 (1985) 651.

14) L. Bonora, P. Pasti and M. Tonin, Phys. Lett. 188B (1987) 335; L. Bonora, M. Bregola, K. Lechner, P. Pasti and M.Tonin, Nucl. Phys. B296 (1988) 877.

15) M. Tonin, preprint DFPD 9/88, Dept of Phys. Univ. of Padova, March 1988.

16) E. Guadagnini and M. Mintchev, preprint IFVP-TH 21/86.

17) I.M. Krichever and S.P. Novikov, Funk. Anal. i Pril. 21 No. 2 (1987) 46.

18) L. Bonora, M. Bregola, P. Cotta-Ramusino and M. Martellini, CERN-TH. 4889/87 to appear on Phys.Lett. B.

19) G.T. Horowitz and A. Strominger, Phys. Rev. Lett. 57 (1986) 519; E.Witten, Nucl. Phys. B268 (1986) 253.

II
NUCLEAR PHYSICS

NUCLEAR COLLECTIVE EXCITATIONS CARRYING ISOSPIN

Renzo Leonardi

Dipartimento di Fisica dell'Università, Trento, Italy

We review the bulk properties of the isovector collective electric modes like the dipole, the monopole, and the quadrupole, determining their strength and their mean excitation energies within schematic forces and sum rules. Their dependence on the symmetry energy, the exchange forces, the neutron and proton radial distributions, their similarities and their differences are illustrated in a simple manner. The peculiarity of the quadrupole excitation in respect to the nonlocality of the residual interaction is discussed.

1. INTRODUCTION

In our lectures we regard nuclei as non-relativistic assemblies of neutrons and protons interacting through the strong attractive nuclear force and the weak electromagnetic Coulomb repulsion between protons.

We know that nuclear force is charge-independent to a high degree. Isospin is shorthand for this charge independence and it entails enormous simplifications through the symmetries that is imposes upon nuclear structure and also great richness through the essentially approximate character of those simmetries.

As a result the isospin quantum numbers are useful to characterize a wide variety of nuclear states both in the discrete low-lying domain and in the high--lying region of collective resonances.

Once one excites a target nucleus it is extremely interesting to analyze the tensorial character of the excitation in respect to the isospin. An isoscalar excitation will be obtained without transfer of isospin; an isovector excitation transfers one unit of isospin to the nucleus and has three independent components; an isotensor (rank 2) excitation transfers two units of isospin and has five independent components and so on.

In this respect an electromagnetic excitation is in general an appropriate combination of an isoscalar and the charge conserving component ($\Delta T_3=0$) of a vector; a (π^+,π^-) excitation is the $\Delta T_3=-2$, $\Delta T=2$ component of an isotensor excitation, etc..

The use of probes which can transfer isospin to the target in a controlled way has proved extremely fruitful for the exploration of several nuclear properties. For example, isoscalar and different charge components of isovector projectiles have been widely used to excite giant collective states and analog resonances within isospin multiplets, thereby relating information extracted from photo- and electro-excitations, muon capture, beta decay, meson excitations, etc..

In what follows we focus on isovector excitations, and we review the kind of problems and the wide variety of nuclear properties one can analyze once *isospin* is transferred to a nuclear target.

Since we are not interested in the details of the strength distribution but rather in the isospin features, our treatment will depart from the conventional "microscopic" spectroscopy: we will deal, however, with the distribution, in energy, of states of fixed isospin by considering their *total strength* and their *centroid energies*. Schematic models and sum rules are the appropriate tools for studying these strengths and centroids and for illustrating the gross structure of some simple isovector modes like the dipole, the quadrupole, the monopole - their common features and their substantial differences and their links with many interesting nuclear properties.

The properties of these simple charge exchange excitations are deeply connected with many interesting and fundamental properties of nuclei - like differences between neutron and proton spatial distribution, isospin properties of the effective schematic interactions, isospin properties of the microscopic forces, structure of the symmetry energy, isovector compressibilities, isospin quantum numbers and selection rules, Coulomb dynamical effects, etc. When spin--isospin modes are taken into account, a further number of interesting problems are opened up. In the following, however, we shall refer to excitations, which do not change the spin quantum numbers.

Mesons have been proved extremely helpful to study charge-exchange excitations. D. Bowman and collaborators[1] did a number of systematic experiments using (π^-,π^0), (π^+,π^0) charge exchange reactions. The isospin quantum numbers and the dynamical properties of resonance-energy pions make the reactions favourable to study the electric isovector resonances like the dipole, the monopole and the quadrupole. These investigations partly confirmed some theoretical expectations based on microscopic models[2,3] and partly contradicted them. For example, the quadrupole mode turned out to be much weaker than expected.

2. ISOVECTOR EXCITATIONS AND SUM RULE CONSTRAINTS

The charge exchange excitation operators have the form $F^{\pm} = \sum_i f_i(r_i, \sigma_i \ldots) \tau_i^{\pm}$ and together with the charge conserving component $F^0 = \sum_i f_i(r_i, \sigma_i \ldots) \tau_i^3$, they constitute the components of the isovector operator $\vec{F} = \sum_i f_i(r_i, \sigma_i \ldots) \vec{\tau}_i$. These three components, when acting on a nuclear target $|T, T_3 = T(>\tfrac{1}{2})\rangle$, lead to the following well-known excitation scheme. According to the scheme of fig. 1, the strength $\langle 0|(F^-)^{\dagger} F^-|0\rangle \equiv \sigma^-$ is split into three fragments $\sigma^-(T-1)$, $\sigma^-(T)$, and $\sigma^-(T+1)$ at mean energies $E^-(T-1)$, $E^-(T)$, and $E^-(T+1)$ respectively.

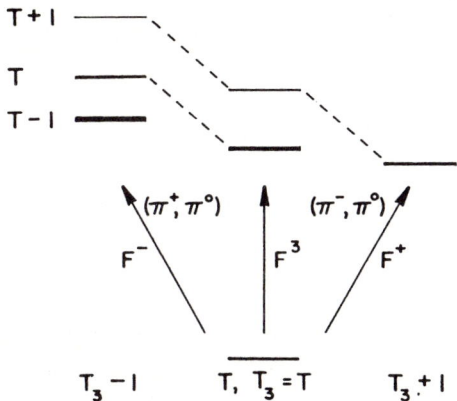

FIGURE 1
Schematic relationship for isovector excitations built on a T≥1 target nucleus

Since our target is fully polarized in the isospin space ($|0\rangle \equiv |T, T_3 = T\rangle$), the strength $\langle 0|(F^3)^{\dagger} F^3|0\rangle \equiv \sigma^0$ is split in only two fragments $\sigma^0(T)$ and $\sigma^0(T+1)$, with mean energies $E^0(T)$ and $E^0(T+1)$ whereas the strength $\langle 0|(F^+)^{\dagger} F^+|0\rangle \equiv \sigma^+$ has only one isospin fragment $\sigma^+(T+1)$ with a mean energy $E^+(T+1)$.

The scheme reflects the superposition of a number of dynamical effects:
(a) the neutron excess combined with the Pauli principle block the strength of the F^+ component in respect to that of the F^- component (fig. 2a and b). Furthermore, the differences in the spatial distribution of neutrons and

protons modify the total strength; in particular, a distribution of neutrons extending to larger radii than the protons reinforces the blocking whereas an opposite distribution diminishes the blocking.

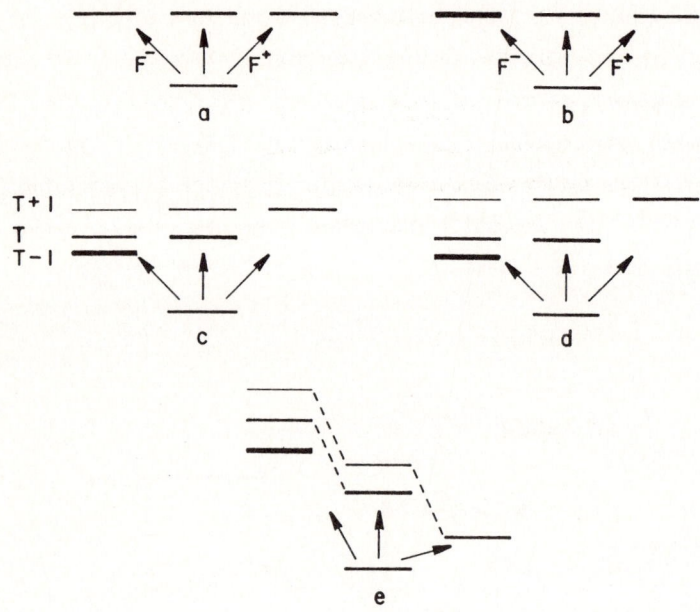

FIGURE 2
Qualitative illustration of the various effects leading to the scheme of fig. 1. (a) Isovector strength ignoring Coulomb shifts, excess neutron effects, and isospin-dependent forces. (b) Effect of neutron excess on the strengths. (c) Effect of neutron excess and antisymmetrization on the energy distribution of the strengths. The various isospin fragments are split and the strength rearranged. (d) Effect of microscopic isospin-dependent forces, which further rearrange and split the strengths. (e) Coulomb shift of the charge exchange modes in respect to the $\Delta T_3 = 0$ mode.

(b) The neutron excess combined with the antisymmetrization and the short-range nature of the nuclear force originate effective isospin-dependent interactions, which force the various isospin fragments to split and further rearrange the total strengths of the excitation modes and their distribution in energy (fig. 2c).

(c) Microscopic isospin-dependent forces contribute further to the energy-splitting and to the redistribution of the strength (fig. 2d).

(d) Coulomb forces remove the degeneracies within the same multiplets shifting up the $F^-|0\rangle$ excitation and lowering down the $F^+|0\rangle$ excitation energy in respect to the $F^3|0\rangle$ excitation (fig. 2e). Further (small) effects on the energies and strengths arise from the interference between Coulomb and nuclear forces (dynamical Coulomb distortions).

The strength distributions of the various modes must satisfy some general relations referred as sum rules: the simplest and most important are listed below.

Let $\sigma(\omega)d\omega$ be the amount of strength in the energy range ω, $\omega+d\omega$ (ω is measured in respect to the target ground state energy), then the following relations hold

$$\int \sigma^-(\omega)d\omega - \int \sigma^+(\omega)d\omega \equiv \sigma^- - \sigma^+ = \langle 0|[F^+,F^-[|0\rangle \tag{1}$$

$$\int \sigma^-(\omega)\omega d\omega + \int \sigma^+(\omega)\omega d\omega = \langle 0|[F^+,[H,F^-,]]|0\rangle \tag{2}$$

$$\int \sigma^0(\omega)\omega d\omega = (1/2)\langle 0|[F^3,[H,F^3]]|0\rangle \tag{3}$$

$$\int \sigma^-(\omega)\omega^2 d\omega - \int \sigma^+(\omega)\omega^2 d\omega = \langle 0|[[F^+,H],[H,F^-]]|0\rangle \tag{4}$$

where the following convention has been used: $[\tau^+,\tau^-]=2\tau_3$. Note that the r.h.s. of these equations involve commutators only. Many other sum rules can be established with the help of anticommutators. In contrast to the latter, the former relations can be easily evaluated within RPA accuracy, (i.e. including RPA correlations) just using as a nuclear ground state $|0\rangle$ an uncorrelated determinantal Hartree-Fock wave function. Where anticommutators are involved, a correlated ground state is necessary to obtain RPA accuracy.

The explicit evaluation of the sum rules can be performed once the nuclear Hamiltonian has been chosen and its ground state determined. Some general remarks are in order here. The commutator of eq. (1) can be simply expressed as

$$\langle 0|[F^+,F^-]|0\rangle = \langle \sum_i f_i^2 \tau_{3i}\rangle = N\langle f^2\rangle_n - Z\langle f^2\rangle_p$$

where $\langle f^2\rangle_n$ and $\langle f^2\rangle_p$ are the mean values of the operator $f_i^2(r_i)$ over the neutrons and protons, respectively. As anticipated the point (a), we can see here how the difference $\sigma^- - \sigma^+$ strongly depends on the spatial distribution of neutrons and protons. Small differences in $\langle f^2 \rangle_n$ and $\langle f^2 \rangle_p$ strongly affect $\sigma^- - \sigma^+$.

For the commutators of eqs. (2-4), one can separate out the contribution which arises from the kinetic energy term K_{in}, of H, from those involving the isospin-dependent part, $V(\tau)$ of the *microscopic* nuclear potential V. (The local isospin-independent part of V does not contribute to the commutators.) In particular,

$$\langle 0|[F^+,[F^-,H]]|0\rangle = \langle 0|[F^+,[F^-,K_{in}]]|0\rangle + \langle 0|[F^+,[F^-,V(\tau)]]|0\rangle$$

where the first part of the commutator is a largely model-independent one-body operator. A similar separation holds for the operator F^3. Note that.

$$\langle 0|[F^+,[K_{in},F^-]]|0\rangle = \langle 0|[F^3,[K_{in},F^3]]|0\rangle = \frac{A}{m}\langle (\nabla f)^2 \rangle.$$

The kinetic energy term of the eq. (4) is

$$\langle 0|[[F^+,K_{in}],[K_{in},F^-]]|0\rangle = \frac{1}{2m^2}\langle \sum_i (\vec{\nabla} f \cdot \vec{\nabla} + \vec{\nabla} \cdot \vec{\nabla} f)^2 \tau_i^3 \rangle.$$

Whereas effective isospin-dependent forces (point b) do not contribute to the commutators of eqs. (2-6), *microscopic* isospin-dependent forces (point c) do, so one can distinguish between these two types of forces by looking at the sum rules (2-4).

3. SCHEMATIC APPROACH TO MEAN ENERGIES AND STRENGTHS

In order to sketch the gross properties of the various isovector components of some simple resonances as the dipole, the quadrupole, the monopole, their common properties, and their fundamental differences, we will choose an oversimplified framework[5].

(1) We disregard the Coulomb interaction (fig. 3a). Coulomb effects will be taken into account properly shifting the energies determined within our framework. This is equivalent to neglecting the effects of Coulomb distortion.

(2) We focus on large T nuclei. In fact, simple isospin geometry arguments (further reinforced by the nuclear dynamics) tell us together with $\sigma^+ \equiv \sigma^+(T+1)$, for large T nuclei $\sigma^- \equiv \sigma^-(T-1)$, $\sigma^0 \equiv \sigma^0(T)$ because the weakness

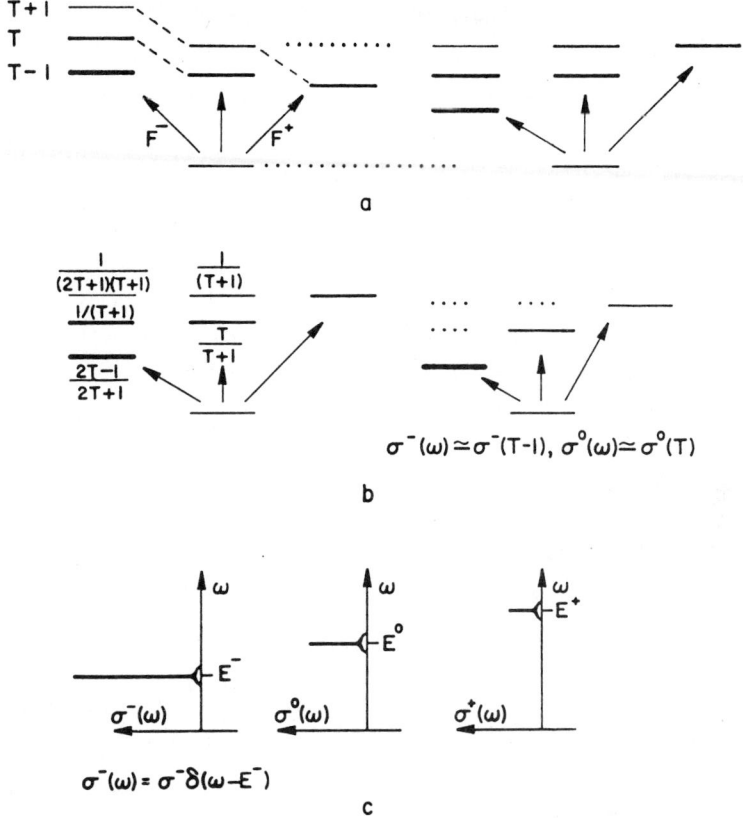

FIGURE 3
Illustration of the various approximations used. (a) Energy distribution (schematic) of the various isospin fragments once the Coulomb shifts are ignored. (b) Leading isovector fragments in the limiting case of large isospin T. (c) Strength distributions for schematic Hamiltonian and closed shell nuclei.

of the strength of the other fragments. As a consequence, the mean energy $\bar{E}^-$ of σ^- coincides essentially with $\bar{E}^-(T-1)$ and the centroid $\bar{E}$ of σ^0 coincides with $\bar{E}^0(T)$. (For $T=T_3$ targets, $\bar{E}^+$ always coincides with $\bar{E}^+(T+1)$). In this limiting case, we can assign a definite isospin to the excitation associated with σ^-, σ^0, and σ^+ respectively (fig. 3b). This is an important point because it opens up the possibility of calculating the strength and the energies of the various isospin fragments within the Random Phase Approximation (RPA) in spite of the fact that the RPA excitations refer to a mixture of isospin fragments[2,5]. Once the T dependence of the various quantities has been determined in the limiting case of large T, many results can be generalized to the other nuclei exploiting the appropriate isospin geometry[5].

(3) We consider nuclei consisting of closed shells so that the excitations produced by an operator F characterized by a given multipolarity, isospin, spin, etc., are an appropriate superposition of single particle-hole excitations.

(4) For each operator $\vec{F}$, we use simple schematic Hamiltonians leading to degenerate unperturbed individual particle-hole excitations; the actions of an external isovector field on the target ground state $|0\rangle$ produces excitations $F^{\pm,0}|0\rangle$, which are eigenstates of the single-particle schematic Hamiltonian. For this approximation the strength distributions of the three modes are simply proportional to three δ-functions (fig. 3c). This simple model illustrates the most important characteristic of the microscopic description of the collective modes and can be generalized to more sophisticated schemes.

For each excitation operator F, some general prescriptions allow us to derive a schematic Hamiltonian to be used in the RPA equations of motion from the following (Skyrme-like) energy functional

$$E = \langle \phi|H|\phi\rangle = \frac{1}{2m}\int \tau_K dV + \int v(\rho)dV + \frac{V_1}{8}\int \frac{\rho_1^2}{\rho}dV + \ldots \qquad (5)$$

where $\tau_K = \sum_i \vec{\nabla}\psi_i \vec{\nabla}\psi_i$, $\rho = \sum_i \psi_i \psi_i$, $\rho_1 = \sum_i \psi_i \vec{\tau}\psi_i$ (= the isovector density). $V_1 \simeq 100-130$ MeV. In (5) we have omitted for simplicity further velocity-dependent terms (current terms), which originate from microscopic exchange potentials.

The schematic Hamiltonian H emerging for an operator $\vec{F} = \sum_i f_i(r)\vec{\tau}_i$ (with the property $\langle \sum_i f_i \cdot \tau_{3i}\rangle = 0$) is[5]

$$H = H_0 + \sum \xi \langle f_V^2 \rangle T\tau_{3i} + \xi \langle \vec{F}\rangle \cdot \vec{F} \qquad (6)$$

where H_0 is a static part of the Hamiltonian and for our purpose can be approximated with a harmonic oscillator Hamiltonian. The second term is a static Lane-type isovector potential affecting differently the neutron and the proton orbitals.

$$\langle f_V^2 \rangle = \frac{1}{N-Z} \langle \sum_i f_i^2 \tau_{i3} \rangle \qquad (7)$$

and

$$\xi = \frac{1}{4} \frac{V_1}{A \langle f^2 \rangle} \qquad (8)$$

where $\langle f^2 \rangle = \frac{1}{A} \langle \sum_i f_i^2 \rangle$. The third term is an appropriate renormalization of the single-particle isovector effects characteristic of the exciting external field F.

For each excitation operator $\vec{F}$ one can use the schematic Hamiltonian H to solve the equations of motion

$$[H, O_F^+]_{RPA} |0\rangle = E_F O_F^+ |0\rangle \qquad (9)$$

by explicitly determining the operators O_F^+ and the energies E_F associated with elementary modes excited by the operator $\vec{F}$.

4. DIPOLE

The dipole excitation induced by the operator $\vec{F} = \frac{1}{2} \sum z_i \vec{\tau}_i$ is perhaps the most widely studied excitation both in the $\Delta T_3 = 0$ and $\Delta T_3 = \pm 1$ charge-exchange channels. A great number of experimental and theoretical works are dedicated to this mode[6]. Even though each isovector operator has its own properties and characteristics, the dipole can be considered as the most representative of them so that we will discuss in some detail the gross behavior of these excitations, the view of illustrating some methods and results which will be helpful in studying the monopole and quadrupole operators.

For the dipole case one can assume that the individual unperturbed particle--hole excitations are degenerated at $\hbar\omega_0$ energy so that the action of the external field $\vec{F}$ will produce an excitation $F|0\rangle$, which is an *eigenstate* of the one-particle Hamiltonian (6), so that we will determine strength and eigenenergies of the excitation operator F simply solving the RPA equation of motion (9). The appropriate schematic Hamiltonian to be used is[5]

$$H = \sum_i \left(\frac{p_i^2}{2m} + \frac{1}{2} m\omega_0^2 r_i^2 + \frac{V_1}{2} \frac{T}{A} \tau_i^3 \frac{\langle r_V^2 \rangle}{\langle r^2 \rangle} \right) + \xi(\vec{D}) \cdot \vec{D} \qquad (10)$$

where $\omega_0 = 41 A^{-1/3}$ and $\vec{D} = \frac{1}{2} \sum_i z_i \vec{\tau}_i$. Once one solve the RPA equation of motion, the following results[5] emerge for the dipole:

$$E^0 = \sqrt{\omega_0^2 + \frac{3}{4m} \frac{V_1}{\langle r^2 \rangle}};$$ (11)

$$\sigma^0 = \frac{A}{8m} \frac{1}{E^0}$$ (12)

$$\sigma^{\pm} = \left((E^0) + \left(\frac{V_1}{2A} T \frac{\langle r_V^2 \rangle}{\langle r^2 \rangle} \right)^2 \right)^{-1/2} \left(\frac{A}{8m} + \frac{V_1 (T \langle r_V^2 \rangle)^2}{12 \langle r^2 \rangle A} \right) \pm \frac{T}{6} \langle r_V^2 \rangle$$ (13)

$$E^{\pm} = \sqrt{(E^0)^2 + \left(\frac{V_1}{2A} T \frac{\langle r_V^2 \rangle}{\langle r^2 \rangle} \right)^2} \pm \frac{T}{2} \frac{V_1}{A} \frac{\langle r_V^2 \rangle}{\langle r^2 \rangle}$$ (14)

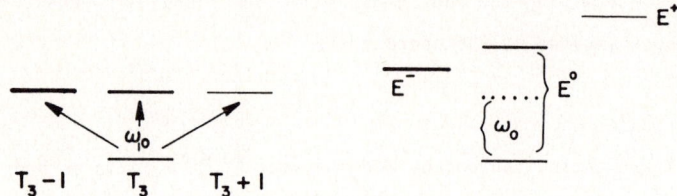

FIGURE 4
Qualitative behavior of the energies and strengths of the dipole mode, in the unperturbed situation (left) and with $V_1 \neq 0$ (right).

Note that E^+ and E^- can be parametrized as follows:

$$E^{\pm} = E^0 \pm \Delta_V + \Delta_T$$ (15)

where

$$\Delta_T = \frac{T}{2} \frac{V_1}{A} \frac{\langle r_V^2 \rangle}{\langle r^2 \rangle}$$ (16)

and, up to $\left(V_1 \frac{T}{A} \right)^2$ order,

$$\Delta_V = \frac{1}{2E^0} \left(\frac{T}{2} \frac{V_1}{A} \frac{\langle r_V \rangle}{\langle r^2 \rangle} \right)^2 = \frac{1}{2E^0} \Delta_V^2 .$$ (17)

Finally, up to $\left(V_1 \frac{T}{A} \right)^2$ order,

$$\sigma_\pm = \frac{1}{E^0}\left(\frac{A}{8m} + \Delta_V \frac{T\langle r_V^2\rangle}{6E^0} - \frac{A\Delta_T}{8m(E^0)^2}\right) \pm \frac{T}{6}\langle r_V^2\rangle. \tag{18}$$

The microscopic meaning of Δ_V and Δ_T is discussed elsewhere[7].

It is rather instructive to analyze these results in terms of sum rules. For the dipole operator and the Hamiltonian (10) the following relations hold:

$$\sigma^- - \sigma^+ = \langle 0|[F^+, F^-]|0\rangle = \frac{T}{3}\langle r_V^2\rangle \equiv f_0; \tag{19}$$

$$E^0\sigma^0 = \frac{1}{2}\langle 0|[F^3,[H,F^3]]|0\rangle = \frac{A}{8m} \equiv \frac{f_1}{2}; \tag{20}$$

$$E^-\sigma^- + E^+\sigma^+ = \langle 0|[F^+,[H,F^-]]|0\rangle = \frac{A}{4m} \equiv f_1; \tag{21}$$

$$(E^-)^2\sigma^- - (E^+)^2\sigma^+ = \langle 0|[[F^+,H],_0H,F^+]]|0\rangle = \omega_0^2 \frac{T}{3}\langle r^2\rangle \equiv f_2. \tag{22}$$

The first sum rule is rather independent (it depends on the model Hamiltonian through the ground state only) and illustrates clearly the blocking effect ($\sigma_- > \sigma_+$) caused by the Pauli principle and the effect of neutron and proton distribution. (This class of sum rules has been introduced in connection with the charge exchange excitation in ref. 4.)

The second sum rule is a particular case of the T-R-K sum rule. Note that within our model one single excited (collective) state at E^0 energy exhausts the sum. Since our microscopic Hamiltonian (5) does not contain any microscopic exchange potential [the exchange potential would generate "current" dependent terms in (5)] the sum rule (20) is just $\langle\frac{1}{2}F^3[,[K_{in},F^3]]\rangle$. Note, however, that our schematic Hamiltonian (10) contains effective isospin-dependent terms: these last (which arise in our model from the antisymmetrization) leave unaffected the sum rule. Similar arguments hold for the energy weigthed sum rule (21).

The fourth sum rule concerns the difference between the square energy weighted strength in the $\Delta T_3 = -1$, $\Delta T_3 = +1$ channels. Once more, since we have disregarded in (5) the "current" terms and further nonlocality terms, the sum rule is just equal to $\langle[[K_{in},F^+],[K_{in},F^-]]\rangle$, which is just proportional to the difference between the kinetic energy of neutrons and the protons. Within our approximations (harmonic oscillator approximation for H_0) the difference between the mean kinetic energy of the neutrons and of the protons is proportional to $\sigma^- - \sigma^+$.

In order to appreciate the powerfulness of these sum rules, we will derive from them relations (15-18), within the hypothesis that $\Delta_T < \Delta_V$. In fact, inserting $E^{\pm} = E^0 \pm \Delta_V + \Delta_T$ in relations (19-22) and using an iterative procedure[7], one obtains from (19-22):

$$\Delta_V = \frac{1}{2}\left(\frac{f_0(E^0)^2 - f_2}{f_1}\right); \tag{23}$$

$$\Delta_T = \frac{1}{2}\frac{\Delta_V^2}{E^0}; \tag{24}$$

$$\sigma_{\pm} = \frac{1}{2}\left(\frac{f_1}{E^0} + \frac{f_1\Delta_V}{E^0} - \frac{f_1\Delta_T}{(E^0)^2} \pm f_0\right); \tag{25}$$

which, up to $V_1 \frac{T}{A}^2$ order, become identical to (13-14) if one uses for E^0 the expression (11).

According to point (1) of Sect. 3, we have disregarded the Coulomb interaction part H_c of H in (10). Relations (1-4) can be evaluated taking into account explicitly H_c. In this case the energies of the left-hand side of the sums involving charge exchange modes include the Coulomb shift, whereas on the right-hand side the Hamiltonians contain H_c. Neglecting Coulomb in (19-22) is equivalent to assuming that the energy shift due to the Coulomb energy in the left-hand side just cancels with the contribution of the commutators arising from the H_c part of H in the rigth-hand side. Within this approximation one can handle the problem of the Coulomb energy, ignoring H_c in the commutators and properly shifting the E^+, E^- energies as emerging from formula (15).

To the extent in which the previous cancellation is not exact, Coulomb distortion play a role. This problem is discussed elsewhere[7].

5. DIPOLE VELOCITY-DEPENDENT CORRECTIONS

As mentioned before, current terms have been neglected in (5). One can easily incorporate their effects in our theory[8,9]. The following results are then obtained for the schematic Hamiltonian

$$H = \sum_i \left(\frac{p_i^2}{2m^*} + \frac{1}{2}\frac{m}{m^*}\omega_0^2 r_i^2 + \frac{1}{2}V_1\frac{\langle f_v^2\rangle}{\langle f^2\rangle}T\tau_{3i} + \frac{1}{4}\frac{V_1}{A}\langle\sum_j f_j\vec{\tau}_j\rangle \cdot f_i\vec{\tau}_i \right.$$

$$\left. + \frac{K_1}{4mA\langle(\vec{\nabla}f)^2\rangle}\langle(\sum_j \vec{\nabla}f_j \cdot \vec{\nabla} + \vec{\nabla}\cdot\vec{\nabla}f_j)\rangle \cdot (\vec{\nabla}f_I \cdot \vec{\nabla} + \vec{\nabla}\cdot\vec{\nabla}f_I)\vec{\tau}_i\right) \tag{26}$$

for operators $\vec{F} = \sum_i f_i(r_i)\vec{\tau}_i$ having the property $\langle 0|\sum_i f_i \tau_{3i}|0\rangle = 0$, where K_1 is an appropriate coupling constant connected with the isovector current generated in (5) by nonlocality terms and m* is an effective mass (for details, see ref. 9).

In the dipole case $2f_i = z_i$ and the explicit formulas for energies and strengths become:

$$E^0 = \sqrt{(1+K)\left(\frac{m}{m^*}\omega_0^2 + \frac{3}{4m}\frac{V_1}{\langle r^2\rangle}\right)} ; \qquad (27)$$

$$\sigma_\pm = \frac{1}{2}\left(\frac{A}{4m}(1+K)\frac{1}{E^0} + \frac{f_0\Delta_V}{E^0} - \frac{A}{4m}(1+K)\frac{\Delta_T}{(E^0)^2} \pm f_0\right) ; \qquad (28)$$

$$\sigma_0 = \frac{1}{8}\frac{A}{m}(1+K)\frac{1}{E^0} ; \qquad (29)$$

$$\Delta_V = \frac{V_1}{2}\frac{T}{A}\frac{\langle r_V^2\rangle}{\langle r^2\rangle}\left(1 + \frac{4}{3}K_1\omega_0^2 m\frac{\langle r^2\rangle}{V_1}\right), \qquad \Delta_T = \frac{1}{2E^0}(\Delta_V)^2 ; \qquad (30)$$

where K, the enhancement factor of the dipole energy weighted sum rule, is linked to m* and K_1 by the relation $K = \frac{m}{m^*} - 1 - K_1$. It is trivial to work out from formulas (11-18) and (27-30) the numerical values of the relevant quantities, for various nuclei. Typical values of the parameters entering the formulas are:

$$\omega_0 = 40A^{-1/3} \text{ MeV}, \qquad V_1 = 100 \div 130 \text{ MeV},$$

$$\frac{\langle r_V^2\rangle}{\langle r^2\rangle} = 1 \div 1.3, \qquad K = 1.2 \div 1.4, \qquad m^* = 1 \div 0.7,$$

$$K_1 = \frac{m}{m^*} - 1 - K, \qquad \langle r^1\rangle^{1/2} = 1.1 A^{1/3}.$$

6. MONOPOLE

The monopole operator is generally expressed as $\vec{F} = \sum_i r_i^2 \vec{\tau}_i$. Differently from the dipole operator, however, for doubly closed nuclei $T \neq 0$, the individual unperturbed particle-hole excitations relevant to the F^- excitation operator are partly degenerate at $2\hbar\omega_0$ energy and partly at $0\hbar\omega_0$. This last group of states which can take a large part of the strength is generally referred to as the isobar analog state, $|IAS\rangle$, whereas the first group originates the giant monopole state in the $\Delta T = -1$ channel, so that in the case of the monopole $\bar{\sigma}^- = \bar{\sigma}_{IAS}^- + \bar{\sigma}_{2\hbar\omega_0}^-$. We can, however, decouple the $|IAS\rangle$ from the giant state with an interesting trick. In fact, if one defines $\vec{F}_M = \sum_i (r_i^2 - \langle r_V^2 \rangle)\vec{\tau}_i$, then F_M^- excites the giant state only; in fact, $\langle IAS|F_M^-|0\rangle = 0$.

This allows us to study the monopole giant state along a path rather similar to the one followed for the dipole. The appropriate schematic Hamiltonian is obtained from (6), putting $f_i = (r_i^2 - \langle r_V^2 \rangle)$. The following results are obtained:

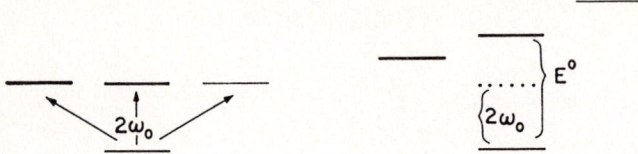

FIGURE 5

Qualitative behavior of the energies and the strengths of the monopole mode, in the unperturbed situation (left) and with $V_1 \neq 0$ (right).

$$E^0 = \sqrt{4\omega_0^2 + \frac{V_1}{m} \frac{r^2}{\langle r^4 \rangle - \langle r^2 \rangle^2}} \; ; \tag{31}$$

$$E^\pm = E_0 \pm \Delta_V + \Delta_T \; ; \tag{32}$$

$$\sigma^0 = \frac{1}{2} \frac{f_1}{E^0} \; ; \tag{33}$$

$$\sigma^\pm = \frac{1}{2} \left(2\sigma^0 + \frac{f_0 \Delta_V}{E^0} - \frac{f_1 \Delta_V}{(E^0)^2} \pm f_0 \right) \tag{34}$$

(where σ^- refers to the giant state $\sigma_{2\hbar\omega_0}$);

$$\Delta_V = \frac{1}{2} \frac{T}{A} \frac{V_1}{2} \frac{(\langle r_V^4 \rangle - \langle r_V^2 \rangle^2)}{(\langle r^4 \rangle - \langle r^2 \rangle^2)} \tag{35}$$

$$\Delta_T = \frac{1}{2} \frac{\Delta_V^2}{E^0} \; ; \tag{36}$$

and f_1, f_0 stand for

$$f_1 = \frac{4}{m} A \langle r^2 \rangle, \quad f_0 = 4T(\langle r_V^4 \rangle - \langle r_V^2 \rangle^2). \tag{37}$$

formulas (34) and (36) are the RPA solutions of eq. (9) up to $\left(\frac{TV_1}{A}\right)$ order.

[The exact solutions, "mutatis mutandis", are similar to those of the dipole case, (13) and (14)].

Sum rules (1) to (4), when evalutated consistently with our Hamiltonian, give

$$\overline{\sigma}_{2\hbar\omega_0} \cdot \overline{\sigma}^+ = f_0 \; ; \tag{38}$$

$$E^- \overline{\sigma}^- + E^+ \overline{\sigma}^+ = f_0 \; ; \tag{39}$$

$$(E^-)^2 \overline{\sigma}^- - (E^+)^2 \overline{\sigma}^+ = f_2 \; ; \tag{40}$$

where the operator $\vec{F}$ is $\sum_i (r_i^2 - \langle r_V^2 \rangle)$, f_0 and f_1 are as defined in (37), and $f_2 = 4\omega_0^2 f_0$.

One can easily incorporate in our Hamiltonian (6) velocity-dependent terms[9]. The appropriate Hamiltonian is obtained from (26) with the choice $f_i = r_i^2 - \langle r_V^2 \rangle$. Their effect in the formulas is:

$$E_0 = \sqrt{(1+K)\left(\frac{4m}{m^*}\omega_0^2 + \frac{V_1}{m}\frac{\langle r^2 \rangle}{\langle r^4 \rangle - \langle r^2 \rangle^2}\right)} \; ; \tag{41}$$

$$\Delta_T = \frac{T}{A}\frac{V_1}{2}\frac{\langle r_V^4 \rangle - \langle r_V^2 \rangle^2}{\langle r^4 \rangle - \langle r^2 \rangle^2}\left\{1 + \frac{K_1 4\omega_0^2 m}{V_1}\frac{\langle r^4 \rangle - \langle r^2 \rangle^2}{\langle r^2 \rangle}\right\} \; ; \tag{42}$$

$$\Delta_T = \frac{1}{2}\frac{\Delta_V^2}{E^0} \; ; \tag{43}$$

$$f_1 = (1+K)\frac{4A}{m}\langle r^2 \rangle \; ; \tag{44}$$

$$f_2 = (1+K)^2 4\omega_0^2 f_0 \; ; \tag{45}$$

where K_1, K and m^* are defined in Sect. 5 (see also ref. 9). Formulas (31) through (45) should be compared with (15) through (26). One realizes that the isospin structure fo the giant monopole resonance is quite similar to the structure of the dipole case. Small differences, however, between the neutron and proton distributions could affect in a different way these two isovector modes. Nonlocality corrections also play a different role in the two cases.

7. QUADRUPOLE

The quadrupole mode corresponds to the excitation operator $\vec{F} = \sum_i x_i x_i \vec{\tau}_i$ (the radial part can be expressed in various different ways).

Similarly to the monopole case, for doubly closed T≠0 nuclei the individual unperturbed particle-hole excitations relevant to the F^- component of the quadrupole operator are partly degenerate at $2\hbar\omega_0$ ($J=2^+$) energy and partly at $0\hbar\omega_0$ ($J=2^+$). In the past, little attention has been paid to this last group of unperturbed states[10] (see fig. 6); we will see, however, that these states could play a rather peculiar role determining the properties of the quadrupole operator. In fact, differently from the monopole case, there is not a simple trick in order to decouple the low energy mode from the giant quadrupole state.

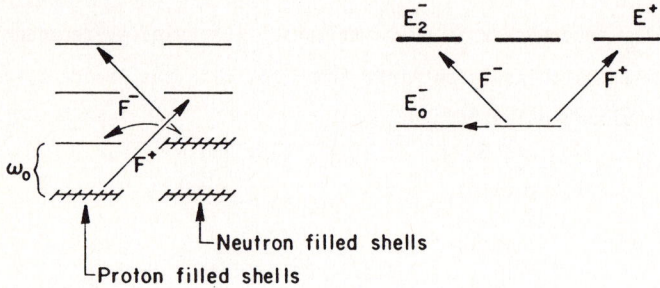

FIGURE 6
Scheme illustrating the origin of the various charge exchange unperturbed excitations in the case of the quadurpole excitation $F^\pm$.

On the other hand, experimental work[1] has shown that no compact isovector quadrupole strength is needed to fit the pion charge exchange data. This result contrasts with the situation for the isovector monopole and dipole resonances whose excitation energies, widths, and strengths are well described by RPA calculations using Skyrme forces. Calculations using the same Skyrme forces predict large compact isovector quadrupole resonances in contrast with the measurements[3].

In the following we discuss these problems. The relevant schematic Hamiltonian for the quadrupole operator is obtained from (6) putting $f_i = x_i y_i$. It is instructive to solve the RPA equation of motion in the limit of $V_1 \to 0$.

One can easily work out the solutions of eq. (9). In fact, exploiting the following commutation relations:

$$[H_0, F^\pm] = iG, \quad [H_0, iG^\pm] = 2\omega_0^2(M^\pm + F^\pm), \quad [H_0, M^\pm] = iG^\pm \tag{46}$$

where

$$M^\pm = \frac{1}{m\omega_0^2} \sum_i p_{y_i} p_{z_i} \tau_i^\pm, \quad G^\pm = \frac{1}{m} \sum_i (y_i p_{i_z} + z_i p_{y_i}) \tau_i^\pm. \tag{47}$$

One can guess that the solutions have the form

$$O^\pm = \alpha F^\pm + \beta M^\pm + \gamma iG^\pm. \tag{48}$$

In fact, inserting $O^\pm$ in eq. (9) one obtains, for the eigenvalues E, the following equations:

$$2\gamma\omega_0^2 = E, \quad 2\gamma\omega_0^2 = \beta E, \quad \alpha + \beta = \gamma E.$$

The system has solutions for E=0 ($\gamma=0$, $\alpha=-\beta$) and for $E=2\omega_0$ ($\alpha=\beta$, $\alpha=\gamma\omega_0$). The corresponding normal modes are:

$$O^-_{E=0} = \alpha_-(F^- - M^-); \tag{49}$$

$$O^+_{E=2\omega_0} = \gamma_+(\omega_0 F^+ + \omega_0 M^+ + iG^+); \tag{50}$$

$$O^-_{E=2\omega_0} = \gamma_-(\omega_0 F^- + \omega_0 M^- + iG^-); \tag{51}$$

where α_-, γ_+, and γ_- are normalization factors.

The excitation operators (49) through (51) excite separately the three unperturbed charge exchange states of fig. 6. In order to get the estimate of the strengths of each of these three modes, one can invert the sum rules (1-4) evaluated within our quadrupole schematic Hailtonian in the limit $V_1 \to 0$. These sum rules are (with the obvious meaning of the symbols)

$$\sigma_0^- + \sigma_0^- - \sigma^+ = \frac{4T}{15} \langle r_V^4 \rangle;$$

$$2\omega_0(\sigma^- + \sigma^+) = \frac{2A\langle r^2 \rangle}{3m} \tag{52}$$

$$(2\omega_0)^2(\sigma_2^- - \sigma^+) = 2\omega_0^2 \left(\frac{4T}{15} \langle r_V^4 \rangle \right).$$

One obtains

$$\sigma_0^- = \frac{1}{2} \left(\frac{4T}{15} \langle r_V^4 \rangle \right) ;$$

$$\sigma_2^+ = \frac{1}{2\omega_0} \left(\frac{A \langle r^2 \rangle}{3m} \right) + \frac{\sigma_0^-}{2} ; \qquad (53)$$

$$\sigma_2^+ = \frac{1}{2\omega_0} \left(\frac{A \langle r^2 \rangle}{3m} \right) - \frac{\sigma_0^-}{2} .$$

It is simple to verify that a large amount of strength is absorbed by the $E_0=0$ state. (As an example for ^{90}Zr $\sigma_0^- \cong \frac{1}{2} \sigma_2^-$, see Table I, first row).

TABLE I
Energies (in MeV) and strength distributions (fm⁴) for the isovector quadrupole mode, for ^{90}Zr.

	E_0^-	σ_0^+	E_2^-	σ_2^+	E_2^+	σ_2^+
$V_1=0$	0	490	18.3	1462	18.3	972
$V_1=120$ MeV	-9	68	23	1319	36.4	407
[ref. 3]	--	--	24.3	1200	31.9	697

The previous simple exercise, done in the limit $V_1 \to 0$, has shown us that:
(i) The operator $F^- = \sum x_i y_i \tau_i^-$ excite two normal modes ($E_0^-=0$, $E_2^-=2\omega_0$).
(ii) The operator $0^\pm$, creating the normal modes, contains nonlocality terms of type $M^\pm$.
(iii) A consistent part of the σ^- strength is absorbed by the 0-energy level.

These results have been obtained in the limiting case $\xi \to 0$. When the interaction is taken into account, i.e., when the full Hamiltonian (6) is used in eq. (9), result (iii) is no longer valid and the strength of the F^- excitation is almost entirely concentrated in the high-energy state $E_2^-|$ (fig. 7); see also Table I, second row. This explains the results of ref. 3 since our Hamiltonian (6) is derived from Skyrme-line functionals. Point (ii), however, suggests to us that, for the quadrupole mode, Hamiltonian (6) is a particular case of a more general one. In fact, since the structure of the excitation operators $O^{\pm}$, a more general schematic Hamiltonian appropriate for the quadrupole operator should include nonlocality terms of the kind $\langle\vec{M}\rangle\cdot\vec{F}$, $\langle\vec{M}\rangle\cdot\vec{M}$, etc. Once time-reversal and parity invariance requirements are satisfied, the more general Hamiltonian is[9]

$$H = H_0 + \sum_i \xi \langle f^2 \rangle T \tau_{3i} + a \langle \vec{F} \rangle \cdot \vec{F} + b \langle \vec{F} \rangle \cdot \vec{M} + c \langle \vec{M} \rangle \cdot \vec{F} + d \langle \vec{M} \rangle \cdot \vec{M} , \qquad (54)$$

where a, b, c, and d are new parameters.

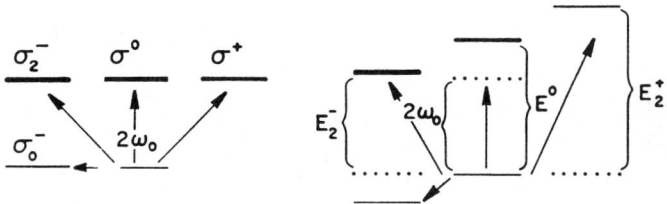

FIGURE 7
Qualitative behavior of the energies and the strengths of the quadrupole mode, in the unperturbed situation (left) and with $V_1 \neq 0$ in the nuclear Hamiltonian (6) (right).

In (54) we have omitted a current-current term proportional to $\langle\vec{G}\rangle\cdot\vec{G}$, which would originate from the current terms disregarded in (5). This term leads to an enhancement factor in the energy weighted sum rules and it is the counterpart of similar terms illustrated, for example, in (26). Our goal here is to discuss the role of the terms proportional to a, b, c, and d.

Within the Hamiltonian (54) the sum rules (1) through (4) becomes:

$$\sigma_0^- + \sigma_2^- - \sigma^+ = \frac{4T}{15} \langle r_v^4 \rangle, \qquad \sigma^0 E^0 = \frac{A \langle r^2 \rangle}{3m} ,$$

$$E_0^- \sigma_0^- + E_2^- \sigma_2^- + E^+ \sigma^+ = \frac{2}{3m} A \langle r^2 \rangle + (a-\xi)\left(\frac{4T}{15} \langle r_v^4 \rangle \right) ,$$

$$(E_0^-)^2\sigma_0^- + (E_2^-)^2\sigma_2^- - (E^+)^2\sigma^+ = 2\omega_0^2\left(\frac{4T}{15}\langle r_V^4\rangle\right) \tag{55}$$

$$+ \frac{4}{3m} A\langle r^2\rangle \left(\frac{4T}{15}\langle r_V^4\rangle\right)(a+b-\xi)$$

$$+\left(\frac{4T}{15}\langle r^4\rangle\right)^3 ((a-\xi)^2+b^2),$$

and E^0 turns out to be

$$E^0 = \sqrt{4\omega_0^2 + (a+b+c+d)\frac{2}{3}A\frac{\langle r^2\rangle}{m}}. \tag{56}$$

Similar equations can be deduced for E_0^-, E_2^-, and E^+. The new parameters a, b, c, and d allow drastic changs in the strength distribution and the mean energies of the quadrupole excitation. The predictions of the Hamiltonian (6) are obtained from (55) and (56) putting $a=\xi$ and $b=c=d=0$.

Let us take the simplifying assumption that the energy of the giant isovector quadrupole resonance in the $\Delta T_3=0$ channel (E^0) is the same as the energy clculated with the traditional schematic Hamiltonian[11] (6). (This implies that the strength is also the same because the sum rule (3) is unchanged). Let us further impose that when the parent nucleus is deformed, the low-lying rotational excitation $\Delta T_3=0$ (the so-called M1 rotational state) remains unchanged at the energy rpedicted by the Hamiltonian (6); then three relations are obtained[10] between the parameters a, b, c, and d, reducing them to one free parameter a:

$$a+b+c+d=\xi, \quad a+d=\xi, \quad ad=bc. \tag{57}$$

The strength parameter a is then constrained by either of the conditions $a \leq 0$ or $a \geq \xi$.

It turns out that within the previous conditions the energies E_0^-, E_2^-, and E^+ predicted by the Hamiltonian (54) remain the same as those predicted by the traditional Hamiltonian (6). We have studied the dependence of the strengths σ_0^-, σ_2^-, and σ^+ on the parameter a. Rather than write down the formal expressions for σ_0^-, σ_2^-, and σ^+, let us illustrate the situation in the practical case of ^{90}Zr.

In Table I we have reported the energies and the strengths for the unperturbed case [$V_1=0$ in (6)] and for $V_1=120$ MeV. One can see in this example a practical case of the behavior already anticipated in the general discussion. The low-lying state goes from a strength of 430 units of the unperturbed situation ($V_1=0$) to a strength of 70 units only ($V_1=120$ MeV). The results for the generalized Hamiltonian (54) are illustrated in Table II in the case of conditions (57).

TABLE II
Quadrupole strength distribution (fm^4) in function of a/ξ, for ^{90}Zr.

a/ξ	0	-0.3	-0.5	-0.8	-1	-1.18	1
$\bar{\sigma}_0$	1232	762	620	397	230	64	68
$\bar{\sigma}_2$	598	759	788	835	875	919	1319
σ_2^+	849	541	427	252	125	2	407
Total Strength	2679	2062	1835	1484	1230	985	1794

The case $\xi=a$ corresponds to the traditional model. As soon as a becomes a bit larger than ξ, the strength $\bar{\sigma}_0$ goes to negative values. Similarly, when a/ξ<-1.20, the strength σ^+ becomes negative so that the physical range of variation of is a≅ξ and -1.20<a/ξ<0. From Table II one can see a number of very different situations. For example, for a/ξ≅-0.5 the total quadrupole strength of the $\Delta T_3=-1$ channel is equally spread between the low-lying and the giant state, whereas in the $\Delta T_3=+1$ channel, the strength remains practically the same as in the traditional case ($\xi=a$), whereas for a/ξ≅-1.2, the strength in the $\Delta T_3=+1$ is entirely suppressed. Also, the strength on the $\Delta T_3=-1$ channel is somewhat suppressed in respect to the $\xi=a$ case.

8. FURTHER REMARKS

Quite recently special interest has been focused on *isotensor* excitations. Isotensor ($\Delta T=2$) excitations must involve at least two nucleons in the process because of charge conservation and for this reason it is a process which is very much influenced by nucleon-nucleon correlations. Hence the excitation operator $\vec{F}$ involved in these processes is a two-body transition operator.

We mention here two processes of this kind. One is the pion double charge exchange ($\pi^{\pm}$, $\pi^{\mp}$) reaction recognized for many years as a useful tool for studying nucleon-nucleon correlations[12]. The pion factories at LAMPF, TRIUMF and SIN opened the possibility of measuring this reaction in nuclei and in the last years many experiments of high quality have been performed so that it is now possible to fulfill the hope of studying correlations in nuclei.

The other one is the double beta decay. For nuclei A≥70 spontaneous double beta decay is energetically allowed and the prediciton of the double beta decay rates (both for two neutrinos and zero neutrinos decay) open up the possibility of exploring the nature of the neutrino and more precisely its Dirac and Majorana masses[13]. Once more these predictions must be based on detailed nuclear structure studies in which nucleon-nucleon correlations play a crucial role.

If we focus on (π^+, π^-) reaction on nuclei the simplest possible transition is induced by the operator $F^- = \sum_{i,j=1}^{A} \tau_i^- \tau_j^-$ where τ^- change a neutron to a proton (Double isobaric analog state).

In this transition the initial and final nuclear wave functions have the same space-spin structure apart from Coulomb distortions. Because this operator is the square of the total isospin lowering operator, and thus only connects states in the same isospin multiplets, once it is applied to a nuclear ground state with isospin T, $T_3=T$ it will give a non vanishing result only to the extent that Coulomb force or other charge-dependent forces mix the state $T'=T$, $T_3'=T-2$ analog of the state $T'=T$, $T_3'=T$, into the $T=T-2$, $T_3=T-2$ final state.

($\pi^{\pm}$, $\pi^{\mp}$) reactions can also excite giant resonances in which two processes of the kind discussed in sects. 2, 4, 6, 7, are involved, sequentially, so that giant resonances can be built on other giant resonances through a two-step process of the kind $F^{\pm}F'^{\pm}$ 0>. To the extent in which giant states are of one-particle one-hole nature (as described in RPA theories) these "double" giant resonances are therefore of two-particle two-hole character. The main question, of course, is how close are such states to the actual eigenstates of the nuclear Hamiltonian, so that they can be observed experimentally as relatively narrow bumps. The experimental results from pion single-charge-exchange experiments ($\pi^{\pm}$, π^o) in which giant resonances are observed, could be of guidance for predicting which of the double giant resonances will be most strongly excited.

For the spontaneous double beta decay the most important excitation operators, at least in the closure approximations, involve both the square of the total lowering isospin operator (Fermi transitions) as in the doubly analogue isobaric excitations and the square of the Gamow-Teller operator $\sum \vec{\sigma}_i \tau_i^{\pm}$, this lead to three different possibilities as suggested by the three different couplings $(\vec{\sigma}_i \otimes \vec{\sigma}_j)_\lambda$, with $\lambda = 0,1,2$.

Before a connection is established between double beta decay experiments and the underlying weak interaction theory, important questions of nuclear structure must be addressed. For an interesting review of this matter we refer to reference 13.

9. CONCLUSIONS

We have investigated, within Hamiltonian (6) and the RPA, the mean energies and the strengths of the isovector dipole, monopole, and quadrupole excitation operators.

For the dipole and the monopole, the results are in line with the existing experimental data. The energy splitting between the various T-fragments (Δ_V and Δ_T) for the dipole and the monopole in the limit of $\langle r_n^2 \rangle = \langle r_p^2 \rangle$ is the same. Differences in neutron and proton rms radii and exchange forces affect differently the two modes [see formulas (30) and (42)].

For the quadrupole, within Hamiltonian (6), a compact isovector quadrupole giant state is predicted in $\Delta T_3 = 0, \pm 1$ channels. The splitting Δ_V and Δ_T, however, turn out to be different from those of the dipole and monopole, even in the limiting case of $\langle r_n^2 \rangle = \langle r_p^2 \rangle$ [use formulas (23) and (24) with the input 52 for f_0, f_1, and f_2].

Measurements in $\Delta T_3 = \pm 1$ channels are inconsistent with a compact giant quadrupole state in these channels. We have studied a possible mechanism which can both decrease and spread the quadrupole strength. This mechanism is connected with the presence of nonlocal components in the nuclear interaction that are absent in, for example, Skyrme-like forces [see Hamiltonian (54)].

More precise data on both the $\Delta T_3 \pm 1$ and $\Delta T_3 = 0$ components of the isovector quadrupole excitation could determine the strength of the nonlocal force components and tests the consistency of $\Delta T_3 = 0$ and $\Delta T_3 = \pm 1$ channels within a defined model Hamiltonian.

ACKNOWLEDGEMENTS

Many of the ideas exposed in this talk have been developed in collaboration with S. Stringari and E. Lipparini.

REFERENCES

1) J.D. Bowman, Proceedings of International Nuclear Physics Conference, Harrogate, U.K., 1986, p. 83 and references therein.

2) A. Bohr and B.R. Mottelson, Nuclear Structure, Vol. II (Benjamin, New York, 1975).

3) N. Auerbach and A. Klein, Nucl. Phys. A345 (1983) 77.

4) R. Leonardi and M. Rosa-Clot, Phys. Rev. Lett. 23 (1969) 874, and Riviste N.C. 1 (1971) 1.

5) R. Leonardi, E. Lipparini and S. Stringari, Phys. Rev. C26 (1982) 2636.

6) Some references are listed below (see refs. 1-5 also):
P. Paul, in Photonuclear Reactions and Applications, Vol. I, 1973, ed. B.L. Berman, p. 401.
R.Ö Akyüz and S. Fallieros, Phys. Rev. Lett. 27 (1971) 1016.
W.A. Sterrenburg, S.M. Austin, R.P. De Vito and A. Galonsky, Phys. Rev. Lett. 45 (1980) 1839.

7) R. Leonardi, Phys. Rev. C14 (1976) 385.

8) R. Leonardi, E. Lipparini and S. Stringari, to be published, and R. Leonardi, Journal de Physique 45 (1984) 319.

9) E. Lipparini and S. Stringari, Nucl. Phys. A371 (1981).

10) R. Leonardi, E. Lipparini and S. Stringari, Phys. Rev. C35 (1987) 1439.

11) A measurement has been reported indicating compact quadrupole strength in the lower fragment of the $\Delta T_3=0$ excitation of ^{90}Y. (R. Zorro, I. Bergqvist, S. Crona, N. Olsson et al., Nucl. Phys., in press).

12) N. Auerbach, AIP Conference Proceedings 163, eds. J. Peterson, D. Strataiman, 1987, p. 134.

13) W.C. Haxton and G.J. Stepehnson, Jr., Progress in Nuclear and Particle Physics, Vol. 12, p. 409.

LARGE SHAPE CHANGES IN NUCLEI

Francisco BARRANCO

Escuela Superior de Ingenieros Industriales, Sevilla, Spain

Enrico VIGEZZI and Ricardo A. BROGLIA

The Niels Bohr Institute, University of Copenhagen, DK-2100 Copenhagen, Denmark and
Dipartimento di Fisica, Università di Milano, and INFN Sez. Milano, Italy

A simple model of large amplitude motion in nuclei is discussed and applied to a variety of phenomena. Superfluidity is found to play an essential role in the tunneling between different configurations.

1. INTRODUCTION

Overwhelming evidence exists, testifying to the fact that nuclei are able to display macroscopic shape changes. The most conspicuous example is provided by the spontaneous fission of heavy nuclei. Of more recent date are the phenomenon of exotic decay [1], and of the sudden depopulation of superdeformed bands at high spins [2].

In the present paper we discuss a model of large amplitude motion, which has been shown to be able to describe phenomena ranging from alpha-decay to fission [3]. Applications to the case of exotic decay and of tunneling between the different configurations of octupole deformed nuclei are discussed.

2. THE MODEL

We shall calculate the changes of shape using a collective Hamiltonian

$$\sum_\lambda \left(-\frac{\hbar^2}{2D_\lambda}\frac{\partial^2}{\partial \epsilon_\lambda^2} + V(\{\epsilon_\lambda\})\right)\psi(\{\epsilon_\lambda\}) = E\psi(\{\epsilon_\lambda\}) \qquad (1)$$

depending on the set of deformation parameters $\{\epsilon_\lambda\}$, of multipolarity $\lambda = 2, 3, \ldots$.

While much effort has been concentrated on the study of the potential energy surfaces, much less is known about the inertial parameters D_λ. In what follows we use a simple model to determine these quantities [3].

We consider a linear sequence of nuclear wavefunctions corresponding to a set of Slater determinants ϕ_i describing different shapes, and assume that the residual interaction connects only nearest neighbours with a constant matrix element v. The Hamiltonian operator for this system is a tridiagonal matrix

$$\begin{bmatrix} \ddots & \vdots & \vdots & \vdots & \vdots \\ \cdots & E_{i-1} & v & \cdots & \cdots \\ \cdots & v & E_i & v & \cdots \\ \cdots & \cdots & v & E_{i+1} & \cdots \\ \vdots & \vdots & \vdots & \vdots & \ddots \end{bmatrix} \begin{bmatrix} \cdots \\ a_{i-1} \\ a_i \\ a_{i+1} \\ \cdots \end{bmatrix} = E_\alpha \begin{bmatrix} \cdots \\ a_{i-1} \\ a_i \\ a_{i+1} \\ \cdots \end{bmatrix} \qquad (2)$$

The factor 1/4 arises because only one of the four possible types of pair jumps between levels of various slopes leads to a level crossing. The collective inertial parameter resulting from eqs. (5) and (8) is $D = \frac{\hbar^2}{(\Delta\epsilon)^2}(2\frac{G}{\Delta^2})$. The quadratic dependence of the inertia on the reciprocal pairing gap found in this formula is well known (cf. e.g. ref. 4).

The above derivation considers only one kind of particle. With both neutron (ν) and proton (π) level crossings one finds

$$D_\lambda = (D_\pi)_\lambda + (D_\nu)_\lambda \quad (\lambda = 2, 3, ...) \tag{9}$$

where

$$(D_i)_\lambda = \hbar^2 \frac{2G}{\Delta_i^2} \left(\frac{dn_i}{d\lambda}\right)^2 \quad (i = \pi, \nu) \tag{10}$$

The next step is to calculate the average distance between neighbouring configurations. This will determine $\Delta\epsilon$ and the distance scale in large amplitude motion. We find the distance between states by deforming the nucleus according to the required shape change, making a criterion to determine when the nucleus has reached a new Hartree-Fock configuration.

We first deform the wavefunction with a single-particle field, $F(\vec{r})$. The wavefunction will be transformed to

$$\phi'_k(\vec{r}) = e^{\vec{\nabla}F\cdot\vec{\nabla}}\phi_k(\vec{r}) \approx \frac{1}{\sqrt{1-\nabla^2 F}}\phi_k(\vec{r} + \vec{\nabla}F) \tag{11}$$

In order to make the new state as low in energy as possible, we shall restrict ourselves to incompressible fields, $\nabla^2 F = 0$. Then F may be expressed as

$$F(\vec{r}) = \sum_\lambda c_\lambda r^\lambda Y_\lambda(\hat{r}) \tag{12}$$

Under the transformation (11) the momentum distribution in the nucleus changes, according to [5]

$$\vec{p} \to (1 + \vec{\nabla}\vec{\nabla}F)\vec{p} \tag{13}$$

We assume that in an Hartree-Fock minimum the momentum distribution of the nucleus is homogeneous and isotropic within the Fermi sphere of radius p_F; the corresponding Wigner function $f(\vec{r}, \vec{p})$ is given by

$$f(\vec{r}, \vec{p}) = \frac{(2\pi)^3}{g} \frac{3}{4p_F^3} \frac{A}{V} \theta(p_F - p) \tag{14}$$

where V is the spatial volume of the system. It is normalized according to

$$\frac{g}{(2\pi)^3} \int_V d\vec{r} \int d\vec{p}\, f(\vec{r}, \vec{p}) = A \tag{15}$$

The Wigner function of the state, subject to the transformation (11-13), is given by

$$f'(\vec{r}, \vec{p}) = f(\vec{r} + \vec{\nabla}F, (1 + \vec{\nabla}\vec{\nabla}F)\vec{p}) \tag{16}$$

The factor 1/4 arises because only one of the four possible types of pair jumps between levels of various slopes leads to a level crossing. The collective inertial parameter resulting from eqs. (5) and (8) is $D = \frac{\hbar^2}{(\Delta\epsilon)^3}(2\frac{G}{\Delta^2})$. The quadratic dependence of the inertia on the reciprocal pairing gap found in this formula is well known (cf. e.g. ref. 4).

The above derivation considers only one kind of particle. With both neutron (ν) and proton (π) level crossings one finds

$$D_\lambda = (D_\pi)_\lambda + (D_\nu)_\lambda \quad (\lambda = 2, 3, ...) \tag{9}$$

where

$$(D_i)_\lambda = \hbar^2 \frac{2G}{\Delta_i^2}\left(\frac{dn_i}{d\lambda}\right)^2 \quad (i = \pi, \nu) \tag{10}$$

The next step is to calculate the average distance between neighbouring configurations. This will determine $\Delta\epsilon$ and the distance scale in large amplitude motion. We find the distance between states by deforming the nucleus according to the required shape change, making a criterion to determine when the nucleus has reached a new Hartree-Fock configuration.

We first deform the wavefunction with a single-particle field, $F(\vec{r})$. The wavefunction will be transformed to

$$\phi'_k(\vec{r}) = e^{\vec{\nabla}F\cdot\vec{\nabla}}\phi_k(\vec{r}) \approx \frac{1}{\sqrt{1-\nabla^2 F}}\phi_k(\vec{r}+\vec{\nabla}F) \tag{11}$$

In order to make the new state as low in energy as possible, we shall restrict ourselves to incompressible fields, $\nabla^2 F = 0$. Then F may be expressed as

$$F(\vec{r}) = \sum_\lambda c_\lambda r^\lambda Y_\lambda(\hat{r}) \tag{12}$$

Under the transformation (11) the momentum distribution in the nucleus changes, according to [5)]

$$\vec{p} \to (1 + \vec{\nabla}\vec{\nabla}F)\vec{p} \tag{13}$$

We assume that in an Hartree-Fock minimum the momentum distribution of the nucleus is homogeneous and isotropic within the Fermi sphere of radius p_F; the corresponding Wigner function $f(\vec{r},\vec{p})$ is given by

$$f(\vec{r},\vec{p}) = \frac{(2\pi)^3}{g}\frac{3}{4p_F^3}\frac{A}{V}\,\theta(p_F - p) \tag{14}$$

where V is the spatial volume of the system. It is normalized according to

$$\frac{g}{(2\pi)^3}\int_V d\vec{r}\int d\vec{p}\, f(\vec{r},\vec{p}) = A \tag{15}$$

The Wigner function of the state, subject to the transformation (11-13), is given by

$$f'(\vec{r},\vec{p}) = f(\vec{r}+\vec{\nabla}F, (1+\vec{\nabla}\vec{\nabla}F)\vec{p}) \tag{16}$$

where E_i are the Hartree energies associated with the different configurations. The solution of the Hamiltonian problem yields eigenfunctions which are linear combinations of the basis determinantal states ϕ_k

$$\Phi_\alpha = \sum_{i=1}^{n} a_i^\alpha \phi_i \qquad (3)$$

The energetics of the process described by this wavefunction is sketched in Fig.1 (top left), which depicts energy levels as a function of some deformation coordinate. Each energy level is a parabola. The bottom is a Hartree-Fock state of some definite energy E_i, and the other points show how the energy changes when the state is deformed. It is energetically very unfavorable to move any distance along the same state. The large-scale shape changes must then come from the jumps from one state to another.

The determinant ϕ_0 describes the configuration associated with the absolute minimum corresponding to the equilibrium shape of the parent nucleus, while ϕ_n is the Slater determinant associated with the touching configuration of the daughter nuclei (cfr. Fig. 1, top right). The collective Hamiltonian (2) may be approximated on a mesh in ϵ-space, with the second derivative operator replaced by the difference operator

$$-\frac{\hbar^2}{2D}\frac{d^2\psi}{d\epsilon^2} = -\frac{\hbar^2}{2D}\frac{\psi(\epsilon_{i+1}) + \psi(\epsilon_{i-1}) - 2\psi(\epsilon_i)}{(\Delta\epsilon)^2} \qquad (4)$$

Here $\Delta\epsilon$ is the step interval between states. Comparing the off-diagonal matrix elements of eqs. (2) and (4) one can make the identification

$$v = -\frac{\hbar^2}{2D(\Delta\epsilon)^2} \qquad (5)$$

The most important part of the nuclear interaction responsible for the off-diagonal matrix elements is the pairing force, which in the BCS approximation is written as

$$H_p = -G \sum_{\nu,\nu'} a_\nu^+ a_{\bar\nu}^+ a_{\bar\nu'} a_{\nu'} \qquad (6)$$

The single particle levels ν and ν' depend on the collective coordinate ϵ. One can group these levels according to whether the single-particle energies increase or decrease with deformation. Under the influence of the interaction (6), pairs of particles are transferred from one type of level to the other, and the system evolves from one configuration to the next, passing through level crossings. This process can be viewed as a pick up of a pair of particles from one type of level in the pairing condensate and a stripping onto the other kind of level in the condensate, without exciting any quasiparticle. In the BCS approximation, each process is associated with an amplitude of the type

$$<\psi_{BCS}|\sum_\nu a_\nu^+ a_{\bar\nu}^+|\psi_{BCS}> = <\psi_{BCS}|\sum_{\nu'} a_{\bar\nu'} a_{\nu'}|\psi_{BCS}> = \sum_\nu U_\nu V_\nu = \frac{\Delta}{G} \qquad (7)$$

The effective residual interaction between different configurations is

$$v = -\frac{G}{4}\left(\sum_\nu U_\nu V_\nu\right)^2 = -\frac{\Delta^2}{4G} \qquad (8)$$

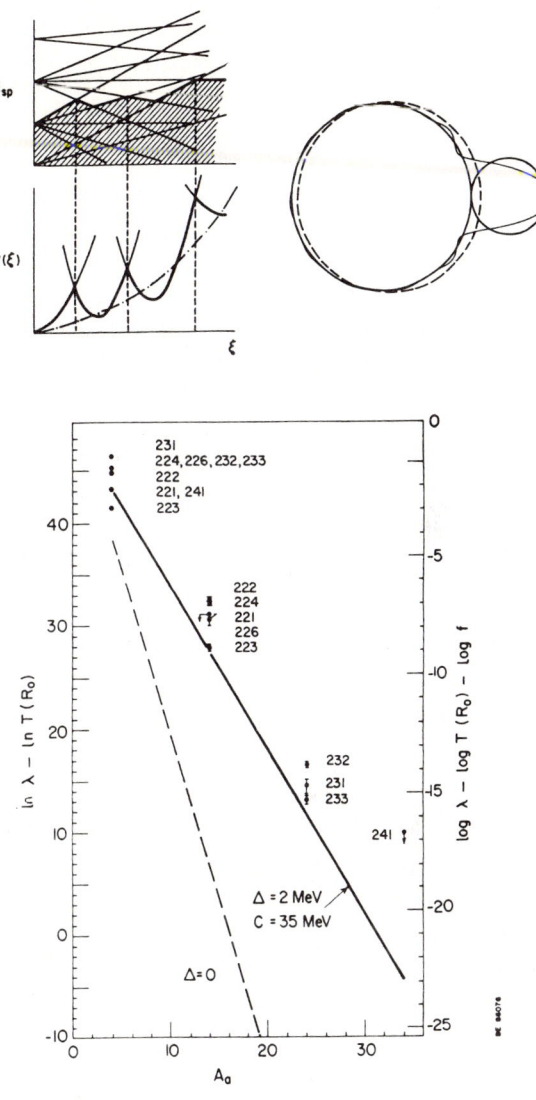

FIGURE 1

Top left: the single-particle energy levels and the energy of the nucleus are shown schematically as a function of the deformation coordinate ξ.
Top right: the shape of ^{223}Ra (dashed line) is compared with the touching nuclei ^{14}C and ^{209}Pb (thick lines). Also shown (thin line) is the shape obtained via an expansion in multipoles up to $L = 10$.
Bottom: the quantities $ln\lambda - lnT$ and $log(P)$ (cf. eqs. (23),(25)), relative to the exotic decay of a fragment of mass A_a, are compared with the experimental data for different parent nuclei.

LARGE SHAPE CHANGES IN NUCLEI

Francisco BARRANCO

Escuela Superior de Ingenieros Industriales, Sevilla, Spain

Enrico VIGEZZI and Ricardo A. BROGLIA

The Niels Bohr Institute, University of Copenhagen, DK-2100 Copenhagen, Denmark and Dipartimento di Fisica, Università di Milano, and INFN Sez. Milano, Italy

A simple model of large amplitude motion in nuclei is discussed and applied to a variety of phenomena. Superfluidity is found to play an essential role in the tunneling between different configurations.

1. INTRODUCTION

Overwhelming evidence exists, testifying to the fact that nuclei are able to display macroscopic shape changes. The most conspicuous example is provided by the spontaneous fission of heavy nuclei. Of more recent date are the phenomenon of exotic decay [1], and of the sudden depopulation of superdeformed bands at high spins [2].

In the present paper we discuss a model of large amplitude motion, which has been shown to be able to describe phenomena ranging from alpha-decay to fission [3]. Applications to the case of exotic decay and of tunneling between the different configurations of octupole deformed nuclei are discussed.

2. THE MODEL

We shall calculate the changes of shape using a collective Hamiltonian

$$\sum_\lambda \left(-\frac{\hbar^2}{2D_\lambda} \frac{\partial^2}{\partial \epsilon_\lambda^2} + V(\{\epsilon_\lambda\}) \right) \psi(\{\epsilon_\lambda\}) = E\psi(\{\epsilon_\lambda\}) \qquad (1)$$

depending on the set of deformation parameters $\{\epsilon_\lambda\}$, of multipolarity $\lambda = 2, 3, \ldots$.

While much effort has been concentrated on the study of the potential energy surfaces, much less is known about the inertial parameters D_λ. In what follows we use a simple model to determine these quantities [3].

We consider a linear sequence of nuclear wavefunctions corresponding to a set of Slater determinants ϕ_i describing different shapes, and assume that the residual interaction connects only nearest neighbours with a constant matrix element v. The Hamiltonian operator for this system is a tridiagonal matrix

$$\begin{bmatrix} \ddots & \vdots & \vdots & \vdots & \\ \cdots & E_{i-1} & v & \cdots & \cdots \\ \cdots & v & E_i & v & \cdots \\ \cdots & \cdots & v & E_{i+1} & \cdots \\ & \vdots & \vdots & \vdots & \ddots \end{bmatrix} \begin{bmatrix} \cdots \\ a_{i-1} \\ a_i \\ a_{i+1} \\ \cdots \end{bmatrix} = E_\alpha \begin{bmatrix} \cdots \\ a_{i-1} \\ a_i \\ a_{i+1} \\ \cdots \end{bmatrix} \qquad (2)$$

Thus, the new shape of the system has been achieved at the cost of a deformed Fermi surface. This gives the system a higher kinetic energy.

A state of lower energy can be constructed by moving the particles above the spherical Fermi surface below. In the classical limit, moving particles in momentum space will not change the density distribution, as long as the volume in momentum space remains constant. Of course, to do this the single-particle basis in the wavefunction has to be transformed to a representation with particles above and below the Fermi sphere. This is not always possible. A given particle wavefunction might be distributed partly above and partly below the Fermi sphere. If there is a nonintegral number of particles above there is no way a spherical Fermi surface could be restored. Consequently, a necessary condition for restoring a spherical Fermi surface is that the integral of the Wigner function outside the Fermi sphere equals an integer. We assume that this is sufficient also for the existence of a single-particle representation that allows a new Hartree-Fock minimum to be reached by moving particles between the orbitals.

The general formula for the criterion is that the following integral be equal to an integer,

$$\frac{g}{(2\pi)^3} \int_V d\vec{r} \int_{p>p_F} d\vec{p}\, f'(\vec{r},\vec{p}) = n \tag{17}$$

Here g is the spin, and possibly isospin, degeneracy of the single-particle levels. Indicating with F_i $(i=1,2,3)$ the eigenvalues of the matrix $\vec{\nabla}\vec{\nabla}F$, one has $F_1+F_2+F_3 = \nabla^2 F = 0$. Moreover, if it is supposed that the transformed Fermi sphere has axial symmetry, F_2 is equal to F_3. It is then possible to show that

$$n = A \int_V \frac{d\vec{r}}{V} \frac{F_1(\vec{r})}{\sqrt{3}} = \frac{A}{\sqrt{3}} <F_1(r)> \tag{18}$$

This formula is very simple for pure quadrupole deformation In fact, writing $F = \epsilon_2 r^2 Y_{20}$, we have

$$\vec{\nabla}\vec{\nabla}F = \sqrt{\frac{5}{4\pi}} \epsilon_2 \begin{pmatrix} 2 & 0 & 0 \\ 0 & -1 & 0 \\ 0 & 0 & -1 \end{pmatrix} \tag{19}$$

and $F_1 = 2\epsilon_2\sqrt{\frac{5}{4\pi}}$. Since F_1 does not depend on $\vec{r}$ in this case, the spatial integration is trivial in eq.(18) and we obtain

$$n = \frac{A}{\sqrt{3}} 2\epsilon_2 \sqrt{\frac{5}{4\pi}} \tag{20}$$

For heavy nuclei, this formula has been demonstrated [5] to be accurate to about 10% in counting the level crossing over large intervals of deformation. It is also surprisingly accurate for finding the deformation of low energy states in light nuclei [3].

3. OCTUPOLE TUNNELING

Several mean field calculations have found octupole-deformed wavefunctions to be most stable for nuclei in the radium region [6-11]. We apply our model to the nucleus ^{222}Ra (cf. also refs. 12 and 13). The shapes are conventionally parametrized by deformation parameters ϵ_2 and ϵ_3, which are related to the shape of the density or the single particle potential by

$$R = R_o \left(1 + \frac{2}{3}\sqrt{\frac{4\pi}{5}}\epsilon_2 Y_{20} + \sqrt{\frac{4\pi}{7}}\epsilon_3 Y_{30} \right) \qquad (21)$$

where R_o is the half-density radius and $Y_{\lambda 0}$ are the spherical harmonics of multipolarity λ. The mean field calculations of ref. 9 predict minima at $\epsilon_3 \approx \pm 0.07$ with a saddle point connecting the two minima at $\epsilon_2 \approx 0.11$ and $\epsilon_3 = 0$.

Making use of eq. (20) we can determine the density of level crossings along the quadrupole coordinate, obtaining $\frac{d(n_\pi + n_\nu)}{d\epsilon_2} \approx 40$. We also find about the same crossing density for octupole deformations, integrating eq. (18) numerically.

Using the values [14] $\Delta \approx \frac{12}{\sqrt{A}} \approx 0.8 MeV$, $G \approx \frac{25}{A} MeV \approx 0.11 MeV$, one obtains $v = 2.9 MeV$ and $D_3 = 276\hbar^2 MeV^{-1}$. To complete the Hamiltonian, the potential from Fig. 2 of Ref. 9 was used. This potential has a barrier between the two minima with a value at zero octupole deformation $V_B = 300$ keV. The two-dimensional Schrödinger equation was solved using a discrete mesh in the $\epsilon_2 \times \epsilon_3$ plane.

The resulting ground state wavefunction is concentrated close to the saddle point of the potential energy surface, while the excited state is peaked around the absolute minima of the potential. Thus the ground state is an even combination of the unperturbed solutions associated with each of the octupole minima, while the first excited state is an odd combination. In this way the system has regained invariance with respect to parity reflections. The energy difference between the two states is found to be 500 keV, which is of the same order of magnitude as that experimentally observed (242 keV) .

4. EXOTIC DECAY

The radioactive decay of heavy nuclei by Carbon and Neon known as exotic decay [1] needs for its description elements taken from both theory of alpha decay and of fission. The decay rate associated with the process

$$B \to a + A \qquad (22)$$

can be written as

$$\lambda = fPT \qquad (23)$$

That is, the particle before its expulsion does not move freely in a fixed potential, but its escape from the nucleus must be considered as composed of two steps, of which the first consists in its release from nuclear matter, and the second in its penetration as a free particle through the Coulomb barrier. The likelihood for the first step is measured by the formation probability P, while the second is controlled by the frequency f with which the particle a hits the Coulomb barrier. Making use of Coulomb penetrabilities T and of standard values for f together with the experimental half-lives, the empirical values of the formation probabilities were calculated. They are displayed in Fig. 1 (bottom).

Approximating the potential energy of the local minima by the parabola

$$V(\xi) = \frac{1}{2}C\xi^2 , \qquad (24)$$

and inserting it in eq. (1), one obtains the Schrodinger equation describing the evolution of the system along the deformation coordinate ξ. The formation probability is given by the square of the ground state wavefunction for $\xi = 1$, that is,

$$P = \left(\frac{\alpha^2}{\pi}\right)^{1/2} e^{-\alpha^2} , \qquad (25)$$

where

$$\alpha^2 = \left(\frac{CD}{\hbar^2}\right)^{1/2} . \qquad (26)$$

The quantity (25) is also shown in Fig. 1, and provides an overall account of the experimental data.

REFERENCES

1) H.J. Rose and G.A. Jones, Nature (London) 307 (1984) 245.

2) P. Twin et al., Phys. Rev. Lett. 57 (1986) 811.

3) G.F. Bertsch, F. Barranco and R.A. Broglia, in *Windsurfing the Fermi Sea*, eds. T.T.S. Kuo and J. Speth, (Elsevier, Amsterdam, 1987) 33; F. Barranco, R.A. Broglia and G.F. Bertsch, Phys. Rev. Lett. 60 (1988) 507.

4) L. Wilets, *Theories of nuclear fission* (Clarendon Press, Oxford, 1964).

5) G.F. Bertsch, Phys. Lett. 95B (1980) 157.

6) G.A.Leander, et al., Nucl.Phys. A388 (1982) 452.

7) W.Nazarewicz, et al., Nucl.Phys. A429 (1984) 269.

8) R.Chasman, J. de Phys. Suppl. 45 (1984) C6-167.

9) P.Bonche, P.H.Heenen, H.Flocard and D.Vautherin, Phys.Lett. 175B (1986) 387.

10) L.M. Robledo, J.L. Egido, J.F. Berger, and M. Girod, Phys. Lett. 187B (1987) 223.

11) L.M. Robledo, J.L. Egido, B. Nerlo-Pomorska, and K. Pomorski, Phys. Lett. 201B (1988) 409.

12) F. Barranco and R.A. Broglia, in *The variety of nuclear shapes*, eds. J.D. Garrett et al. (World Scientific, Singapore, 1988) 253.

13) F. Barranco, E.Vigezzi, R.A.Broglia and G.F. Bertsch, Phys. Rev. C 38 (1988) 1523.

14) A. Bohr and B.R. Mottelson, *Nuclear Structure*, vol. II (Benjamin, Reading, 1975), p.645; R.A. Broglia, D.R. Bes and B.S. Nilsson, Phys. Lett. 50B (1974) 213.

Temperature Dependence of the Lifetime of Giant Dipole Resonances

R. A. Broglia

Dipartimento di Fisica, Universita' di Milano

and INFN sez. Milano, via Celoria 16, 20133 Milano

and

The Niels Bohr Institute, University of Copenhagen

DK-2100 Copenhagen Ø, Denmark

and

A. Bracco

Dipartimento di Fisica, Universita' di Milano

and INFN sez. Milano, via Celoria 16, 20133 Milano

The frequency of the giant dipole resonance (GDR) can be determined from the Thomas-Reiche-Kuhn and the polarization sum rules. The resulting expression:

$$\hbar\omega_{dip} = \sqrt{\frac{b_{sym}}{<r^2>}} \qquad (1)$$

depends on the symmetry coefficient appearing in the mass formula, and on the nuclear dimensions through the mean square radius.

In a spherical nucleus, the dipole resonance will be present as a single peak, while in quadrupole deformed nucleus it will break into three components with frequencies [1]:

$$\omega_k = \omega_{dip}(1 - \beta \cos\frac{3\gamma}{k}) \qquad (2)$$

associated with vibrations along the three principal axis of the system.

Because the nuclear deformation, characterized by the parameters β and γ, depends on both temperature and angular momentum, the frequencies ω_k also do. Note that the centroid energy (1) is not expected to change with temperature.

By rotating a liquid drop around an axis fixed in space with increasing rotational frequency, the interplay of the Coriolis force and the nuclear tension will produce an oblate nucleus, then to a prolate system, leading finally to a saddle shape and, at the critical value of the angular momentum to the fissioning of the nucleus. Because of the shell structure, this sequence of shapes may be inverted. Also the nucleus may acquire super- and hyperdeformed configurations, where the ratio between largest and smallest axis of deformation is 2:1 and 3:1, respectively.

The above scenario indicates that at low rotational frequencies, the energy splitting of the different magnetic dipole components will increase markedly with angular momentum, eventually faster by simultaneously heating the nucleus and thus reducing the surface tension and the barriers between the different configurations. A saturation effect is expected when the critical value of the angular momentum is reached.

Absent from this picture are the effects of both quantal and classical fluctuations. In fact, throughout the above discussion the surface of the nucleus was assumed to be static. However, due to the zero point motion associated with collective multipole vibrations, the surface fluctuates [2]. These fluctuations take place around the equilibrium deformation of the mean field, determined by the interplay of Coulomb repulsion, surface tension and shell effects.

Furthermore, at finite temperature the system has a finite probability to be in an ensemble of configurations associated with different shape deformation parameters, of which the most important are expected to be β and γ [3-5].

In the quantal case, the fluctuations can be viewed as doorway states of the giant resonances to the compound nucleus, and thus to statistical equilibrium. Thermal fluctuations, being controlled by Boltzmann factors, are also associated with thermodynamic equilibrium.

Both quantal and thermal fluctuations eventually lead to thermodynamic equilibrium and thus to the damping of the sharp lines in which the dipole strength breaks, under the influence of static deformations.

The simultaneous treatment of both the quantal and the thermal damping mechanisms, is a central problem in the physics of hot finite many-body systems. A first step in dealing with the solution of it can be carried as follows (cf. ref. [3] and references therein). For a given temperature and angular momentum, the potential energy surface in the (β, γ)-plane is constructed making use of the Nilsson-Strutinsky method [1]. From it, the free energy is calculated.

Next, one calculates at each point in the (β, γ)-plane the different components of the dipole vibration, making use of the random phase approximation, which leads to three or five lines according to whether the rotation takes place along the symmetry axis or not. Furthermore, each component is coupled to the fluctuations of the surface around that equilibrium position. This is done by allowing the particle or the hole of the different components in the associated RPA wavefunction to bounce inelastically off the surface and excite a multipole vibration, thus leading to a 2p-2h doorway state. The coupling of those states to the compound nucleus is done, in an average way, through a small imaginary component of the energies involved in the process. Although the resulting strength function is insensitive to the actual value of the averaging parameter, there is experimental evidence indicating it to be ≤ 30 keV [6].

The damping mechanism described above is the same and the only, as that operating in the damping of giant resonances at zero temperature, where the doorway hypothesis has been checked.

The resulting (quantal) width is found to be rather independent of temperature [7]. A possible reason for this result is, that the strong cancellation found to operate between the particle and the hole contributions to the quantal amplitude, may lead to an effective harmonic coupling between the dipole vibration and the surface modes.

Carring out the thermal average of the strength functions by weighting them at each point of the (β, γ)-plane with the corresponding Boltzmann factor, leads to a strength function which contains both quantal and thermal fluctuations on somewhat equal footing. A rough estimate of the total damping width can be given by:

$$\Gamma = \sqrt{\gamma_Q^2 + (2\Delta\omega_0)^2} \qquad (3)$$

where $\Delta\omega_0$ is the spread in vibrational frequency of the selected K-component over the whole grid in the (β, γ)-plane.

Increasing the temperature one would expect a continuous increase of the component of Γ arising from thermal fluctuations. The opposite is in fact true, at least from some temperature which can roughtly be estimated at ≈ 2 MeV. This phenomenon is known as motional narrowing [8,9] and has a simple explanation.

If the dipole has no time to perform a number of vibrations at any given point in the (β, γ)-plane, and feel the deformation as static, before the compound nucleus has changed deformation jumping to another point in (β, γ)-plane, the effect of deformation will be averaged out and the spread in frequency $\Delta\omega_0$ reduced by the factor $\Delta\omega_0 \tau$, ratio between the time $\frac{1}{\Delta\omega_0}$ needed to feel the full spread in frequency associated with the variety of quadrupole shapes, and the actual available time τ (hopping time) that the system remains in a given configuration.

Theory and experiment are set on top of each other in Fig. 1. They display the full width at half maximum (FWHM) of the GDR of the compound nucleus in the mass region

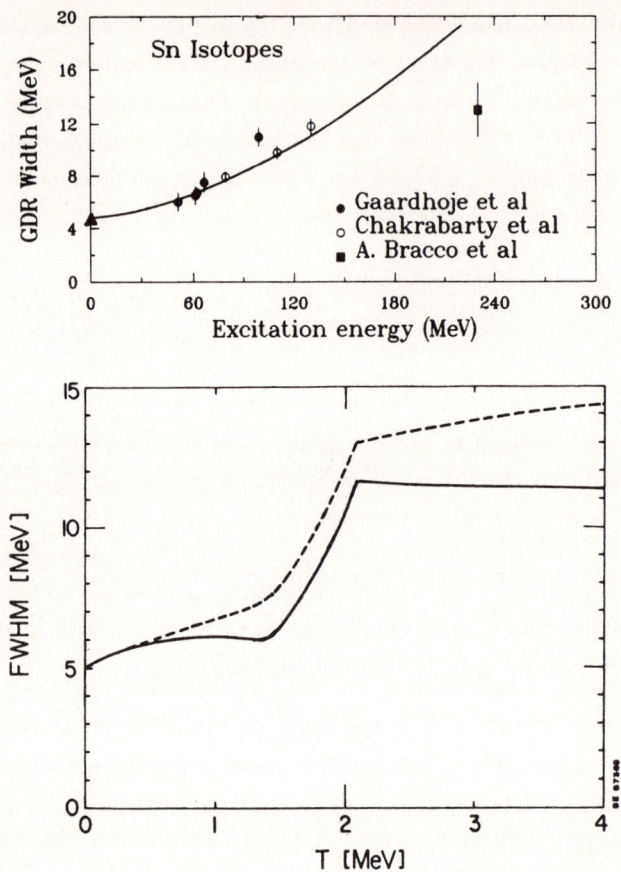

Fig. 1. In the upper part of the figure (a), the FWHM of the GDR in $^{108-110}$Sn extracted fron refs. [10] (filled circles), [11] (open circles), and [12] (filled square) are shown. The ground state GDR width obtained from the (γ,n) raction is also indicated by the triangle. The continuous curve corresponds to the expression (4). In the lower part of the figure (b), the theoretical calculation of the width of ^{110}Sn is displayed. The continuous line takes into account motional narrowing [9], while the dashed curve does not [3].

of Sn as a function of excitation energy E* ($0 \leq E^* \leq 240 MeV$), and covering a range of temperature $0 \leq T \leq 3$ MeV. The curve in figure is the plot of the phenomenological law:

$$\Gamma = 4.8 + 0.0026(E)^{1.6} (MeV) \tag{4}$$

which was determined by fitting the GDR widths in the excitation energy region 0-130 MeV [11].

It should be noted that the data in the figure correspond to different values of the angular momentum, which also increases with increasing excitation energy. Moreover, the data point at E* = 240 MeV (T $\approx$ 3.2 MeV) corresponds to a maximum angular momentum of $\approx$ 60 $\hbar$, which is about the same as that associated with the point at E* = 130 MeV (T$\approx$ 2 MeV). This fact should be taken into account in seeking an explanation for the large deviation the empirical law (4) derived from the low excitation energy data, and the most recent E* = 240 MeV data point.

The calculations displayed in the lower part of Fig. 1 as a continuous line were carried out using a "realistic" model, taking into account both quantal and thermodinamic fluctuations, as discussed in references 8 and 9. They include the mechanism of motional narrowing. Excluding this effect lead to the results shown as a dashed curve.

In comparing theory to experiment it has to be noted that the calculations were carried for a fixed value of excitation energy and angular momentum and for a given isotope (^{110}Sn), while the experimental data reflect an average of excitation energies and angular momenta as well isotopes.

Although the present experimental and theoretical results do not allow to give a clear cut answer to the question whether the phenomenon of motional narrowing is operative in determining the lifetime of GDR, a whole field of research is being opened by asking such question.

References

1) A. Bohr and B.R. Mottelson, *Nuclear Structure* vol. I, (Benjamin, New York, 1969)p.372
2) G.F. Bertsch, P.F. Bortignon, and R.A. Broglia, Rev. Mod. Phys. **55** (1983)287
3) M. Gallardo, M. Diebel, T. Dossing, and R.A. Broglia, Nucl. Phys.**A443** (1985)415
4) S. Levit, Proc. Int. Conf. Harrogate, 25-30 Aug., 1986, Eds. J.L. Durrel et al., vol 2, p227
5) A. Goodman, *The Variety of Nuclear Shapes*, eds. J. Garret et al., (World Scientific, 1988)345
6) A. Richter et al. Darmstädt preprint and to be published.

7) P.F. Bortignon, R.A. Broglia, G.F. Bertsch, and J. Pacheco, Nucl. Phys. **A460** (1987)149
8) R.A. Broglia, W.E. Ormand, and M. Borromeo, Nucl. Phys.**A482** (1988)141c.
9) B. Lauritzen, R.A. Broglia, W.E. Ormand, and T. Døssing Phys. Lett. **B.207**(1988)238
10) J.J. Gaardhøje *et al.* Phys. Rev. Lett. **56** (1986)1783.
11) D.R. Chakrabarty, S. Sen, N. Alamanos, P. Paul, R. Schicker, J. Stachel, and J.J. Gaardhøje, Phys. Rev.**C36**(1987)1886.
12) A. Bracco *et al.*, XXVI International Winter Meeting on Nuclear Physics, Universitá degli studi di Milano, Supplemento n. 63,(1988)280, and to be published.

The Darmstadt effect

Ettore Remiddi
Dipartimento di Fisica, Università di Bologna
INFN, CNAF e Sezione di Bologna.

The name Darmstadt effect indicates the presence of a set of narrow positron peaks (sometimes referred to as Darmstadtons) repeatedly found since 1981 in the collision of heavy ions (such as U+U) at the GSI facility of Darmstadt. When the positron is interpreted as occurring in the e^+e^- decay channel of a structure X, one finds for X the following mass spectrum and widths (both in keV) [1] :

$$M_X = \quad 1873 \pm 10 \qquad \Gamma_X = \quad 40$$
$$1782 \pm 20$$
$$1646 \pm 10 \qquad\qquad\qquad 25$$
$$1498 \pm 20$$

Those results summarizes the work done in the years 1981-85 by two groups, using two different apparatus at GSI, the EPOS and the Orange spectrometers, which have sometimes observed the accompanying electron too. A more recent experiment [2] in the super HILAC at LBL has found a correlated back to back $\gamma\gamma$ signal, with

$$M_X = 1062 \pm 1 \qquad \Gamma_X = 2.5 \text{ keV}.$$

The very first question is obviously whether the structures are real ones or spurious, i.e. due to fluctuations or to some subtle artifacts of the elaborate detector apparatus; as we have no competence on such matters, waiting for the experimental clarification, we will discuss only the problems which the peaks pose within our current knowledge of QED and nuclear or particle physics.

One of the motivations for producing heavy ion collisions is the study of the laws of QED in intense external electric fields. Let us recall the familiar Balmer formula

(1) $$E_n = -\frac{m(Z\alpha)^2}{2n^2} ,$$

which gives the energy levels of a nonrelativistic electron in the Coulomb field of a point charge Ze; at the same nonrelativistic approximation, the electron velocity is $\beta_n = \frac{(Z\alpha)}{n}$ (in $c = 1$ units). Note that E_n increases quadratically with $Z\alpha$, so that for $Z\alpha > 1$ the binding energy esceeds the rest mass. For large $Z\alpha$, however, the problem becomes relativistic, and one has rather to use the Dirac equation, which gives the relativistic formula

(2) $$E(n,j) = \frac{mc^2}{\sqrt{1 + \left(\frac{Z\alpha}{n-|k|+\sqrt{k^2-(Z\alpha)^2}}\right)^2}} ,$$

where $k = j \pm \frac{1}{2}$ and j is the total angular momentum (orbital and spin) of the electron. For $Z\alpha > 1$, or $Z > 137$, eq.(2) return a complex value, indicating that the one particle treatment is inadequate in that limit. Eq.(2), further, holds only in the idealized case of a pointlike nucleus; taking into account the form factor of an actual nucleus, the problem requires a more complicated numerical treatment, but the net result remains very similar; the value of Z for which the binding energy of the electron exceeds its rest mass is increased from $Z > 137$ to $Z > 150$; when $Z > 173$, further, the binding energy exceeds twice the rest mass. In those conditions it is energetically convenient for the Coulomb field of the nucleus to create an e^+e^- pair, a phenomenon referred to as sparking of the vacuum [3]. The highest Z nucleus observed so far has $Z = 109$, while the highest Z target is Curium, with $Z = 96$, well below the $Z \simeq 173$ threshold; to get higher values one has to go to the scattering of two high Z nuclei; the detection of vacuum sparking was indeed one of the motivations for producing heavy ion collisions. It is to be noted, however, that there are several mechanisms for e^+e^- production such reactions (the kinetic energy of the colliding ion is about 6 MeV/nucleon, i.e. > 1 GeV/nucleus, while the threshold for an e^+e^- pair is just $1 MeV$), so that the interpretation of the produced electrons or positrons as actually due to vacuum sparking is not model independent. Some of the unexpected results are shown in Fig.(1) (taken from Ref. [1]); a very narrow peak, stubbornly sitting at $380 keV$ of kinetic energy, independent from the charge of the colliding ions, two features which do not fit with our current knowledge of nuclear and atomic physics.

Let us discuss widths first. The heaviest nuclei have radii of less than 10f (1f $=10^{-13}$ cm); the speed of the ion beam is $\beta \simeq 0.1$, so that the transit time τ_R of the two nuclei near to each other on a classical Rutherford trajectory is of the order of magnitude $\tau_R \simeq 10f/0.1c = 3 \times 10^{-22}$ sec (presumably somewhat larger, see later); on the other hand the radius of the ground state orbit of a bound electron at $Z \simeq 173$ is $\simeq 300f$, its speed being essentially c; if the time τ_b for binding an electron is of the order of the time needed for a complete revolution around the orbit, $\tau_b \simeq 10^{-21}$ sec; if the interaction time is just 3 times larger than τ_R, there is enough time for the formation of the bound state electron.

The peaks, however, are observed also at $Z < 173$, and their very smooth Z dependence (even if many ore data would be needed to extract a clear Z-pattern) seems to contradict the theoretical model, which predicts a Z^n behaviour, with n as high as 20. A much more serious problem is further posed by the narrowness of the peak. Recall that $1s^{-1} = 6.6 \times 10^{-16} eV/\hbar c$; a time τ_b corresponds therefore to a non surprising width of 1MeV or higher (the typical width of heavy nuclei resonances), while the width of the peaks is in the $20 - 40 keV$ range.

Two opposite attitudes are here possible:

i) the peaks can be optimistically interpreted as the very indication that exteremely long-lived hyperheavy nuclear states have been formed;

ii) they are too narrow to be related to nuclear phenomena.

To support attitude i), one must stretch as much as possible the interaction time, within the usual tunnel effect description of hyperheavy nuclei formation. In the actual case, the incoming nucleus has a kinetic energy $E_c \simeq 1440 MeV/nucleus$ (corresponding to $\simeq 6 MeV/nucleon$). If the charges of the nuclei are Z_1, Z_2, that corresponds to a minimum distance R given by $Z_1Z_2/R \simeq E_c$, or $R \simeq 7f$. The capture and the formation of a metastable state can be described by invoking the existence of a somewhat ad-hoc short range behaviour of the two nuclei potential $U(r)$, increasingly repulsive, for large r, up to $r \simeq R$ (the Coulomb barrier, which is very strong because Z_1, Z_2 are both large), suddenly decreasing in an attractive well, due to nuclear forces, for $r \leq R$. If the incoming nucleus succeeds in penetrating

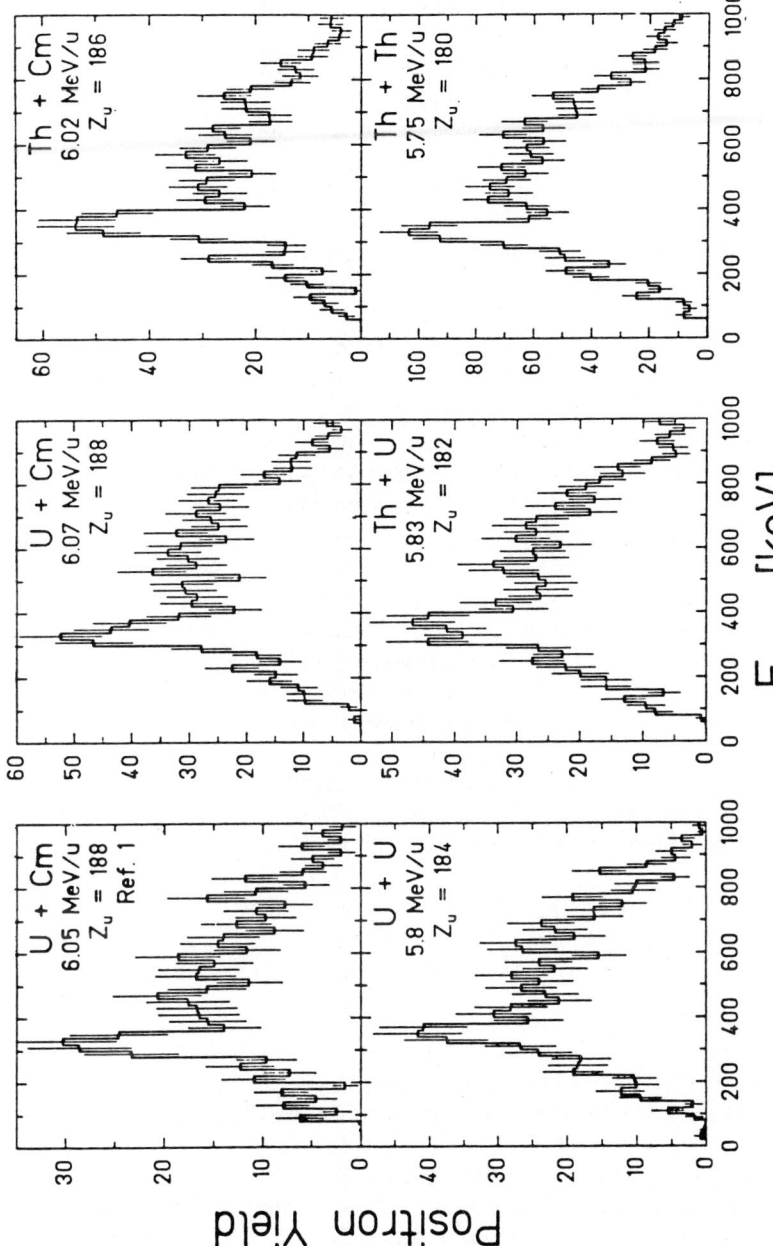

FIG.1 - Positron energy spectra for the five collision systems and bombarding energies indicated.(taken from Ref.[1]).

the Coulomb barrier, a metastable state is formed, which then decays by tunnel effect in a time

$$\tau_d \simeq \frac{R}{v} e^{2\frac{\sqrt{2}m}{\hbar} \int_b^R dr\sqrt{U(r)-E_c}}$$

with $(R/v) \simeq 10^{-22}$sec; ad-hoc adjustments of the sofar unspecified potential might give the required stretching of the interaction time, its unusual length being related to the unusual height of the Coulomb barrier (which is bilinear in the charges of the ions, both unusually large).

If the above sketched pattern can't be turned into a satisfactory quantitative description, one must turn to the last attitude ii), trying to intepretate the peaks as due to the decay of some presumably newly discovered particle.

That last (but by no means less interesting) interpretation faces immediately a preliminary problem: in the decay of a particle, it is the reconstructed total invariant mass of its decay particles to be peaked, not the momentum of the decay products themself (electrons and positrons, in out case), which are on the contrary expected to be Doppler broadened; a peak in the momentum distribution of the decay products implies a somewhat unusual feature, i.e. that the decaying particles are all produced with the same velocity in the laboratory (practically at rest, in the considered case).

Even ignoring that problem, the intrerpretation of the peaks as particles of mass in the 1-2 MeV range is not a simple task: the whole range from the electron mass $m_e \simeq 0.5$MeV to to the muon mass $m_\mu \simeq 106$MeV is indeed completely empty. Even theoretical predictions falling in that range are quite rare. The almost unique candidate might be the axion [4]; purposedly introduced to keep small the CP violation due to $G_{\mu\nu}\tilde{G}_{\mu\nu}$ in the standard model, it is expected to have a mass of the order $M(\text{axion}) \simeq 25N_f(x + 1/x)$keV, where N_f is the number of quark families, x an unknown parameter. The Darmstadtons, while too light to be normal particles, are too heavy to be axions (or at least "standard" axions); indeed Darmstadton masses are recovered for x or $\frac{1}{x>10}$; but in that model the heavy quarkonium decay rates into an axion and a photon are given by

$$\Gamma(\psi(3100) \to a\gamma) \sim x^2, \qquad \Gamma(\Upsilon(9460) \to a\gamma) \sim x^{-2} ;$$

with either choice of $x, 1/x$, one of the two rates turns out to be too large for not having been observed already.

Independently from the theoretical motivations, one can also ask whether the existence of the Darmstadtons is compatible with our current information and understanding of elementary particles, looking at their indirect or direct effects. Limits on indirect effects come from a_e, the anomalous magnetic moment of the electron. If the Darmstadtons can decay into electron-positron pairs, the electron can emit Darmstadtons and virtual Darmstadtons must contribute to a_e. Experimentally, a_e is known with a very high precision, with an absolute error Δa_e of the order of a few parts in 10^{-12}, corresponding to a relative error of few parts in 10^{-9}; on the other hand the theoretical (QED) value of a_e is known with essentially the same precision, and the two numbers agree [5], so that little is left for the Darmstadton contribution. Quantitatively [6], one finds from the comparison with a_e an upper limit on the electron-Darmstadton coupling which in turns gives a lower limit ($\sim 6 \times 10^{-14}$sec) on its lifetime.

Concerning direct observation, the Darmstadtons are too heavy to be produced in positronium decay; to get more energy at disposal, one has to perform an electron beam dump experiments, looking at the invariant mass of electron-positron pairs corresponding to the decay of a Darmstadton "radiated" by the incoming electron. Various limits are so obtained for the combined mass and lifetime values; but no Darmstadtons have yet been found [6].

Conclusions are easily drawn: the theory cannot give more stringent predictions without further experimental hints; the original Darmstadt experiment should be repeated by other groups, systematically varying energy and Z of the ion beam as well as the Z of the target, until the new data will help to solve the problem, either disproving the existence of the Darmstadtons or showing a clear pattern upon which a theaory can be built and tested.

References.
[1] Cowan, T. et al., *Phys. Rev. Lett.* 54, 1761 (1985); *Phys. Rev. Lett.* 56, 444 (1986).
[2] Danzmann, K. et al., *Phys. Rev. Lett.* 59, 1885, (1987).
[3] J. S. Greenberg and W. Greiner, *Physics Today* 25, August 1982.
[4] R. D. Peccei, *Proc. of the XXIII Inernational Conference on High Energy Physics*, 3,S.C. Loken ed., World Scientific (1987).
[5] T. Kinoshita, preprint, july 1988 CERN-TH.5097/88.
[6] M. Davier, *Proc. of the XXIII Inernational Conference on High Energy Physics*, 25,S.C. Loken ed., World Scientific (1987).

THE OPTICAL MODEL IN THE NUCLEON (ANTINUCLEON) - NUCLEUS SCATTERING

F. Iazzi and B. Minetti

Dipartimento di Fisica, Politecnico di Torino
INFN, Sezione di Torino

G. Puddu

Dipartimento di Scienze Fisiche, Universita' di Cagliari
INFN, Sezione di Cagliari

1. INTRODUCTION

The elastic scattering of a nucleon on a nucleus shows remarkable analogies with the diffusion of light by an obstacle. In both the cases the angular distribution gives rise to a characteristic diffraction behavior.

This simple observation is the physical background of the Optical Model.

By an Optical Model Potential (OMP) we mean a potential that represents the interaction between a nucleon (antinucleon) or group of nucleons and a nucleus. When inserted in the Schrodinger equation it gives the differential cross-section and polarization for elastic scattering, the reaction cross-section and some other less important quantities.

It is called the Optical Model since the replacement of the many-body nucleon-nucleus potential is analogous to description of the the propagation of light in a refractive and absorbing medium by an index of refraction.

It is not possible to treat elastic scattering without considering the accompanying absorption processes. To a good approximation they can, however, be all lumped together and treated as a process that removes particles from the incident beam. Then it is not necessary to distinguish between the various absorption processes and consider them in details. It is sufficient to consider a complex potential, like a complex index of refraction, in order to take into account the nucleon-nucleus absorption.

The OMP can be obtained in several ways. Basically it can be calculated from the nucleon-nucleon interaction in nuclear matter; this is difficult and involves a detailed knowledge of the internal structure of the nucleus and its influence on the fundamental nucleon-nucleon interaction.

Alternatively the OMP can be found by postulating a form of potential with a number of free parameters and adjusting such parameters to optimize the fit to the experimental data.

Optical potentials can also be classified by the range of experimental data they are designed to fit. Global potentials give fairly good overall fit to the scattering from many nuclei over a wide range of energies. More precise fits can be obtained adapting the OMP to few particular nuclei over a limited range of energies; but in this case the global behavior of the OMP can be lost.

There can be some ambiguities in the OMP, which are familiar in phenomenological analysis: sometimes it can be found that several potentials fit the same data equally well. It is usually thought that one of these is the "physical" potential, namely that one obtained by a microscopic calculation.

It is important to identify the "physical" potential, that can be used with more confidence in situations different from those from which it was obtained.

It is possible that even among microscopic potentials there are ambiguities: different types of calculations could give different potentials that nevertheless give equally good fits to the data.

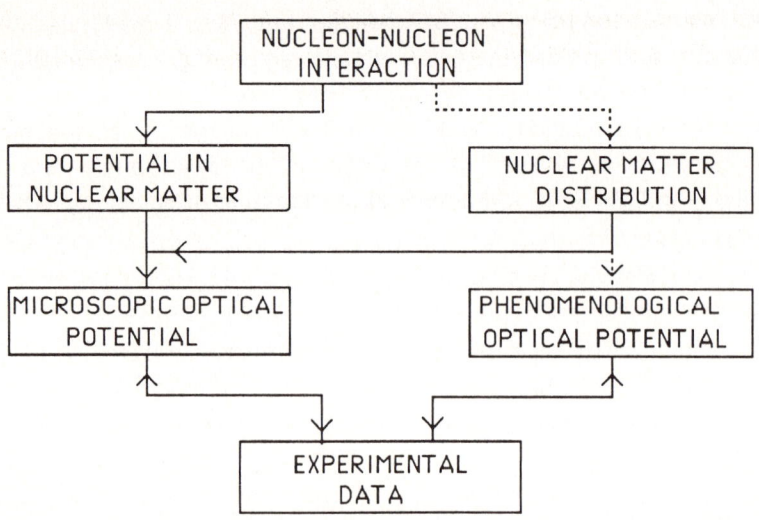

FIGURE 1

Logical Scheme in the Optical Model Analysis

Then the OMP is of its nature a theoretical construction, then care is necessary in describing it as "physical".

The logical path between the elementary interaction and the experimental data can be summarized in fig. 1.

In conclusion the real part of the OMP is mainly responsible of the elastic scattering while the imaginary part takes into account all the absorption processes.

To explain the polarization effects in the scattering of particles with spin greater than zero, the OMP must include at least one term dependent on the spin, usually proportional to **L · S**. (**L** is the angular momentum and **S** is the spin of the incoming particle).

2. EVALUATION OF THE OBSERVABLE QUANTITIES FOR AN ASSUMED POTENTIAL

The main difficulty in comparing the OMP results and the experimental data is that of calculating the matrix elements for each partial wave for an assumed form of the two-body potential.

The problem is therefore that of finding a solution of the wave equation for the interaction, having the required asymptotic form. When the wave function is known the observables are easily evaluated.

2.1 Scattering by a local OMP

Here we summarize the main results; the details about the numerical methods, the error discussion and the automatic search techniques can be found in Refs. 1 and 2

The wave equation for the system reads

$$[(-\hbar^2/2m)\Delta + U]\Psi = E\Psi \qquad (2.1)$$

where $m = m_i \cdot m_t/(m_i + m_t)$ is the reduced mass and $E = E_{LAB}\, m_t/(m_i + m_t)$ is the center of mass kinetic energy, m_i and m_t being respectively the masses of the incident particle and target nucleus. E_{LAB} is the kinetic energy of the incident particle in the laboratory frame.

The OMP can be written as a sum of a central and spin-orbit part in addition to a Coulomb term.

$$U(r) = V_{COUL}(r) - U_C(r) - U_{SO}(r)\, \mathbf{L \cdot S} \qquad (2.2)$$

The total wave function can be expanded as a sum over radial, angular and spin function

$$\Psi(r) = \sum_{l=0}^{\infty} \frac{1}{r} \psi_l^j(r) \cdot C_{m\lambda\mu}^{jls} \cdot i^l \cdot Y_l^{\lambda}(\theta,\varphi) \cdot \chi_\mu^s \qquad (2.3)$$

where $C_{m\lambda\mu}^{jls}$ are the Clebsch-Gordan coefficients and $Y_l^{\lambda}(\theta,\varphi)$ are spherical harmonics.

The spin of the incident particle can couple to the orbital angular momentum to give the total angular momentum $j = l + s$

If the expression (2.3) is inserted in (2.1), a set of coupled radial equations is obtained:

$$\{d^2/dr^2 - l(l+1)/r^2 + (k^2/E) \cdot [E - (V_{COUL}(r) - U_C(r) - U_{SO}(r) \cdot [(j(j+1) - l(l+1)$$
$$- s(s+1)/2]\} \psi_l^j(r) = 0 \qquad (2.4)$$

where:

$$k = (2mE/\hbar^2)^{1/2} = 0.218728 \ (mE)^{1/2} \ fm^{-1}$$

is the wave number (m is expressed in atomic mass units; E in MeV).

The radial wave functions must vanish at the origin and their asymptotic form must be matched to a linear combination of regular and irregular Coulomb wave functions at a point R_m beyond which the nuclear potential can be neglected.

$$\psi_l^j(r) \propto F_l(\eta,r) + G_l(\eta,r) + S_l^j[F_l(\eta,r) - iG_l(\eta,r)] \qquad (2.5)$$

$F_l(\eta,r)$ and $G_l(\eta,r)$ are respectively the regular and irregular Coulomb wave functions and

$$\eta = m \, Z_i Z_t \, e^2/\hbar^2 k) = 0.157454 \ Z_i Z_t (m_i/E_{LAB})^{1/2}$$

The S_l^j are complex matrix elements of unit amplitude for a purely refracting potential, $((Re S_l^j)^2 + (Im \, S_l^j)^2 \leq 1)$. Only the outgoing wave is modified by the interaction and ψ_l^j reduces to the incident wave when the interaction is zero ($S_l^j = 1$)

All the physical measurable quantities are expressed in terms of these matrix elements.

For spin zero particle the scattering amplitude is given in terms of the Legendre polynomials:

$$A(\theta) = (1/(2ik)) \sum_l [(2l+1)(S_l - 1) \, e^{2i\sigma_l} P_l(\cos\theta)] + f_C(\theta) \qquad (2.6)$$

where $f_C(\theta)$ is the Coulomb scattering amplitude

$$f_c(\theta) = -\eta \, \text{EXP}[-i\cdot\eta\cdot\ln(\text{sen}^2(\theta/2)) + 2i\sigma_0] / (2k \, \text{sen}^2(\theta/2))$$

and the coulomb phase shifts are given by
$$\sigma_l = \arg \Gamma(l + 1 + i\eta) \quad \text{or} \quad \exp(2i\sigma_l) = \Gamma(l + 1 + i\eta) / \Gamma(l + 1 - i\eta)$$

The differential elastic cross-section is then

$$\frac{d\sigma_{el}}{d\Omega} = |A(\theta)|^2 \tag{2.7}$$

The reaction cross-section is given by

$$\sigma_{reac} = (\pi/k^2) \sum_l [(2l + 1)(1 - |S_l|^2)] \tag{2.8}$$

For neutral incident particles, $f_c(\theta)$ and σ_l are zero and the elastic cross-section is finite:

$$\sigma_{el} = (\pi/k^2) \sum_l [(2l + 1)|1 - S_l|^2] \tag{2.8}$$

The total cross-section in this case is

$$\sigma_{tot} = (\pi/k^2) \sum_l [(2l + 1)(1 - \text{Re } S_l)] \tag{2.8}$$

Two important relations can be derived from the above equations. The optical theorem

$$k\,\sigma_T = 4\pi \cdot \text{Im } A(0°) \tag{2.9}$$

and the Wick's limit

$$\frac{d\sigma_{el}(0°)}{d\Omega} \geq \left(\frac{k\,\sigma_{tot}}{4\pi}\right)^2 \tag{2.10}$$

In the case of incident particle of spin one half there are two scattering amplitudes

$$A(\theta) = f_c(\theta) + (1/2ik) \sum_l [(l+1)\cdot S_l^{l+1/2} + l\cdot S_l^{l-1/2} - (2l+1)] \, e^{2i\sigma_l} P_l(\cos\theta)$$

$$B(q) = (1/2ik) \sum_l [S_l^{l+1/2} - S_l^{l-1/2}] \, e^{2i\sigma_l} P_l^1(\cos\theta) \tag{2.11}$$

where $P_l^1(\cos\theta)$ is the associated Legendre polynomial. The differential elastic cross-section is

$$\frac{d\sigma_{el}}{d\Omega} = |A(\theta)|^2 + |B(\theta)|^2$$

and the polarization of the scattered particle is

$$P(\theta) = \frac{2\,\text{Im}(A\cdot B^*)}{|A|^2 + |B|^2}$$

in the case of unpolarized incident particles.

The integrated reaction, elastic and total cross-sections are:

$$\sigma_{reac} = (\pi/k^2) \sum_l [(l+1)\cdot(1-|S_l^{l+1/2}|^2) + l\cdot(1-|S_l^{l-1/2}|^2)]$$

$$\sigma_{el} = (\pi/k^2) \sum_l [(l+1)\cdot|1-S_l^{l+1/2}|^2 + l\cdot|1-S_l^{l-1/2}|^2)] \qquad (2.12)$$

$$\sigma_{tot} = (\pi/k^2) \sum_l [(l+1)\cdot(1-\mathrm{Re}\, S_l^{l+1/2})^2 + l\cdot(1-\mathrm{Re}\, S_l^{l-1/2})^2]$$

In the case of particle of unit spin the equations are more complicated and are given in Refs. 1 and 2.

2.2 Scattering from a non local OMP

Detailed studies of the theory of nuclear matter have shown that the OMP is non local in character. This means that the potential acting on a particle located at a point **r** depends not only on **r** but also on the values of the wave function throughout all space and so takes into account of the finite size of the incident particle and of the dispersive nature of the nuclear matter.

The term $U\Psi$ in the Schrodinger equation (2.1) is replaced by

$$U\Psi \longrightarrow \int U(\mathbf{r},\mathbf{r'})\, \Psi(\mathbf{r'})\, d^3\mathbf{r'} \qquad (2.13)$$

where $U(\mathbf{r},\mathbf{r'})$ is the non local OMP.

We are now faced with an integro-differential wave equation that can be reduced to a radial form by expanding the wave functions $\Psi(\mathbf{r})$, $\Psi(\mathbf{r'})$ and the non local OMP in spherical harmonics.

Details of the calculations are given in Ref. 3, where the kernel $U(\mathbf{r},\mathbf{r'})$ has the separable form

$$U(\mathbf{r},\mathbf{r'}) = U_N((\mathbf{r}+\mathbf{r'})/2) \cdot H(\mathbf{r}-\mathbf{r'}) \qquad (2.14)$$

After integration of the radial wave equation, the matrix elements S_l^i and the observables are obtained as in the case of a local OMP.

If the functions $H(\mathbf{r}-\mathbf{r'})$ has the form

$$H(\mathbf{r}-\mathbf{r'}) = (1/(\pi a^2)^{3/2})\, \exp\left[-\left(\frac{\mathbf{r}-\mathbf{r'}}{a}\right)^2\right] \qquad (2.15)$$

and the range of non locality a is small, the integro-differential equation reduces to an ordinary differential one by a suitable expansion in power of a. The problem is then reduced to a local one but for a particle of variable mass[1]

$$m(r) = m/[1-(a^2 m/2\hbar^2) \cdot U_N(r)]$$

This is the effective mass approximation, and the Schrodinger equation reads

$$(1/8)\left[\mathbf{p}^2 \frac{1}{m(r)} + \mathbf{p}\frac{2}{m(r)}\mathbf{p} + \frac{1}{m(r)}\mathbf{p}^2\right]\Psi(r) = [E-V_{COUL}(r)+U_N(r)]\,\Psi(r) \qquad (2.16)$$

where **p** is the well known differential operator.

Unfortunately the effective mass approximation can be applied only if the range of non locality is very small.

When this assumption is no longer valid an equivalent local potential can be evaluated in the Gaussian approximation[3].

An equivalent local potential is defined as the local potential $U_L(r)$ whose local wave function has the same asymptotic behavior of the non local one $U_N(r)$. In this case the matrix elements S_i^j and the cross-section are the same.

Perey and Buck[3] have shown that, if $U_N(r)$ is approximately constant inside the nucleus, the equivalent local OMP is

$$U_L(r) = U_N(r) \cdot \exp[-(ma^2/2\hbar^2) \cdot (E - U_L(r))] \qquad (2.17)$$

As it can be shown the local equivalent OMP is energy dependent also when the non local one is energy independent. This means that the energy dependence found for the phenomenological OMP's can be ascribed to its non-local character.

3. PHENOMENOLOGICAL ANALYSIS OF THE NUCLEON-NUCLEUS ELASTIC CROSS-SECTIONS

When the analytical form of the OMP is defined, the free parameters can be determined by minimization of the quantity

$$\chi^2 = \sum \{[\sigma_{OMP}(\theta_i) - \sigma_{EXP}(\theta_i)]/\Delta\sigma(\theta_i)\}^2 \qquad (3.1)$$

where $\sigma_{OMP}(\theta_i)$ and $\sigma_{exp}(\theta_i)$ are the theoretical and experimental differential elastic scattering cross-sections at the angle θ_i and $\Delta\sigma(\theta_i)$ is the experimental error.

Two sources of errors[2] in the measured values of θ can be included in the minimization process.

The first one is related to the detector angular acceptance α. In (3.1) the $\sigma_{EXP}(\theta_i)$ can be replaced by

$$\sigma_{EXP}(\theta_i) \longrightarrow [\sigma_{EXP}(\theta) + \sigma_{EXP}(\theta - \delta\theta) + \sigma_{EXP}(\theta + \delta\theta)]/3$$

where $\delta\theta = \alpha/(2\sqrt{2})$.

The second one is the experimental uncertainty $\Delta(\theta_i)$ of the scattering angle θ_i. The square of the experimental error $\Delta\sigma(\theta_i)$ can be replaced by

$$\Delta(\theta_i) \longrightarrow [\Delta\sigma(\theta_i)]^2 + [d\sigma(\theta)/d\theta]^2 \cdot (\Delta\theta_i)^2$$

3.1 Choice of the phenomenological form of the OMP

In principle any radial dependence of the central and spin-orbit part of the OMP can be accepted provided the range of the OMP be of the same order of magnitude of the nuclear radius.

In practice any *a priori* knowledge of the nuclear structure must be used to define the OMP.

To take into account the absorption processes the central and spin-orbit term of the OMP appearing in (2.2) must be splitted in a real and imaginary part

$$U_C(r) = V_C(r) + iW_C(r) \qquad U_{so}(r) = V_{so}(r) + iW_{so}(r)$$

The Coulomb potential $V_{COUL}(r)$ is usually taken to correspond to a uniform charge density within a spherical nucleus of radius R_C.

$$V_{COUL}(r) = (Z_i Z_t e^2 / 2R_c) \cdot (3 - r^2/R_c^2) \qquad \text{for } r \leq R_C$$
$$= Z_i Z_t e^2 / r \qquad \text{for } r \leq R_C$$

The exact form of $V_{COUL}(r)$ has been found to have little effect on the final results. This potential has no free parameters and the radius R_c can be given by[4]

$$R_c = (1.149\ A^{1/3} + 1.788 \cdot A^{1/3} - 1.163\ A^{1/3})\ fm$$

The real part of the OMP can be extimated as the sum of all the elementary interactions $V(r-r')$ of the incident particle and the individuals nucleons in the target.

Referring to fig. 2 and supposing a continuum distribution of nucleons in the target nucleus with density $\rho(r)$, we have

$$V_C(r) = \int \rho(r) \cdot v(r-r') \cdot d^3 r' \tag{3.2}$$

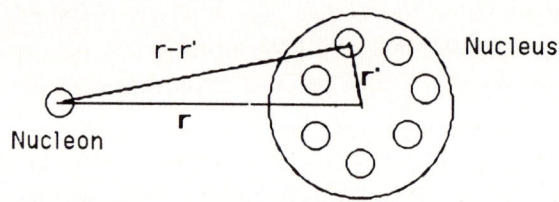

FIGURE 2

Sketch of the geometrical derivation of phenomenological OMP

Since the elementary interaction is of short range, it can be approximated by a Dirac function. If the nuclear deformation is neglected, the real part of the central OMP is proportional to $\rho(r)$.

Usually the nuclear density is approximated by a Saxon-Woods form factor

$$SW(r,R,a) = [1 + \exp((r-R)/a)]^{-1}$$

and the real central OMP can be written

$$V_C(r) = V \cdot SW(r,R_R,a_R) \qquad R_R = r_r A^{1/3} \qquad (3.3)$$

This expression depends on three free parameters, (V, r_R, a_R).

The imaginary part is usually assumed as a sum of a surface term proportional to $(-d/dr)SW(r,R_S,a_S)$ and a volume term proportional to $SW(r,R_V,a_V)$. The first term is expected to be important in the low energy range where the incoming particle has a mean free path less than the nuclear radius. The second term is important at higher energies where the incident particle has a large chance to be absorbed inside the nucleus. We can write

$$W_C(r) = W_S(-4a_S) \cdot (d/dr)SW(r,R_S,a_S) + W_V \cdot SW(r,R_V,a_V)$$
$$R_S = r_S A^{1/3} \qquad R_V = r_V A^{1/3} \qquad (3.4)$$

with six free parameters $(W_S, r_S, a_S, W_V, r_V, a_V)$.

The spin-orbit potential in (2.2) takes into account the spin dependence of the OMP and explains very simply the separation of the $j = l \pm 1/2$ states in nuclei and the increase of this separation with l. The derivative form factor of the Thomas term for spin-orbit forces in atoms gives the spin-orbit potential in the form

$$V_{SO}(r) = (\hbar/m_\pi c)^2 \cdot (V_{SO}/r) \cdot (-d/dr)SW(r,R_{SO},a_{SO})$$
$$W_{SO}(r) = (\hbar/m_\pi c)^2 \cdot (W_{SO}/r) \cdot (-d/dr)SW(r,R_{SO},a_{SO})$$
$$R_{SO} = r_{SO} A^{1/3} \qquad (3.5)$$

Here the π meson Compton wavelength $\lambda_\pi = (\hbar/m_\pi c) \approx \sqrt{2}$ fm is a convenient scaling factor.

This potential has four free parameters $(V_{SO}, W_{SO}, r_{SO}, a_{SO})$.

3.2 Dependence of the OMP from the isospin

As it is well known the elementary nucleon-nucleon interaction depends on the isospin. We expect that also the OMP contains a term dependent on the isospin. Lane[5] has shown that the central part of the OMP can be written as a sum of an isoscalar term $V_0(r)$ and an isovectorial term $V_1(r)$:

$$U_C(r) = U_0(r) + U_1(r) \cdot (4/A) \cdot (t \cdot T) \qquad (3.6)$$

In the above equation t, T are the isospin of the incident nucleon and target respectively. The review article of Satchler[6] discusses the argument in detail. The isospin interaction splits the radial part of the potential into two diagonal terms which are responsible for the proton and neutron scattering, and a non diagonal term responsible for the (p, n) quasi elastic scattering.

The diagonal terms are

$$V_C(r) = (V_0 \pm \varepsilon V_1) \cdot SW(r,R_R,a_R) + \Delta V_C \qquad (3.7)$$

where $\varepsilon = (N-Z)/A$ is the nuclear asymmetry coefficient. The + and - signs are for protons and neutrons respectively.

The last term represents the Coulomb correction term. It can be interpreted as a consequence of the energy dependence of the OMP.

The OMP become as more attractive as the kinetic energy of the incoming proton is reduced. Let us suppose a linear dependence of the local potential V on the kinetic energy T of the incident particle inside the nucleus, so that
$$V = V^* - \alpha \cdot T$$
where α is a constant. For protons of energy E
$$T = E + V - V_{COUL}$$
Eliminating T from the last two equations, we get
$$V = V^*/(1+\alpha) - [\alpha/(1+\alpha)] \cdot E + [\alpha/(1+\alpha)] \cdot V_{COUL} = V_0 + \beta V_{COUL}$$
$$V_0 = V^*/(1+\alpha) - [\alpha/(1+\alpha)] \cdot E \tag{3.8}$$
$$\beta = \alpha/(1+\alpha) = -\partial V/\partial E = -\partial V_0/\partial E \qquad \beta > 0$$
This expression can be generalized to all the values of r. Thus we have
$$V(r,E) = V_0(r,E) - \frac{\partial V_0(r,E)}{\partial E} \cdot V_{COUL}(r) \tag{3.9}$$
The last Coulomb correction term is greater than zero for protons and is zero for neutrons.

By comparison with (3.7) we have
$$\Delta V_C = \frac{\partial V_0(r,E)}{\partial E} \cdot V_{COUL}(r) \tag{3.10}$$
The coulomb potential is a slowly variable function and can be replaced by a mean value
$$V_{COUL}(r) \longrightarrow \overline{V}_{COUL} = \int_0^{R_C} V_{COUL}(r) \cdot 4\pi r^2 \cdot dr \Big/ \int_0^{R_C} 4\pi r^2 \, dr \sim (6/5) \, Z_t e^2/R_C$$
If we assume $R_C = r_C A^{1/3}$, the Coulomb potential can be replaced by a term proportional to $Z_t A^{-1/3}$ and the radial dependence of ΔV_C is like that of the real potential $V_0(r,E)$.

In conclusion the real potential can be written as
$$V(r,E) = [V_0(E) \pm \frac{N-Z}{A} \cdot V_1(E) + \delta \cdot Z_t/A^{1/3}] \cdot SW(r, R_R, a_R) \tag{3.11}$$
Sign + and $\delta > 0$ are to be taken for protons, the sign - and $\delta = 0$ for neutrons.

3.2 Ambiguities in the OMP

As seen in the introduction, in many cases there are different OMP that give equally good fits to the experimental data. The multidimensional function

χ^2 defined in (3.1) is a surface in the space of the free parameters and can show several local minima. It is often difficult to identify the OMP with physical significance.

This fact make necessary to choose an analytic form of the phenomenological OMP coherent with some physical hypothesis derived from the general properties of the nuclear matter.

Some ambiguities are continuous, so that all sets of parameters in a particular region give equally good fits. Other ambiguities are discrete in the sense that only a series of particular values of the parameters gives acceptable fits while intermediate values do not.

The ambiguities are more serious if the experimental data are insufficient or affected by large errors.

The most important types of parameter ambiguities are discussed below, other types are discussed in Ref. 1.

a) Normalization ambiguities

In many cases the experimentally measured cross-sections include an over-all normalization error depending on the calibration of the experimental apparatus. It can be shown that, by analyzing the same set of data with different normalization, the parameters of the OMP depend on the normalization factor chosen. In fig. 3 some results relative to 12 MeV deuteron scattering on ^{60}Ni are reported [7].

It is possible to determine the normalization factor by a suitable change of the χ^2 defined in (3.1), but this procedure can introduce additional uncertainties into the values of the OMP.

It can be shown (fig. 3) that some pairs of parameters such as (V, r_R) or (W, a_V) vary in opposite ways. This corresponds to the $V \cdot r_R^n$ = constant and $W \cdot a_V$ = constant ambiguities.

b) Valley ambiguities

The χ^2 function can be thought as a surface in a space of many dimensions and the best fit corresponds to the deepest minima. In practice sometimes these minima are very broad and shallow, so that it is possible to vary several parameters at once without any appreciable change in χ^2. An example of this is the $V \cdot r_R^n$ = constant ambiguities in which the χ^2 remains almost constant provided that V and r_R are varied at the same time keeping constant $V \cdot r_R^n$, with n of the order of two or three.

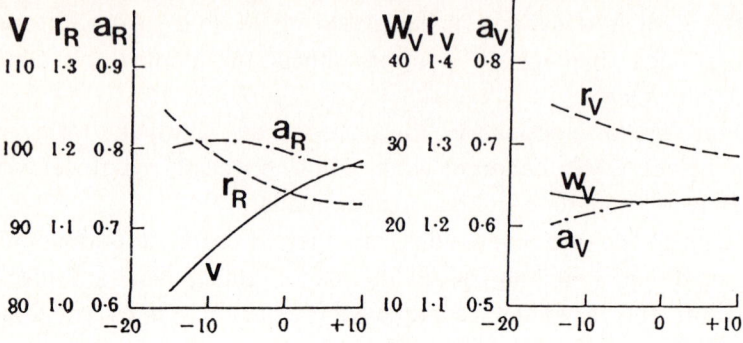

FIGURE 3
Best values of the parameters $V, r_R, a_R, W_V, r_V, a_V$ of the OMP as a function of the normalization of the experimental data

Many effects of the valley ambiguities can be removed by fixing some form-factor parameters to a suitable average values at the beginning of the analysis. The fits obtained in such a way are only slightly inferior to those obtained with all the parameters free and they allow the systematic study of the potential depths with the energy or the atomic mass number.

c) Potential-depth ambiguities

If the real part of the OMP is steadily varied, and at each depth all the other parameters are adjusted to minimize χ^2, this quantity passes through a whole set of minima of the same order of magnitude. This happens because for a given partial wave all the wave functions have the same asymptotic form and hence the same matrix element S_l^j. The value of the imaginary part of the OMP also increases quite regularly, and can be possible to find an empirical relationship between the values of V and W_V for each potential. A more detailed analysis of this ambiguity is given in[1].

3.3 Systematic analyses of differential elastic cross-sections

The results of many analyses are reported in Ref. 1 and in the references enclosed therein. Only some typical results are reported here.

One of the most comprehensive and detailed analysis of a wide range of protons and neutrons data on medium and heavy nuclei for energies less than 50 MeV has been made by F. D. Becchetti and G. W. Greenlees[8] using a phenomenological OMP of the form

$$U(r) = V \cdot SW(r, R_R, a_R) +$$

$$+ i [W_V \cdot SW(r, R_I, a_I) + 4a_I W_S \cdot (-d/dr)SW(r, R_I, a_I) +$$
$$+ (\hbar/m_\pi c)^2 \cdot V_{SO} \cdot (-d/dr)SW(r, R_{SO}, a_{SO})$$
$$R_R = r_R A^{1/3}, \quad R_I = r_I A^{SO}, \quad R_{SO} = r_{SO} A^{1/3} \tag{3.12}$$

They obtained excellent over-all fits with a wide range of different assumptions about the dependence of the parameters on the atomic mass number A and the energy E.

An important hypothesis is the $A^{1/3}$ dependence assumed for all the radius parameters. Although this assumption is peculiar for the nuclear density, the elementary nucleon-nucleon interaction can change this A dependence. Any departures from this dependence are reflected in the A dependence of the others parameters.

The best parameters for protons were found to be :
$V = (54 - 0.32 \cdot E + 24 \cdot (N-Z)/A + 0.4 \cdot Z/A^{1/3})$ MeV
$r_R = 1.17$ fm $a_R = 0.75$ fm
$W_V = (0.22 \cdot E - 2.7)$ MeV, or zero, whichever is greater
$W_S = (11.8 - 0.25 \cdot E + 12 \cdot (N-Z)/A)$ MeV, or zero, whichever is greater
$r_I = 1.32$ fm $a_I = (0.51 + 0.7 \cdot (N-Z)/A)$ fm
$V_{SO} = 6.2$ MeV $r_{SO} = 1.01$ fm $a_{SO} = 0.75$ fm

The best parameters for neutrons were found to be :
$V = (56.3 - 0.32 \cdot E - 24 \cdot (N-Z)/A)$ MeV
$r_R = 1.17$ fm $a_R = 0.75$ fm
$W_V = (0.22 \cdot E - 1.56)$ MeV, or zero, whichever is greater
$W_S = (13 - 0.25 \cdot E - 12 \cdot (N-Z)/A)$ MeV, or zero, whichever is greater
$r_I = 1.26$ fm $a_I = 0.58$ fm
$V_{SO} = 6.2$ MeV $r_{SO} = 1.1$ fm $a_{SO} = 0.75$ fm

A recent analysis [9] has been performed by R. L. Varner et al. for proton and neutron scattering on medium and heavy nuclei. They use some new features in the parametrization of the OMP but they retain the same radial dependence used by Becchetti and Greenlees.

The potential radii are not simply proportional to $A^{1/3}$. The energy dependence of the imaginary potential exhibits a smooth transition from surface absorption at high energies to volume absorption at low energies avoiding the linear energy dependence.

The parameters a_R, a_I, V_{SO}, a_{SO} were found independent of A, Z and energy E. The A, Z and E dependence of the others parameters is given by

$$V_R = V_0 \pm V_t \cdot (N-Z)/A + V_e \cdot (E-E_c) \qquad + : \text{protons}, - : \text{neutrons}$$
$$R_0 = r_0 \cdot A^{1/3} + r_0^{(0)}$$

$$E_C = 6e^2Z/5R_C = 1.73 \cdot Z/R_C \text{ MeV} \qquad \text{for (p,p)}$$
$$E_C = 0 \qquad \text{for (n,n)}$$
$$R_C = r_C \cdot A^{1/3} + r_C^{(0)}$$
$$R_{SO} = r_{SO} \cdot A^{1/3} + r_{SO}^{(0)}$$
$$W_V = W_{V0} \cdot (1 + \exp\{[W_{VE0}-(E-E_C)]/W_{VEW}\})^{-1}$$
$$W_S = [W_{S0} \pm W_{St} \cdot (N-Z)/A] \cdot (1 + \exp\{[(E-E_C) - W_{SE0}]/W_{SEW}\})^{-1}$$
$$R_I = r_I A^{1/3} + r_I^{(0)}$$

The parameters of the global fit are:

V_0=52.9 MeV	V_t=13 MeV	V_e=-0.3	
r_0=1.25 fm	$r_0^{(0)}$=-0.24 fm	a_0=0.69 fm	
V_{SO}=5.9 MeV	r_{SO}=1.39 fm	$r_{SO}^{(0)}$=-1.43 fm	a_{SO}=0.65 fm
W_{V0}=10 MeV	W_{VE0}=35 MeV	W_{VEW}=15 MeV	W_{S0}=9 MeV
W_{St}=14 MeV	W_{SE0}=29 MeV	W_{SEW}=23 MeV	
r_I=1.32 fm	$r_I^{(0)}$=-0.41 fm	a_I=0.72 fm	
r_C=1.24 fm	$r_C^{(0)}$=0.12 fm		

The main characteristics of this new parametrization are:
a) The real potential shows an isovector strength V_1 about half of the Becchetti one. This arises from the simultaneous analysis of proton and neutron data.
b) The parametrization of the radii requires a constant term.
c) The imaginary potential has a smooth energy dependence.
d) The spin-orbit potential depth is independent on E and there is not need of an imaginary component

Neutron data were also analyzed by F. G. Perey and B. Buck[3] with a non local potential of the form
$$U(r,r') = U(p) \cdot H(r,r') +$$
$$+ [(V_{SO}+iW_{SO}) \cdot (\hbar/m_\pi c)^2 \cdot (1/r) \cdot (-d/dr)SW(r,R_{SO},a_{SO})] \cdot \delta(r-r')$$
with $p = |r+r'|/2$

The potential $U(p)$ varies with p in the same way as the corresponding local potential in (3.12) depends on r. The function $H(r-r')$ is given by (2.15).

Perey and Buck were able to fit neutron data up to 25 MeV with the above non local potential with energy independent parameters.

They found

$V = 71$ MeV $\quad W_S = 15$ MeV $\quad W_V = 0$ $\quad V_{SO} = 14.4$ MeV

$r_R = r_I = 1.22$ $\quad a_R = a_{SO} = 0.65$ fm $\quad a_I = 0.47$ fm

The range of non locality was found $a = 0.85$ fm.

Another interesting main feature of this analysis is that an energy independent non local potential obtained by a fit of the elastic differential cross-sections reproduces fairly well also the reaction and total cross-sections as a functions of E and A, except at low energies where the contribution of the compound nucleus can be important.

This means that the energy dependence of the local OMP is at least partially fictitious, reflecting only the fact that the OMP is essentially a non local one.

4. MICROSCOPIC THEORY OF THE OMP

In the previous section it has been shown that a phenomenological OMP, with suitably adjusted parameters, can give a good overall account of the experimental data on elastic scattering.

This OMP is generally smooth, local and slowly varying function of the energy. It is useful if we are only interested in a correct evaluation of the elastic scattering cross-section for practical purposes.

Unfortunately this approach can't give physical informations about the nuclear structure and the fundamental (nucleon-nucleon) interaction into the nuclear matter. Thus it becomes necessary to calculate the OMP from more fundamental physical data.

This microscopic approach is quite difficult because the exact evaluation of the OMP is equivalent to the complete and exact solution of the many body problem.

The formal theory[10] of nuclear reactions gives an expression for the OMP but the numerical calculations are difficult.

A similar approach by the Bruckner's nuclear matter calculations[11] gives definite results but it is difficult to apply to finite nuclei.

Many other attempts were made to evaluate the OMP but they cannot be reported here. This section is concerned only with simple approaches to the problem, to emphasize the main features of the microscopic OMP's.

4.1 The theory of Greenlees

G. W. Greeenlees and collaborators[12] start from a variational principle equivalent to the Schrodinger equation and found the OMP as the sum of all its constituent interactions $v(|r-r_i|)$ with the individual nucleons in the target. If $\psi(\xi)$ is the normalized wave functions describing the ground state of the target nucleus, and if ξ denotes the spatial, spin and isospin coordinates of all the nucleons, the real part of OMP can be written

$$U(r) = \int \psi^*(\xi) \sum_{i=1}^{A} v(|r-r_i'|) \psi(\xi) d\xi = \int \rho(r') v(|r-r'|) d^3r' \quad (4.1)$$

where $\rho(r)$ is the density of nucleons in the nucleus.

The general form of the elementary nucleon-nucleon interaction can be written as

$$v(\rho) = v_d(\rho) + v_\tau(\rho) (\tau \cdot \tau_i) + v_\sigma(\rho) (\sigma \cdot \sigma_i) +$$
$$+ v_{\sigma\tau}(\rho) (\tau \cdot \tau) \cdot (\sigma \cdot \sigma_i) + [v_t(\rho) + v_{t\tau}(\rho) (\tau \cdot \tau_i)] S_{12} +$$
$$+ v_{ls}(\rho) (1/\hbar) [(r-r') \times (p-p') \cdot (\sigma \cdot \sigma_i)] \quad (4.2)$$

where $\rho = r - r_i$ and S_{12} is the tensor force operator.

For zero spin nuclei, only the first, second and last term contributes to $U(r)$ in 4.1. Using the expression of the neutron ($\rho_n(r)$) and proton ($\rho_p(r)$) densities in terms of the nuclear wave function $\psi(\xi)$, the OMP real part can be written

$$U = U_R + U_I + U_{SO} \quad (4.3)$$

where the central term is

$$U_R = \int [\rho_p(r') + \rho_n(r')] \cdot v_d(|r-r'|) d^3r' \quad (4.4)$$

the isotopic spin dependent term is

$$U_I = \tau_z \int [\rho_p(r') - \rho_n(r')] \cdot v_\tau(|r-r'|) d^3r' \quad (4.5)$$

the spin-orbit term is given by an infinite series whose leading term is

$$U_{SO} = -(1/\hbar) \cdot (4\pi/3r) \cdot (d/dr)[(\rho_p(r) + \rho_n(r)] \cdot \int v_{SO}(\eta) \eta^4 d\eta \cdot l \cdot \sigma \quad (4.6)$$

This expression has the same form of the last term in (4.12), with a suitable expresion for V_{SO}. Here τ_z is +1 for neutrons and -1 for protons. To use these expression in an actual calculation it is necessary to chose an explicit form for the nuclear densities and for the nucleon-nucleon interaction.

As a first approximation it can be assumed

$$\rho_p(r) = (Z/A) \rho_m(r) \qquad \rho_n(R) = (N/A) \rho_m(r)$$

with $\rho_m(r) \propto SW(r, R_m, a_m)$ and $R_m = r_m A^{1/3}$.

If also $v_\tau(\rho) = -\zeta \cdot V_R(\rho)$ with ζ a constant parameter indipendent on r, the isospin term reduces to

$$U_I(r) = \zeta \cdot U_R(r) \cdot [(N-Z)/A] \cdot \tau_z$$

wich is of the form commonly used in phenomenological analyses.

The Greenlees OMP was derived under some drastic assumptions: the effect of polarization of the target nucleus by the incident particle was neglected, the incident nucleon was treated as indistinguishable and the wave function was not antisimmetrized with respect to the exchange of the incident nucleon with a nucleon in the target nucleus. If we take into account the antisymmetrization, additional exchange terms must be introduced in (4.1) and the resulting potential will be non local.

Furthermore eq. (4.1) is also inadequate because it gives an energy indipendent OMP.

The justification of the theory of Greenlees has been given by Kidway and Rook [1,13] They found an expression that is the same as that of Grenlees (4.1) except for the substitution of the t-matrix for nucleon-nucleon interaction $t(r-r')$ in place of $v(r-r')$. The difference vanishes for an unperturbed system for which $t(r-r') = v(r-r')$.

A formal theory to obtain the t-matrix in the nuclear matter starting from the free nucleon-nucleon interaction was given by Kerman, Mc Manus and Thaler [14] with the Green-function formalism.

In spite of the various approximations the theory of Greenlees was successfully applied to the experimental data [12].

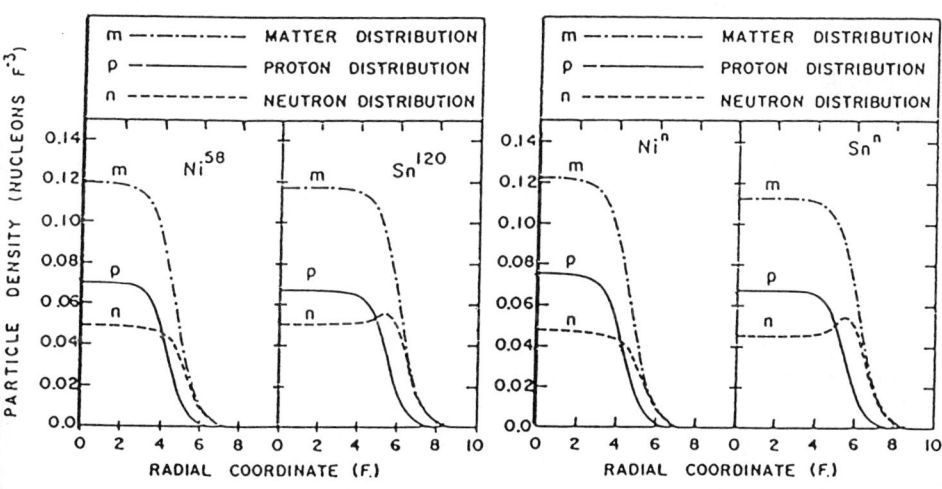

FIGURE 4
Radial behavior of nuclear matter, neutron and proton densities determined from proton (Ni^{58}, Sn^{120}) and neutron (Ni^n, Sn^n) analysis.

The analysis was performed taking as free parameters the radius r_m and the diffuseness a_m of the nuclear matter distribution, toghether with the strengths of the central and spin-orbit potentials, the strengths W_S and W_V, the radius r_I and the diffuseness a_I of the surface and volume imaginary optical potentials.

With these eight parameters the experimental data can be fitted as accurately as with the purely phenomenological ten-parameter potential (3.12).

This calculation was improved by assuming the protons and the neutrons to have different density distributions, the distribution of protons being taken from electron scattering data and the parameters of the neutron distributions being treated as variables.

The result [12] for medium and heavy nuclei consistently indicated that the neutron r.m.s radii exceed those of the proton by about 0,5 fm

$$<r_n^2> - <r_p^2> \approx 0.5 \text{ fm}$$

Some density distributions are given in fig. 4.

A subsequent analysis given by Friedman[15] using a density dependent effective interaction in (4.1) reduced this value to about 0.2 fm.

4.2 The OMP in the Impulse Approximation

Starting from the formal theory of nuclear reactions, Giannini and Ricco[16] derived an expression for the generalized OMP.

Let us consider a system consisting of one nucleon plus a nucleus with mass number A.

The single particle Schrodinger equation can be written

$$(-\hbar^2/2m) \cdot \Delta\psi(r) + \int V(r,r') \cdot \psi(r') \cdot d^3r' = E \cdot \psi(r) \qquad (4.7)$$

the OMP $V(r,r')$ being the representative of the one-body operator

$$V(r,r') = <g_0|U_{OP}|g_0> \qquad (4.8)$$

and g_0 being the part of the total wave functions wich corresponds to the nucleus in its ground state.

Starting from Watson's multiple scattering theory [17], the optical potential operator U_{OP} can be expressed by a multiple-scattering expansion whose first term is

$$U_{OP} = \sum_{i=1}^{A} \tau_i + ...$$

Here τ_i is the single scattering t-matrix for bound nucleons.

When the energy of the incident particle is sufficiently high, only the first term of the series can be retained (single-scattering approximation) and the operators τ_i can be approximated by the scattering t-matrix for free nucleons t_j (impulse approximation).

The OMP is then
$$V(r,r') = \sum_{i=1}^{A} \langle g_0|t_i|g_0\rangle \tag{4.9}$$

In the coordinate space the corresponding expression of $V(r,r')$ in given in[16]

For medium and heavy nuclei the spin- and isospin-indipendent part of eq, (4.9) reads

$$V_C(r,r') = (2\pi)^{-3} \cdot \int d^3p \, d^3q \cdot e^{i\mathbf{p}\cdot(\mathbf{r},\mathbf{r}')} \cdot e^{i\mathbf{q}\cdot(\mathbf{r},\mathbf{r}')/2} \cdot$$
$$\cdot t_E((\mathbf{p}+\mathbf{q})/2,(\mathbf{p}-\mathbf{q})/2) \cdot F(q) \tag{4.10}$$

where $F(q) = \int \rho(r) \cdot e^{-i\mathbf{q}\cdot\mathbf{r}} \cdot d^3r$ is the nuclear ground state form factor and $t_E(k_f,k_i)$ is the off-shell nucleon-nucleon scattering amplitude. k_f and k_i are the initial and final momenta of the nucleon-nucleon c. m. system and $\mathbf{p}=k_i+k_f$, $\mathbf{q}=k_f-k_i$.

If the t-matrix in eq. (4.10) does not depend on $\mathbf{p}$, i.e. the t-matrix depends only on the momentum transfer $\mathbf{q}$ and the off-shell effects are neglected, the OMP is local and is given by
$$V(r,r') = V_L(r) \cdot \delta(r-r')$$
with
$$V_L((r) = \int d^3r \cdot e^{i\mathbf{q}\cdot\mathbf{r}} \cdot t_E(q) \cdot F(q) = \int d^3r' \cdot \rho(r') \cdot t_E(|r-r'|) \tag{4.11}$$

The expression (4.11) gives the OMP in terms of an effective interaction. If the t-matrix is written in the Born approximation we have $t_E(|r-r'|) = v(|r-r'|)$ and one obtain the Greenlees formula (4.4).

The non local effects in the OMP are related to the off-shell behaviour of the t-matrix. The simplest way to take into account these effects is to factorize the t-matrix:
$$t_E((\mathbf{p}+\mathbf{q})/2,(\mathbf{p}-\mathbf{q})/2) = g(p) \cdot t_E(q) \tag{4.12}$$
The OMP becomes a separable one :
$$V(r,r') = U((r+r')/2) \cdot H(r-r')$$
with
$$H(r-r') = (2\pi)^{-3} \cdot \int d^3r \, e^{i\mathbf{p}\cdot(\mathbf{r}-\mathbf{r}')} \cdot g(p)$$
$$U((r+r')/2) = \int d^3q \cdot e^{i\mathbf{q}\cdot(\mathbf{r}+\mathbf{r}')/2} \cdot t_E(q) \cdot F(q) \tag{4.13}$$
If a Gaussian form for the off-shell factor $g(p)$ is assumed
$$g(p) = e^{-a^2p^2/2}$$
one obtains the Perey-Buck non locality factor
$$H(r-r') = (\pi a^2)^{-3/2} \cdot e^{-(r-r')^2/a^2}$$

The non local OMP is in general energy dependent because of the energy dependence of the t-matrix. Neverthess this dependence is weaker than for the equivalent local potentials and in many cases can be neglected.

If the off-shell effects can be neglected, the local OMP (4.11) can be expressed[18] in terms of the nucleon-nucleon scattering amplitude

$$M(q) = - (2\pi^2 m/\hbar^2) \cdot t_E(q).$$

On the other hand $F(q)$ is strongly forward peaked, while $M(q)$ changes much more slowly. In this case only $M(0)$ is important and we have the forward scattering approximation.

$$U_L(r) = - ((4\pi \hbar^2)/m) \cdot M(0) \cdot \rho(r) \qquad (4.14)$$

This approximation gives the real part of the OMP proportional to the density.

$$V(r) = - ((4\pi \hbar^2)/m) \cdot \text{Re } M(0) \cdot \rho(r) \qquad (4.15)$$

as used in phenomenological analyses.

The same approximation gives an expression for the imaginary part of the OMP[18]. Using the optical theorem, the imaginary part of the forward scattering amplitude is given by

$$\text{Im } M(0) = (k/4\pi) \cdot <\sigma_{TOT}>$$

where k is the wave number of the incident particle inside the nucleus at the position r and $<\sigma_{TOT}>$ is the average total nucleon-nucleon cross section at the same position. The average has to be done because the Pauli principle inside the nucleus can forbid some final states.

The imaginary part of the OMP becomes

$$W(r) = (1/2)\hbar \cdot v \cdot [\rho_p(r) \cdot <\sigma_p> + \rho_n(r) \cdot <\sigma_n>] \qquad (4.16)$$

where $<\sigma_p>$ and $<\sigma_n>$ are the total cross sections of the incident particle on a proton and a neutron respectively and v is the velocity of the incident particle at the position r.

The velocity v inside the nucleus can be approximated by [1,18]

$$(1/2) \cdot m \cdot v^2 = E - V(r) - V_{COUL}(r)$$

The expression (4.16) can also be deduced from semi-classical arguments [1] taking into accounts the relation between the OMP and the complex refractive index of the nuclear matter.

5. THE OMP IN THE ANTINUCLEON-NUCLEUS SYSTEM

Although the largest number of measurement is related to the nucleon-nucleus system, recently very accurate measurements of antiproton-nucleus elastic scattering on ^{12}C, ^{40}Ca, ^{16}O, ^{18}O, ^{208}Pb have been performed[19].

Like the nucleon-nucleus the p̄-nucleus elastic scattering cross-sections have a diffractive behavior but a special characteristic is the more pronounced diffraction structure. It indicates a strong absorptive potential which has its origin in the annihilation of antinucleons on nucleons, which is a dominant effect.

The angular distributions are like those of a black disk with a diffuse edge[23]. This means that the real OMP will influence only the details of the angular distributions. As a consequence of the strong absorption the OMP will be determined only on the nuclear surface. The inner behavior could be determined only by the elastic scattering at very high energy where the mean free path of the antiprotons in the nuclear matter is longer.

5.1 Phenomenological analysis of the p̄-nucleus elastic scattering.

A detailed phenomenological analysis of the p̄-nucleus elastic scattering has been performed by S. Janouin et al[20]. They adopt the standard OMP (3.12) assuming only the volume absorption term. Also the spin-orbit interaction has been neglected, since from N̄-N elementary interaction it is expected to be weak because of the cancellation of ρ-and ω-meson contribution[21]. This assumption is also justified by the low values in the polarization measurements[22]. The OMP has six free parameters, a typical best fit is shown by fig. 5 and the corresponding parameters are given in the following table[20].

target	energy [MeV]	V_0 [MeV]	r_{0V} [fm]	W_0 [MeV]	r_{0W} [fm]	$a_V=a_W$ [fm]	χ^2	σ_R [barn]
^{12}C	46.8	25	1.22	61	1.17	0.56	0.70	616
^{40}Ca	47.8	9	1.4	143	1.03	0.63	0.68	1243
^{208}Pb	48.3	0.0		22	1.38	0.50	0.58	3458
^{12}C	179.7	44	1.05	184	0.935	0.56	0.86	510
^{16}O	178.4	35	1.2	79	1.2	0.52	0.71	581
^{18}O	178.4	38.5	1.05	150	0.98	0.62	0.89	660
^{40}Ca	179.8	40.5	1.1	111	1.1	0.63	0.63	1035
^{208}Pb	180.3	60	1.097	105	1.13	0.70	1.07	2710

For such strongly absorbed particles, one expects that all the OMP's with the some behavior at the nuclear surface give the some best fit. Therefore it is interesting to test which radial region can be considered to be well mapped by the analysis of the experimental data. This can be done by moving a localized perturbation of the OMP systematically through the nucleus and studying the effect by looking at the ratio χ^2/χ^2_{MIN} versus the distance of the center of the perturbation. The conclusion[20] is that this region is located near the nuclear radius, then the OMP can be determined only at the nuclear surface.

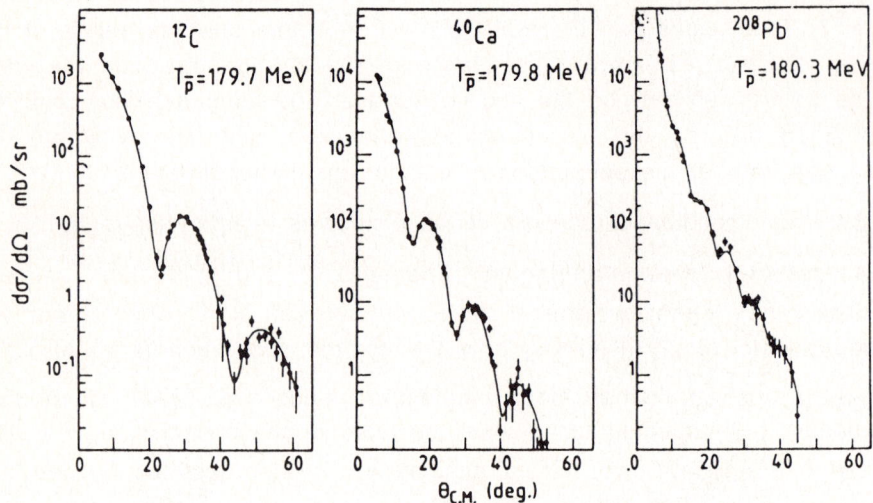

FIGURE 5
OMP calculations of the antiproton-nucleus scattering

A different phenomenological analysis has been performed by Friedman and Lichtenstadt[25] using a model-independent potential. This OMP, as any arbitrary function of r, can be expanded as a series of spherical Bessel function.

$$U(r) = \sum_{n=1}^{N} a_n \cdot j_0(n\pi r/R_c) \qquad r < R_c$$
$$= 0 \qquad r \geq R_c$$

The cut-off radius, R_c, is chosen to be sufficiently larger than the nuclear radius. The complex coefficients a_n are obtained by a standard χ^2 minimization so as to obtain a potential which best fit the experimental cross-sections. This method[26] allows also an easy evaluation of the errors in the determinations of the OMP's.

The behavior with r of the OMP's found with this analysis are given in fig.6. It is noticeable to see how the error bands growth as r approaches the inner part of the nucleus.

The OMP's before described reflect a purely phenomenological approach. To go up a simple folding model can be envisaged to obtain informations about the elementary effective interaction in the nuclear medium. This approach is essentially the Greenlees model where an effective (phenomenological) $\bar{p}$-bound-nucleon interaction is folded into the nuclear density distribution to obtain the optical potential.

$$V(r) = - V_0 \int v_{eff}(|r-r'|) \cdot \rho_m (r') \cdot d^3 r'$$

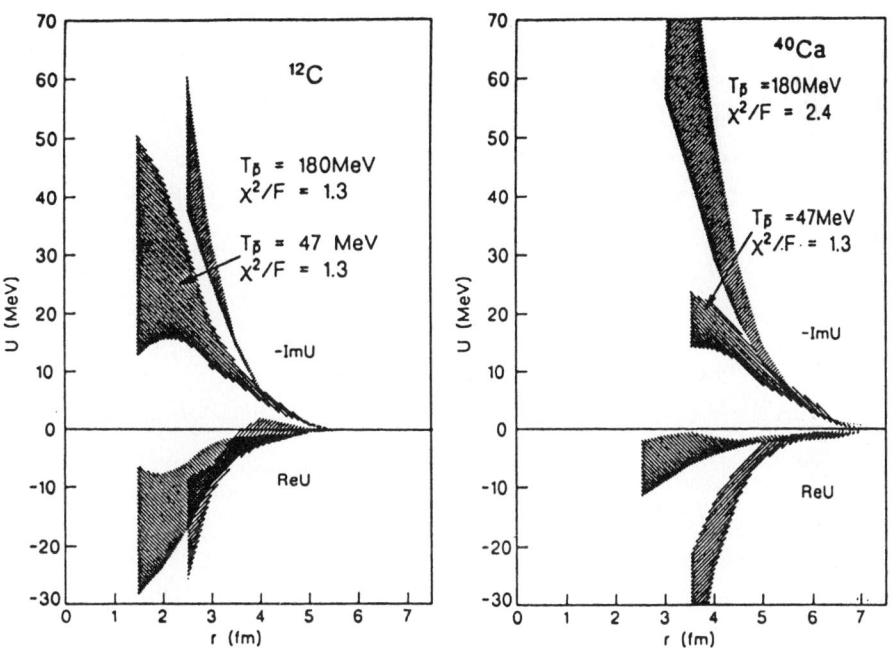

FIGURE 6
Radial and energy dependence of the model-independent OMP for antiproton-nucleus scattering

Janouin and collaborators[20] used a Yukawa interaction $\exp(-\mu r)/(\mu r)$ and, after some considerations about the r.m.s. radii of the OMP and of the elementary interaction, they showed that it is reasonable to assume $\mu \approx 0.6$ fm for both the real and the imaginary part of the potential.

Friedman and Lichtenstadt[25] used a Gaussian interaction with a small linear dependence from the nuclear density

$$v_{eff}(|\mathbf{r-r'}|) = [1 - \gamma \cdot \rho(r')^{2/3} \exp[-(r-r')^2/a^2]$$

They found the same χ^2 for any values of γ ranging from zero to 1 fm^2.

This means that the $\bar{p}$-nucleus effective interaction can be assumed to be density independent; the opposite result was found for the nucleon-nucleus system. The values of the 'a' parameters for nuclei up to the ^{40}Ca are

$$a_R \approx 1.5 \text{ fm} \qquad a_I \approx 1 \text{ fm}$$

for the real and imaginary part of the OMP.

5.2 Microscopic analysis of the $\bar{p}$-nucleus elastic scattering

Elastic scattering of antiprotons from nuclei can supply informations about the elementary $\bar{N}$-N interaction in the low density nuclear surface. The microscopic procedure uses a folding technique of the effective interaction with the target ground state density.

A systematic analysis of the elastic $\bar{p}$-nucleus scattering with microscopic OMP's has been performed by Adachi and Von Geramb[27].

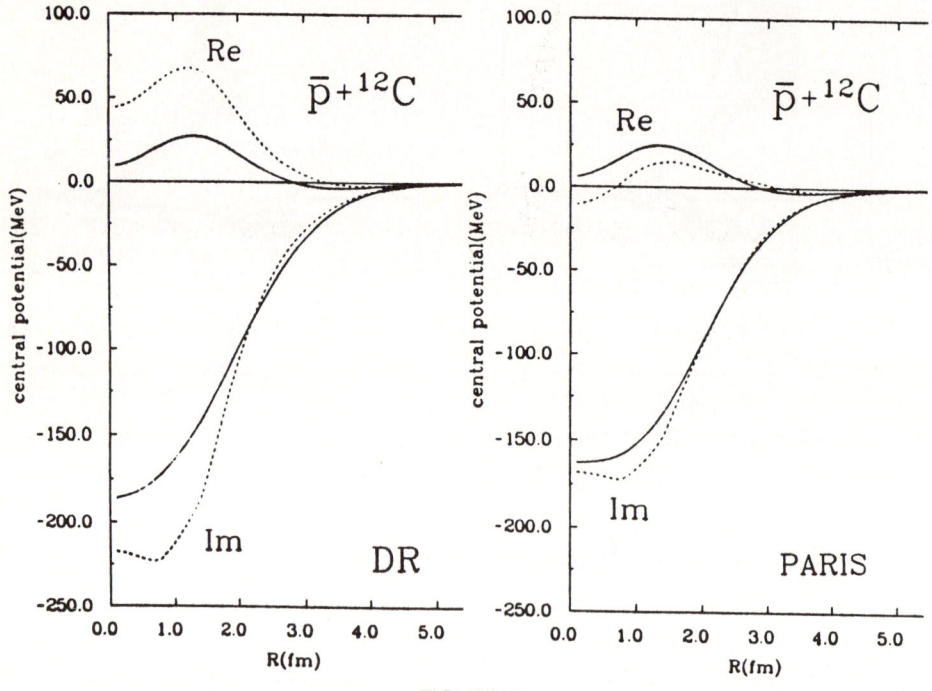

FIGURE 7
Real and imaginary part of the microscopic OMP for antiproton scattering on ^{12}C at energies of 47.5 MeV (dashed line) and 180 MeV (solid line) with DR and P potentials

In their microscopic approach, the effective interactions between the incident particle and a target nucleon are built from the Bruckner reaction matrix[28]. The obtained reaction matrix depends on the momenta of the incident antinucleon and of the hit nucleon. To use it in finite nuclei Adachi and Von Geramb construct an effective interaction with the assumption that there is a local potential which reproduces in an average way the reaction matrix. The effective interaction consists of a central, spin-orbit and tensor part like the interaction nucleon-nucleus (4.2). The analogy with the nucleon-nucleus system is evident.

As an input of two-body antinucleon-nucleon potential, Dover-Richard (DR)[29] and Paris (P)[30] potentials were used.

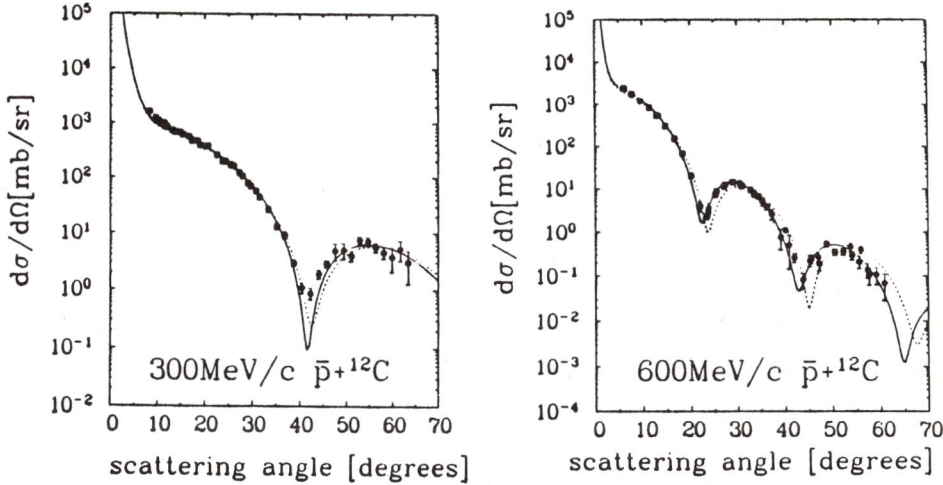

FIGURE 8
Microscopic OMP calculations with the DR (solid line) and P (dashed line) potential for antiproton-^{12}C elastic scattering

The DR potential consists of a real potential arising from t-channel meson exchange and a purely phenomenological, spin- and isospin-independent, annihilation potential with a Saxon-Woods form factor. The P potential is obtained by the G-parity transformation of the Paris nucleon-nucleon potential.

In fig. 7 the central OMP's for $\bar{p}$ scattering on ^{12}C at 47.5 MeV and 180 MeV are shown. The real central potential has an attractive pocket in the surface region and becomes repulsive in the inner region. The imaginary part is negative, *i. e.* absorptive and roughly follows the density distribution. As the energy increases the imaginary potential becomes deeper in the surface region.

Fig. 8 shows a typical microscopic calculation compared with the experimental data. As it can be seen the agreement is quite good.

In conclusion, there is a very remarkable difference between the phenomenological and microscopic OMP's in the antinucleon-nucleus system and this fact has no analogies in the nucleon-nucleus system. This difference lies mainly in the radial behavior of the real central OMP in the core region.

Unfortunately the present experimental data are insensitive to the inner part of the potential and therefore the shape of the OMP in the inner nuclear region is more of theoretical importance than of practical need to reproduce experimental data.

6. CONCLUSION

Although the OMP is used from some tenth of years, the physical interest remains inaltered also at the present time where more fundamental microscopic theories of nuclear matter (quarks effects in nuclei) are developed.

Phenomenological analyses of the nucleon-nucleus experimental data have greatly improved and the spline technique now gives very precise fits. The existence of extensive and accurate global potentials that fit the data for many nuclei over a range of energies, allows the evaluation of any quantity of interest to be put in more sophisticated programs used in many branches of applied physics.

Microscopic theories of the OMP have now been developed to the stage where they can give, without parameter adjustement, potentials that are in good overall agreement with the experimental data. From these potentials it is possible to extract many physical informations about the nuclear matter with an effort considerably less than that requested from more sophisticated theories.

Work in the area of semi-phenomenological theories, like the folding model, is proving very successful and should facilitate the unification of the microscopic and phenomenological theories.

These developements in the theory of the OMP show the need of more extensive and accurate experimental data and this should stimulate further experimental work.

The case of antinucleon-nucleus scattering, where microscopic and phenomenological OMP's are very different, is an example of the present situation and of the future needs.

REFERENCES

1) P. E. Hodgson, The optical model of elastic scattering (Clarendon Press, Oxford, 1963)

   P. E. Hodgson, Nuclear reactions and nuclear structure (Clarendon Press, Oxford, 1971)

2) M. A. Melkanoff, T. Sawada, J. Raynal, Nuclear Optical Model Calculations, in : Computational Physics, Vol 6, Nuclear Physics (Academic Press, 1966)

3) F. G. Perey and B. Buck, Nucl. Phys 32 (1962) 353

4) L. R. B. Elton, Nuclear charge distribution, in: Nuclear Radii, ed. H. Shopper (Springer-Verlag, Berlin, 1967)

5) A. M. Lane, Nucl. Phys. 35 (1962) 676, Phys. Rev. Lett. 8 (1962)171
   A. M. Lane and J. M. Soper, Nucl. Phys. 37 (1962) 663

6) G. R. Satchler, in: Isospin in Nuclear Physics, ed. D. H. Wilkinson (North Holland, 1969)

7) J. K. Dickens and F. G. Perey, Phys. Rev. B138 (1965) 1080

8) F. D. Becchetti and G. W. Greenlees, Phys. Rev. 182 (1969) 1190

9) R. L. Warner, T. B. Clegg, T. L. McAbee and W. J. Thompson, Phys. Lett. 185B (1987) 6

10) H. Feshbach, Ann. Rev. of Nucl. Science 8 (1958) 49
    R. H. Lemmerr, Fundamentals in Nuclear Theory, ed. A. de Shalit and C. Villi, (International Atomic Energy Agency, Vienna, 1967)

11) K. A. Bruckner, Phys. Rev. 96 (1954) 508
    K. A. Bruckner and C. A. Levinson, Phys. Rev. 97 (1955) 1344

12) G. W. Greenlees, G. J. Pyle and J. C. Tang, Phys. Rev. 171 (1968) 1115
    G. W. Greenlees, W. Makofske, and G. J. Pyle, Phys. Rev. C1 (1970)1145
    G. W. Greenlees, V. Hnizdo, O. Karban, J. Lowe and W. Makofske, Phys. Rev. C2 (1970) 1063

13) H. Kidwai and J. R. Rook, Nucl. Phys. A169 (1971) 417

14) A. K. Kerman, H. McManus and R. M. Thaler, Ann. of Phys. 8 (1959) 551

15) E. Friedman, Phys. Lett. B29 (1969) 213

16) M. M. Giannini and G. Ricco, Ann. of Phys. 102 (1976) 458

17) M. L. Golberger and K. M. Watson, Collision theory (John Wiley, New York, 1964)
    H. Feshbach, A. Gal and J. Hufner, Ann. Physics 66(1971)20
    M. M. Giannini, Nuovo Cimento A3 (1971)365

18) B. Sinha, Phys Reports 20 (1975) N. 1

19) D. Garreta, et al., Phys. Lett. 135B (1984) 266, Phys. Lett. 139B (1984) 464, Phys. Lett. 149B (1984) 64, Phys. Lett. 151B (1985) 473

20) S. Janouin et al., Nucl. Phys. A451(1986) 541

21) C. B. Dover and J. M. Richard, Phys. Rev. D17 (1978)1770

22) R. Birsa et al., Phys. Lett. 155B (1985) 437

23) J. Lichtenstadt et al., Phys. Rev. C32 (1985)1096

24) J. G. Cramer and R. M. De Vries, Phys, Rev. C22 (1980) 91

25) E. Friedman and J. Lichtenstadt, Nucl. Phys. A455 (1986) 573

26) E. Friedman and C. S. Batty, Phys. Rev. C17 (1978) 34

27) H. V. Von Geramb, K. Nakano, and L. Rikus, Lett. Al Nuovo Cimento 42 (1985) 209
    S Adachi and H. V. Von Geramb, Nucl. Phys. A470 (1987) 461

28) J. Hufner and C. Mahaux, Ann. of Phys. 73 (1972) 525
    J. P. Jeukenne, A. Lejeune and C. Mahaux, Phys. Report 25C (1976) 83

29) C. B. Dover and J. M. Richard, Phys. Rev. C21 (1980) 1466

30) J. Cote', M. Lacombe, B. Loiseau, B. Moussallam and R. Vinh Mau, Phys. Rev. Lett. 48 (1982) 1319

CHARM, BEAUTY AND NUCLEI

Tullio BRESSANI
Dipartimento di Fisica Sperimentale dell'Universita' di Torino
Istituto Nazionale di Fisica Nucleare, Sezione di Torino, Italy

and

Felice IAZZI
Dipartimento di Fisica del Politecnico di Torino
Istituto Nazionale di Fisica Nucleare, Sezione di Torino, Italy

1. INTRODUCTION

The title could be changed in the equally cryptic one "Supernuclei" or in the perhaps more clear one "Hypernuclei with charm and beauty", i. e. nuclear systems in which a Λ_c^+ or a Λ_b replaces a nucleon. The rather complete freedom in the choice of the definition of the matter is due to the fact that, at moment, there is no experimental evidence for the existence of these systems, but only theoretical speculations, started by Tyapkin[1] in 1975. A recent overview of the subject is due to Starkov and Tsarev[2], and we will inspire to their work on several points.

The starting idea is the similarity between the quark structure of Λ_s(uds), Λ_c^+(udc) and Λ_b(udb). Since we know from a long time the existence of nuclear systems in which the Λ_s has replaced a nucleon in a nucleus, we may infer "by similarity" the possible existence of analogous systems for Λ_c^+ and Λ_b. Alternatively, if we describe a Λ_s-hypernucleus as a system in which an s-quark substitutes one of the many u- and d- quarks of an ordinary nucleus[3], we may think of systems in which a c- or b- quark makes the substitution.

As a counterpart of these simple and nice considerations, there is the constatation that, whereas the Λ_s-hypernuclei were identified about five years after the discovery of the Λ_s, there is no experimental evidence for the existence of Λ_c^+-hypernuclei, after more than thirteen years from the discovery of the Λ_c^+. An analogous pessimistic conclusion can be drawn considering the total number of Λ_s that were studied at the moment of the discovery of the Λ_s-hypernuclei, and the total number of Λ_c^+ that are up to now collected and studied.

These facts may led us to the conclusion that or the basic idea is wrong or that there are experimental limitations with the existing machines and techniques. We believe that the second possibility is more grounded, we will describe why and we will finally suggest a (rather remote) experimental possibility offered by a new machine, the ARES (Acceleratore Ricircolato per Elettroni, Superconduttore) facility proposed at Frascati[4].

2. THEORETICAL ASPECTS

In Table I we have reported the known static properties of Λ_s and Λ_c^+. For Λ_b we report the mass given by Basile et al.[5], even if disputed, and a guess for the lifetime.

TABLE I

Mass and lifetime of Λ_s, Λ_c^+ and Λ_b

	Λ_s	Λ_c^+	Λ_b
M(MeV)	1115.6	2281.2	5425
τ(sec)	2.6×10^{-10}	2.3×10^{-13}	$\approx 10^{-13}$

The lifetime of Λ_c^+ is much longer than the typical nuclear time (10^{-22}s) but of the same order of the nuclear lifetime for γ-deexcitation. We may then expect isomeric states and a conversion $\Lambda_c \to \Lambda_s$. The mass difference between Λ_c^+ and the nucleon (≈ 1341 MeV) is very much greater than that between Λ_s and the nucleon (≈ 176 MeV) and obviously even worse is the case of Λ_b. Then only this consideration may suggest that all the approximations and symmetries that are applied for the theoretical analysis of the Λ_s-hypernuclei, and that are valid, cannot at all hold for baryons with heavy flavours.

Indeed Λ_c^+ and possibly Λ_b could stick to a core of nucleons, in spite of the mass difference, due to the specificity of the elementary interactions. Unfortunately there is no experimental information on the strength of the interaction $\Lambda_c^+(\Lambda_b)$-N due to the impossibility of getting beams with heavy flavours. We recall in fact that even the lifetimes of the D and B mesons are in the order of 10^{-13} s, and then they travel only a very reduced distance.

A guess on the strength of the interaction $\Lambda_c^+(\Lambda_b)$-N can be done only by means of analogies based on the "universality of constants", the exchange of the relevant mesons, the mass of the constituents and, last but not least, the Coulomb interaction which is present for Λ_c^+, <u>not</u> for Λ_s and Λ_b. Starkov and Tsarev[2,5] obtained a prediction for the binding energy $B_{\Lambda i(i=s,c,b)}$ of a Λ_i for a Λ_i-hypernucleus in the ground state, defined as:

$$B_{\Lambda i}(g.s) = M_{core} + M_{\Lambda i} - M_{Hy\Lambda i} \quad (1)$$

where M_{core} is the mass of the nucleus $^{(A-1)}Z$ in the ground state, $M_{\Lambda i}$ the mass of the Λ_i particle and $M_{Hy\Lambda i}$ the mass of the hypernucleus $^{A}_{\Lambda i}Z$. Their results are shown by Fig. 1, and one can notice how the binding energy, as a function of the mass number A, shows a saturation for all the Λ_i. Of particular importance is the effect of the Coulomb interaction for Λ_c^+ that restricts the existence of Λ_c^+- hypernuclei to a reduced range of masses around A=70.

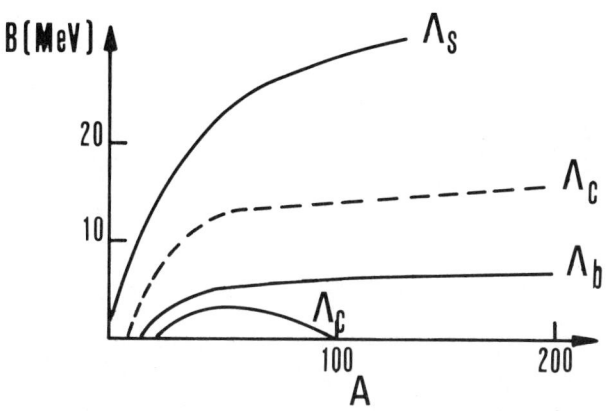

FIGURE 1
Calculation of the binding energy $B_{\Lambda i}$(g.s) for Λ_s, Λ_c^+ and Λ_b in nuclei as a function of the mass number A, taken from Ref. 2. For Λ_c^+ the dashed line indicates the calculation without the Coulomb interaction.

The same authors predict also several other properties of the nuclei with heavy flavours, like the absence of spin dependent interactions, the dominance, in the energy spectra, of states with high angular momentum, but we will not report them here. Our conclusion is that, even if there are many "ad-hoc" hypotheses and assumptions, the existence of nuclei with heavy flavours seems possible.

3. POSSIBLE METHODS OF PRODUCTION

We may imagine two methods of producing Λ_i in a nucleus:
i) by nonflavoured beams
ii) by flavoured beams

In the first case we may notice that, using a high energy proton beam, the inclusive spectrum of Λ_c^+ produced on quasi-free nucleons of the nucleus, shown by Fig. 2 is shifted towards high momenta; the probability of sticking to the nucleus is then extremely small. If we consider instead the spectrum of Λ_s, we observe that it is shifted towards low momenta, there is a non-negligible chance of having Λ_s with momenta lower than the Fermi momentum of nucleons ($\approx$ 250 MeV/c), and then we may observe the production of Λ_s-hypernuclei. As a matter of fact, the only experiment up to now performed[6] using emulsions exposed to 250 GeV protons gave negative results. Only two ambiguous candidates of Λ_c^+-hypernuclei were reported.

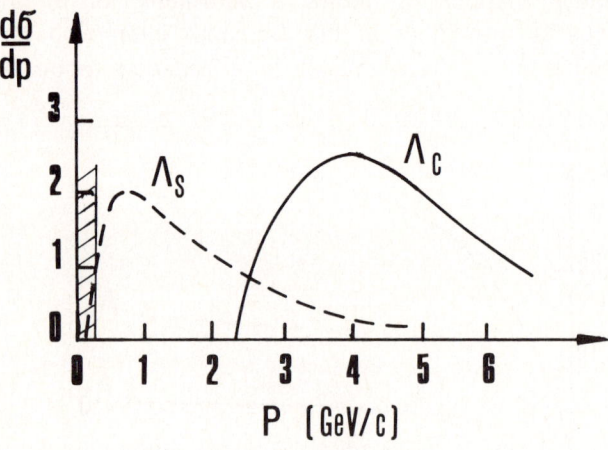

FIGURE 2
Inclusive cross-section for the production of Λ_s (dashed line) and Λ_c^+ (solide line) in pp collisions, in the rest frame of one of the protons taken from Ref. 5. The ordinate scale is arbitrary. The shadowed area indicates the momentum range useful for Λ_s- and Λ_c- hypernuclei production.

In analogy to the Λ_s-hypernuclei, for whose production and study the "winning" elementary reaction was the "strangeness exchange" one:

$$K^- + n \to \Lambda_s + \pi^- \tag{2}$$

one can ask whether there are similar features in reactions induced by flavoured beams, like:

$$D^+ + p \to \Lambda_c^+ + \pi^+ \tag{3}$$

and

$$B^- + n \to \Lambda_b + \pi^- \tag{4}$$

We recall[3] that the main advantage of (2) is in the kinematics, shown by Fig. 3; at $\theta\pi_{Lab} = 0°$ and at $p_{k\text{-}Lab} = 530$ MeV/c (the "magic" momentum) the Λ_s is produced at rest in the target nucleus, with a maximum of probability of sticking. In the same figure we have reported also the kinematics for the reactions (3) and (4). For (4) we used for $M_{\Lambda b}$ the value 5425 MeV[5] and another guessed value of 5700 MeV. We may observe that even the kinematics with flavoured particles shows a "magic" momentum, not too far from that of (2), a part the case $M_{\Lambda b} = 5425$ MeV.

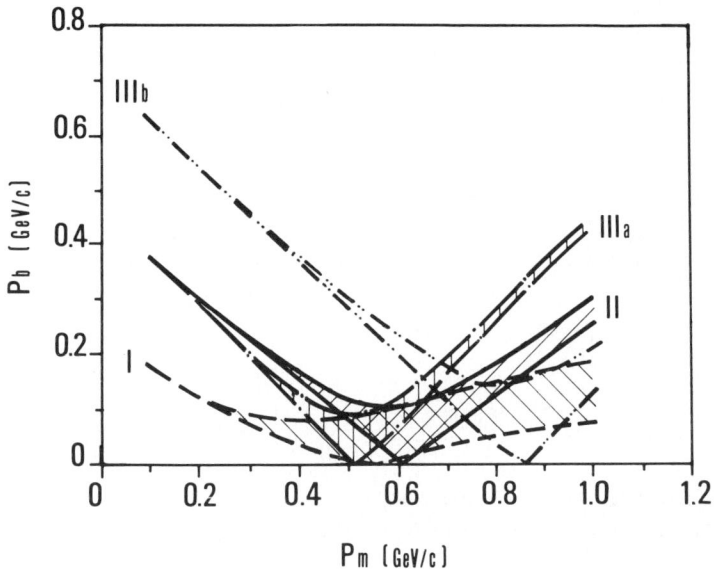

FIGURE 3

Kinematics in the forward direction for the reactions $K^- + n \to \Lambda_s + \pi^-$ (I, dashed lines), $D^+ + p \to \Lambda_c^+ + \pi^+$ (II, continuous lines) and $B^- + n \to \Lambda_b + \pi^-$, with $M_{\Lambda b} = 5700$ MeV (III$_a$, dot-dashed lines) and 5425 MeV (III$_b$, two dots-dashed lines). The curves which limit the hatched areas downwards are for a π emission angle of 0°, those which limit upwards for 5°. P_m indicates the momentum of the meson in the lab. frame, P_b that of the baryon.

Unfortunately, as mentioned before, the present techniques of production do not allow the use of flavoured beams, due to the lifetime of D and B. Even the possibility of using a two-step mechanism, in the same nucleus or in two enough "near" (<100 μm) nuclei, in which the first step is the production of D by means of high-energy (250 GeV) protons, the second a reaction like (3), is hopeless if we look at the inclusive spectrum of the produced D, shown by Fig. 4. The momenta of the produced D did not cover the region of the magic momentum for the reaction (3). Again the opposite happens for the case of K, where the inclusive spectrum has just a maximum near the magic momentum for (2).

In conclusion, the known features of the experimental methods that can be foreseen with the existing machines explain the lack of experiments up to now performed.

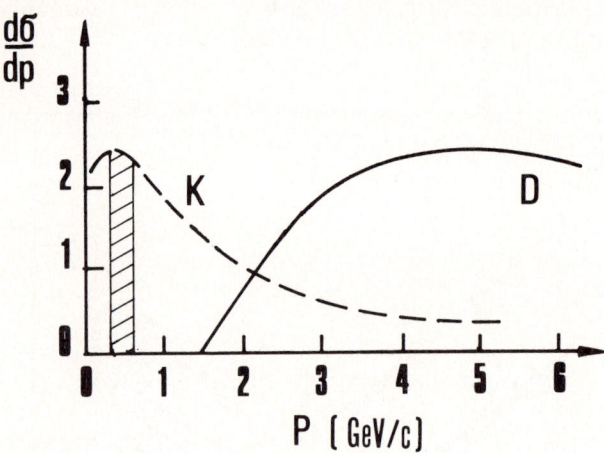

FIGURE 4
Inclusive cross-section for the production of K and D mesons in pp collisions in the rest frame of one of the protons, as taken from Ref. 5. The ordinate scale is arbitrary. The shadowed area indicates the region around the magic momentum.

4. A NEW EXPERIMENTAL POSSIBILITY

A few months ago a proposal for a new machine, called ARES, was put forward[4]. The project is based on one racetrack superconducting accelerator, able to accelerate bunches of electrons and positrons at various energies up to a maximum of 7.5 GeV following different cycles and modes of operations. Fig. 5 shows a scheme of the proposed complex taken from Ref. 4, in which details of the expected performances are obviously given. In a few words, the machine is a single-pass collider, able to reach a luminosity of 10^{33} cm^{-2} s^{-1} and spanning c.m. energies from $\approx$ 4.0 to 15.0 GeV. The luminosity is optimised at the Y(4S) resonance (10.58 GeV), since the main physics program is a careful study of the B-$\bar{B}$ system. Indeed at 4.0 GeV (c$\bar{c}$ system) one can still get a luminosity of 0.35 x10^{33} cm^{-2} s^{-1}

The technical realization of ARES is very ambitious, and in particular the high luminosity is reached by means of a very reduced size of the beams at the intersection point ($\sigma \approx$ 1µm).

This is the most important parameter that allows to think of a direct production of Λ_c^+-hypernuclei by means of the reactions (3) and (4) . One could imagine to put a nuclear target, in form of an annulus, of thickness ≈100μm and internal radius ≈ 30μm just around the interaction region.

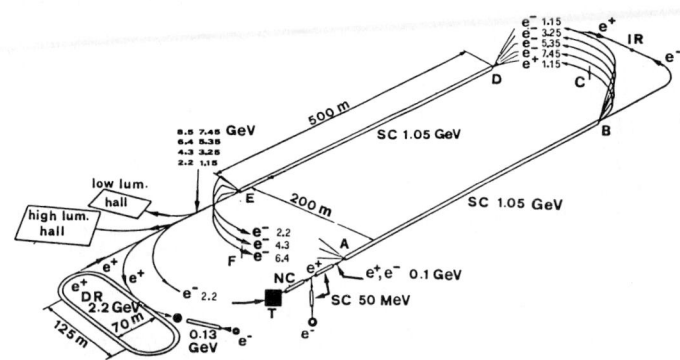

FIGURE 5
Scheme of the ARES accelerator complex, as taken from Ref. 4.

One of the main purposes of the machine is that of producing abundantly D-$\bar{D}$ and B-$\bar{B}$ pairs, in clean experimental conditions. The numbers are ≈10^6 D and ≈10^5 B per day. By this way one could exploit the reactions (3) and (4), around the magic momentum, for the study of the two-body reactions on a target nucleus $^A_Z X$:

$$D^+ + {^A}X \rightarrow {^A_{\Lambda_c^+}}(X-1) + \pi^+ \qquad (5)$$

and

$$B^- + {^A}X \rightarrow {^A_{\Lambda_b}}X + \pi^- \qquad (6)$$

In order to estimate the cross section $d\sigma/d\Omega_{Hyp}$ for (5) and (6) it is possible to take the analogy with the Λ_s-hypernuclei production case, where it is known[8] that an order of magnitude can be given by the expression

$$(d\sigma/d\Omega)_{Hyp} = (d\sigma/d\Omega)_{free}\, N(A,\sigma_1,\sigma_2)\, |F(q)|^2 \qquad (7)$$

in which $(ds/d\Omega)_{free}$ is the forward differential cross-section for the free reactions (3) and (4) respectively, $N(A,\sigma_1,\sigma_2)$ is the effective nucleon number of the nucleus, defined as:

$$N(A, \sigma_1, \sigma_2) = \frac{1}{\sigma_2 - \sigma_1} \int d^2b \{\exp[-\sigma_1 T(\mathbf{b})] - \exp[-\sigma_2 T(\mathbf{b})]\} \quad (8)$$

and F(q) is a form factor accounting for the trasformation of a nucleon of a nucleus in a charmed baryon of a supernucleus and depending from the momentum transfer and from the wave-functions of the initial and final state. In (8) σ_1 is the total cross-section for D(B)-N, σ_2 the total cross-section for π-N and

$$T(\mathbf{b}) = A \int_{-\infty}^{+\infty} dz \rho(\mathbf{b}, z) \quad (9)$$

in which $\rho(\mathbf{b},z)$ is the nuclear density normalised to one. The physical meaning of the relations (8) and (9), due to Koelbig and Margolis for the incoherent hadron production, can be easily understood by looking at fig. 6.

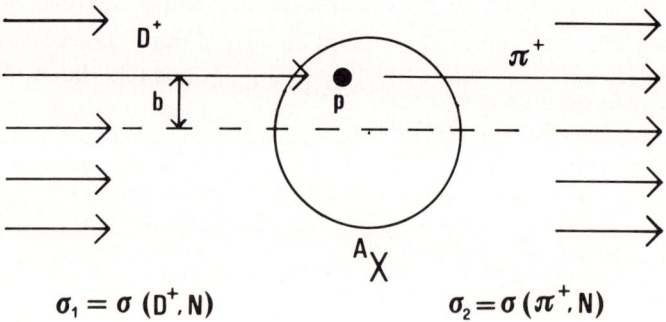

FIGURE 6
Simple representation of the model used in order to evaluate the effective nucleon number.

Unfortunately in the case of reactions (5) and (6), $(d\sigma/d\Omega)_{free}$ and σ_1 are unknown, and F(q) is very difficult to evaluate, even in a rough model. Indeed we may notice that there is a compensation of the effects due to

$(d\sigma/d\Omega)_{free}$ and σ_1 in (7), since a large value of the D(B)-N cross-section produces a decrease in the value of $N(A, \sigma_1,\sigma_2)$ and vice-versa. This fact is well evidenced in $\bar{p}$ -Nucleus reactions[10], for which σ_1 is very large. Following these considerations, we assume for $(d\sigma/d\Omega)_{Hyp}$ a tentative value of 10 mb/sr.

The rate of production of Λ_c^+-hypernuclei through the subsequent reaction (3) and (5), may then be evaluated in the order of a few events/day. The rate of production of D^+ could be enhanced by a factor 3 at E_{cm} = 3.77 GeV, corresponding to the J" resonance, decaying predominantly in $\bar{D}$-D. On the other side the momentum of the D from the decay of the J" is 0.242 GeV/c, somehow far from the magic momentum and corresponding to a decay length of 36 µm. The rate of production of Λ_b-hypernuclei through the subsequent reactions (4) and (6) can be evaluated an order of magnitude lower than that of Λ_c-hypernuclei, under the same hypotheses.

From the experimental side, a search for $\Lambda_{c(b)}$-hypernuclei could be done by putting the annular reaction target in a general purpose 4π facility for c and b physics that is under active study[11].

The first problem that has to be investigated carefully with the machine physicists is whether the annular target may induce problems on the colliding beams at the focus.

5. CONCLUSIONS

From the above considerations we may draw the conclusion that an experiment for the search of $\Lambda_{c(b)}$-hypernuclei at the ARES facility is difficult, but not "a priori" impossible. Such an experiment would have to be incorporated in a general purpose 4π facility that would be built at the machine.

From the physics point of view, the experiment could be the only one to give direct information on the D(B)-N interaction, absolutely unknown at present. Furthemore this investigation could shed light on the behaviour of c- and b- quarks in Nuclei, that could be considerably different than that of light quarks.

REFERENCES

1) A. A. Tyapkin, Sov. J. Nucl. Phys. 22 (1975) 181
2) N. I. Starkov and V. A. Tsarev, in Proc. of the 1986 INS International Symposium on Hypernuclear Physics, Tokyo, August 20-23, 1986, eds. H. Bando, O. Hashimoto and K. Ogawa, INS University of Tokyo, p. 247
3) see e.g. T. Bressani in Hadronic Physics at Intermediate Energy, eds. T. Bressani and R. A. Ricci (North Holland, Amsterdam, 1986), p. 259
4) U. Amaldi and G. Coignet, in Proc. on Heavy-Quark Factory and Nuclear-Physics Facility with Superconducting Linacs, Courmayeur 14-18 December 1987, eds. E. De Sanctis, M. Greco, M. Piccolo and S. Tazzari, Conf. Proc. Vol. 9 (Italian Physical Society, Bologna, 1988), p. 59, quoted in the following as Courmayeur '87

5) M. Basile et al., Lett. Nuovo Cimento 31 (1981) 97 and Nuovo Cimento 68A (1982)289
6) N. I. Starkov and V.A. Tsarev, Nucl. Phys. A450 (1986), 507
7) Yu. A. Batusov et al., JETP Lett. 33 (1981) 56
8) G.C. Bonazzola et al., Phys. Rev. Lett. 34 (1975), 683
9) K. S. Koelbig and B. Margolis, Nucl. Phys. B6 (1968), 85
10) T. Bressani et al., Europhys. Lett. 2 (1986), 587
11) M. Piccolo, in Courmayeur '87, p. 401

III

SUBNUCLEAR PHYSICS

MEASUREMENT OF THE $\bar{P}$ ANNIHILATION CROSS-SECTIONS AT VERY LOW ENERGIES

Evandro LODI RIZZINI

Dipartimento Automazione Industriale - Universita' degli Studi di Brescia - Via Valotti 9 - 25060 Brescia - Italia

1. INTRODUCTION

Whith the advent of LEAR, it became possible to study in detail the interaction mechanisms of antinucleons whith nucleons and nuclei. In this talk, I focus on the measurement of $\bar{N}N$ annihilation cross sections in the very low-energies regime (defined here as <100 MeV/c projectile momentum).
The knowledge of the antiproton-proton ($\bar{p}p$) and antiproton-nuclear ($\bar{p}A$) annihilation cross-section at very low energy ($p_{\bar{p}}$ < 100 MeV/c) is of great interest for the understanding of the antinucleon-nucleon ($\bar{N}N$) interaction. For istance, the low-energy limit ($\beta \to 0$) of $\lim \beta \, \sigma_{ann}$ is an important input for potential calculation and, if better known, may allow the existence of baryonia close to threshold to be ruled out or proved. Moreover, the measurement of partial-annihilation channels allows definite bounds to be put on the contribution of partial waves. Below 100 MeV/c annihilation and elastic scattering are the only possible reaction channels.
In 1949, before the discovery of the antiproton, Fermi and Yang noted that certain repulsive nucleon-nucleon forces could become attractive in the $N\bar{N}$ system. They also predicted that there may exist many bound states due to this effect. These observations set the stage for studying the $\bar{N}N$ system in terms of a potential model framework.
The real (nonannihilation) part of the $N\bar{N}$ potential can be derived from its NN counterpart. In the conventional picture of the NN interaction, a potential is generated by the exchange of light, nonstrange scalar (S*, δ, ϵ), pseudoscalar (π, η, η') and

vector (ρ, φ, ω) mesons. The resulting potentials are usually reffered to as "one-boson-exchange potentials" (OBEP), or "two-pion-exchange potentials" (TPEP) in the case of two-pion exchange. This description is only valid for the medium and long-range parts of the potential ($\geqslant 0.8$ fm). Individual contributions to the potential from each meson exchanged include spin-spin, tensor, spin-orbit, and isospin-dependent terms, where applicable. The sum of the contributions from all exchanged mesons gives the real potential $V(NN) = \sum_m V_m$ where V_m denotes the contribution of meson m, and m=$\{\pi, 2\pi, \eta, \omega$, ect.$\}$. The long-range part is mostly due to pion exchange, the pion being the lightest of the mesons. The short-range "hard-core" repulsion is due mainly to ω exchange.

The $N\bar{N}$ potential can be obtained from that of the NN system since the two only differ at the N-m-N ($\bar{N}$-m-$\bar{N}$) interaction vertices of the relevant Feynman diagrams. This introduces a factor G_m for each meson exchanged, where G_m is the G parity of meson m. This technique is commonly reffered to as the "G-parity transformation". The resulting $N\bar{N}$ potential can then be written as $V(NN) = \sum_m G_m V_m$. Contributions from mesons with odd G parity (π, ω, φ, δ) have apposite signs in the $N\bar{N}$ and NN cases. In particular, the strong, short-range NN repulsion due to ω exchange now becomes a strong, short-range attraction. The naive implication of this observation is that with such a strong attractive potential there should be many bound states in the $N\bar{N}$ system.

The first attempt at incorporating annihilation into the $N\bar{N}$ potential was made by Bryan and Philips. They used the OBEP of Bryan and Scott applied the G-parity transformation, and added a phenomenological, purely imaginary potential, which is independent of spin, isospin, and energy. The authors, noted that imaginary potential is very strong at short distances and greatly attenuates the wave functions there, acting somewhat like the NN repulsive core. Also, this attenuation suppresses the short-range forces in the real part of the potential and essentially eliminates sensitivity to therm.

Subsequent improvements to the NN potentials, in particular due to the Paris group, have evolved. Dover and Richard[1] (DR) introduced the G-parity transformed version of this potential, along with a complex phenomenological spin, isospin, and energy-independent annihilation potential (model DR I), figure 3A.

The Paris group[2] subsequently fit the available $p\bar{p}$ data, including angular and polarization distributions, and provided a much more flexible form for the annihilation potential. This form has no real part as in the Dover and Richard model, but does include an explicit energy dependence. It was introduced to account for the fact that one expects more annihilation channels to open up with increased energy. Six parameters were adjusted to give the best fit to the data, and are different for isospin 0 and 1. The I=0 and I=1 forms of the potential W differ by up to 20% at very low energies ($E_{lab} \simeq 0$), with W(I=0) being larger. The effects at higher energies ($E_{lab} \simeq 100$ MeV) are not as strong. Figure 3B shows the cross sections for $\bar{p}p$ and $\bar{n}p$ for the Paris model near the energy region of interest here. It has characteristics similar to that of the DR I model discussed above.

In the following we discuss the possibility of extending annihilation cross-section measurements to energies as low as possible. For simplicity we consider only the case of a hydrogen target, in an experimental apparatus like OBELIX[3] at LEAR (CERN).

Let us consider antiprotons of an initial kinetic energy E_0 of $T_{\bar{p}} = 6$ MeV (or a momentum of $p_{\bar{p}} = 106.3$ MeV/c).

Antiprotons from LEAR loss energy due to beryllium and mylar windows, thin scintillation counters and some centimeters of air before the entrance in the gas target. At this point the kinetic energy of antiprotons is spread around E_i see fig.1. Tipical values (E_0, E_i) may be (110, 97 MeV/c), (105, 90 MeV/c), (100, 83 MeV/c), with a spread of ± 4MeV/c for E_i.

In fig.2 we see the time of flight (S1-TARGET 150 cm) for antiprotons of different momenta. The measurement of this time of flight plus the deceleration time allows for a good determination of $\bar{p}$ momentum at the annihilation point ($\sim \pm 2$ MeV/c), using a barrel of scintillation counters (TOF) around the target for the detection of the emitted charged particles. It should be possible to measure the annihilation cross-section in this way down to 40 MeV/c (1MeV) with a good momentum resolution, having a time (≈ 1 ns) and space resolution (≈ 1 cm), for the annihilation point, see fig.2 The $\bar{p}$ momenta distribution in the target is obtained from

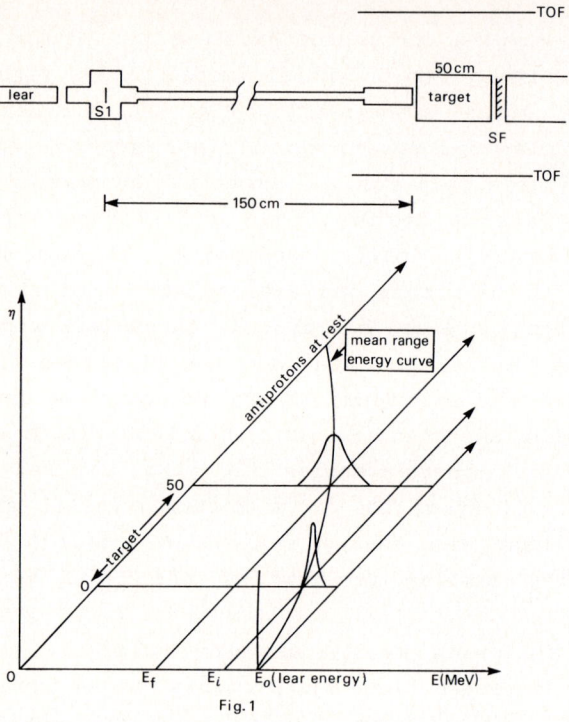

Fig. 1

S1 - SF - TOF   Scintillation Counters

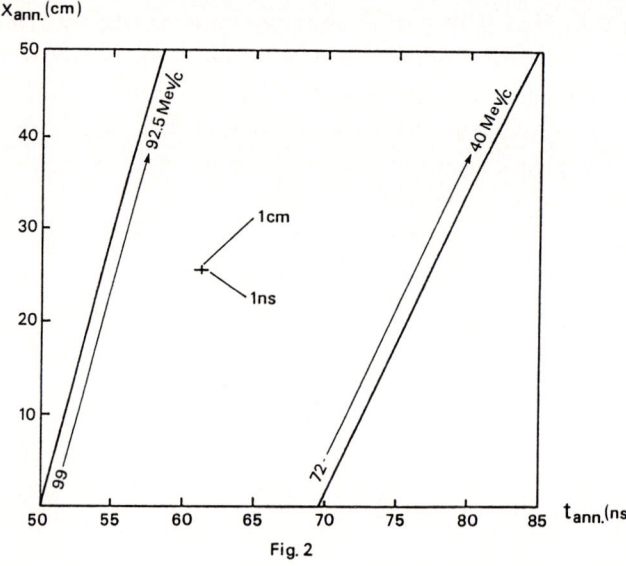

Fig. 2

S1 - SF time of flight (non annihilated antiprotons).

The annihilation rates are estimated using the following parametrization for the annihilation cross section:

$$\sigma_{ann} = 38 + \frac{35}{p_{\bar{p}} [GeV/c]} \text{ mb}$$

For $p_{\bar{p}} < 0.1$ GeV/c this reduces essentially to

$p_{\bar{p}} \sigma_{ann} \approx 38$ mb   or $\beta \sigma_{ann} \approx 37$ mb.

The number of annihilations is given by

$$N_{ann} = N_{\bar{p}} L \rho \, dx \, \sigma_{ann}.$$

The annihilation rate per 10 cm length and 10^6 $\bar{p}$ for gaseus H_2 at NTP is hence

$$R_{ann} \approx \frac{20}{p_{\bar{p}} [GeV/c]} \, s^{-1},$$

On the other hand, the annihilation rate per time interval Δt,

$$R_{ann} = N_{\bar{p}} L \rho c \Delta t \, \beta \sigma_{ann} = N_{\bar{p}} L \rho c \Delta t \times 37.$$

The LEAR beam intensity ($10^6 - 10^7$ $\bar{p}$/s) insure for good statistics in few hours.

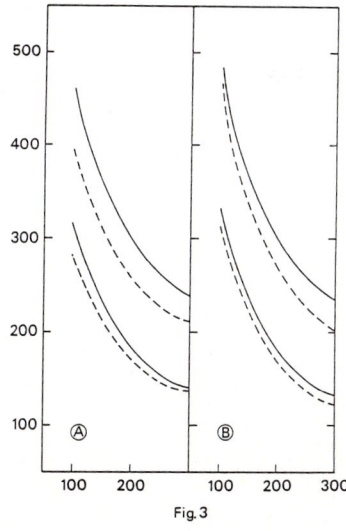

Fig. 3

1) C.B. Dover and J.M. Richard, Phys.Rev C 21, 1466 (1980).

2) J.Cote et al., Phys. Rev. Lett. 48, 1319 (1982).

3) OBELIX proposal (PS 201) CERN.

PROBING THE STANDARD MODEL AT THE SPS COLLIDER.
EXPERIMENTAL RESULTS FROM UA1

Alessandro BETTINI

Dipartimento di Fisica and I.N.F.N., Padova, Italy

1. INTRODUCTION

In these lectures a review of the experimental results which have been obtained by the UA1 experiment at the CERN $p\bar{p}$ collider is presented. When relevant, results from the parallel UA2 experiment are quoted for comparison. Both experiments collected data in a series of data taking periods between 1982 and 1985 at centre of mass energies $\sqrt{s}$ = 546 GeV and, mostly, at $\sqrt{s}$ = 630 GeV. The total integrated luminosity is between 0.7 and 0.8 events per picobarn (pb^{-1}).

The experimental results will be compared with the predictions of the Electroweak theory in Sections 4.2 - 4.6 and 5.2 and of the Quantum Cromo-Dynamics in Chapter 5. The observation of the W in all the three expected leptonic channels (W → eν_e, W → $\mu\nu_\mu$, W → $\tau\nu_\tau$) and of the Z^0 in both the electron and muon channels (Z^0 → e$^+$ e$^-$, Z^0 → μ^+ μ^-) will be reviewed as well as the measurements of their masses and decay properties. Complete agreement will be found with the predictions of the Electroweak theory. The production properties of the intermediate vector bosons (production cross sections and differential cross sections) will be found in agreement with the predictions of QCD.

As discussed in Section 5.3 and Chapters 7 and 8, the data put significant limits on extensions and/or modifications of the standard model; in particular limits are obtained on the masses of the top quark and of the charged lepton of a possible fourth generation and stringent upper limits on the number of neutrino species. Lower limits on the masses of hypothetical new objects as the supersymmetric particles are also obtained.

Space does not allow to cover the physics of the hadronic jets, that are interpreted as the way in which the constituents appear to the experimenters; we will only quote (Chapter 8) the analogous of the experiment of Geiger and Marsden at these energies and the limits on the diameter of the quark. The limited space does not allow to cover other interesting topics as the discovery of the $B^0 - \bar{B}^0$ oscillations[1].

A new source of antiprotons, ACOL, is now in operation at CERN and will allow an increase in luminosity by an order of magnitude. Both UA1 and UA2 experiments are in the process of being substantially upgraded to provide more stringent tests of the standard model in the coming years. In the same period the Fermilab Tevatron Collider, that started operation $\sqrt{s}$ = 1.8 TeV, will be operational. Certainly collider physics will continue to provide existing results in the near future.

2. THE INTERMEDIATE VECTOR BOSONS. EXPECTATIONS AT THE $P\bar{P}$ COLLIDER

Beta decays of nuclei, originally observed at the end of the last century, are interpreted as beta decays of neutrons or protons bound in the nucleus:

a. $\quad n \to p + e^- + \bar{\nu}_e \quad ; \quad p \to n + e^+ + \nu_e$

We know today that the beta decays of the nucleons are not yet elementary processes but are due to the beta decay of quarks bound in the nucleons:

b. $\quad d \to u + e^- + \bar{\nu}_e \quad ; \quad u \to d + e^+ + \nu_e$

The intermediate vector bosons W+ and W- are predicted by the Electroweak theory to be the mediators of these processes. At the lowest order the Feynman graphs are shown in Fig. 1.

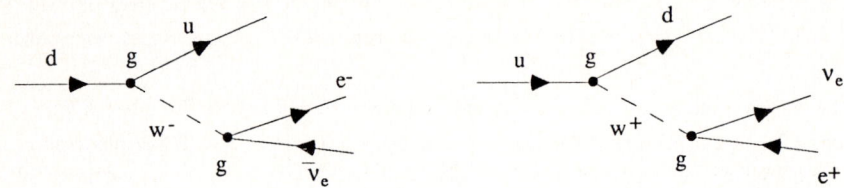

FIGURE 1

In both graphs the quarks in the initial and final states have the same colour. The corresponding matrix element is:

$$M = \left[\frac{g}{\sqrt{2}} \bar{\nu}_e \gamma^\lambda (1-\gamma^5) e \right] \frac{1}{M_W^2 - q^2} \left[\frac{g}{\sqrt{2}} \bar{d}_c \gamma_\lambda (1-\gamma^5) u \right] \qquad (1)$$

where g is the coupling constant, M_W the mass of the W, q^2 the square of the four-momentum transfer and the particle symbols represent the corresponding (relativistic) spinors. Universality of the coupling i.e. the same g in both lepton and quark current terms is obtained, as shown by Cabibbo, introducing substantial mixing amongst the charge -1/3 quarks (d_c is the Cabibbo rotated down quark).

Note that the propagator factor $1/M_W^2 - q^2$ is the probability that a W moves from one vertex to the other of the graphs of Fig. 1. When the energy is very small with respect to the W mass $(q^2 \ll M_W^2)$ the W cannot move significant distances and the interaction is effectively in a point. We are in the limit the four-field Fermi point interaction. The propagator is effectively a constant: $1/M_W^2$, and we can identify the Fermi constant G_F:

$$\frac{G_F}{\sqrt{2}} = \frac{g^2}{8M_W} \qquad (2)$$

Note that g has the same order of magnitude as the fine structure constant α (the coupling of the electromagnetic interaction), G_F is much smaller because M_W is large.

We finally remind that the current terms have a V - A structure as is immediately visible in (1). This implies that the W has spin-parity $J^P = 1^-$ and that it couples only with **fermions** (quarks and leptons) with **negative** helicity (spin antiparallel to the velocity) and **antifermions** with **positive** helicity. The test of this important aspect (**parity violation**) of the theory will be described in Sect. 4.4.

W+ and W- can be produced from an hadronic initial state by the graphs shown in Fig. 2.

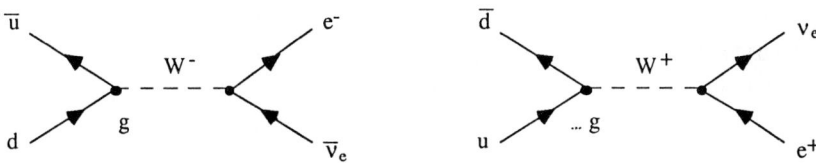

FIGURE 2

Of course different final states as $\nu_\mu \mu$, $\bar{u}d$ etc are possible. Note that also now the quark and the antiquark in the initial state must have the same colour. Due to these graphs the cross sections of the processes:

c. $\bar{u}d \to e^-\bar{\nu}_e$; $\bar{d}u \to e^+\nu_e$

have a resonance at $\sqrt{\hat{s}} = M_W$, where $\sqrt{\hat{s}}$ is the energy in the centre of mass of the quark antiquark system.

$$\sigma(\bar{u}d \to e^-\bar{\nu}_e) \alpha \frac{\Gamma_i \Gamma_f}{\left(\sqrt{\hat{s}} - M_W\right)^2 + \left(\frac{\Gamma}{2}\right)^2} \qquad (3)$$

where Γ is the total width, Γ_i and Γ_f the partial widths in the initial and final states respectively. SU(2) x U(1) gives the following definite predictions:

$$M_W^2 = \frac{A^2}{1-\Delta r}/\sin^2\vartheta_W \qquad , \qquad A^2 = \frac{\pi\alpha}{\sqrt{2}G_F} \qquad (4)$$

From the values of the fine structure constant and of the Fermi constant, known with high accuracy form low energy precision measurements, we obtain[2]:

$$A^2 = (37.2810 \pm 0.0003)^2 \text{ GeV}^2 \tag{5}$$

Δr contains the higher order radiative corrections, it is calculated by the Electroweak theory[3]:

$$\Delta r = 0.0711 \pm 0.013 \tag{6}$$

The Weinberg angle is known from low energy measurements, mainly from the ratio of neutral current and charged current neutrino reactions[4]:

$$\text{sen}^2\vartheta_W = 0.231 \pm 0.004 \text{ (stat)} \pm 0.005 \text{ (syst)} \tag{7}$$

With these values the expected value of the mass of the W is:

$$M_W = 80.3 \pm 0.9 \text{ GeV}$$

The Electroweak theory gives also a prediction for the total width:

$$\Gamma_W \approx 2.6 \text{ GeV}$$

(the precise value depends on the unknown mass of the top quark).

In practice processes c) cannot take place because quarks are not found free in Nature, but only inside the hadrons. We will then consider a high energy proton beam colliding against a high energy antiproton beam. Both beams can be considered in a first approximation as wide band parton $(q, \bar{q}, g)$ beams.

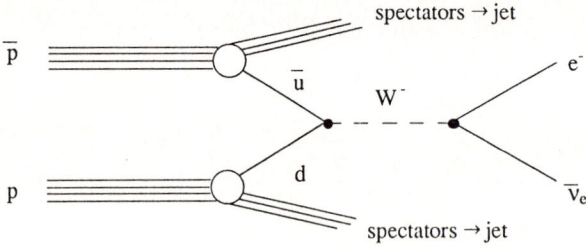

FIGURE 3

In the rare case when an $\bar{u}$ quark of the antiproton closely encounters a d quark of the same colour of the proton, the process depicted in Fig. 3 happens: a W is produced that immediately (after a time $1/\Gamma_W$) decays for example in $e\nu_e$. The process has a sharp resonance at $\sqrt{\hat{s}} = M_W$. In a good approximation we can forget the motion of the quarks transverse to the beam; if x_d is the fraction of the proton momentum taken by the d quark, and $x_{\bar{u}}$ the fraction of the antiproton momentum taken by the $\bar{u}$ the energy squared in the parton centre of mass is:

$$\hat{s} = x_d \cdot x_{\bar{u}} \cdot s \tag{8}$$

where s is the energy squared in the $p\bar{p}$ centre of mass.

When an $\bar{u}$ and a d annihilate into a W the proton and antiproton remain in a coloured unstable state; each of them must rearrange its state "hadronizing" and giving origin to a jet of hadrons that move at small angle with original direction: two jets (spectator jets) at low p_T (p_T is the momentum transverse to the beams direction) are observed.

The calculation of the cross section of process d) was done by G. Altarelli at al.[5] with QCD; knowledge of the parton distribution functions in the nucleon (structure functions) is needed. These functions have been measured by deep inelastic lepton scattering at low energies and must be evolved to high energies taking into account QCD scale violations. Fig. 4 shows the relevant quark structure functions. u_v and d_v are the valence quark, s is the "sea" contribution, that contains both quarks and antiquarks. Notice that at the CERN collider energy $\sqrt{s}$ = 630 GeV, the value of x at resonance is $x = M_W / \sqrt{s} = 0.13$, a value where the valence quark completely dominate.

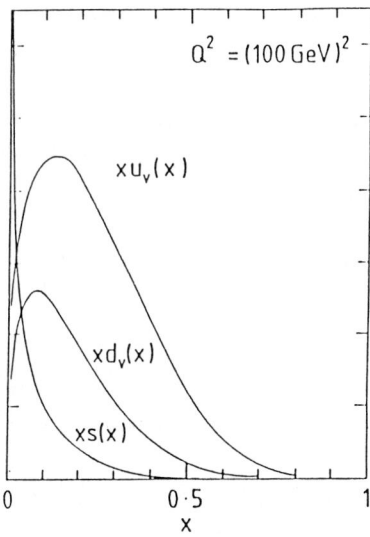

Fig. 4

We are then entitled to think that the interacting quark was in the proton, the interacting antiquark in the antiproton. This is important to be able to observe the asymmetry predicted by V - A (parity violation).

The expectation for the cross section from QCD calculation at $\sqrt{s} = 630$ GeV is:

$$\sigma(\bar{p}p \to W \to e\nu) = 570^{+180}_{-100} \text{ pb} \qquad (9)$$

a value that is of the order of 10^{-8} of the total cross section.

We turn now to the neutral vector boson, Z^0. It is needed because neutral currents are observed, i.e. processes of the type:

$$\nu_\mu N \to \text{hadrons } (+\text{no } \mu)$$

The relevant Feynman graphs are:

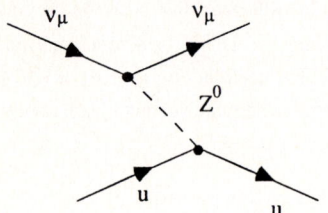 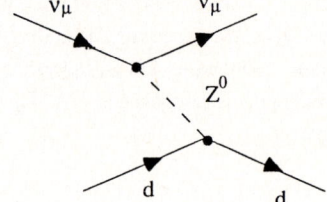

FIGURE 5

and again allow the production of Z^0 in $\bar{p}p$ collisions:

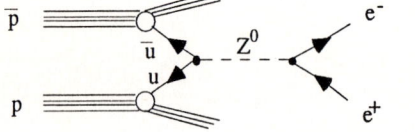 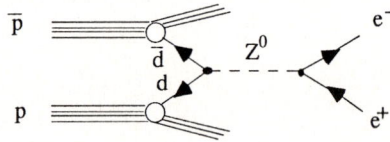

FIGURE 6

Again the quark and antiquark in the initial state must have the same colour and again other final states are possible ($\mu^+\mu^-$, $\tau^+\tau^-$, $u\bar{u}$, $d\bar{d}$, etc). Note that in this case two different processes ($u\bar{u}$ and $d\bar{d}$ annihilation) contribute to the Z^0 production.

SU(2) x U(1) gives clear predictions for the mass

$$M_Z = \frac{M_W}{\cos\vartheta_W} = 91.6 \pm 0.7 \text{ GeV}$$

and the total width

$\Gamma_Z \approx 2.6$ GeV

Again the expected value of the cross section is more uncertain (QCD calculation)[5]:

$$\sigma(\bar{p}p \to Z^0 \to e^+e^-) = 54^{+17}_{-11} \text{ pb} \qquad (10)$$

expected at $\sqrt{s} = 630$ GeV, a value an order of magnitude smaller than for $W \to e\nu$.

In summary the production of the intermediate vector mesons at the $p\bar{p}$ collider is a rare process ($10^{-8} \div 10^{-9}$); their observation requires a rejection power against background at least of 10^{10}. The more frequent final states of the IVB's are $q\bar{q}$ states (about 70%), experimentally observed as two hadronic jets. Unfortunately the background due to gluon-gluon, gluon-quark and quark-antiquark scatterings is very high, two order of magnitude higher than the signal. An accurate measurement of the jet-jet invariant mass should allow to observe resonance peaks at the W and Z^0 masses, but the experimental resolution $\sigma (M_{jj}) \sim 8$ GeV is still too poor to allow a clear identification of the peaks (UA2 has indeed observed a shallow hump in correspondence to the non resolved W and Z masses[6], UA1 is building a "compensated" calorimeter that will largely increase its accuracy in jet energy measurements).

In a much more favorable situation are the leptonic final states. For the W:

$$W \to \begin{cases} e\nu_e \\ \mu\nu_\mu \\ \tau\nu_\tau \end{cases}$$

where both the charged lepton (e, μ or τ) and the neutrino have a **high** p_T (momentum component perpendicular to the beams) and are **isolated** (meaning that no other energetic particle is produced near to their direction). For the Z:

$$Z \to \begin{cases} e^+e^- \\ \mu^+\mu^- \\ \tau^+\tau^- \end{cases}$$

the signature being here **two isolated high p_T leptons** of opposite charge.

In both cases the debris of the "spectators" are present in the final state ("underlaying event") but can be easily distinguished because their p_T is much smaller (typically smaller than 1 GeV). Occasionally, as we will see, the initial quark or antiquark can radiate a gluon: a high p_T jet appears in this case in the final state in addition to the two leptons.

The "**isolation**" of a track can be defined quantitatively. The UA1 definition is in the space (ϕ, $\eta = -\ln \text{tg } \vartheta/2$) where ϕ is the azimuth around the beam axis and ϑ the angle with the beam; the pseudorapidity variable η is used because the track density in η is much more uniform than in ϑ. A track is defined to be isolated if the energy flux in a cone of given semiaperture

$\Delta R = [\Delta\phi^2 + \Delta\eta^2]^{1/2}$ around the track is less than a given value E_{max}. The specific value of ΔR and E_{max} depend on the particular case considered, as we will see later.

As we will see all the before mentioned final states (with the exception of $Z^0 \to \tau\tau$) have been clearly observed with a very small background. Let us now consider the measurements of the IVB's masses and the relative accuracy. In the Z^0 case we observe an $e^+ e^-$ pair or a $\mu^+ \mu^-$ pair and measure the energy E_1 and E_2 of the leptons and the angle ϑ between their directions. The invariant mass of the system is:

$$m = \left(4E_1 E_2 \mathrm{sen}\frac{\vartheta}{2}\right)^{1/2} \tag{11}$$

and the related resolution:

$$\frac{\sigma_m}{m} = \frac{1}{2}\left[\left(\frac{\sigma_1}{E_1}\right)^2 + \left(\frac{\sigma_2}{E_2}\right)^2 + \left(\frac{\sigma_\vartheta}{\mathrm{tg}\frac{\vartheta}{2}}\right)^2\right]^{1/2}$$

The angle ϑ is generally big and $\mathrm{tg}\frac{\vartheta}{2}$ has value around one; as the directions of the charged tracks are accurately measured, σ_ϑ is of the order of 10^{-2} and the contribution of the angle measurement resolution is negligible. As on the other hand E_1 and E_2 are of the same order:

$$\frac{\sigma_m}{m} = \frac{1}{\sqrt{2}} \frac{\sigma_E}{E}$$

We must at this point distinguish the electron from the muon case. The energy of the electron is measured by the calorimeters with a resolution $\sigma_E / E = 16\% / \sqrt{E}$ (with E in GeV) i.e. $\sigma_E / E = 2\%$ for m = 90 GeV (E = 45 GeV). In conclusion the resolution of the measurement of the Z mass in the electron channel is $\sigma_m = 1.4$ GeV. The muons, on the other hand, are not observed by the calorimeters, their energy (better their momentum) is measured by a magnetic spectrometer. The resolution is typically $\sigma_p / p = 0.5\% \cdot p$ (with p in GeV) that is 22.5% at 45 GeV (the sagitta of 1 m long track in 0.7 T magnetic field is 6 mm). The corresponding resolution for the Z^0 mass is $\sigma_m = 14$ GeV an order of magnitude worse than the electron channel.

The resolution values quoted above refer to the single measurement and can of course be reduced increasing the statistical sample. The ultimate resolution will be determined by the systematics that will be discussed in Sect. 4.2.

The measurement of the W mass is different because only the charged lepton is directly observed. The charged lepton momentum in the W centre of mass frame is obviously $p_e = M_W/2$. If ϑ^* is the angle between the charged lepton and the beam and we neglect for the moment the transverse motion of the W, the transverse momentum $p_T^e = (M_W / 2) \mathrm{sen}\vartheta^*$, is the same in the

W rest frame and in the laboratory frame. From the known decay angular distribution of the W $dn/d\vartheta^*$, one can obtain the p_T^e distribution simply with a coordinate transformation:

$$\frac{dn}{dp_T^e} = \frac{dn}{d\vartheta^*} \cdot \frac{d\vartheta^*}{dp_T^e} = \frac{1}{\left[\left(\frac{M_W}{2}\right)^2 - p_T^{e\,2}\right]^{1/2}} \cdot \frac{dn}{d\vartheta^*} \tag{12}$$

As one can see the "Jacobian" of the transformation has a peak (the "Jacobian peak") at $M_W/2$; the presence of this peak in the data gives direct evidence for the W, its position allows a determination of its mass. The transverse motion of the W smears the peak and can alter its position in a way dependent on dn/dp_T^W. To avoid the consequent model dependence of the evaluation of the mass the momentum of the neutrino should be determined. This can be achieved by measuring the momenta of all the charged and neutral particles produced in the interaction. As the neutrino does not interact with the detector, its momentum can be obtained, applying momentum conservation, from the overall momentum unbalance. The "measurement" of the "missing momentum" is achieved by UA1 with a calorimeter that is hermetic. Unfortunately complete hermeticity cannot be achieved: particles at small angle with the beams escape detection; in practice only the transverse components of the missing momentum $\left(\vec{p}_T^{\,miss}\right)$ are determined. An independent signature of W events is the presence of the Jacobian peak in the p_T^{miss} distribution.

We can now define the "transverse invariant mass":

$$m_T^{ev} = \left[2p_T^e p_T^{miss}(1 - \cos\phi)\right]^{1/2} \tag{13}$$

where ϕ is the angle between the two leptons in the transverse plane. The distribution of m_T^{ev} has a peak only a few GeV lower than M_W. It is only in these few GeV that the model dependence of dn/dp_T^W enters.

3. THE DETECTOR

The discussion of the previous chapter allows us to lay down the main characteristics to demand to the detector. It must be able to identify the elementary objects produced by the collision and to measure their energy, momentum and charge. The elementary objects are: a) the charged leptons, the **electron** (identified by its characteristic shower in a heavy material), the **muon** (identified by its ability to penetrate deeply into matter) and the **tau** (identified by the shape of the jet produced by its fast (lifetime = 0.33 ps) decays into hadrons; b) the **neutrinos** (identified by an overall momentum unbalance); c) **quarks** and **gluons** experimentally measured as hadronic **jets** (identified by the tracking chambers and by the calorimeters).

The UA1 apparatus consists of different layers of different detectors (Fig. 7). The innermost is a cylindrical drift chamber around the beam pipe 5.8 m long, 2.3 m in diameter; it is called the

Central Detector[7] (C.D.) and provides a three-dimensional picture of the charged tracks, down to 0.2° with respect to the beams. It is immersed in a uniform dipole magnetic field horizontal and perpendicular to the beams (B = 0.7 T); it provides the momentum measurement with a typical resolution $\Delta p/p = 0.005\ p$ (p in GeV).

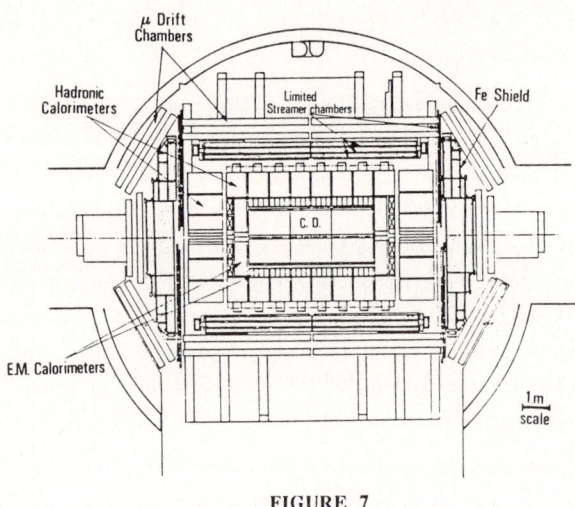

FIGURE 7
The UA1 detector

Outside the lateral surface of the Central Detector one finds the central electromagnetic[8] calorimeter, which covers the polar angle interval between 25° and 155° and 2π in azimuth. It consists of two half cylinders on the two sides, each composed of 24 elements, the "gondolas". A gondola is a lead-scintillator sandwich of 26.6 radiation lengths with independent read-out of four longitudinal samplings of 3.3, 6.6, 10.1 and 6.6 r.l. respectively. The energy deposited in the gondola is measured with an accuracy $\Delta E / E = 16\% / \sqrt{E}$ (E in GeV). Each gondola covers the angle interval $\Delta\vartheta = 5°$, $\Delta\phi = 180°$.

The two end-cap calorimeters, the "bouchons" consist of 32 "petals" of lead-scintillator sandwich. They are 27 r.l. thick, segmented four times in depth. The energy resolution is the same as that of the gondolas. With the gondolas they provide coverage down to 5° to the beams, additional forward located calorimeters provide coverage down to 0.2°[8].

The electromagnetic calorimeters are located inside the magnetic field. The yoke of the dipole magnet is outside the gondolas and is made of iron plates separated by plastic scintillator plates to form the central hadronic calorimeter[9]. Similar structures outside the bouchons from the forward hadronic calorimeters. The calorimeters are read-out in two separated longitudinal samples. The central one is 5 absorption lengths thick, the forward ones 7 absorption lengths. Their granularities are $\Delta\vartheta \times \Delta\phi = 15° \times 18°$ and $5° \times 10°$ respectively. Their energy resolution is $\Delta E / E = 0.8 / \sqrt{E}$ (E in GeV). Electron and photon showers in the electromagnetic calorimeters are completely absorbed, hadrons penetrating in the heavy material of the calorimeters interact giving origin to a hadronic shower; a good fraction of their energy is deposited in the electromagnetic calorimeter, the rest in the hadronic one. Rarely tails of the hadronic shower emerge from the back of the calorimeters. Iron shields 60 cm thick on the sides, more than 1 m thick in the forward direction completely absorb these tails. Limited streamer tubes detectors[10] layers instrument the shields allowing the tracking of penetrating charged particles, the muons. On the sides the iron is magnetized. Outsides the iron shielding two planes of drift chambers[11], separated by a 60 cm distance, give accurate measurements of the muon track coordinates. Each plane measures twice each of two orthogonal coordinates. The muon detection system covers the full interval in azimuth and between 5° and 175° in polar angle, it allows also to determine the deflection of the track in the magnetic field in the yoke and in the lateral walls thus providing and independent measurement of the muon momentum.

Electrons are identified by the characteristic longitudinal development of the shower and by the presence of an associated charged track. The principal background is due to one (or more) $\pi^°$ near to a medium energy $\pi^{\pm}$; the electromagnetic shower of the photons from the $\pi^°$ overlapped to the hadronic shower of the charged π can mimic a truly electromagnetic shower. A comparison of the total energy of the shower with the momentum of the charged track as measured in the C.D. eliminates part of this background. This background is of course normally near or into the hadronic jets. In conclusion high p_T, isolated electrons are easy to identify (discriminating power against jets ~40.000), much more difficult at lower p_T or near to the jets, impossible inside jets.

Muons are identified by the presence of a charged track in the muon chambers associated with a charged track in C.D. The principal background is due to K or π decays into a muon in C.D., when the kink of the track is not observed. In these cases the track momentum is overevaluated and an oppositely directed "missing momentum", interpreted as a neutrino, is generated. The background is controlled by an accurate study of the track quality in the C.D. and comparing the two measurements of its momentum. The event sample containing one muon is 80% pure at $p_T = 20$ GeV, 25% pure at $p_T = 6$ GeV. In conclusion isolated high p_T muon can be easily separated from background but this becomes increasingly difficult as p_T decreases. On the other hand it is possible to identify muons even inside jets, opening the route to heavy flavour (bottom and top) physics.

A **jet** appears clearly as a localized energy deposit in the calorimeters; all the hadronic events with total transverse energy E_T bigger than 20-30 GeV consist on two or more jets, as the example shown in Fig. 8. This observation leads immediately to interpret jets as manifestations of the quarks and of the gluons. On the other hand the identification of the energy and direction of the jets with those of the parton is subject to uncertainties both experimental and theoretical; the most important are the non linear response of the calorimeter at low energies, the different response to a $\pi^°$ (electromagnetic) and to a $\pi^{\pm}$ (hadronic) of the same energy, the uncertainties on the soft tails of the fragmentation, the subtraction of the contribution of the underlaying event.

Tau leptons are identified[12], in their hadronic decays, by the narrow, low multiplicity shape of the jet.

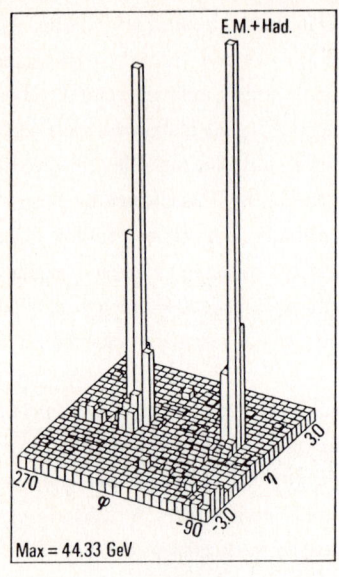

FIGURE 8
"Lego plot" of a two jet event. Energy deposit in the calorimeter cells in the φ, η space.

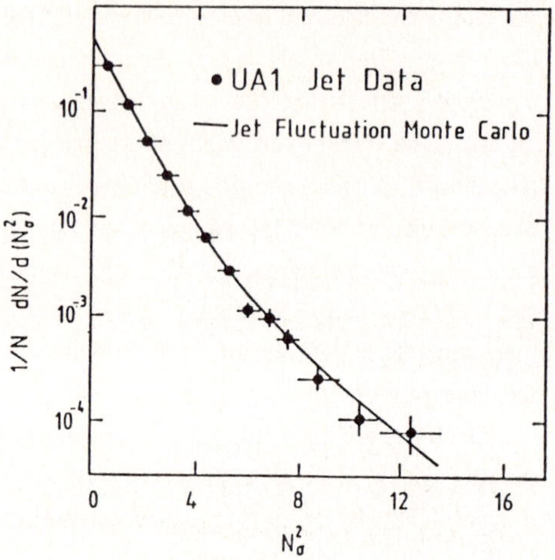

FIGURE 9

The presence of **one or more neutrinos** (or other unknown non interacting particles) can be inferred by measuring the total momentum vector of the interacting particles and obtaining values different from zero (outside the errors). In practice even providing a coverage down to a fraction of a degree, it is impossible to measure the energy of the secondaries at small angle with the beams; the longitudinal component of the total momentum can not be measured. On the other hand undetected particles give a negligible contribution to the transverse momentum allowing its determination with good accuracy. If E_i is the energy deposited in the i cell of the calorimeter and $\vec{u}_i$ the versor of the cell from the interaction point we can define the energy (momentum) vector in the cell: $\vec{E}_i = E_i \vec{u}_i$ and its component in the transverse plane $\vec{E}_{iT}$; the missing transverse momentum is then:

$$\vec{p}_T^{\,miss} = -\sum_i \vec{E}_{iT} - \vec{p}_T^{\,\mu}$$

where $\vec{p}_T^{\,\mu}$ is the transverse momentum of the muons (that deposit very small energy in the calorimeters).

Events with high $\vec{p}_T^{\,miss}$ are the of outstanding importance, as they can signal the presence of new physics; the knowledge of the resolution function in the $\vec{p}_T^{\,miss}$ measurement is as a consequence extremely relevant. The study of a sample of events where neutrinos are not expected allowed UA1 to determine this function that is a gaussian with:

$$\sigma\left(p_T^{miss}\right) = 0.7\sqrt{|E_T|} \tag{14}$$

where $|E_T|$ is the scalar sum of the transverse energies. It is shown, as a function of $N_\sigma\left(p_T^{miss}\right) = 0.7\sqrt{|E_T|}$ i.e. the number of standard deviations, in Fig. 9 where the data are compared with the expectations due to the fluctuations in the measurement of the energy of the jets. As it can be seen by the agreement, the resolution function is completely under control.

4. THE INTERMEDIATE VECTOR BOSONS

4.1   THE DATA SAMPLE

In a series of data taking periods between 1982 and 1985 UA1 collected data at $\sqrt{s}$ = 546 GeV for an integrated luminosity of 118 nb^{-1} and at $\sqrt{s}$ = 630 GeV for 568 nb^{-1}. The total of 686 nb^{-1} corresponds to about 50 x 10^9 $\bar{p}p$ collisions.

The main triggers, simultaneously operating, have been the following:

    1. One ore more electromagnetic clusters with E_T bigger than a given threshold (depending on running conditions, typically 10 GeV);

    2. One or more hadronic jets with E_T bigger than a given threshold (typically 25 GeV);

    3. Left-right E_T unbalance and at least one jet with E_T > 25 GeV;

    4. One or more muons;

    5. Total E_T bigger than a given threshold;

    6. Other physics triggers.

4.2 $W \to e\nu$ AND $Z^0 \to e^+ e^-$

In both cases an isolated high p_T electron is searched for. The selection criteria[13] are tuned to obtain a high rejection against jets ($\pi^0 \pi^\pm$ overlap) and a reasonable high efficiency for electrons. The main cuts are the following:

    1) One electromagnetic cluster with calibrated E_T > 15 GeV;

    2) A charged track in the C.D. associated in position and angle with the electromagnetic cluster with momentum consistent with p_T > 15 GeV;

    3) Longitudinal development of the shower, as measured in the 4 samplings, compatible with electromagnetic nature;

    4) Small (< 0.6 GeV) energy deposit in the first hadronic compartment;

    5) Isolation, requiring both p_T (C.D. measurement) and E_T (calorimeter measurement) in ΔR < 0.7 to be smaller than 3 GeV.

The efficiency of all these cuts is around 70% for electrons from IVD's decay. Having selected one electron we now look for missing p_T to select W's, for a second electron to select Z^0's.

In the W selection $p_T^{miss} > 15$ GeV is requested, giving the final sample of 299 events (the overall selection efficiency is around 60%). The Z^0 sample is obtained requiring a second electromagnetic cluster with less tight selection criteria (isolation with higher threshold and transverse energy of the electron > 8 GeV); it consists of 33 events (selection efficiency ~70%).

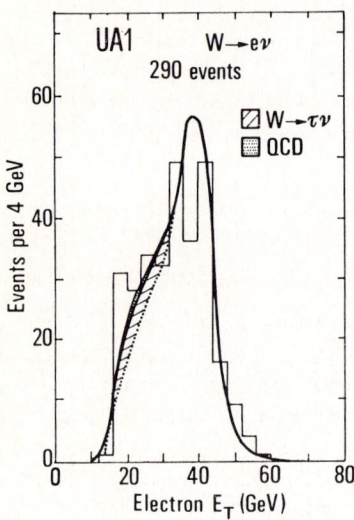
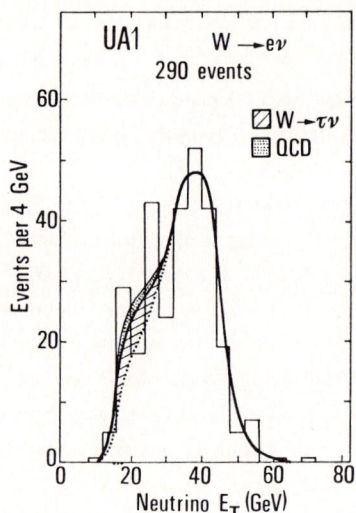

FIGURE 10   **FIGURE 11**

Shaded parts show expected background contributions from jet fluctuations and from $W \to \tau\nu$. Curves are predictions for $M_W = 82.7$ GeV.

Fig. 10 and 11 show the distributions of p_T^e and p_T^ν respectively. The Jacobian peaks are clearly seen out of the background dominant at low values. The background in the W sample due to jets misidentification is evaluated extrapolating the background dominated low p_T distribution into the sample region. It amounts to 3% of the sample. A second background source comes from the W itself through the decay $W \to \tau\nu$, when the τ decays into one electron or misidentified hadrons. Knowing the decay properties of the W it can be reliably calculated and it amounts to 6% of the data. In conclusion the sample of 299 W candidates contains 26.1 ± 2.6 background events. These events are all at low values of p_T^e and p_T^ν. For the determination of M_W a further cut: $p_T^e > 30$ GeV, $p_T^\nu > 30$ GeV is applied to obtain a background free sample of 149 events.

The transverse mass (M_T) distribution is shown in Fig. 12. To extract M_W from the data[14], Monte Carlo calculations are done with different values of M_W and fixed value of Γ_W near to expected one. The simulation takes into account the expected transverse and longitudinal motions of the W (and associated QCD uncertainties). The best fit is shown in Fig. 12. The same calculation is used to evaluate systematic errors, obtaining[14]:

UA1 $M_W = 82.7 \pm 1.0$ (stat) ± 2.7 (syst) GeV   (15)

For comparison UA2 finds (with similar statistics)[15]

UA2 $M_W = 80.2 \pm 0.8$ (stat) ± 1.3 (syst) GeV   (15)

Both values agree perfectly with the expectations (see Chapter 2). Note that the error is dominated by the systematics due to uncertainty on the absolute energy scale.

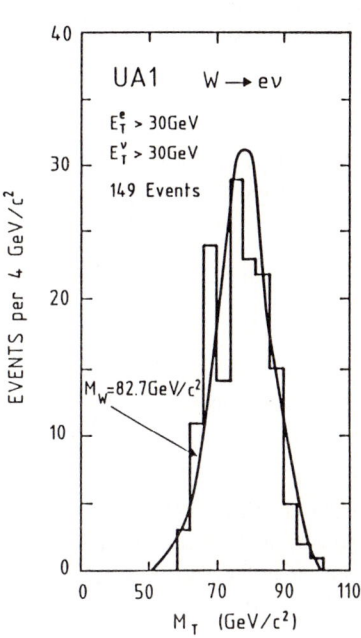
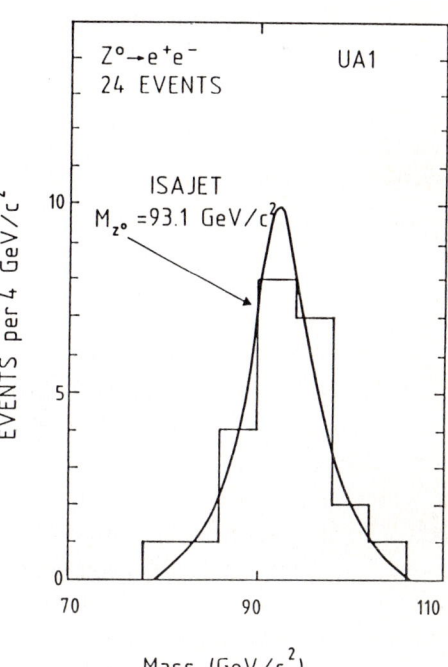

FIGURE 12
Transverse mass distribution for W candidate events with well measured electron and leptons transverse momenta bigger that 30 GeV.

FIGURE 13
$e^+ e^-$ invariant mass distribution for Z^0 candidates with well measured electrons.

Turning now to the width, we note that for both experiments the width of the resolution function is comparable with the expected natural one Γ_W. The uncertainties on the resolution function in these conditions allow only to put upper limits on Γ_W[14,15]:

UA1 $\Gamma_W < 5.4$ GeV   (16)
UA2 $\Gamma_W < 7.0$ GeV   (16)

at 90% c.l.

Let us now consider the 33 events Z^0 sample that is practically free from background (evaluated number of background events is 0.2 ± 0.02). To determine M_Z a subsample of 24 events with both electrons accurately measured is selected. The two-electron invariant mass distribution is shown in Fig. 13. From the fit we obtain[14]:

UA1 $M_Z = 93.1 \pm 1.0 \pm 3.1$ GeV   (17)

For comparison[15]:

UA2 $M_Z = 91.5 \pm 1.2 \pm 1.7$ GeV   (17)

Both are in perfect agreement with expectations.

Again the dominant error is that on the energy scale, again both experiments give only upper limits for the width; at 90% c.l.[14,15]:

UA1 $\Gamma_Z < 5.2$ GeV   (18)
UA2 $\Gamma_Z < 5.6$ GeV   (18)

The uncertainties on the energy scale is evaluated to be ±3% for UA1 and ±1.6% for UA2. Both experiments measure the energy deposited by the particles in sampling calorimeters that are sandwiches of active material (scintillator) and passive material (lead or iron). The quantity measured is the total charge collected at the anodes of a number of photomultipliers. This charge results as the final product of a long series of processes: energy loss in scintillator, light emission from the modules of the scintillator; partial collection of the light by the light guides, photoelectric effect at the photocathodes, amplification of the electrons in the P.M. Of course the conversion factor from collected charge to energy lost by the particle must be obtained calibrating the calorimeter cells on an electron and pion beam at different energies. Unfortunately the "calibration constants" are subjected to vary with time due for example to ageing and radiation damage; it is then necessary to monitor continuously their values with light sources and radiative sources. At the hadron colliders a direct, in situ, calibration tool, as that given by $e^+ e^-$ elastic scattering at the $e^+ e^-$ machines, is not in fact available. The complex procedure outlined leads to the residual uncertainty quoted above. The UA2 experiment, having smaller calorimeter cells, was able to recalibrate part of them periodically in a test beam; on the contrary UA1 could not move its large calorimeter cells to a test-beam area.

One of the main target of the upgrade program of both experiments is to obtain a much better control of the systematic. In particular UA1 has developed a new technique, where the ionization charge is directly collected without amplification to this aim.

4.3 COMPARISON WITH THE STANDARD MODEL

Three independent definitions of the Weinberg parameter $\text{sen}^2\vartheta_W$ can be given. The first:

$$\text{sen}^2\vartheta_W = 1 - (M_W/M_Z)^2 \tag{19}$$

is free from radiative corrections and, from the experimental point of view, insensitive to the energy scale error.

The second definition (from (5)):

$$\text{sen}^2\hat{\vartheta}_W = (1 - \Delta r) / A \cdot M_W^2 = (38.65 / M_W)^2 \tag{20}$$

contains the radiative correction term Δr, that is, as we will see, in particular sensitive to the top mass M_t.

The third value is obtained from the ratio of neutral and charged current neutrino interactions. The weighted average of the two must precise experiments[14] is:

$$\text{sen}^2\vartheta_W^\nu = 0.232 \pm 0.004 \text{ (stat)} \pm 0.003 \text{ (theor)} \tag{21}$$

The three quantities could be a priory different. Their equality is a definite prediction of the Electroweak theory. Experimentally from (19) we have:

UA1 $\qquad \text{sen}^2\vartheta_W = 0.211 \pm 0.025$ (22)

UA2 $\qquad \text{sen}^2\vartheta_W = 0.232 \pm 0.027$ (22)

and from (20)[14]:

UA1 $\qquad \text{sen}^2\hat{\vartheta}_W = 0.218 \pm 0.005 \pm 0.014$ (23)

UA2 $\qquad \text{sen}^2\hat{\vartheta}_W = 0.232 \pm 0.003 \pm 0.008$ (23)

The equality of all the values constitutes a stringent test of the standard model. The agreement is shown in graphical form in the (M_Z - M_W, M_Z) plane in Fig. 14. A global fit to the neutrino data and to the W and Z masses gives for the radiative correction term[16]:

$$\Delta r = 0.077 \pm 0.037 \tag{24}$$

This value is consistent with the theoretical one (6); this last value could be different for two main reasons: 1) for high value of the top mass ($M_t > 80$ GeV), Δr decreases becoming negative for $M_t \sim 250$ GeV; this is due to the fact that the graph shown in Fig. 15 gives negative contribution to the W mass proportional to $M_t^2 - M_b^2 \cong M_t^2$ and there is no similar contribution to the Z^0 mass. The result is[3]:

$$\Delta r = 0.0711 - \frac{3\alpha}{16\pi} \left(\cos^2 \vartheta_W / \mathrm{sen}^4 \vartheta_W \right) \cdot \frac{M_t^2}{M_W^2} \tag{25}$$

2) new fermion doublets with large mass splitting give similar contribution, decreasing Δr.

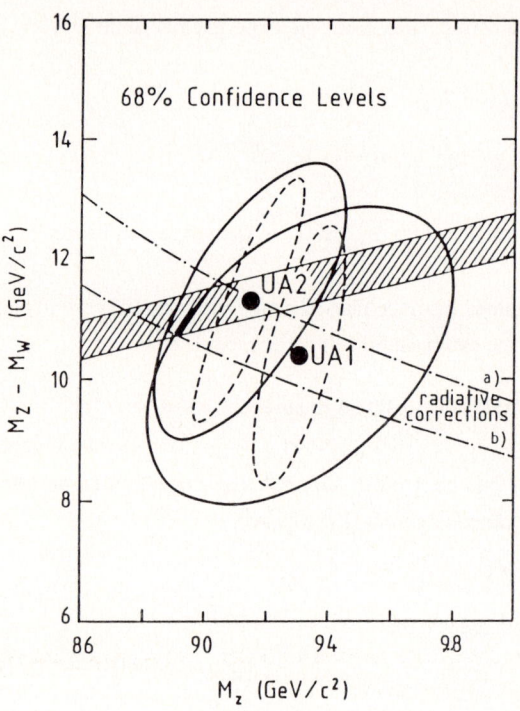

FIGURE 14
Dotted ellipses are 68% c.l. confidence contours taking account only statistical uncertainties, solid ellipses are the same including systematics. Dashed band is the prediction from low energy data. Curve a) S.M. prediction with $\rho = 1$ with radiative corrections, curve b) the same without radiative corrections.

The value (24) with expression (25) allows us to put an upper limit on the mass of the top: $M_t < 180$ GeV (in the hypothesis of the three families).

One of the most important targets of the improved UA1 and UA2 experiments is the precise measurement of Δr. This will allow to limit the mass of the top, if not yet found, or test the number of families, if M_t will be known.

In the minimal Standard Model, considered so far, an isospin doublet of Higgs fields is assumed. This implies equality of the coupling strength of neutral and charged currents. Their ratio:

$$\rho = M_W^2 / \left(M_Z^2 \cos^2 \vartheta_W \right) \tag{26}$$

may be different from 1 for more complicated isospin structures of the Higgs fields. Using (20) to evaluate $\text{sen}^2 \vartheta_W$ we have:

UA1[14] $\rho = 1.009 \pm 0.028 \pm 0.020$ (27)
UA2[15] $\rho = 1.001 \pm 0.028 \pm 0.006$ (27)

both perfectly compatible with 1.

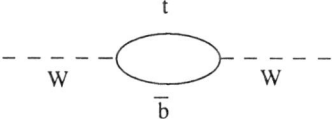

FIGURE 15

4.4 THE PARITY VIOLATION

The V-A coupling of the W implies parity violation of the weak interactions. As a consequence a charge asymmetry will be observed; as we will now see electrons will tend to be emitted forward with respect to incident protons, positrons backward. The charge asymmetry is apparent at CERN collider energies because, as we mentioned in Chapter 1, W production is mainly due to the valence quarks and antiquarks and we may identify the direction of the initial quark with that of the proton beam, the direction of the antiquark with the $\bar{p}$ beam.

V-A coupling fixes the helicity states of the leptons that couples to the W. With reference to Fig. 16

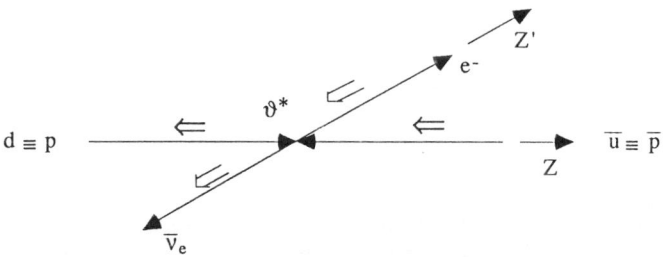

FIGURE 16

we see that the initial state has $J = 1$, $J_z = -1$, i.e. the W is fully polarized with respect to the beam direction. The final state is, on the other hand, $J = 1$, $J_{z'} = -1$. The transition element must then be proportional to the rotation matrix element $d^1_{-1,-1}$, i.e.:

$$\frac{d\sigma}{d\Omega} \propto \left[d^1_{-1,-1} \right]^2 = \left[\frac{1}{2}(1 + \cos \vartheta^*) \right]^2 \tag{28}$$

where ϑ^* is the angle between e^- and p in the W^- centre of mass, or between e^+ and $\bar{p}$ in the W^+ centre of mass.

Experimentally p_L^ν is not measured, so the transformation to the W centre of mass is not trivial. The problem is solved imposing to the ev invariant mass to be equal to M_W. The problem has two solutions for p_L^W but in the great majority of the events only one is physical. Keeping these events and further requiring that the electron charge is well measured we are left with 149 events. The (corrected) angular distribution for this sample is shown in Fig. 17 compared with the prediction of the V-A structure. Notice that non conservation of parity is observed, but V-A is not proven. In fact V+A would give exactly the same prediction.

More generally assuming an arbitrary spin J for the W it can be shown[17] that the following predictions hold:

$$
\begin{array}{lll}
J = 1 \text{ and } V \pm A & : & \langle \cos \vartheta^* \rangle = \dfrac{1}{2} \\
J = 0 & : & \langle \cos \vartheta^* \rangle = 0 \\
J > 1 & : & \langle \cos \vartheta^* \rangle \le \dfrac{1}{6}
\end{array}
\qquad (29)
$$

From the data in Fig. 17 we can calculate:

$$\langle \cos \vartheta^* \rangle = 0.43 \pm 0.07$$

proving both $J = 1$ and $V \pm A$ coupling (maximum helicity).

The situation is different for the Z^0, that couples to fermions, f, of both helicities. The Z^0 $f\bar{f}$ vertex is in fact proportional to:

$$\frac{g}{\cos \vartheta_W} \gamma_\mu \frac{1}{2} \left(C_V^f - C_A^f \gamma_5 \right)$$

where C_V^f and C_A^f are the vector and axial couplings of fermion f respectively. For the final state $e^+ e^-$, $C_A^e = -1/2$, $C_V^e = -1/2 + 2 \mathrm{sen}^2 \vartheta_W = -0.03$. Should $\mathrm{sen}^2 \vartheta_W$ be equal to 1/4, $C_V^e = 0$ and the coupling would have been purely axial, the angular distribution symmetric. The smallness of C_V^e produces a slightly asymmetric angular distribution. To calculate $d\sigma/d\cos\vartheta^*$ we must remind that two processes contribute:

$$u\bar{u} \to Z^0 \to e^+ e^- \qquad ; \qquad d\bar{d} \to Z^0 \to e^+ e^-$$

For each of them:

$$\frac{d\sigma}{d \cos \vartheta^*} = \left(C_V^{i\,2} + C_A^{i\,2} \right) \left(C_V^{e\,2} + C_A^{e\,2} \right) (1 + \cos^2 \vartheta^*) + 8 C_V^i C_A^i C_V^e C_A^e \cos \vartheta^* \qquad (30)$$

where i, meaning initial state, can be u or d. The last term gives the asymmetry, depending on C_V^i, C_A^i that are known functions of the Weinberg angle. Fig.18 shows the experimental distribution, where a small asymmetry is observed. If $N^+(N^-)$ is the number of events with $\cos\vartheta^* > 0 \ (<0)$ we have:

$$A = \frac{N_+ - N_-}{N_+ + N_-} = 0.30 \pm 0.15 \tag{31}$$

This value can be compared with the prediction based in the weighted average of the contribution of the two processes (30) obtaining a measurement of the Weinberg angle.

$$\mathrm{sen}^2\vartheta_W = 0.18 \pm 0.04 \tag{32}$$

We have still another independent way to measure this important quantity, that will give another check when statistics will be more adequate.

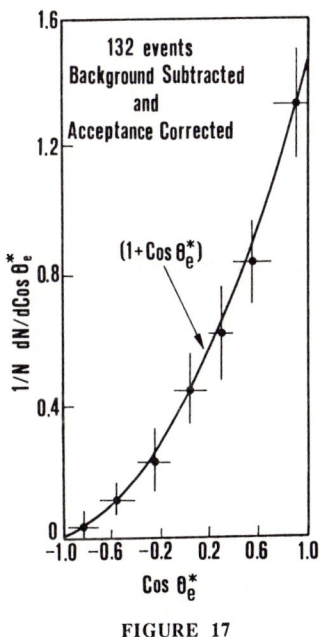

FIGURE 17

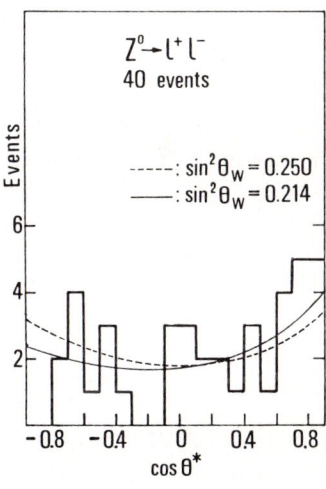

FIGURE 18

4.5 $W \to \mu\nu$ AND $Z^0 \to \mu^+\mu^-$

Universality requires that equal rates of the IVB's are observed in the electron and muon channels. Experimentally, as discussed above, electron and muons are detected in completely different ways with different backgrounds. This allows a powerful check of the results.

To select the W → μν sample[18,14] an isolated track is required in the external drift chambers matching a high momentum track in the C.D. with p_T^μ > 15 GeV and a missing momentum p_T^{miss} > 15 GeV. The resulting sample consists on 57 events with negligible background. The reduced events number with respect to the electron channel is due to the limited solid angle covered by the muon trigger. As described in Chapter 3 the momentum of the muon is measured with a sizeable error, about 25% at p = 50 GeV. More precisely the measured, gaussian, quantity is 1/p, hence we consider the inverse transverse mass $1/M_T$. Its distribution is shown in Fig. 19. Fitting the expected shape of the distribution we obtain[14]:

$$M_W = 81.8^{+6.0}_{-5.3}(\text{stat}) \pm 2.6(\text{syst}) \text{ GeV} \tag{33}$$

where the systematic error is due to the uncertainties in the correction of the Central Detector distortions.

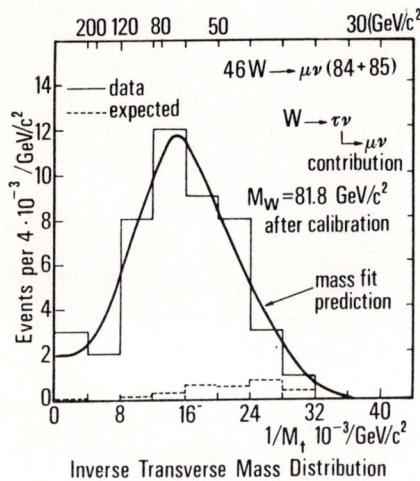

FIGURE 19
Inverse transverse mass distribution for W → μν candidates. Solid curve is the mass fit prediction with M_W = 81.8 GeV.

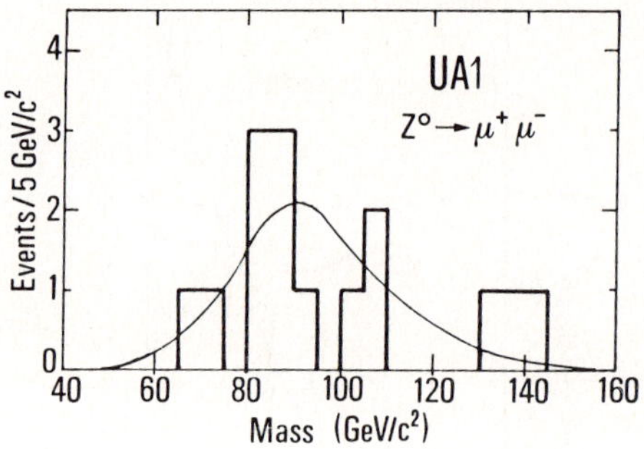

FIGURE 20
The dimuon invariant mass distribution, before the use of momentum balance, for the 18 events for which both μ tracks are well measured. Three events are in over flow. Curve is the fit with M_Z = 94.1 GeV.

Requiring the presence of a second muon with p_T^μ > 15 GeV and such that the dimuon invariant mass $M_{\mu\mu}$ > 40 GeV UA1 gets[14] a sample of 18 background free Z^0 events. The $M_{\mu\mu}$ distribution is shown in Fig. 20. Note the asymmetry due to the fact that the resolution is Gaussian in 1/p. The fit gives:

$$M_Z = 94.1^{+8.4}_{-6.6}(\text{stat}) \pm 2.8\,(\text{syst})\,\text{GeV} \tag{34a}$$

By imposing energy-momentum conservation to the whole event a more precise value is obtained:

$$M_Z = 90.7^{+5.2}_{-4.8} \pm 3.2\,\text{GeV} \tag{34b}$$

The muon channel allows independent measurements of the W and Z masses. They are in complete agreement with the (more precise) values obtained in the electron channel.

4.6 $W \rightarrow \tau\nu$, THE MISSING ENERGY ROUTE

The hermeticity of the UA1 detector allows a reliable determination of the transverse components of the missing momentum vector $\vec{p}_T^{\,mis}$. As we have seen in Chapter 3 the resolution function on p_T^{mis} is gaussian with $\sigma = 0.7\sqrt{|E_T|}$. Events with high p_T^{mis} are extremely interesting because they contain high p_T neutrinos or, possibly, other non-interacting particles (like photinos, $\tilde{\gamma}$) that should be evidence for new physics. Fig. 21 summarizes the expectations.

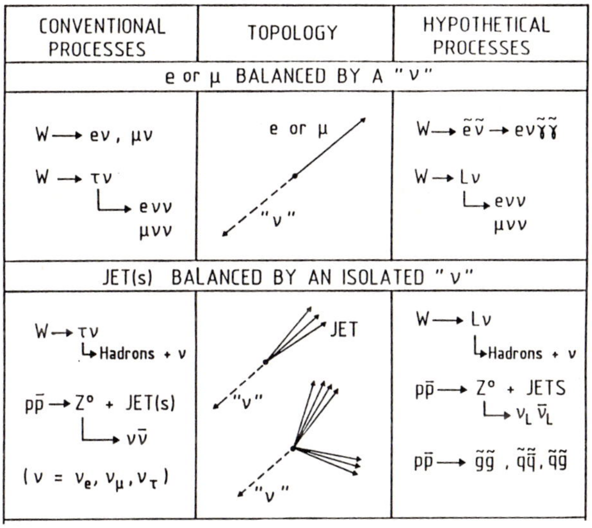

CONVENTIONAL PROCESSES	TOPOLOGY	HYPOTHETICAL PROCESSES
e or μ BALANCED BY A "ν"		

$W \rightarrow e\nu, \mu\nu$
 $W \rightarrow \tau\nu$
     $\hookrightarrow e\nu\nu$
     $\mu\nu\nu$
	e or μ /
	   /
	 /"ν"
		$W \rightarrow \tilde{e}\tilde{\nu} \rightarrow e\nu\tilde{\gamma}\tilde{\gamma}$
		 $W \rightarrow \tilde{L}\nu$
		     $\hookrightarrow e\nu\nu$
		     $\mu\nu\nu$

JET(s) BALANCED BY AN ISOLATED "ν"		

$W \rightarrow \tau\nu$
     $\hookrightarrow$ Hadrons + ν
 $p\bar{p} \rightarrow Z^0 + \text{JET(s)}$
     $\hookrightarrow \nu\bar{\nu}$
 ($\nu = \nu_e, \nu_\mu, \nu_\tau$)
	JET
	 "ν"
	 "ν"
		$W \rightarrow \tilde{L}\nu$
		     $\hookrightarrow$ Hadrons + ν
		 $p\bar{p} \rightarrow Z^0 + \text{JETS}$
		     $\hookrightarrow \nu_L \bar{\nu}_L$
		 $p\bar{p} \rightarrow \tilde{g}\tilde{g}, \tilde{q}\tilde{q}, \tilde{q}\tilde{g}$

FIGURE 21

The first two topologies, high p_T electron or muon and high p_T^{mis} at large angle, are those previously considered for $W \rightarrow e\nu$ and $\mu\nu$. Unknown processes as W decays in scalar electron or muon (predicted by supersymmetry) could contribute (see Sect. 7.2). In the next row "monojet" events are considered: they contain high p_T hadronic jets opposed to p_T^{mis}. We expect contributions from $W \rightarrow \tau\nu$ with hadronic decay of the τ; to monojets and multijets we expect also

In conclusion the W is observed also in the τν final state with a value of the mass fully consistent with expectations.

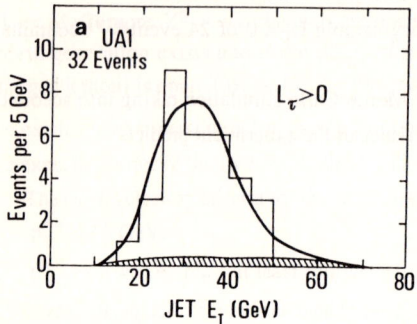
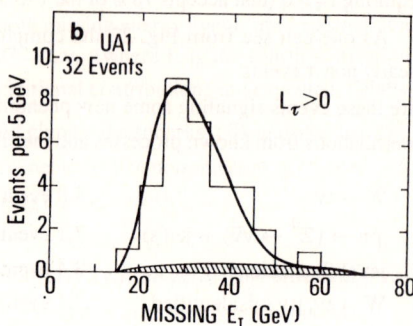

FIGURE 24
a) Jet transverse energy distribution for events passing the τ selection cut $L_\tau > 0$ (histogram) compared with the prediction of the Monte Carlo for tau's from W decay (28.7 events, upper curve) and the predicted background processes (2.7 events, shaded).
b) Missing transverse energy distribution for events passing the τ selection cut $L_\tau > 0$

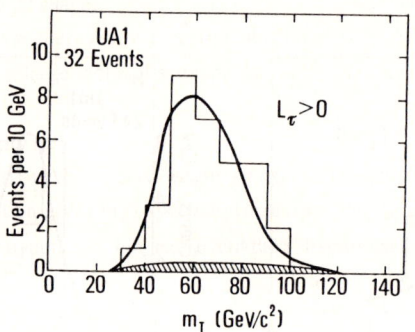

FIGURE 25
Transverse mass distribution for events passing the τ selection cut $L_\tau > 0$ (histogram) compared with the prediction of the Monte Carlo for tau's from W decay and background processes.

5. PRODUCTION PROPERTIES OF W AND Z

5.1 THE CROSS SECTIONS

The cross sections for W production followed by the decays into eν, μν and τν and for Z^0 production followed by the decays into e^+e^- and $\mu^+\mu^-$ are calculated from the events in the sample discussed in the previous chapter subtracting the background contributions and correcting for detector acceptance and for selection criteria inefficiency. The total luminosity, continuously monitored during the data taking runs, is known with ±15% uncertainty that gives a corresponding systematic uncertainty on cross section values. Fig. 26 shows the measured cross sections[12] compared with the theoretical QCD predictions of Altarelli et al.[5]; the uncertainties on these arise from the uncertainties on the structure functions, from the ambiguities on the choice of the Q^2 scale and from the not yet evaluated contribution of higher order terms. The UA2 results[21] are also in good agreement. To give the reader a feeling of the sensitivity of the theory to the input structure functions, we note that if we use the new measurement of $F_2^{\mu p}(x)$ from BCDMS[22] the predicted cross sections increase by 20% in even better agreement with the measured values.

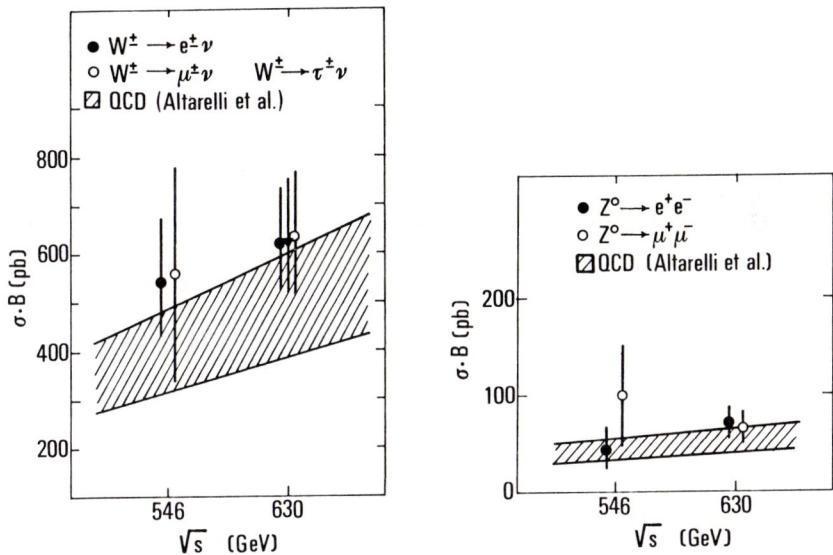

FIGURE 26
W and Z production cross sections times leptonic branching ratio as measured by UA1 compared with QCD predictions (shaded area).

5.2 TEST OF THE UNIVERSALITY

Independently on QCD calculations the ratio of the cross sections in the different leptonic channels gives a test of universality of the leptonic coupling at the energy scale M_W/M_Z.

The square roots of the rations of cross sections give the ratios of the coupling constants. For the charged currents (W)[20]:

$g_\mu/g_e = 1.00 \pm 0.07$ (stat) ± 0.04 (syst)   (37a)

$g_\tau/g_e = 1.01 \pm 0.10$ (stat) ± 0.06 (syst)   (37b)

and for the neutral currents (Z)[20]:

$k_\mu/k_e = 1.02 \pm 0.15 \pm 0.04$ (38)

The uncertainty on the luminosity obviously cancels in the ratios.

5.3 NUMBER OF COLOURS AND NUMBER OF NEUTRINO SPECIES

The theoretical calculation mentioned in Sect. 5.1 depends also on: 1) the value of the top mass (assumed to be $M_t = 40$ GeV); varying M_t the changes of the cross sections are small compared to other uncertainties; 2) the number of colours N_c; 3) the number of neutrino species N_ν. The number of colours enters in the calculation of the cross sections in two ways: the probability for a quark to meet an antiquark of the same colour in the initial state is $1/N_c$, the total widths of the IVB depend on N_c (the number of decay channels depends on N_c). We have:

$$\sigma B_{ev}(\bar{p}p \to W \to e\nu) = \frac{1}{N_c} \sum_{(q\bar{q})} \sigma(q\bar{q} \to W) \frac{\Gamma(W \to e\nu)}{\Gamma_W^{tot}(N_c)}$$

In a first approximation the branching ratio can be evaluated by counting the final states:

$$\frac{\Gamma(W \to e\nu)}{\Gamma_W^{tot}} \approx \frac{1}{N_L + N_c \cdot N_Q}$$

where N_L and N_Q are the numbers of lepton and quark families respectively. The correct calculation for $N_c = 2$ and $N_c = 4$ (the two cases nearer the expected $N_c = 3$ value) gives:

$$\frac{\sigma \cdot B(N_c = 2)}{\sigma \cdot B(N_c = 3)} = 1.9 \quad ; \quad \frac{\sigma \cdot B(N_c = 4)}{\sigma \cdot B(N_c = 3)} = 0.6 \qquad (39)$$

both values being inconsistent with data[20] at 90% c.l. we have a further test of QCD assumption $N_c = 3$.

Up to now we have seen that the collider data are in very good agreement with Electroweak theory and with QCD. The Standard Model does not make any prediction on the number of fermion families nor on the masses. Two important questions are open: what is the top mass and how many families are realized in Nature. We will start from the second question observing that it is equivalent to ask how many are the neutrino species, N_ν. An indirect limit exists based in the comparison of the He/H abundance ratio in the Universe and the calculations in the Big-Bang model. Two calculations give upper limits[23] $N_\nu < 4$ and $N_\nu < 6$ valid for neutrino masses not bigger than 1 MeV.

A more direct evaluation, valid for $M_\nu \ll M_Z/2$, can be obtained from the measurement of Γ_Z. In fact, in the assumption that charged leptons and quarks of further generations are so heavy that do not contribute to Z^0 decays, the total width of the Z^0 is:

$$\Gamma_Z = \Gamma_Z(3 \text{ gen}) + (N_\nu - 3) \Gamma_Z(\nu\bar{\nu}) \qquad (40)$$

where $\Gamma_Z(3 \text{ gen})$ is the total width expected for 3 generations and $\Gamma_Z(\nu\bar{\nu}) = 170$ MeV is the expected contribution of each neutrino species.

Unfortunately a measurement of Γ_Z does not exist yet (it will come from LEP) but only the upper limits (18). From them we have

$$N_v \leq 16 \qquad (41)$$

Much more sensitive limits are obtained starting from the observation that the number of $W \to l\nu$ where l means a charged lepton, and $Z^0 \to ll$ decays are sensitive to additional open channels like $Z^0 \to \nu\bar{\nu}$. As originally suggested by N. Cabibbo [24] the ratio

$$R = \frac{\sigma(W \to l\nu)}{\sigma(Z \to ll)} = \frac{\sigma_W}{\sigma_Z} \frac{BR(W \to l\nu)}{BR(Z \to ll)} \qquad (42)$$

is free from many theoretical uncertainties and from the experimental uncertainty in the luminosity.

The R parameter can be written as the product of three factors:

$$R = \frac{\Gamma(W \to l\nu)}{\Gamma(Z \to ll)} \cdot \frac{\Gamma_Z}{\Gamma_W} \cdot \frac{\sigma_W}{\sigma_Z} \qquad (43)$$

The first factor, the ratio of the partial leptonic widths is well known from Electroweak theory; the second factor, the ratio of the total widths, can be reliably calculated when all the different decay channels are known. Let us assume for the moment that charged leptons and quarks of possible heavier families are heavy enough that their contributions to W and Z^0 widths can be neglected and that all the neutrinos of these families are light enough that no phase space suppression takes place in the Z^0 decay ($M_\nu \ll M_Z/2$). Γ_W then depends only on M_t (it is a decreasing function of M_t) and Γ_Z depends essentially only on N_v (it is an increasing function of N_v). In principle Γ_Z depends also on M_t up to $M_t = M_Z/2$ where the $t\bar{t}$ channel closes; we can ignore this dependence, given the experimental lower limit $M_t > 44$ GeV (see Chapter 7).

The third factor in (43), the cross section ratio, is the most uncertain as its calculation is based on QCD. The main uncertainties are due to the structure functions and the Weinberg angle. The best estimate is[20]:

$$\frac{\sigma_W}{\sigma_Z} = 3.23 \pm 0.10 \qquad (44)$$

The R parameter has been measured by UA1[20] and UA2 [21]:

UA1 $\qquad R = 9.1^{+1.7}_{-1.2} \qquad ; \qquad R < 11.5$ (90% c.l.) $\qquad (45)$

UA2 $\qquad R = 7.2^{+1.7}_{-1.2} \qquad ; \qquad R < 9.5$ (90% c.l.) $\qquad (45)$

The results are consistent and can be combined giving:

$$R = 8.4^{+1.2}_{-0.9} \quad ; \quad R < 10.1 \ (90\% \ c.l.) \quad (46)$$

The upper limit on R can be translated in an upper limit, function of M_t, on Γ_Z using formula (43) hence in an upper limit (again a function of M_t) on N_ν using (40). The result is shown in Fig. 27 where $\sigma_W/\sigma_Z = 3.13$ as been taken (the less constraining value at one standard deviation from the most probable value in (44)).

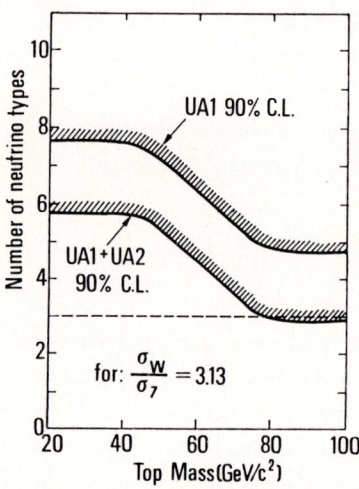

FIGURE 27
Limits (90% c.l.) from UA1 and UA1 + UA2 data on the number of neutrinos as a function of M_t.

Taking into account the above mentioned lower limit on M_t we conclude that[20]:

$N_\nu < 5.7$ at 90% c.l.

that, being N_ν necessarily integer, means:

$$N_\nu \leq 5 \quad (47)$$

The limit becomes even more stringent if M_t is bigger. In particular if $M_t > 75$ GeV the existence of a fourth neutrino flavour (a fourth generation) is highly unlikely.

If we now allow for the decay $W \to L\nu_L$, where L is the fourth generation lepton and with mass $M_L > 41$ GeV (as we will see in Sect. 7.1) the limit becomes worse by one neutrino species.

A similar method has been used to place an upper limit on N_ν at e^+e^- machines. The process $e^+e^- \to Z^0\gamma \to \nu\bar{\nu}\gamma$ (with virtual Z^0) gives origin to the peculiar final state: one γ + missing energy. The cross section is perfectly calculable and is a function of N_v. The upper limit obtained combining the data of the experiments ASP, MAC and CELLO[25] is $N_v < 4.9$ (90% c.l.). Again little space is allowed for not yet known neutrino flavours.

5.4 W TRANSVERSE MOTION

Experimentally events with a W emitted at moderate or high transverse momentum (p_T^W) are observed. In all of them p_T^W is balanced by the transverse momentum of one or (more rarely) more jets. In the QCD interpretation the production process of the W via quark-antiquark annihilation discussed in Chapter 2 is only the lowest order process; at higher orders gluon radiation from one of the initial quarks is expected giving exactly the observed topology.

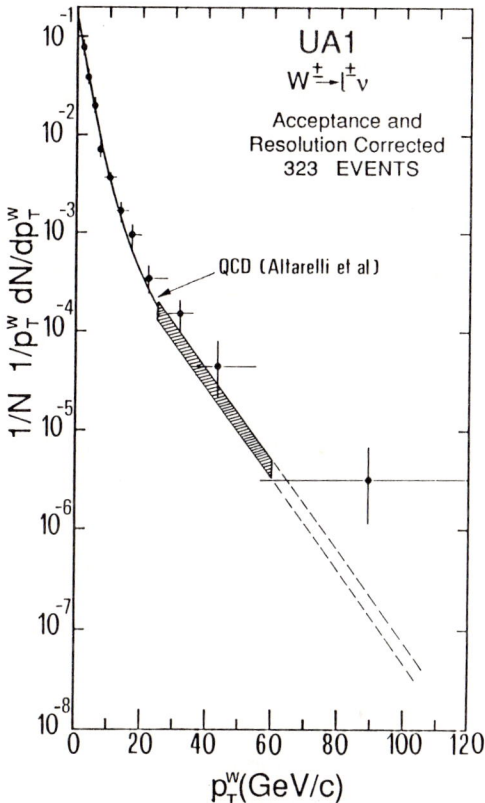

FIGURE 28

W transverse momentum distribution for both $W \to e\nu$ and $W \to \mu\nu$ events.

At sufficiently high values of p_T^W perturbative QCD calculations are feasible and in fact have been done by G. Altarelli et al.[5] and a quantitative test of the theory is possible.

$$I = \left[\left(\sum E_T / 3 \right)^2 + \left(\sum p_T / 2 \right)^2 \right]^{1/2}$$

in a cone of $\Delta R = 0.7$. The study of this variable allows the optimization of the selection criteria.

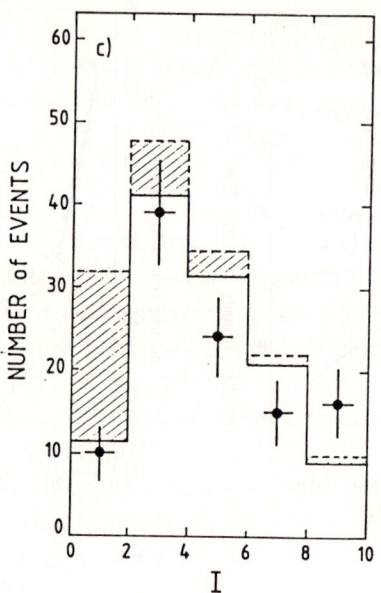

FIGURE 31
Isolation distribution for the sample described in the text. Histogram gives the predictions not including top. Shaded area gives the expected contribution for top with $M_t = 30$ GeV.

Fig. 31 shows the distribution of I for the sample with at least two jets, $p_T^\mu > 12$ GeV, E_T of the highest energy jet > 12 GeV and transverse invariant mass $M_T (\mu \nu) < 40$ GeV. The histogram is the result of the simulation and is seen to account perfectly for the data.

As the data do not require any top contribution and as the top production is a decreasing function of M_t we can put a lower limit on M_t. Again with a Monte Carlo calculation we evaluate the expected contribution of the two top channels for different values of M_t. The results for $M_t = 30$ GeV is shown hatched in Fig. 31. It is clear that this value is not compatible with the data. Fig. 32 shows the experimental upper limits on t the $t\bar{t}$ and $t\bar{b}$ cross sections compared with the expected values as functions of M_t. The crossing points give the lower limits (at 90% and 95% c.l.) for M_t. We note that the contribution to top production of W decays is estimated with high reliability (and is shown dashed in Fig. 32), on the contrary the QCD calculation for $t\bar{t}$ production is uncertain probably by a $\pm 50\%$. This problem is illustrated in Fig. 33, where the ratios between various QCD cross section calculations (with different choices of structure functions and Q^2 scale) and a particular one (σ_0) are reported as a function of M_t. The

energy. The cross section is perfectly calculable and is a function of N_v. The upper limit obtained combining the data of the experiments ASP, MAC and CELLO[25] is $N_v < 4.9$ (90% c.l.). Again little space is allowed for not yet known neutrino flavours.

5.4 W TRANSVERSE MOTION

Experimentally events with a W emitted at moderate or high transverse momentum $\left(p_T^W\right)$ are observed. In all of them p_T^W is balanced by the transverse momentum of one or (more rarely) more jets. In the QCD interpretation the production process of the W via quark-antiquark annihilation discussed in Chapter 2 is only the lowest order process; at higher orders gluon radiation from one of the initial quarks is expected giving exactly the observed topology.

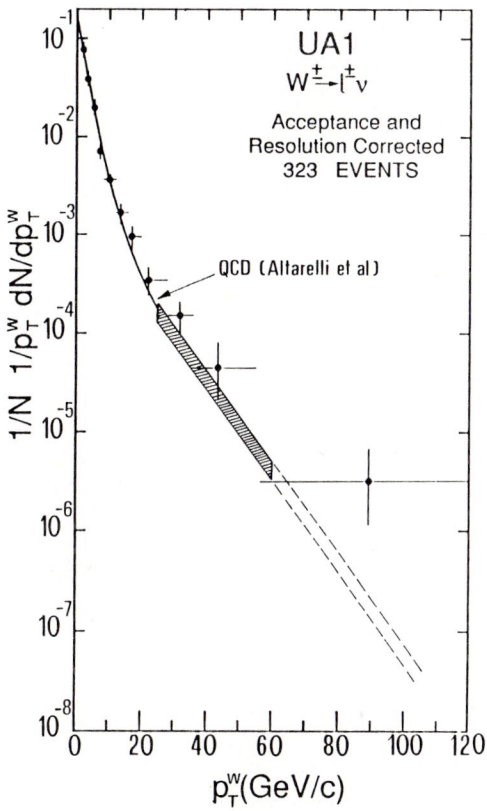

FIGURE 28
W transverse momentum distribution for both $W \to e\nu$ and $W \to \mu\nu$ events.

At sufficiently high values of p_T^W perturbative QCD calculations are feasible and in fact have been done by G. Altarelli et al.[5] and a quantitative test of the theory is possible.

Fig. 28 shows the invariant differential cross section[26] compared with the QCD calculations. Data (323 W leptonic decays) have been corrected for detection and selection efficiency and resolution. We observe very good agreement with the theory. At very high p_T^W the agreement is not perfect: for $p_T^W > 80$ GeV 0.2 ± 0.06 events are expected, 2 are observed. This is not of course an inconsistency but a closer observation shows that their topology (both events contain two jets with invariant mass near to M_W) is very unlikely in the QCD calculation. More statistics is needed to clarify this point, **the only found where the agreement with the S.M. is not perfect.**

6. THE SEARCH FOR TOP

The only not yet observed quark of the three known families is the top quark. Its existence is required to cancel flavour-changing neutral currents as necessary, in the specific case, to explain the experimental upper limit for the branching ratio of the B^0 into $\mu^+\mu^-$ anything of 10^{-3}. The theory des not give any prediction for M_t, but the theoretical upper limit $M_t < 180$ GeV can be obtained, as shown in Sect. 4.3.

An experimental lower limit, $M_t > 25$ GeV, comes from the non observation of top at the highest energy e^+e^- collider, TRISTAN[27].

At a $p\bar{p}$ collider two processes are expected to give rise to top[28]: through W decay into $t\bar{b}$ (or $\bar{t}b$) and through direct strong (QCD) production of $t\bar{t}$. To have a not too unfavourable signal to background ratio, semileptonic decays of the t:

$$t \to \mu b \nu \qquad \text{or} \qquad t \to e b \nu$$

that are expected to have branching ratios of ~1/9, must be considered. In both cases the expected topology is: one high p_T lepton (μ or e), two hadronic jets (one for $\bar{b}$ produced with t, one for b of the decay of the t) and p_T^{mis}. Fig. 29 shows, on the left scale, the expected number of events (before any selection cut).

The principal physical background is due to the processes:

$$p\bar{p} \to b\bar{b}g \quad ; \quad b \to \mu \nu c \quad ; \quad \bar{b} \to \text{jet} \quad ; \quad g \to \text{jet}$$
$$\bar{p}p \to c\bar{c}g \quad ; \quad c \to \mu \nu s \quad ; \quad \bar{c} \to \text{jet} \quad ; \quad g \to \text{jet} \qquad (48)$$

Fig. 30 shows that the signal is expected to be one order of magnitude less than the background. It is then necessary to apply selection criteria to enhance the relative top contribution. We must also observe that p_T^{mis} is in this case much smaller than for $W \to l\nu$ (its value is in fact of the same order of the experimental resolution) and that p_T^l is of the order of 10 - 20 GeV, a region where the identification of both electron and muon is difficult; accurate control of both the instrumental and the physical backgrounds is mandatory. Note also that the topology under study contains jets and that the lepton from the t decay is often not far from the b jet; we will then start considering the μ channel that is experimentally cleaner.

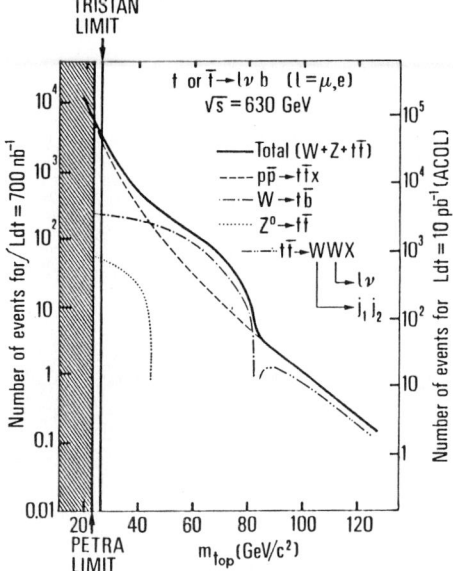

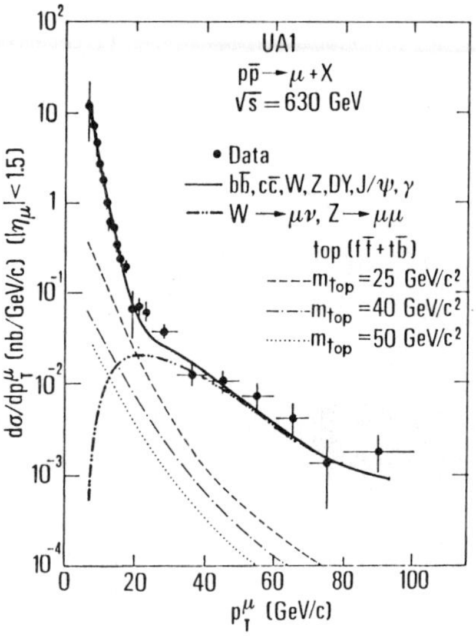

FIGURE 29
Expected numbers of events from top semileptonic decay as functions of M_t.

FIGURE 30
Inclusive μ production cross section and expected top contribution.

The t decay into μ b ν is much more "esothermic" than the muonic decay of the b or the c (due to the high value of M_t). As a consequence μ's from t are much more isolated than those from the background (48). The isolation variable is then of great importance to enhance signal to background ratio.

Simulation calculations have been done to evaluate the expected contributions of the physical and instrumental backgrounds. Their results have been checked on a sample of events selected to consist essentially of background. The selection criteria (one μ and two jets, $10 < p_T^\mu < 15$ GeV, transverse energy of the highest energy jet bigger than 12 GeV) provide 858 events. The Monte Carlo calculations include contributions from known processes and predict 380 ± 5 events from processes (48), 235 ± 22 events from π and K decays, 24 ± 1.5 events from W, Z, J/ψ etc. for a total of expected 639 ± 23 events in reasonable agreement with the observations. More important, the distributions of all the relevant kinematical variables have been checked and agreement between data and simulation always observed, showing that the background is fully understood.

Let us now define an isolation variable I that takes into account both the C.D. measurements (p_T) and the calorimeter measurements (E_T)

$$I = \left[\left(\sum E_T/3\right)^2 + \left(\sum p_T/2\right)^2 \right]^{1/2}$$

in a cone of $\Delta R = 0.7$. The study of this variable allows the optimization of the selection criteria.

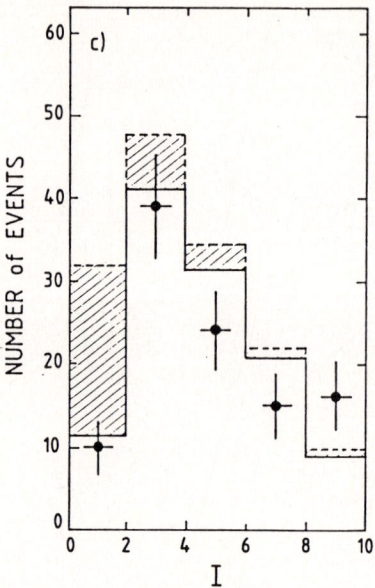

FIGURE 31
Isolation distribution for the sample described in the text. Histogram gives the predictions not including top. Shaded area gives the expected contribution for top with $M_t = 30$ GeV.

Fig. 31 shows the distribution of I for the sample with at least two jets, $p_T^\mu > 12$ GeV, E_T of the highest energy jet > 12 GeV and transverse invariant mass $M_T (\mu\nu) < 40$ GeV. The histogram is the result of the simulation and is seen to account perfectly for the data.

As the data do not require any top contribution and as the top production is a decreasing function of M_t we can put a lower limit on M_t. Again with a Monte Carlo calculation we evaluate the expected contribution of the two top channels for different values of M_t. The results for $M_t = 30$ GeV is shown hatched in Fig. 31. It is clear that this value is not compatible with the data. Fig. 32 shows the experimental upper limits on t the $t\bar{t}$ and $t\bar{b}$ cross sections compared with the expected values as functions of M_t. The crossing points give the lower limits (at 90% and 95% c.l.) for M_t. We note that the contribution to top production of W decays is estimated with high reliability (and is shown dashed in Fig. 32), on the contrary the QCD calculation for $t\bar{t}$ production is uncertain probably by a $\pm 50\%$. This problem is illustrated in Fig. 33, where the ratios between various QCD cross section calculations (with different choices of structure functions and Q^2 scale) and a particular one (σ_0) are reported as a function of M_t. The experimental upper limit curve (95% c.l.) is obtained from the complete information of the muon and electron (not discussed here) channels. The most pessimistic limit[22] is obtained from the lowest theoretical prediction:

$M_t > 44$ GeV   (95% c.l.)   (49)

This limit puts the top search out of the reach of LEPI and SLC. The increase in statistic expected with ACOL and upgrade of UA1 and UA2 will allow both experiments to push the limit up to 70 GeV or, hopefully, to discover the top.

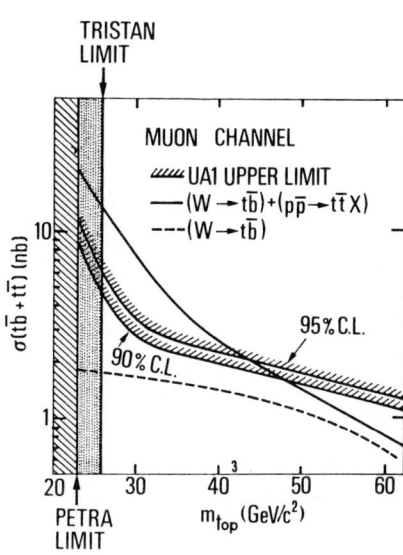
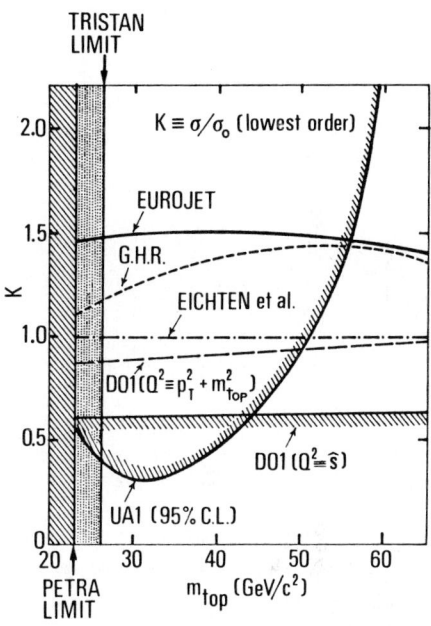

FIGURE 32
Confidence level contours in the top cross section versus M_t plane from μ channel. Solid curve refers to Monte Carlo (EUROJET) calculation.

FIGURE 33
Sensitivity of the mass limit to the ratio $K = \sigma(t\bar{t})/\sigma_0$ where σ_0 is the lowest order cross section calculation using EHLQI structure functions and $Q^2 = M_t^2 + p_T^2 (K = 1)$. The choice of structure function D01, $Q^2 = \hat{s}$ gives the lowest cross section.

7. SEARCH FOR EXTENSIONS AND MODIFICATIONS OF THE STANDARD MODEL

7.1   SEARCH FOR HEAVY LEPTON

As mentioned before the Standard Model does not specify the number of quark and lepton families. Direct searches of the fourth generation charged lepton L in $e^+ e^-$ collisions have placed a lower limit on its mass $M_L > 22.7$ GeV (90% c.l.). At the $\bar{p}p$ collider the heavy lepton is expected to be present in the W decay $W \to L\nu_L$ (with universal coupling), giving a contribution to the high

E_T^{mis} sample (when the L decays semihadronically). As discussed in Sect. 4.5 no such contribution is necessary as the data are completely explained by standard sources. As the number of events expected from the heavy lepton is a decreasing function of its mass, we can put a lower limit on M_L.

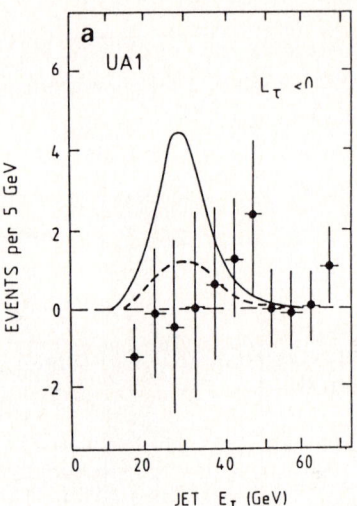

FIGURE 34
Jet transverse energy distribution for background-subtracted $L_\tau < 0$ events. Solid line is the expected contribution of heavy lepton with $M_L = 35$ GeV, dashed line for $M_L = 55$ GeV.

The heavy lepton contribution was evaluated with a Monte Carlo calculation with full detector simulation, including trigger and selection criteria acceptance[19]. The simulation shows that the process populates preferentially the region $L_t < 0$, E_T (jet) < 40 GeV that is then considered for the analysis. In this region we expect 16 events for $M_L = 25$ GeV, 7.6 events for $M_L = 45$ GeV and 1.4 events for $M_L = 65$ GeV. The additional contributions to the E_T (jet) distribution, after subtraction of standard sources, is shown in Fig. 34 for $M_L = 35$ GeV and $M_L = 55$ GeV. Taking into account the contribution of Z^0 + jet followed by $Z_0 \to \nu_L \bar{\nu}_L$ we derive the limit [19]

$M_L > 41$ GeV   (90% c.l.)   (50)

The upgraded experiments UA1 and UA2 with the increased luminosity of ACOL are expected to move the limit to about 70 GeV where the channel $W \to L\nu_L$ starts closing.

7.2 SEARCH FOR SUPERSYMMETRIC PARTICLES

Theoretical speculations predict the existence for each known particle of a supersymmetric partner. In many models the lightest supersymmetric particle, the photino ($\tilde{\gamma}$), is stable and behaves much like a neutrino, non interacting with matter. "SUSY" particles are as a consequence expected in final states with large missing p_T. Again the non observation of any unexpected contribution in the large p_T^{mis} sample allows us to put limits on the SUSY particles masses. (SUSY theory gives no prediction for the masses but specifies the coupling constants).

Given the richness of the hypothesized supersymmetric zoo a specific model must be used. We will assume that the 5 lower mass squarks $(\tilde{q})$ have the same mass $M_{\tilde{q}}$ and no contribution from $\tilde{H}, \tilde{Z}, \tilde{W}, \tilde{t}$. We will also assume the photino to be massless and we will consider the strong production of **squarks** and **gluinos** $(\tilde{g})$ through the following processes:

$$\bar{p}p \to \tilde{g}\tilde{g} + \text{anything}$$
$$\to \tilde{q}\tilde{q} + \text{anything}$$
$$\to \tilde{g}\tilde{q}(\tilde{g}\tilde{q}) + \text{anything}$$

that have cross sections calculable by QCD for given values of $M_{\tilde{g}}$ and $M_{\tilde{q}}$. Gluinos and squarks are not stable; in the case $M_{\tilde{q}} > M_{\tilde{g}}$, the decays are $\tilde{q} \to q\tilde{g}$, $\tilde{g} \to q\bar{q}\tilde{\gamma}$, in the case $M_{\tilde{q}} < M_{\tilde{g}}$ the decays are $\tilde{q} \to q\tilde{\gamma}$, $\tilde{g} \to \bar{q}\tilde{q}$ or $q\tilde{\bar{q}}$. In both cases the event topology consists on a number of hadronic jets from fragmentation of quarks and on p_T^{mis} from the $\tilde{\gamma}$.

Fig. 35 shows the bounds[30] obtained in the plane $M_{\tilde{q}}$, $M_{\tilde{g}}$ for $M_{\tilde{q}} \gg M_{\tilde{g}}$ the bound is $M_{\tilde{g}} > 53$ GeV, for $M_{\tilde{g}} \gg M_{\tilde{q}}$, $M_{\tilde{q}} > 45$ GeV.

These limits are expected to be moved to about 100 GeV with the increased luminosity of ACOL and to 200 GeV at the Tevatron. This shows the importance of the available energy.

Supersymmetric leptons, if light enough, should be produced by W (and Z^0) decays. In particular the scalar electron $\tilde{e}$ should be observed in the decay:

$$W \to \tilde{e}\, \tilde{\nu}$$
$$\quad\ \hookrightarrow e\tilde{\gamma}$$

This process is completely calculable, for given values of the masses, because the couplings are predicted by the theory. Its topology is similar to the standard $W \to e\nu$ decay with a high p_T isolated electron and opposite p_T^{mis} (from the $\tilde{\nu}$ and $\tilde{\gamma}$). The difference is in the p_T^e distribution (no Jacobian peak) and in the ϑ^* distribution (substantially no asymmetry). As these distributions are found in perfect agreement with the standard theory the lower limits shown in Fig. 36 can be placed on the masses $M_{\tilde{\nu}}$ and $M_{\tilde{e}}$[13].

As a final observation we remind the reader that some theories postulate the existence of further IVB's analogous of W and Z at higher masses. No such event having been observed the following limits on the masses can be placed:

$$M_{W'} \geq 232 \text{ GeV} \quad ; \quad M_{Z'} \geq 188 \text{ GeV} \qquad \text{at 90\% c.l.} \qquad (51)$$

if W' and Z' have the same coupling of W and Z.

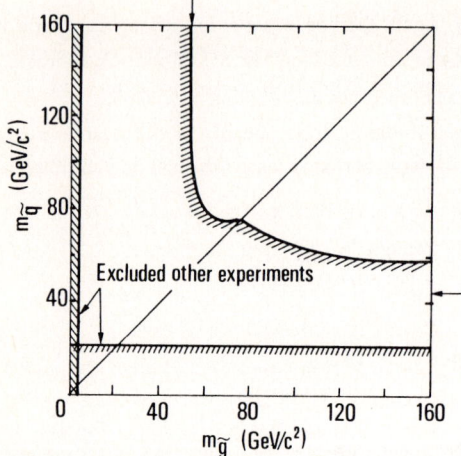 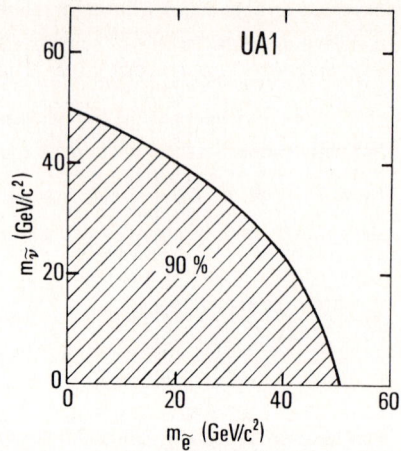

FIGURE 35
Limits (at 90% c.l.) on squark and gluino masses. The arrows indicate the asymptotic values of the 90% c.l. contour as the squark or gluino mass becomes infinitely large.

FIGURE 36
Limits (90% c.l.) on scalar electron and scalar neutrino masses.

8. LIMITS ON THE DIAMETER OF QUARKS

One of the very first observations of UA1 and UA2 was the clear evidence of jets in the high E_T events, where E_T is the scalar sum of the transverse energies of the particles emerging from the collision. In practice all the events with E_T larger than, say, 20 GeV contain jets. The great majority of the events contains two, opposite in azimuth jets; in about 10% of the cases a third jet is present, more rarely a fourth one.

We obviously interpret these events as due to hard scattering of two partons ($q, \bar{q}, g$) that then fragment in the final state in jets, occasionally gluon radiation gives origin to a further detectable jet. The spectators give also origin to jets, but at low p_T. The situation is depicted in Fig. 37; Fig. 8 gives an examples of a two-jet event.

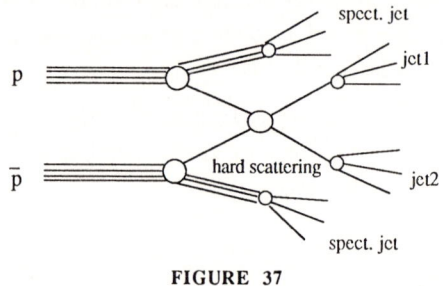

FIGURE 37

The jet identification is relatively easy at high p_T at collider energies. The following step is the identification of the energy and direction of the jet with those of the parton that has originated it. This involves systematic distortions that are calculated through Monte Carlo calculations. The procedure leaves of course uncertainties of both theoretical and experimental nature. Not withstanding these limitations jet physics is extremely interesting being a window on the elementary interactions at the parton level. Jet physics will not be discussed in these lectures, we will only mention one important result that provides a lower limit on the diameter of the quark.

If quarks are composite objects, they are expected to have a non zero diameter. The interaction between their constituents will be mediated by some mediator of high mass and will reduce to a contact term g^2/Λ_c at low energy. The diameter of the quark in this hypothesis is $1/\Lambda_c$.

From elementary optics it is well known that to resolve structures of diameter D the wavelength of the "light" should be $\lambda \leq 2D$. The resolving power achievable with a beam of momentum $p = 2\pi/\lambda$ is then $D_{min} = \pi/p$.

In the scattering of partons $p \approx \sqrt{\hat{s}}/2$ and, typically, $\sqrt{\hat{s}} = (1/6)\sqrt{s}$. At $\bar{p}p$ energy $\sqrt{s}$ the minimum observable substructure diameter is then:

$$D_{min} \approx \pi/p = 2\pi/\sqrt{\hat{s}} \approx 1/\sqrt{s}$$

We can select a sample of events dominated by the process[31]:

$$q\bar{q} \to q\bar{q} \to jet + jet$$

taking events with only two high E_T jets and with high invariant dijet mass, M_{jj}. In this way partons carrying a great fraction of the proton (or antiproton) momentum are selected, that are dominantly (valence) quarks and not gluons. The directions of the jets are well measured and are expected to be very near to the partons directions (the same is not true for the energies). We can then reliably measure the quark-quark scattering differential cross section. In a first approximation for point like quarks this is expected to be the Rutherford cross section:

$$d\sigma/d\cos\vartheta \cong \frac{\alpha_s^2}{\hat{s}}(1-\cos\vartheta)^{-2} \tag{52}$$

Any quark structure is expected to show up as an increased counting rate at large angles. This is exactly the same experiment as done on atoms as targets by Geiger and Marsden in 1909[32]. Before proceeding we define the useful variable:

$$\chi = (1+\cos\vartheta)/(1-\cos\vartheta) \tag{53}$$

that has a constant distribution for Rutherford scattering. Fig. 38 shows the measured $d\sigma/d\chi$ for dijet events with $240 < M_{jj} < 300$ GeV.

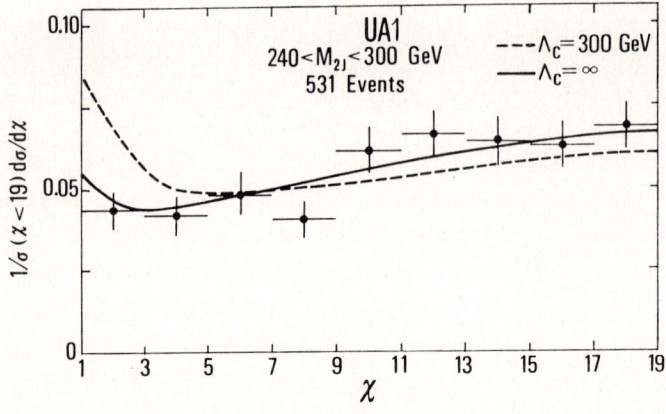

FIGURE 38
The angular distribution $(1/\sigma)(d\sigma/d\chi)$ for the very high mass $(240 < m_{2j} < 300$ GeV$)$ jet pairs shown as a function of χ, where $\chi = (1 + \cos\theta)$ and θ is the subprocess CM scattering angle. The solid curve represents the QCD prediction which corresponds to $\Lambda_c = \infty$, and is the best fit to the data. The dotted curve, which corresponds to $\Lambda_c = 300$ GeV, is clearly excluded by the data.

The continuous line gives the behaviour expected from QCD (that is similar but not exactly equal to Rutherford) and the expectations including and additional contact term with $\Lambda_c = 300$ GeV, clearly excluded by the data. On this basis the limit is:

$$\Lambda_c > 415 \text{ GeV} \quad (95\% \text{ c.l.}) \tag{54}$$

corresponding to a quark diameter:

$$D_q < 0.5 \text{ am} \quad (1 \text{ am} = 10^{-18} \text{ m}) \tag{55}$$

These are the most stringent limits on quarks dimensions. To have better limits **higher centre of mass energies** are needed. A limit $\Lambda_c > 2$ TeV ($D_q < 0.1$ am) is expected from Tevatron.

9. CONCLUSIONS

UA1 and UA2 working at the CERN $\bar{p}p$ collider have definitively shown that **precision physics** can be done with a hadronic collider.

The **standard model** has been verified in the force **mediators** (vector) **sector**. The spinor sector ($J = 1/2$) is also O.K. but we **still miss the top quark** ($M_t > 44$ GeV). The not yet discovered families cannot be many as indicated by independent limits: $N_\nu \leq 5$ or less if top is heavy, $M_L > 41$ **GeV**, mass splitting of the quark doublet increasing going to heavier families and on the other hand near to the limits allowed by the agreement between M_W and $\sen^2\vartheta_W$ for the third family.

Physics beyond the Standard Model has been searched for but no evidence for has been found. In particular limits on the masses of SUSY particles of the order of $M_Z/2$ have been established. No evidence of quark structure down to 0.5×10^{-18} m has been found.

In the next few years the upgraded UA1 and UA2 experiments are expected to collect 10 pb^{-1} integrated luminosity, an increase more than tenfold with respect to the present samples. The upgraded calorimetry will allow a precise measurement of M_W/M_Z, while M_Z will be precisely measured at LEP. More tight tests of the Standard Model will be done.

The top quark will also be discovered if it is lighter than 70 GeV. The limits for new phenomena will be pushed to higher energy but, in this particular sector, the higher centre of mass energy of the Tevatron collider will be winning.

The scalar sector (Higgs) is on the other hand probably outside the reach of the existing hadron colliders.

Other reviews of collider results are listed in ref. 33.

REFERENCES

1. C. Albajar et al., Phys. Lett. B186 (1987) 247.

2. Particle Data Group, Phys. Lett. 170B (1986) 1.

3. A. Sirlin, Phys. Rev. D22 (1980) 971.
   W.J. Marciano, Phys, Rev., D20 (1979) 274.
   W.J. Marciano and A. Sirlin, Phys. Rev. D29 (1984) 945 and D22 (1980) 2695.
   F. Antonelli et al., Phys. Lett. 91B (1980) 90.
   M. Veltman, Phys. Lett. 91B (1980) 95.

4. H. Abramowicz et al., Phys. Lett. 57 (1986) 298.
   F. Bergsma et al., CERN-EP/87-140 (1987). To be published in Z. Phys. C.

5. G. Altarelli et al., Nucl. Phys. B246 (1984) 12 and Z. Phys. C27 (1985) 617.

6. R. Ansari et al., Phys. Lett. 168B (1987) 452

7. M. Barranco Luque et al., Nucl. Inst. and Meth. 176 (1980) 175.
   M. Calvetti et al., Nucl. Inst. and Meth. 174 (1980) 285.
   M. Calvetti et al. IEEE Trans. Nucl. Sci. NS-30 (1983) 71.

8. C. Cochet et al., Nucl. Inst. and Meth. A243 (1986) 45.
   B. Andortet et al., Nucl. Inst. and Meth. 176 (1980) 115.
   C. Bacci et al., Nucl. Inst. and Meth. 200 (1982) 195.

9. M. Corden et al., Nucl. Inst. and Meth. A238 (1985) 273.

10. G. Bauer et al., Nucl Inst. and Meth. A253 (1987) 179.
    A. Bettini et al., Nucl Inst. and Meth. A253 (1987) 189.
    G. Bauer et al., Nucl Inst. and Meth. A260 (1987) 101.

11. K. Eggert et al., Nucl Inst. and Meth. 176 (1980) 223.

12. C. Albajar et al., Phys. Lett. 185 (1987) 233.

13. G. Arnison et al., Nuovo Cimento Lett. 44 (1985) 1.
    G. Arnison et al., Europh. Lett. 1 (1986) 327.

G. Arnison et al., Phys. Lett. 166B (1986) 484.

14. UA1 Collaboration, CERN-EP/87-154 (1987).

15. R. Ansari et al., Phys. Lett. 186B (1987) 440.
    R. Ansari et al., Erratum Phys. Lett. 190B (1987) 238.

16. U. Amaldi et al., Phys. Rev. D36 (1987) 1385.

17. M. Jacob, Nuovo Cimento Lett. 9 (1978) 826.

18. G. Arnison et al., Phys. Lett. 134B (1984) 469.

19. C. Albajar et al., Phys. Lett. 185B (1987) 241.

20. C. Albajar et al., Phys. Lett. 198B (1987) 271.

21. R. Ansari et al., Phys. Lett. 194B (1987) 158.

22. R. Voss, Rapporteur talk on Charged Lepton Interactions at 1987 International Symp. on Lepton and Photon Inter. at High Energies (Hamburg, July 1987) p. 1.

23. G. Steigman et al., Nature 300 (1981) 142.
    J. Ellis et al., Phys. Lett. 167B (1986) 457.

24. N. Cabibbo, "What Next ?" Third Topical Workshop on Proton-Antiproton Collisions held at Rome, Jan. 1983. Proceedings CERN 83-04 (1983) 567.

25. For a review see M. Davier: "Search for New Particles", Proceedings of the XXIII Int. Conf. on HEP, Bekely 1986, p. 25.

26. C. Albajar et al., Phys. Lett. 193B (1987) 389.

27. K. Amako et al.: "New Result from VENUS";
    A. Miyamoto et al.: "New Results from TOPAZ";
    Y. Sakai et al.: "New Results from AMY".
    Presented at " Physics in Collision VII" Tzukuba 1987.

28. C. Albajar et al. "Study of Heavy Flavour Production in Events with a Muon Accompanied by Jet(s) at the CERN Proton-Antiproton Collider", CERN-EP/87-18 and "Search for New Heavy Quarks at the CERN Proton-Antiproton Collider", CERN-EP/87-190. To be published on Z. Phys. C.

29. S. Komamiya in Proc. 1985 Intern. Symp. on Lepton and Photon Interactions at High Energies (Kjoto 1985).

30. C. Albajar et al., Phys. Lett. 198B (1987) 261.

31. G. Arnison et al., Phys. Lett. 177B (1986) 244.

32. Geiger and Mardsen. Proc. Roy. Soc. 82 (1909) 495.

33. C. Rubbia, Rev. Mod. Phys. 57 (1985) 699 (Nobel Lecture).
    Lectures at the Cargese Summer School 1987 by L. Di Lella, Proton-Antiproton Collider Physics: Experimental Aspects. CERN-EP/88/02 (1988) and by M. Jacob, " $p\bar{p}$ Collider Physics, present and prospects", CERN-TN-4813 (1987).
    T. Müller, Produiction Properties of Intermediate Vector Bosons W and Z at the CERN $p\bar{p}$ Collider, CERN-EP/88-48 (1988).

HADRONIC TOTAL CROSS SECTION MEASUREMENTS AT HIGH ENERGIES

Rino CASTALDI

INFN, Sezione di Pisa, Pisa, Italy

1. Introduction

The total cross section is one of the most fundamental parameters of hadron scattering processes and therefore it is one of the first measurements that must be performed as soon as a new accelerator makes available a new energy domain. The expectation that at high energies the strong interaction mechanism that controls hadron scattering should become simpler is perhaps only a prejudice which however has always made people believe that an asymptotic behavior can be finally reached and that the underlying dynamics can be more easily understood.

The unexpected rise of the proton-proton total cross section[1,2] discovered in 1973 at the CERN ISR has brought about a considerable change in the phenomenological interpretation of the asymptotic features of hadron scattering. Up to that time, in fact, total cross sections were believed to approach constant values with increasing energy. Actually cosmic-ray data[3] already suggested that the p-p total cross section could asymptotically rise with energy, furthermore, some theoretical models[4,5] had also considered the possibility of indefinitely rising hadronic cross sections. Nonetheless, the generally accepted picture was that the hadronic total cross sections should reach asymptotically a finite limit, and the rising p-p cross section measured at the ISR certainly came as a big surprise. Afterwards, fixed target experiments at Fermilab found[6] that this behavior is a general feature of all hadrons as can be seen from the compilation of the total cross section data shown in figure 1 (where ISR and SPS Collider data are not included).

The p-$\bar{\mathrm{p}}$ total cross section measurement performed at the CERN SPS Collider[7] at $\sqrt{s}$=546 GeV confirmed that up to this energy σ_{tot} rises as $\ln^2 s$, i.e. at the maximum rate compatible with the Froissart bound. The Collider measurement, together with a compilation of the lower energy p-p and p-$\bar{\mathrm{p}}$ total cross section data, is shown in figure 2.

Moreover the value of the p-$\bar{\mathrm{p}}$ elastic cross-section σ_{el} found at the SPS Collider has shown (see figure 3) that the ratio σ_{el}/σ_{tot} also grows with energy. The rise of this ratio indicates a clear violation of "Geometrical Scaling" and implies that not only the radius but also the opacity of the nucleon is increasing with energy. Once again the attractive phenomenological picture of the nucleon scattering derived from the ISR data, in which an asymptotic regime of "Geometrical Scaling" seemed to have already been reached, cannot

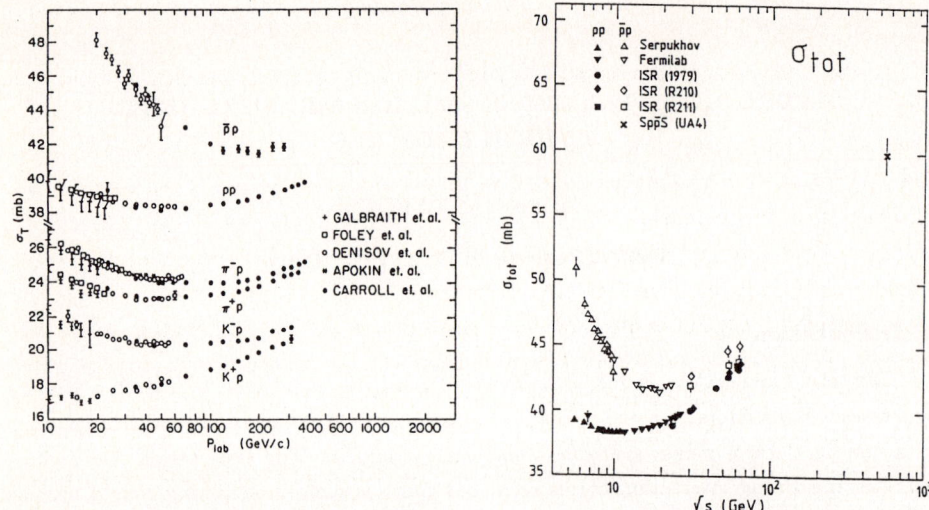

Figure 1: Compilation of hadron-proton cross sections above 5 GeV (ISR and SPS Collider data are not included).

Figure 2: Total cross section data for p-p and p-p̄ scattering.

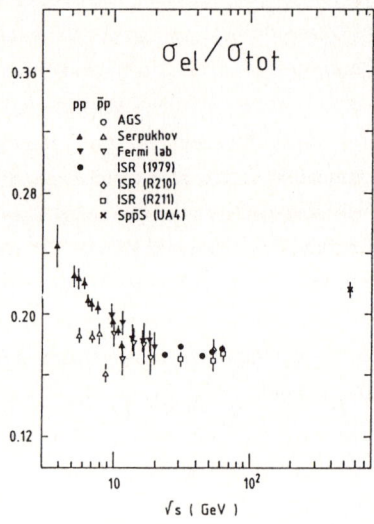

Figure 3: Energy dependence of the ratio σ_{el}/σ_{tot} for p-p and p-p̄ scattering.

survive with the SPS Collider results.

In addition, the recent measurement[8], still at the SPS Collider, of the ratio ρ of the real to the imaginary part of the elastic scattering amplitude in the forward direction indicates that new surprises in the asymptotic behavior of hadron scattering will certainly arrive from the forthcoming measurements in the new energy range ($\sim$ 2 TeV) opened by the Tevatron Collider at Fermilab.

However, the main discussion here will not concern the phenomenological interpretation of the many available measurements of total cross sections at high energies, although it should be very interesting. As a matter of fact many extensive reviews already exist in the literature[9] (for instance detailed discussions on this subject can be found in a review[10] by G. Sanguinetti and myself).

The aim of this lecture will rather be the discussion of the different experimental techniques used so far to measure the total cross sections at high energies, trying to point out the many and subtle difficulties often arising in these measurements. As a matter of fact these experiments, rather easy to conceive in principle, are in practice difficult to perform at high energy due to the accuracy required to make them physically significant. Only consistent results obtained by different experiments, possibly with different methods, can set the scale error attached to the measurement, thus ensuring that unforeseen systematic errors are not affecting the observed values.

2. Total Cross Section Measurements at a Fixed Target Accelerator

The "transmission experiment" technique is the most frequently used to measure total cross sections at a fixed target accelerator. In this kind of experiments the beam intensity is measured before and after the target and the total cross section is derived, in the approximation of an infinitely thin target, from the relation

$$\sigma_{tot} = \frac{1}{N} \ln(I_0/I) \tag{1}$$

where I_0 and I are the incident and the transmitted beam intensity respectively and N is the number of nuclei/cm^2 in the target.

The layout of a tipical transmission experiment[6] is shown in figure 4. The incoming particles are defined and the beam intensity measured by means of Čerenkov counters ($\check{C}_i$),

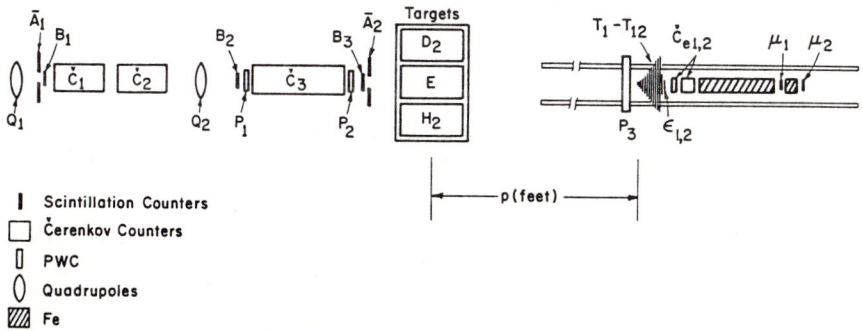

Figure 4: Layout of a typical transmission experiment. This scheme corresponds to a detector actually used to measure total cross sections at Fermilab[6].

scintillation counters (B_i) and multiwire proportional chambers (P_i). The transmission counters T_i (usually covering the range $0.01 \leq |t| \leq 0.1$ GeV^2 with a radius increasing along the beam direction) determine the number of particles transmitted through the target.

Data is usually taken with liquid hydrogen or deuterium targets (as shown in the layout). An empty target made with an identical vessel mounted on the same frame is used to measure the beam absorption in the vessel material. The muon contamination in the beam is removed using iron absorbers and muon counters downstream the transmission counters.

Ideally the intensity of transmitted non-interacting particles could only be measured by an infinitesimal counter. The finite size of the transmission counters only leads to the determination of a partial cross section for each counter

$$\sigma_i = \sigma_{tot} - \int_0^{\Delta\Omega_i} \frac{d\sigma_{tot}}{d\Omega} d\Omega \qquad (2)$$

where $\Delta\Omega_i$ is the solid angle covered by the transmission counter T_i. The total cross section σ_{tot} is then derived with an extrapolation of the partial cross sections σ_i to zero solid angle.

This technique allows rather precise measurements of total cross sections and point-to-point errors of ± 0.1 % have been obtained with a scale error of about ± 0.4 %.

To obtain such an accuracy a rather careful extrapolation has to be performed, taking into account both multiple scattering and finite beam size; in addition the target density must be very well known (≤ 0.1 %) and refined corrections, not only for Coulomb scattering (and Coulomb-nucleon interference) in the target, but also for accidentals and dead-time in the electronics must be applied.

3. Total Cross Section Measurements at a Collider Machine

Transmission techniques, similar to the one previously described, cannot be used for total cross section measurements at colliding-beam machines. Three different methods can be used instead, by measuring simultaneously two of the three following quantities: total interaction rate, forward elastic rate and machine luminosity. The three experimental procedures, extensively used at hadron Colliders, are the following:

a) The most direct method[2] requires the measurement of the total interaction rate R_{tot} (by means of a detector with $\sim 4\pi$ solid-angle coverage) and of the beam luminosity L. The total cross section is then directly determined by the ratio

$$\sigma_{tot} = \frac{R_{tot}}{L} \qquad (3)$$

b) The second method[1] derives the value of the total cross section, via the optical theorem, extrapolating to $t = 0$ the measured rate of elastic scattering at small angle. Neglecting spin effects one can write

$$\sigma_{tot} = \sqrt{\frac{16\pi}{1+\rho^2} \frac{d\sigma_{el}}{dt}}\Big|_{t=0} = \sqrt{\frac{16\pi}{1+\rho^2} \frac{1}{L} \frac{dR_{el}}{dt}}\Big|_{t=0} \qquad (4)$$

where L is again the measured luminosity of the beams and ρ is the ratio of the real to the imaginary part of the elastic scattering amplitude in the forward direction.

c) In the two previous cases an absolute calibration of the machine luminosity is needed. However, when an accurate luminosity calibration method is not available, the simultaneous measurement of low-t elastic scattering and of the total interaction rate allows the determination of σ_{tot} independent of the machine luminosity[7,11]. As a matter of fact, combining the two expressions (3) and (4) one can eliminate the luminosity L and therefore the total cross section can be obtained simply from the ratio of the counting rates

$$\sigma_{tot} = \frac{16\pi}{1+\rho^2} \frac{\frac{dR_{el}}{dt}|_{t=0}}{R_{tot}} \qquad (5)$$

4. Total Cross Section Measurements at the ISR

Method a) was first exploited at the CERN ISR by the Pisa-Stony Brook Collaboration. The layout of the experiment is shown in figure 5. The detector consisted of two telescopes of scintillation counter hodoscopes (H) covering 30° aperture cones downstream from the interaction region and a central box made of two layers of scintillation counters (L, L_s) covering angles larger than 40°. Each telescope consisted of two pairs, (H_1, H_2) and (H_3, H_4), of circular hodoscopes in coincidence. Small hodoscopes (TB), positioned downstream from the H-hodoscopes in each arm, completed the almost 4π trigger. Furthermore in each telescope the angular distribution of charged particles produced in the interaction was detected by additional hodoscopes ($H_i\theta$), finely split into polar angle bins.

The great majority of p-p interactions was detected by the coincidence between the hodoscopes in the two arms. The resolving time of the trigger was kept rather wide and the beam-beam events were then identified off-line from the single-beam background by means of time-of-flight informations. However, corrections must be applied to the event rate to take into account losses due to the incomplete solid angle coverage, mainly in the forward direction inside the beam pipe. The losses of elastic events (ranging from ~ 0.5 mb up to

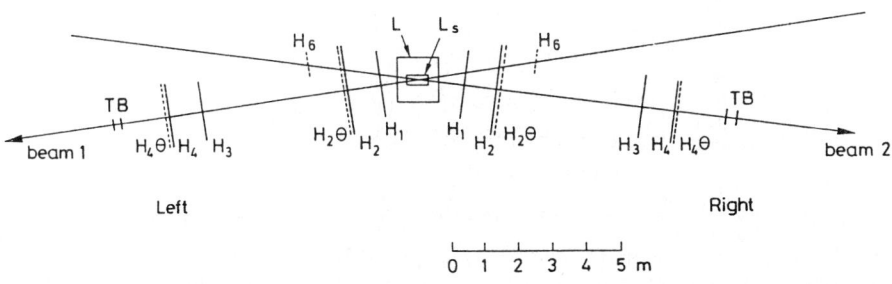

Figure 5: Layout of the proton-proton total cross section measurement performed at the ISR by the Pisa–Stony Brook Collaboration[2].

2 mb depending on the ISR energy) are easily calculated while the small losses of inelastic events ($\leq$ 0.5 mb) can be extrapolated from the measured angular distributions.

Method b) was first used, again at the CERN ISR, by the CERN-Roma Collaboration. The general layout of the detector is shown in figure 6. The experimental apparatus consisted of small scintillation counter hodoscopes placed 9 m from the intersection region inside special thin-wall movable sections of the ISR vacuum chamber (called "Roman pots").

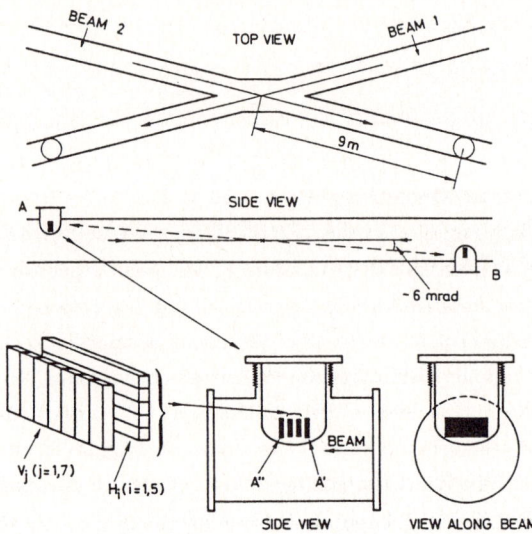

Figure 6: Layout of the proton-proton total cross section measurement performed at the ISR by the CERN–Roma Collaboration[1].

Each pot could be displaced vertically towards the beam when, after injection, stable beam conditions were obtained. Protons elastically scattered in the vertical plane (at angles around 5 mrad) were detected in coincidence by the trigger counters (A', A" and B', B" respectively) of pots A and B. Two hodoscopes of horizontal (H_i) and vertical (V_i) finger counters measured the angles of scattered particles and elastic events were identified by collinearity plots. The kinematic correlation of elastically scattered protons at a given angle, as seen in the (H,V) matrix of the opposite hodoscopes resulted in a rather prominent peak over a small background caused by inelastic processes. However, in order to keep the width of the collinearity peak smaller than the dimensions of the hodoscopes, all data had to be taken with the ISR working in Terwilliger scheme[12]. In these operating conditions, the transverse dimensions of the crossing region were particularly small: the distribution of the interaction points was approximately gaussian in shape with r.m.s. values of $\sim$ 3 mm in the radial direction and of $\sim$ 1.5 mm in the vertical direction. As a consequence, the collinearity peak observed in one hodoscope, corresponding to the shadow of the interaction volume as

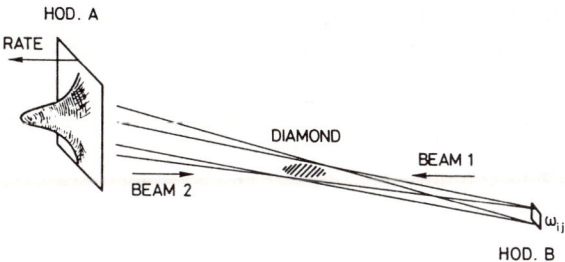

Figure 7: Distribution of hits in hodoscope A when one element of hodoscope B is selected.

projected by one element of the other side hodoscope, is narrow enough to be well contained inside the detector (see figure 7), and the unavoidable edge losses can thus be kept rather small and the appropriate corrections easily applied. Great care must be devoted to the evaluation of the solid angle covered by the counter hodoscopes taking into account dead space and edge effects between two adjacent counters. With the use of movable pots the apparatus could easily detect elastic events in a t region as small as ~ 0.01 GeV^2 at each available ISR energy and it was therefore possible to extrapolate in a rather safe way the elastic rate to $t = 0$. An independent measurement of the machine luminosity (see later) thus allowed, by means of equation 4, the determination of the p-p total cross section over the full range of the ISR energies.

The combined method c) was first suggested[13] and then exploited[11] by the Pisa-Stony Brook and the CERN-Roma Collaborations running simultaneously the two previously described apparata in the same intersection region of the ISR. Both experiments had indeed measured, at the beginning of the CERN ISR operation, an unexpected remarkable rise of the proton-proton total cross section. At that time, however, systematic errors as large as 2 % could not be excluded due to the uncertainty in the calibration of the beam displacement scale during the luminosity measurement. For this reason, although the good agreement between the results of the two experiments indicated that the luminosity measurement could not have been affected by large systematic uncertainties (a systematic error on the luminosity would change the scale of the measured cross section in a different way in the two experiments), the two groups decided to join and perform a combined experiment[11] exploiting simultaneously the three methods a), b) and c), thus obtaining three different determinations of the p-p total cross section. Consistent results were found, which allowed to achieve a precision of $\pm$ 0.6 % in the measurement. This new experiment definitely confirmed that indeed hadronic cross sections rise with energy at a rate compatible with $\ln^2 s$ over the ISR energy range[†].

[†] Other precise measurements of the p-p total cross section at the ISR have been performed also by other experiments[14] exploiting the three methods previously described and obtaining always consistent results.

Only few years ago the antiproton cooling technique[15] successfully carried out at CERN allowed to replace one of the two proton beams in the ISR with a beam of antiprotons. A direct precise comparison of p-p and p-p̄ processes at the same energies, by means of the same detectors, became therefore possible. In particular the classic experiments Pisa-Stony Brook and CERN-Roma have been repeated and both the p-p and p-p̄ total cross sections have been measured by the new editions of the two previously described experiments.

The experimental techniques are exactly the same, and the same are the methods used to measure the total cross sections: the R210 experiment[16] measured the total interaction rate using an almost 4π detector [method a)] while R211 measured[17] elastic scattering at very small angles and derived the total cross section value through the optical theorem [method b)]. These new measurements, however, had the advantage of a direct calibration of the ISR beam displacement scale by means of special vertical scrapers (see later) installed on both beam lines upstream from the interaction point. The over-all precision (systematic and statistical errors combined) achieved in the total cross section measurements was as good as 1 % both for the proton-proton and the proton-antiproton reactions.

5. Luminosity Measurement at the ISR

In order to derive the value of the total cross section, both the pioneer Pisa-Stony Brook and the CERN-Roma experiments had to make use† of the measured ISR luminosity L (a problem common to all experiments aimed at measuring absolute values of cross sections).

The absolute calibration of the luminosity was performed in both cases by a simple method (v.d.M.) suggested by S. van der Meer[19]. This technique consists in measuring the interaction rate as a function of the relative vertical displacement δ of the two continuous beams.

At the ISR the value of the luminosity depends only on the vertical (z-axis) overlap of the beams (in addition of course to the currents I_1 and I_2 circulating in the two rings) since in this machine the two beams cross in the horizontal plane at a fixed angle ($\alpha = 14.77°$). All the information is condensed in the "effective beam height" h_{eff} defined as the inverse of the normalized vertical density overlap integral:

$$\frac{1}{h_{eff}} = \frac{\int_{-\infty}^{+\infty} i_1(z) i_2(z) dz}{\int_{-\infty}^{+\infty} i_1(z) dz \int_{-\infty}^{+\infty} i_2(z) dz} \tag{6}$$

where $I_{1,2} = \int_{-\infty}^{+\infty} i_{1,2}(z) dz$. Indeed the luminosity L can be expressed in terms of the ISR parameters as

$$L = \frac{I_1 I_2}{K} \frac{1}{h_{eff}} \tag{7}$$

---

† Actually also a different approach had been tried[18] at the lowest ISR energies (23 and 31 GeV) by the CERN-Roma Collaboration to fix the absolute value of the elastic cross section measuring the elastic scattering at very small momentum transfer, where the Coulomb cross section, a well-known electromagnetic scattering process, is dominant. However this approach can hardly be used as the energy increases.

where $K = \beta ce^2 \sin\frac{\alpha}{2}(1-\beta^2\sin^2\frac{\alpha}{2})^{-\frac{1}{2}} \simeq \beta ce^2 \tan\frac{\alpha}{2} = \beta \cdot 0.9972 \cdot 10^{-28}$ (A^2·cm·s).

In the v.d.M. method the ISR luminosity can be obtained measuring the interaction rate $R_M(\delta) = \sigma_M L(\delta) = \sigma_M(\frac{I_1 I_2}{K})(1/h_{eff})$ of a monitor counter M, triggered by beam-beam events that correspond to an inclusive cross section σ_M. If the vertical displacement $\delta = z_1 - z_2$ between the centroids z_1 and z_2 of the two beams is varied in small and precise steps, by steering the ISR beams vertically, the value of the integral

$$\int_{-\infty}^{+\infty} R_M(\delta)d\delta = \frac{\sigma_M}{K}\int\int_{-\infty}^{+\infty} i_1(z)i_2(z+\delta)dzd\delta = \frac{\sigma_M}{K}I_1 I_2 \qquad (8)$$

can be easily measured. Since the beam currents are known to better than 0.1 % the above relation leads to the determination of the monitor calibration constant σ_M.

Once σ_M is known, the ISR luminosity can be obtained, at any time, from the beam-beam interaction rate of the monitor using the relation $L = R_M/\sigma_M$.

An example of the beam displacement curve as measured in the early days of the ISR operation by the Pisa-Stony Brook experiment is shown in figure 8. Although this method is rather simple in operation, a number of precautions have to be adopted when a high precision measurement has to be achieved in the v.d.M. luminosity calibration.

All the v.d.M. curves have to be swept by constantly increasing (or decreasing) the vertical displacement δ in order to avoid the effects of magnet hysteresis. Moreover, perturbations in the vertical beam displacement, due to cross-talk between different experimental areas, must be avoided by performing the v.d.M. calibration in one interaction area at a

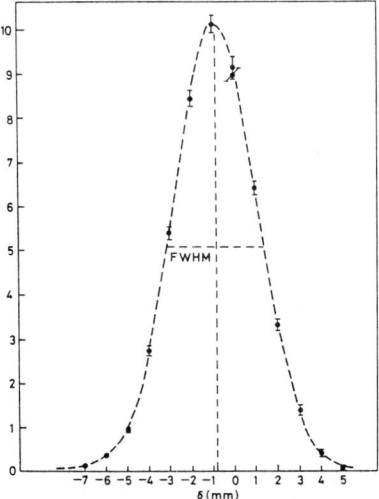

Figure 8: Example of beam displacement curve[2] at $\sqrt{s}=53$ GeV. The full curve is a Gaussian fit to the data. FWHM = 4.50 mm.

time. Corrections must be applied in order to compensate for the difference between the actual and the nominal values of the parameters (radial average position and width) of each ISR stack. In order to minimize the number of accidental single-beam coincidences with respect to beam-beam events, the calibration measurements should be performed with neither too low nor too high currents circulating in the ISR (a value of ~ 4 A was found to satisfy this requirement). A good calibration of the beam displacement is moreover needed since the scale error on the machine luminosity depends linearly (see equation 8) on the absolute scale value of δ.

In a more recent[20] calibration of the beam displacement, performed by means of vertical scrapers, the accuracy was found to be better than 0.5 %, and this has shown that indeed a scale error of such a small magnitude can be achieved also on the luminosity measurements at the ISR.

6. Total Cross Section and Luminosity Measurements at the p-p̄ Collider

The v.d.M. method, successfully used at the ISR, could in principle be easily generalized to the luminosity measurement at the p-p̄ Collider[21].

In this case the two beams are bunched and collide head-on. Therefore at the Collider the value of the luminosity depends on the "effective area" of the collision of the two bunches and equation 6 should be generalized to

$$\frac{1}{a_{eff}} = \frac{\int\int_{-\infty}^{+\infty} i_1(x,y)i_2(x,y)dxdy}{\int\int_{-\infty}^{+\infty} i_1(x,y)dxdy \int\int_{-\infty}^{+\infty} i_2(x,y)dxdy} \qquad (9)$$

where $i_1(x,y), i_2(x,y)$ are the current density profiles of the two beams.

It should be noticed that the value of the luminosity does not depend on the length of the two bunches, since particles in two beams colliding head-on always cross each other, no matter how long the collision volume is. Following the v.d.M. method, the interaction rate $R_M(\delta_x, \delta_y)$ of a monitor counter M can be measured while the two beams are displaced by small amounts δ_x and δ_y along the x and y directions. In analogy with equation 8 the value of the integral

$$\int\int_{-\infty}^{+\infty} R_M(\delta_x,\delta_y)d\delta_x d\delta_y = \sigma_M f_0 N_p N_{\bar{p}} \int\int_{-\infty}^{+\infty} \frac{d\delta_x d\delta_y}{a_{eff}(\delta_x,\delta_y)} = \sigma_M f_0 N_p N_{\bar{p}} \qquad (10)$$

determines the monitor calibration constant σ_M, where f_0 is the revolution frequency of the machine (43.4 KHz for the SPS Collider) and N_p and $N_{\bar{p}}$ are the numbers of protons and antiprotons in the two colliding bunches. Consequently the machine luminosity can be evaluated at any later time simply from the monitor interaction rate ($L = R_M/\sigma_M$).

Up to now, however, a v.d.M. scan in the vertical and horizontal planes, in principle possible by means of a pair of electrostatic separators, was in practice never used in the luminosity measurement of a physics run although extensive tests have been carried out[22].

The proton and antiproton bunch profiles are usually measured in the vertical and horizontal planes by means of rapidly moving wires[23] scanning the beams† in both directions. A 50 μm thick berillium wire (or a 25 μm thick carbon wire) is stretched on a 170 mm long fork (see figure 9). The fork, driven by a dc motor, moves the wire through the beams at a speed of 4.3 m·s^{-1} and therefore during a revolution time the wire is displaced of $\sim$ 0.1 mm. The flying wire, when crossed repeatedly by the bunch, acts as a target for the production of secondary particles and therefore the profiles of the two beams can be reconstructed simply by measuring the interaction rates using scintillation counters placed near the beam pipe in the downstream proton and antiproton directions.

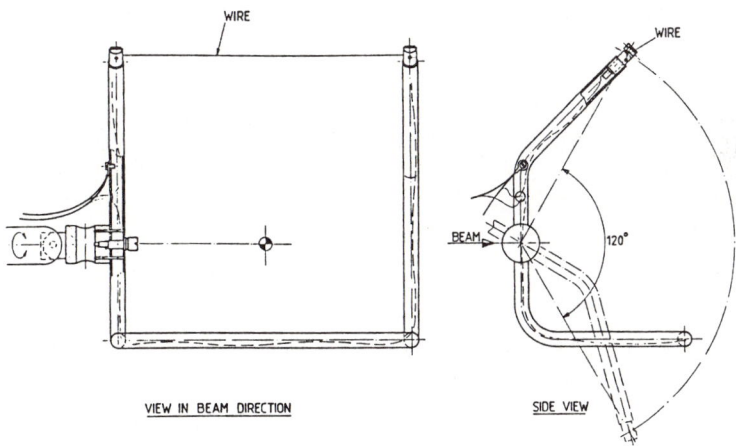

Figure 9: Schematic diagram of the flying wire assembly used[23] to measure the beam profiles at the SPS Collider.

From these measurements, performed at different locations of the machine, the luminosity at the crossing points can be extrapolated using the relation

$$L = \frac{f_0}{2\pi}\sqrt{\frac{\beta_x}{\beta_x^*}\frac{\beta_y}{\beta_y^*}}\frac{N_p N_{\bar{p}}}{\sqrt{(\sigma_{px}^2 + \sigma_{\bar{p}x}^2)(\sigma_{py}^2 + \sigma_{\bar{p}y}^2)}} \quad (11)$$

where the σ's are the measured r.m.s. beam sizes for proton and antiproton bunches while β/β^* is the ratio of the betatron function values at the location of the wire scan and at the crossing point, in the vertical and horizontal planes respectively.

When there are more bunches of protons and antiprotons circulating in the machine, of course each individual bunch profile must be measured separately and the luminosity must be evaluated as the sum of the proper combinations of colliding bunches in a given interaction region.

---

† Other methods, for instance using synchrotron light measurements, have also been attempted[24].

Up to now however a precise absolute measurement of the Collider luminosity with an accuracy better than 10 % remains a very difficult task. Therefore the only experimental procedure that can be used to measure with high accuracy the value of the p-p̄ total cross section at the Collider is the luminosity-independent method previously described (equation 5). Indeed the UA4 experiment was specifically designed[7,25] to measure, using the combined method, the total cross section at the CERN p-p̄ Collider.

The UA4 experimental apparatus is conceptually very similar to the one used at the ISR by the Pisa-Stony Brook and the CERN-Roma Collaborations. However, in practice, both UA4 detectors, the one for the low-t elastic scattering measurement and the 4π detector that measures the total inelastic rate, must achieve quite a greater degree of sophistication. As a matter of fact, since the higher Collider energy causes a remarkable shrinkage in the polar angle distribution of elastically scattered particles, a very good angular resolution (~ 0.05 mrad, approximately one order of magnitude better than at the ISR) is required in a very forward region ($\theta \approx 0.5$ mrad). As a consequence, telescopes of precision drift chambers must be placed at quite large distance from the interaction point, and therefore after the machine quadrupoles. Moreover data must be collected in dedicated Collider runs in which a special optics features a high value of the betatron function β in the intersection region. In fact the usual low-β optics, while providing higher luminosities, is not suitable for low-t elastic scattering measurements due to the larger angular spread of the beams (proportional to $1/\sqrt{\beta}$). In these special high-β runs the typical β values at the crossing point in both the horizontal and the vertical plane were 100 m, and the transverse beam size and the beam angular divergence at the crossing point had r.m.s. values of ~ 1.4 mm and ~ 0.014 mrad respectively.

The layout of the elastic scattering apparatus is shown in figure 10. Elastic events were detected in a system of four telescopes placed symmetrically above and below the machine plane. Particles scattered within the telescope acceptance must enter the quadrupoles before detection, and therefore the trajectory deflections are to be calculated with the standard transfer matrices using the precisely known strength and effective length of the magnets. The effective distance of the telescopes from the crossing point became thus 35 and 40 m in the horizontal and vertical plane respectively, still very close to the geometrical distance of 37 m. The detectors within each telescope were placed inside two "Roman pots" 6 m apart. The detectors could thus be moved vertically towards the beam plane once stable beam conditions were reached. It was found that for safe operation the typical distance from the edge of the sensitive region of the chambers to the beam axis was ranging from 15 to 25 mm (about 15 times the r.m.s. size of the beam itself).

Each pot contained a trigger counter, a hodoscope of vertical finger scintillation counters and a wire chamber; the chamber had four drift planes with three drift cells, suitably staggered to resolve the up-down ambiguity. The drift planes were followed by a proportional plane measuring the horizontal coordinate by means of the charge-division method. Tracks

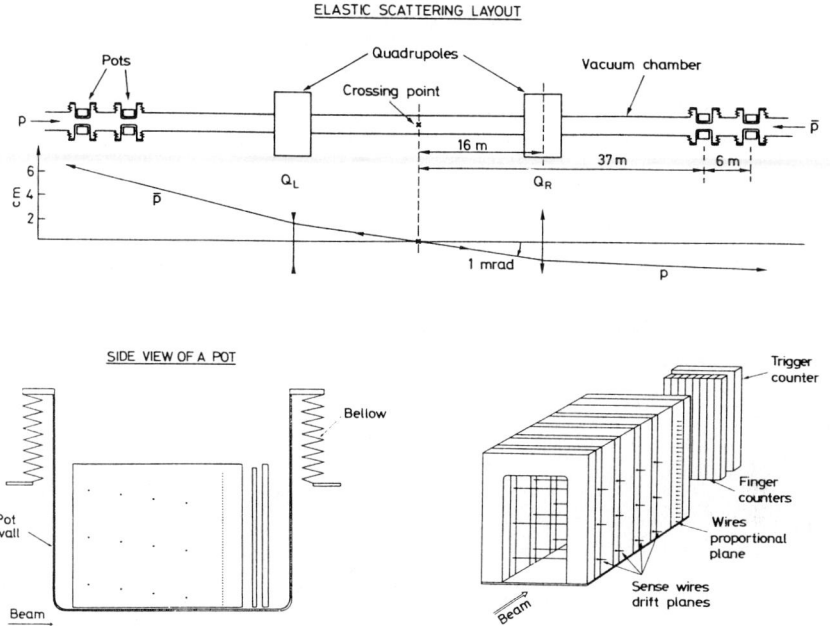

Figure 10: Sketch of the UA4 elastic scattering set up. A detail of the detector inside the "pot" is shown together with a perspective view.

were reconstructed with an accuracy in the vertical and in the horizontal coordinates of 130 μm and 400 μm respectively.

In figure 11 the elastic differential t-distribution measured at the SPS Collider in the range $0.03 \leq -t \leq 0.5$ GeV^2 is shown. In the region $0.03 \leq -t \leq 0.15$ GeV^2 the distribution is well described by a simple exponential e^{bt} with a constant slope parameter $b = 15.2 \pm 0.2$ GeV^{-2} (shown as a solid line in the figure). The value of the elastic rate extrapolated to $t = 0$ (the optical point) can be easily determined using this "linear" fit with a statistical error of 1%. "Quadratic" fits like e^{Bt+Ct^2} have also been tried but, since the data in this t range force the value of C to be almost zero, they give a value for the extrapolated forward rate which is only 0.8% higher than the rate determined by the "linear" fit. In any case this implies a systematic uncertainty in the extrapolation smaller than the intrinsic statistical error. Once the $t = 0$ extrapolation has been performed, the total elastic rate can also be determined by integrating the t-distribution shown in figure 11.

During the data-taking, the measurement of the inelastic rate was performed at the same luminosity as for the elastic rate by simultaneously enabling both triggers with a suitable prescaling factor applied to the inelastic trigger. The measurement of the inelastic rate requires a trigger as inclusive as possible and a clear discrimination of beam-beam events from background interactions.

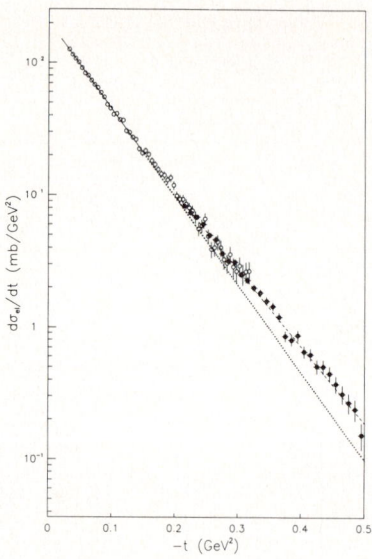

Figure 11: Differential cross section of p-p̄ elastic scattering at $\sqrt{s}$=546 GeV.

The use of time-of-flight information from the trigger counters, highly effective at the ISR for the rejection of background interactions, is not powerful enought to fully discriminate beam-beam interactions in a bunched machine such as the Collider because the background is bunched as well, and therefore in-time with the good beam-beam events. Consequently tracking chambers are needed to reconstruct the vertex of the interaction to an accuracy at least comparable to the length of the interaction region.

One arm of the apparatus used to measure the total inelastic rate is shown in figure 12. Two identical systems of three telescopes (D_1, D_2, D_3) placed simmetrically on the left and right sides of the crossing region covered polar angles corresponding to a pseudo-rapidity range 2.5≤| η |≤5.6. Each telescope consists of six drift chamber planes followed by a plane of trigger counters (T_1, T_2, T_3). The coordinate along the drift wire is measured by means of a delay line close to the drift wire itself. The region $-1.7 \leq \eta \leq 1.7$ was covered by the UA2 central detector[26], used by the UA4 experiment during the high-β runs. To obtain the total inelastic rate the fraction of events escaping detection due to the incomplete angular coverage, mostly in the forward direction, must be carefully determined by means of empirical extrapolations of the measured angular distributions. The losses were found to be less than 1% of the total inelastic rate.

The result of the UA4 experiment, obtained combining the low-t elastic scattering and the integrated elastic rate with the total inelastic rate according to equation 5, gives $(1 + \rho^2)\,\sigma_{tot}$ = 63.3 ± 1.5 mb.

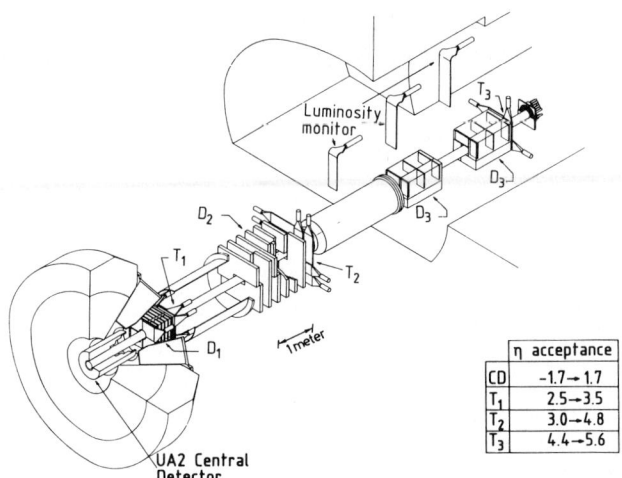

Figure 12: Layout of one arm of the UA4 inelastic scattering detector. The η acceptance of the trigger hodoscopes is shown in the table.

Recently UA4 measured[8], in a special Collider run with a very high-β optics, the ratio ρ between the real and the imaginary part of the forward elastic scattering amplitude making use of the classical Coulomb interference method. The result was $\rho = 0.24 \pm 0.04$, an unexpectedly high number. Using this value for ρ, the total p-$\bar{\text{p}}$ cross section comes out $\sigma_{tot} = 60\pm1.5$ mb, 50% higher than the value found at ISR energies. As a consistency check, UA4 also derived a less accurate value of σ_{tot} from equation 4, using the measurement of the machine luminosity obtained from the determination of the beam profiles with the wire scan technique. The same technique was also used by the UA1 Collaboration[27] in a short Collider run with very low integrated luminosity. The resulting total cross section measurements had quite large errors and are not shown in figure 2.

The relevance of the ρ measurement at the Collider energy comes out from the fact that the real part of the elastic scattering amplitude is related to the imaginary part via dispersion relations; on the other hand the total cross section can be related through the optical theorem to the imaginary part of the elastic scattering amplitude in the forward direction. As a consequence the energy behaviors of ρ and σ_{tot} are mutually constrained by dispersion relations. The extremely high value of ρ measured by UA4 at the Collider arouses therefore great interest in the behavior of total cross sections at even higher energies and make very important a precise measurement of the p-$\bar{\text{p}}$ total cross section at the Tevatron Collider.

The experimental techniques that must be used at this new machine are quite similar to those already used at the CERN SPS Collider. However a precise measurement of low-t elastic scattering may be even more difficult because of the higher center of mass energy that requires efficient particle detection at even smaller scattering angles.

The two experiments[28,29], CDF and E710, that are facing these difficult problems at the Tevatron Collider have just started the data-taking. Hopefully very soon important and exciting results will become available on the behavior of the hadronic total cross sections in this new energy domain.

Acknowledgements

I would like to thank G. Bellettini and P.G. Verdini for their constructive criticism in reading the manuscript and C. Vannini for the many evaluable contributions at all stages of this work.

References

1) U.Amaldi et al.; Phys. Lett. 44B (1973) 112.
2) S.R.Amendolia et al.; Phys. Lett. 44B (1973) 119,
   S.R.Amendolia et al.; Nuovo Cimento 17A (1973) 735.
3) G.B.Yodh, T.Pal and S.J.Trefil; Phys. Rev. Lett. 28 (1972) 1005.
4) W.Heisenberg; Kosmische Strahlung, Springer-Verlag, Berlin (1953) 148.
5) H.Cheng and T.T.Wu; Phys. Rev. Lett. 24 (1970) 1456.
6) A.S.Carroll et al.; Phys. Rev. Lett. 33 (1974) 928 and 932.
7) M.Bozzo et al.; Phys. Lett. 147B (1984) 392.
8) D.Bernard et al.; Phys. Lett. 198B (1987) 583.
9) M.Block and R.Chan; Rev. Mod. Phys. 57 (1985) 563,
   M.Giffon and E.Predazzi; Riv. Nuovo Cimento 7 (1984) 1,
   G.Alberi and G.Goggi; Phys. Rep. 74 (1981) 1,
   M.Jacob and G.Giacomelli; Phys.Rep. 55 (1979) 1,
   E.Predazzi; Riv. Nuovo Cimento 2 (1979) 1 and 6 (1976) 217,
   U.Amaldi, M.Jacob and G.Matthiae; Ann. Rev. Nucl. Sci. 26 (1976) 385,
   G.Giacomelli; Phys. Rep. 23 (1976) 123.
10) R.Castaldi and G.Sanguinetti; Ann. Rev. Nucl. Part. Sci. 35 (1985) 351.
11) U.Amaldi et al.; Phys. Lett. 62B (1976) 460 and Nucl. Phys. B145 (1978) 367.
12) K.M.Terwilliger, Proc. Int. Conf. on High-Energy Accelerators,
    CERN, Geneva (1959) 53.
13) CERN-Rome and Pisa-Stony Brook Collaborations; Memorandum,
    CERN/ISRC/73-14 (1973)
14) K.Eggert et al.; Nucl. Phys. B98 (1975) 93,
    L.Baksay et al.; Nucl. Phys. B141 (1978) 1.
15) D.Möhl et al.; Phys. Rep. 58 (1980) 73 and references therein.
16) G.Carboni et al.; Nucl. Phys. B254 (1985) 697.
17) N.Amos et al.; Nucl. Phys. B262 (1985) 689.
18) U.Amaldi et al.; Phys. Lett. 43B (1973) 231.

19) S.van der Meer; Calibration of the effective beam height in the ISR; Internal report; CERN ISR-PO/68-21 (1968).
20) K.Potter, S.Turner; IEEE Trans. Nucl. Sci., Vol. NS-22, n.3, (1975) 1589.
21) C.Rubbia; CERN p-p̄ Note 38 (November 14, 1977).
22) R.Dubois et al.; CERN/SPS/85-5 (DI-MST)
23) J.Bosser et al.; Nucl. Instr. and Methods A235 (1985) 475.
24) J.Bosser et al.; CERN/SPS/83-15 (ABM).
25) R.Battiston et al.; Nucl. Instr. and Methods A238 (1985) 35.
26) M.Dialinas et al.; Orsay report LAL-RT 83-14 (1983).
27) G.Arnison et al.; Phys. Lett. 128B (1983) 336.
28) G.Apollinari et al.; CDF note 623, October 1987,
   F.Abe et al.; Fermilab Pub-88/25-E, submitted to Nucl. Instr. and Methods.
29) N.Amos et al.; Nucl. Instr. and Methods A252 (1986) 263,
   M.Bertani et al.; Int. Journal of Mod. Phys. A, Vol. 2, n.4 (1987) 891.

WINDSURFING ON THE COLOURLESS SEA

Francesco-Luigi Navarria

Dipartimento di Fisica dell'Università, I-40126 Bologna

INFN, Sezione di Bologna, I-40126 Bologna

ABSTRACT

The first evidences for quark-partons are recalled together with the quarks' properties which are determined by sum rules. Some recent data on $R = \sigma_L/\sigma_T$ and $F_2(x, Q^2)$ are then discussed with emphasis on the BCDMS results.

1 INTRODUCTION

History is perhaps just a different intonation

of the same metaphores (Stevie Wonder)

The evidence for substructure has always been derived from two kinds of observations. One is the measurement of static properties and in particular spectroscopy: the observation of excited states is a proof that the system under investigation is composite. The other is the scattering experiment, elastic and inelastic. Both methods yield information about the target structure, but, in the inelastic case, the probe may scatter onto the target constituents and break it up, providing striking evidence for them.

More than 20 years have passed since the first experimental evidence of the existence of hard chunks inside the nucleon from deep-inelastic scattering (DIS) of electrons [1.1] and the static quark model (QM) of Gell-Mann and Zweig [1.2] is just a few years older. The parton model (PM) [1.3], the quark-parton model (QPM) [1.4] and the theory which stems out of it, namely quantum chromodynamics (QCD) [1.5], have been the subject of hundreds of papers, reviews and books. Yet the quarks have never been found free outside the particles they constitute in spite of the ingenuity of quark hunters and of the sparse positive claims that died away later on [1.6]. A fact which, though naturally embedded in the framework of the standard theory, is nevertheless frustrating, expecially to experimentalists, whose profession is mainly to find out facts before advancing explanations. The confinement of quarks inside hadrons sets a completely different scenario compared to previous layers of matter at larger distances and could stop the indefinite regression to smaller and smaller structures, typical of russian dolls within dolls. For this reason the evidence for quarks is not direct like that for leptons or other particles. However, the indirect evidence that strongly interacting particles (hadrons) are made up from fractionally charged spin-1/2 point-like fermions (quarks) and that the strong force is carried by coloured massless vector particles (gluons) appears overwhelming to the majority of physicists. For instance, the QPM predicts nicely the outcome of e^+e^- collisions into $q\bar{q}$ pairs or pairs of jets and the QCD corrections show up clearly as larger number of jets in one event, in a way analogous to bremmstrahlung in QED. Seen in retrospective it is impressive how well the naive QM reproduces the spectrum of s-wave baryon and meson states and more recently the case for the reality of quarks has received support from the variety of energy levels above the J/ψ and Υ particles. Other aspects are intriguing and some of them will be discussed in the following sections. QCD makes specific predictions which are not so easy to verify with good precision experimentally: the running of the strong coupling constant, α_s, with Q^2 for instance is far from being pinned down at the moment. Some of the observable quantities, such as the nucleon and meson structure functions, are not predicted by the present theory. Only their Q^2 evolution can be confronted with the theory, which is usually quite hard to do within one single experiment for lack of lever arm in Q^2, given the observed logarithmic dependence of scaling violations and, at low Q^2, the presence of non perturbative effects. In addition, from time to time, new data and new unexpected effects have, at least temporarily shaken some of the pillars on which QCD or QPM stand, e.g. the EMC-effect [1.7] and very recently, the polarized EMC deep inelastic data [1.8]. Both effects can eventually be reconciled with the standard wisdom, and looking from another perspective, it is amazing how long this sector of the standard theory has stood the test of experiments [1.9].

This paper gives a brief review of the old days with emphasis on how long it took to convince both experimentalists and theorists about what today seems to most of us a trivially obvious explanation. Then, after a short discussion of various sum rules, which shed some light on the properties of quarks, some of the most recent data on nucleon structure functions are presented.

2 THE FIRST EVIDENCE FOR QUARKS

Le plus important secret de la vie est d'attendre

ce qu'on n'a pas prevu (Emile Chartier dit Alain)

Quarks, historically, could have been 'discovered' very early indeed. Evidence that the nucleons are not elementary is as old as the observation in 1933 of an anomalous magnetic moment. The present day data [2.1], $\mu_p = (2.7928444 \pm 0.0000011)\ \mu_N$ and $\mu_n = (-1.91304308 \pm 0.00000054)\ \mu_N$, with $\mu_N = e\hbar/2m_p c$, nuclear magneton, expected for a Dirac particle, can in fact be reproduced within few percent precision assuming that the nucleon comprises three quarks with third integral charges and mass 1/3 of its mass [2.2]. However, it took many more years, the discovery of many more particles, of resonances and of strangeness, plus the measurement of their static properties, before a QM could be postulated. After the papers of Gell-Mann and Zweig [1.2], the attitude of physicists was quite divided. In fact the quark hypothesis, with its fractionally charged particles, was thought by some as simply a means of explaining the data, not necessarily having any direct physical basis [2.3]. Others instead saw things in a different way, thinking that the reality of quarks was necessary for the model to be worthwhile [2.4]. Among experimentalists some started frantically searching for quarks in all possible interactions, in cosmic rays and in tiny droplets of matter - some theorists tried very hard to reconcile the masses and energies which were resulting with non relativistic approximations - others dismissed the whole thing as not well founded theoretically. After a short time it became clear that quarks were not easy to find as independent particles. On the other hand, the regularity of the hadron spectrum was not sufficient to convince all physicists of the necessary existence of u,d,s quarks. Apart from fractional charges and baryon numbers, one had to swallow the absence of multiple quark states with the exception of $q\bar{q}$ and qqq, the problem with statistics of the Δ^{++} and Ω^- states which, together with the π^0 lifetime, eventually led to the introduction of colour [2.5] and, most of all, to justify the absence of free quarks themselves. All of a sudden, the deep-inelastic electron-proton scattering experiments at SLAC changed the scenario rather dramatically, providing direct evidence for the scattering of the e.m. probe off hard subunits inside the nucleon [1.1]. Elastic electron scattering off nuclei and nucleons [2.6] had already been a popular tool for quite some time, providing information on nucleon form factors in the space-like region. In elastic scattering the only kinematic variable is $Q^2 = -q^2 = 4EE'\ sin^2(\theta/2)$, with E,E' and θ the incoming and scattered lepton energy and the scattering angle in the laboratory, respectively. Elastic nucleon form factors used to decrease roughly as Q^{-4}, showing a forward diffraction peak and hinting towards the existence of a structure. It was a kind of pre-Rutherford era and nobody could tell whether it was a cheesecake or a pomegranate. The theoretical discussion of nucleon form factors was in terms of pole dominance or vector meson dominance, whilst a common parametrization was the so-called dipole form factor

$$G_M(Q^2)/\mu = G_E(Q^2) = (1 + Q^2/0.71)^{-2} \quad (2.1).$$

The electric and magnetic form factors, G_E and G_M, are, in the $Q^2 \ll M^2$ limit, the Fourier transforms of the charge and magnetic moment distributions of the nucleon respectively. They are extracted from the measured cross section

$$(d\sigma/d\Omega)_{el} = (d\sigma/d\Omega)_{Mott} \cdot [G_M^2(Q^2)Q^2/2M^2 tg^2(\theta/2)+$$
$$+(G_E^2(Q^2) + G_M^2(Q^2)Q^2/4M^2)/(1+Q^2/4M^2)] \, E'/E \qquad (2.2)$$

where $(d\sigma/d\Omega)_{Mott} = (4\alpha^2 E'^2/Q^4)cos^2(\theta/2)$ is the Mott cross section for scattering off point-like spin 1/2 particles and M is the proton mass.

The breakthrough came with a new high intensity high energy machine, the 20 GeV 2 mile Stanford linac which started in 1961 and with new powerful detectors, the 20 and 8 GeV spectrometers. The inclusive inelastic measurements at SLAC started in summer 1967 and the first results were presented by J.I. Friedman at the Vienna conference one year later [1.1]. In the inelastic case another kinematic variable, the energy loss, $\nu = (p \cdot q)/M = E-E'$ is available and the ratio $x = Q^2/2M\nu$ can vary between 0 and 1 (for elastic scattering x=1); p and q are the nucleon and virtual photon four momenta [see Fig. 2.1]. Alternatively one may use another variable, $W^2 = M^2 + 2M\nu - Q^2$, the mass squared of the final hadronic system and this was the variable actually used in the first plots [see Fig. 2.2]. In the continuum above the resonance region (W $\simeq$ 2,3 GeV) the cross section divided by the point-like cross section is almost constant with Q^2 varying between 0.5 and 2.5 GeV^2, whilst for elastic scattering from eqs. 2.1 and 2.2 one would have expected a decrease of a factor of about 50. The independence of the structure function from Q^2 at fixed x, and only approximately at fixed W^2, is equivalent to the onset of the $sin^{-4}(\theta/2)$ regime at large angles in the Rutherford experiment. They both signal the scattering off point-like centres within the target. Notice that since W^2 is not a scaling variable, it isn't the best choice to display the effect! Quoting from Panofsky's Vienna rapporteur's talk: 'theoretical speculations are focussed on the possibility that these data might give evidence on the behaviour of point-like charged structures within the nucleon' [1.1].

The double differential cross section measured in DIS of unpolarized electrons off protons at SLAC can be expressed [1.4] in terms of two structure functions, W_1 and W_2, as

$$d^2\sigma/d\Omega dE' = (d\sigma/d\Omega)_{Mott} \cdot [W_2(Q^2,\nu) + 2W_1(Q^2,\nu)tg^2(\theta/2)] \qquad (2.3)$$

and the approximate constancy of the cross section divided by the point-like cross section implies that the structure functions are not functions of Q^2 and ν separately, but just of their ratio, x. In the PM x is interpreted as the fraction of proton momentum carried by the parton [1.3] and the cross section may be seen as the incoherent sum of the contributions to the scattering of the individual partons. The approximate constancy of the cross section ratio implies therefore the scattering off point-like constituents, forgetting for

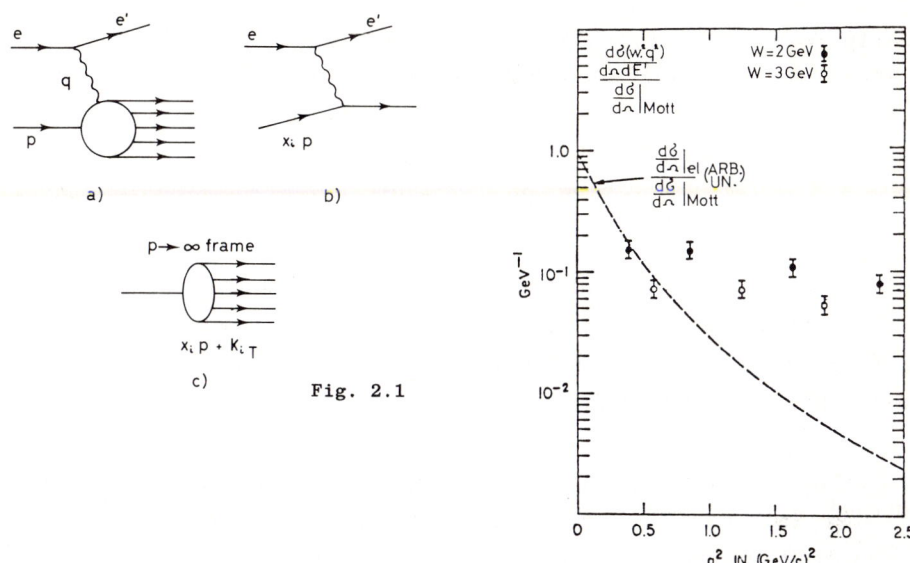

Fig. 2.1

Fig. 2.2

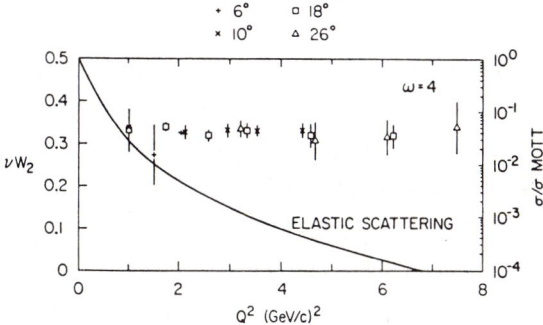

Fig. 2.3

2.1 Kinematics of DIS: a) Lowest order diagram for DIS of electrons on nucleons; b) Feynman diagram of DIS in the PM; c) the infinite momentum frame.

2.2 $(d^2\sigma/d\Omega dE')/\sigma_{Mott}$ plotted as a function of Q^2 in W bins, presented by the SLAC-MIT group at the Vienna Conference in 1968. First evidence for point-like constituents in the nucleon.

2.3 SLAC-MIT data for νW_2 as a function of Q^2 for fixed $\omega = 1/x = 4$. For comparison the elastic form factor is plotted on a logarithmic scale along the right vertical axis.

the moment the complications due to the fact that they are tightly bound, and the question is whether there is any relation (e.g. a unitary transformation) between current partons and constituent quarks. In fact the theoretical situation persisted in being rather confuse for at least two years and some theorists attributed the scaling behaviour of structure functions

$$2MW_1(Q^2,\nu) \to 2xF_1(x)$$
$$\nu W_2(Q^2,\nu) \to F_2(x)$$

at large ν/Q^2 ($\geq 1.5 GeV^{-1}$) to diffractive vector meson production [2.7]. On the other hand, Bjorken had already emphasized in 1967 at Varenna the power of leptons in attacking directly the problem of elementary constituents within the nucleon and later conjectured that for large Q^2, ν one should have scaling [2.8,2.9]. In the mean time Feynman had conjectured a parton model for hadron-hadron reactions and for DIS with emphasis on the infinite momentum frame, without publishing much of it [2.10], and probably was the first one to see the meaning of these early results in the way they are presently understood [1.3,2.11]. In more detail, Bjorken scaling can be connected to the scattering off point-like objects as follows. Setting $G_E = G_M = 1$ in eq. 2.2, i.e. considering the scattering of the electron on a muon, one can rewrite it as

$$d\sigma/d\Omega dE' = (d\sigma/\Omega)_{Mott} \cdot (1 + Q^2/2M^2 tg^2(\theta/2)) \cdot \delta(\nu - Q^2/2M)$$

and the corresponding structure functions are trivially given by

$$2MW_1(Q^2,\nu) = Q^2/2M\delta(\nu - Q^2/2M) = x\delta(1-x)$$
$$\nu W_2(Q^2,\nu) = \nu\delta(\nu - Q^2/2M) = \delta(1-x).$$

Now take free point-like objects inside the proton. For DIS, if Q^2 and ν are large compared to the parton mass, m_i, and to the parton transverse momentum, k_T, the momentum of the i-th parton will be approximately a fraction of the proton momentum,

$$p_i = x_i \cdot p$$

and, since the virtual photon scatters on it, one has $\nu_i = (p_i \cdot q)/m_i = \nu$. The parton mass is $m_i = x_i M$ and the structure function is

$$2m_i W_1^{(i)}(Q^2,\nu_i) = e_i^2 Q^2/2m_i \cdot \delta(\nu_i - Q^2/2m_i) = e_i^2 x\delta(x_i - x)$$

with e_i charge of the parton in terms of the proton charge (and similarly for W_2), implying that the virtual photon must have the right value of x to be absorbed by a parton with momentum fraction x_i. Assuming the incoherence of the scatterings off the partons, one has Bjorken scaling, i.e. no mass scale is explicitly present (compare with the 0.71 GeV^2 of the dipole formula),

$$2MW_1(Q^2,\nu) = \sum_N P(N) \sum_{i=1}^N \int_0^1 dx_i f_N(x_i) 2m_i W_1^{(i)}(Q^2,\nu) =$$
$$= \sum_N P(N) \sum_{i=1}^N e_i^2 x f_N(x) = 2xF_1(x)$$

$$\nu W_2(Q^2,\nu) = \sum_N P(N) \sum_{i=1}^N e_i^2 x f_N(x) = F_2(x),$$

with P(N) probability that the proton has N partons and $f_N(x_i)$ probability that the i-th parton carries a fraction x_i of the proton momentum. Since in the derivation it is assumed that the virtual photon scatters off spin 1/2 particles, in the PM one has the Callan-Gross relation [2.12]

$$F_2(x) = 2xF_1(x) \qquad (2.4).$$

Eq. (2.4) is the master formula of the PM. Working in the infinite momentum frame, it can be shown that relativistic time dialtion slows down the rate of parton interactions with one another. Hence the virtual photon interacts essentially with a free particle [see Fig. 2.1] and this may be rephrased in the incoherence assumption, i.e. the summing of probabilities, rather than amplitudes, of individual scatterings. This impulse approximation is complicated here by the fact that quarks are confined, i.e. there is a final state interaction, which eventually recombines everything into colourless objects. The dressing up however happens over a much longer time scale compared to the initial kick [2.11]. For a discussion of this point in connection with the EMC effect see [2.13]. In 1969 data at $10°$ were added [2.14] reaching higher Q^2 and at the 1970 Kiev conference the data were completed at $18°$ and $26°$ [2.15], showing better evidence for scaling at some values of x (see Fig. 2.3). These new data resulted in the possibility of separating the two structure functions via measurements of the cross section at fixed (x, Q^2) and different angle

$$d^2\sigma/dx dQ^2 = 4\pi\alpha^2/Q^2 \cdot cos^2(\theta/2) \cdot E'/xE \cdot [F_2(x,Q^2) + Q^2/2M^2x^2 \cdot 2xF_1(x,Q^2) \cdot tg^2(\theta/2)]$$

obtaining the first measurement of

$$R = \sigma_L/\sigma_T = [1 + 2Mx/\nu] \cdot F_2/2xF_1 \quad - \quad 1,$$

the ratio between the cross sections for absorption of longitudinal and transverse virtual photons (see Fig. 2.4). In addition the neutron structure function could be separated from the proton one combining data from hydrogen and deuterium targets. The first measurements of R clearly showed that this quantity is closer to zero, expected in the QPM for spin 1/2 partons, than to infinity, expected for spin 0 partons [2.12]. In fact for spin 1/2 quarks σ_L and hence R should be zero by angular momentum conservation and only introducing the Fermi motion of the quarks (and the radiation of gluons in QCD) the expected value of R becomes different from zero but remains small. In any case a decrease of R with Q^2 is expected, e.g. taking into account the quark Fermi momentum k_T, one has $R = 4 \leq k_T^2 \geq /\, Q^2$ [1.3]. Spin 1/2 constituents implied that the most economical assumption was to identify them with the quarks of Gell-Mann and Zweig. Structure function data could then be understood, as we shall see in the next section, with three valence (V) quarks and with a sea (S) of $q\bar{q}$ pairs, the latter infinite in number. At Kiev and later

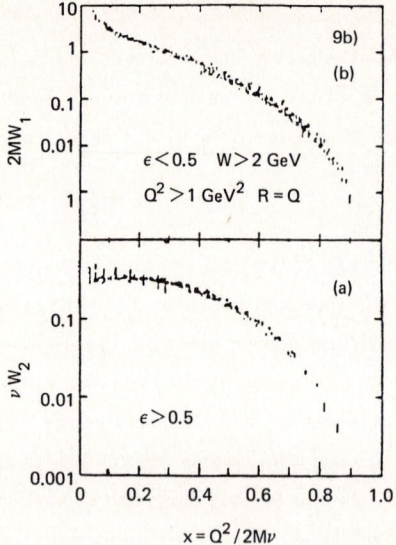

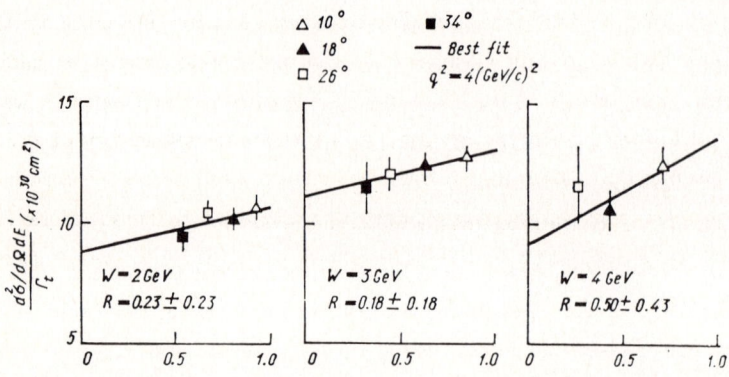

Fig. 2.4

Fig. 2.5

2.4 νW_2 and $2MW_1$ as determined from the double differential cross-section assuming $R = \sigma_L/\sigma_T = 0$. a) $\epsilon \geq 0.5$, b) $\epsilon \leq 0.5$. $\epsilon = 1/[1 + 2(1 + \nu^2/Q^2)tg^2(\theta/2)] \simeq 2(1-y)/[1+(1-y)^2]$ is the virtual photon polarization.

2.5 Measurement of $R = \sigma_L/\sigma_T$. The data are plotted as a function of ϵ.

with more statistics, another piece of evidence came into the puzzle. Copious multihadron production away from the vector meson tails had been observed at Adone [2.16]. The data were in favour of a $\sim 1/E^2$ decrease of the cross sections, similar to e^+e^- annihilations into point-like particles (μ pairs) and with a magnitude comparable to it. Again the data hinted at point-like electrically charged constituents, pair produced in e^+e^- annihilations and subsequently decaying to observable final states. So the idea that there are elementary constituents in the nucleon can be said to have been accepted as from 1970. To identify dynamic partons with static quarks and to tell charge, spin and colour of the partons had to wait longer and was a matter of debate for several years. Charge and colour could indeed have been measured at Adone, but the statistics in each experiment were low and $R = \sigma_{multihadron}/\sigma_{\mu\mu}$ was assessed only later, above the charm threshold.

In 1973 the first ν and $\bar{\nu}$ total cross sections were published by the Gargamelle group [2.17] and in 1975 the first structure functions [2.18]. In DIS of ν and $\bar{\nu}$ the double differential cross section is given by

$$d^2\sigma^{\nu,\bar{\nu}}/dxdQ^2 = G^2/2\pi \cdot [(1+(1-y)^2)/2 \cdot 2xF_1^{\nu,\bar{\nu}}(x,Q^2)+$$
$$+(1-y-Mxy/2E)(F_2^{\nu,\bar{\nu}}(x,Q^2) - 2xF_1^{\nu,\bar{\nu}}(x,Q^2)) \pm (y-y^2/2)xF_3^{\nu,\bar{\nu}}(x,Q^2)] \quad (2.5)$$

with $G \simeq 10^{-5}/M^2$ the Fermi coupling constant, xF_3 a new structure function which appears because of the parity violating character of the weak interaction and $y = \nu/E$ the inelasticity. From the Gargamelle data, the total cross sections were found to be proportional to the incident energy, $E_{\nu,\bar{\nu}}$, a fact which is explained naturally assuming that the ν and the $\bar{\nu}$ scatter on point-like objects, as it can be seen integrating (2.5) in the scaling limit [2.6]. In addition, the structure functions showed a similar behaviour in the e.m. and weak case and, after taking into account the e.m. and weak charges of the partons, were in fact found to be the same. The inelasticity distributions in ν and $\bar{\nu}$ scattering reflect the angular distributions in the rest frame of the neutrino parton system. The observation was essentially flat y distributions in νN and $(1-y)^2$ behaviour in $\bar{\nu}N$, which is naturally interpreted as the scattering off spin 1/2 partons. Pictorially (the $\rightarrow$ denotes momentum, the $\Rightarrow$ helicity in the limit in which masses are negligible)

ν $\overset{\Leftarrow}{\rightarrow}$ $\overset{\Rightarrow}{\leftarrow}$ spin 1/2 parton $\quad J_z = 0 \quad$ flat angular or y distribution
$\bar{\nu}$ $\overset{\Rightarrow}{\rightarrow}$ $\overset{\Rightarrow}{\leftarrow}$ spin 1/2 parton $\quad J_z = 1 \quad (1-y)^2$

whilst for $\nu, \bar{\nu}$ on spin 0 partons one expects a (1-y) dependence. The agreement of the data with this picture, after taking into account a small momentum fraction carried by antipartons, is found to be quite good [2.19]. Upon integration of (2.5), one expects naively that the ratio of total cross sections, $R_\nu = \sigma_{\bar{\nu}}/\sigma_\nu$, is $1/3 \leq R_\nu \leq 3$, the experimental value being close to 0.5 [2.19].

Further confirmations of the quark-parton picture came in the following years with the observation of 2-jet events in e^+e- annihilations [2.20], with the discovery of heavy flavours [2.21] and with the observation of jets in hadron-hadron collisions [2.22].

3 PROPERTIES OF QUARKS AND GLUONS

> Ce que le tableau represente cela depend de
> celui qui le regarde (J.A. McNeill Whistler)

3.1 Structure functions and parton densities [3.1]

Assuming that the nucleon contains partons of flavour f = u,d,s,c and neglecting scaling violations as a first approximation (or working at fixed Q^2), the structure functions for charged lepton scattering on protons are

$$2xF_1^{lp} = F_2^{lp} = \sum_f e_f^2 x[f(x) + \bar{f}(x)] \quad (3.1)$$

where f(x) etc. are the probabilities that a parton of flavour f with charge e_f carries a fraction x of the proton momentum. The validity of the Callan-Gross relation [2.12] is consistently assumed when using eq. (2.4). By isospin invariance the u[d] parton distribution in the neutron is the same as the d[u] in the proton, with s and c unchanged. For an isoscalar target one therefore has

$$2xF_1^{lN} = F_2^{lN} = 1/2(e_d^2 + e_u^2)x[u(x) + \bar{u}(x) + d(x) + \bar{d}(x)]$$
$$+ e_s^2 x[s(x) + \bar{s}(x)] + e_c^2 x[c(x) + \bar{c}(x)] \quad (3.2)$$

For neutrino scattering on protons one has

$$2xF_1^{\nu p} = F_2^{\nu p} = 2x[\bar{u}(x) + d(x) + s(x) + \bar{c}(x)]$$
$$2xF_1^{\bar{\nu} p} = F_2^{\bar{\nu} p} = 2x[u(x) + \bar{d}(x) + \bar{s}(x) + c(x)]$$
$$xF_3^{\nu p} = 2x[d(x) + s(x) - \bar{u}(x) - \bar{c}(x)]$$
$$xF_3^{\bar{\nu} p} = 2x[u(x) + c(x) - \bar{d}(x) - \bar{s}(x)] \quad (3.3)$$

For the neutron again one exchanges u with d and viceversa, so that for isoscalar targets it follows

$$2xF_1^{\nu N} = F_2^{\nu N} = 2xF_1^{\bar{\nu} N} = F_2^{\bar{\nu} N} = \sum_f x[f(x) + \bar{f}(x)]$$
$$xF_3^{\nu N} = xF_3^{\bar{\nu} N} = 1/2[xF_3^{\nu p} + xF_3^{\nu n}] = \sum_f x[f(x) - \bar{f}(x)] \quad (3.4)$$

3.2 The charge of the quarks

The charge of the quarks may be extracted comparing the structure functions measured in DIS of charged leptons and neutrinos. Taking the ratio of (3.2) and (3.4) we have

$$F_2^{lN}/F_2^{\nu N} = 1/2 \cdot (e_u^2 + e_d^2) \cdot [1 - 3/5 \cdot x(s + \bar{s} - c - \bar{c})/F_2^{\nu N}]. \quad (3.5)$$

The correction term is small and decreases with x: using $2s/(\bar{u} + \bar{d}) = 0.5$ from opposite sign dimuon events [3.2] and the measured charm sea distribution [3.3], one has

~ 6% correction at x = 0.03. The data on iron, after correction for a small non-isoscalarity, and on carbon are shown in Fig. 3.1; they are consistent with eq. (3.5) within 5 to 10 % normalization shifts, which are compatible with the errors quoted by the groups, if one compares the measurements in iron by the WA1 [3.4], the CCFRR [3.5] and the EMC [3.6] collaborations (see Fig. 3.1). The statistically very precise data of the BCDMS collaboration in carbon [3.7] show however a systematic trend with x when compared with other experiments, which will be discussed later.

The agreement with eq. (3.5) and in particular with $1/2(e_u^2 + e_d^2) = 5/18$ as predicted by the naive QPM implies that both charged leptons and neutrinos scatter from the same point like constituents of the nucleon, provided that third integral electric charges are assigned to them. In addition, the detailed form of the structure functions in terms of quark-parton densities, eqs. (3.1) to (3.4), is found to be valid. In other words the constancy of the ratio (3.5) with x measures the mean square charge of the u,d quarks following a variation of the absolute value of the $F_2's$ by about two orders of magnitude. Taking into account normalization uncertainty and other systematic effects and taking the average of the measurements of the EMC [3.8] and WA25 [3.9] collaborations in deuterium, one has $1/2(e_u^2 + e_d^2) = 0.29 \pm 0.02$. The charges of the different flavour partons can be extracted assuming that there are three valence quarks in the proton, two u-flavoured and one d-flavoured, as in the static model, so that $\int u_V dx = 2$ and $\int d_V dx = 1$, which is supported by the data (see next section). From eq. (3.1) one has

$$F_2^{lp} - F_2^{ln} = (e_u^2 - e_d^2)x[u_V(x) - d_V(x)]$$

and the Gottfried sum rule [3.10] follows

$$I = \int_0^1 dx \cdot [F_2^{lp} - F_2^{ln}]/x = e_u^2 - e_d^2$$

Fig. 3.2 shows $F_2^{lp} - F_2^{ln}$ measured in deep inelastic electron [3.11] and muon [3.8] scattering. The integral is measured between 0.02 and 0.8 and is extrapolated over the rest of the x-region; in particular assumptions of well behaved integrand as $x \to 0$ have to be made. The result is $I = 0.24 \pm 0.11$ [3.8], where the error is mainly systematic, compatible, even if lower, with the QM model prediction 1/3. Solving from the sum and the difference of the square charges of the partons, one obtains

$$|e_u| = 0.64 \pm 0.05 \qquad |e_d| = 0.41 \pm 0.08$$

consistent with the values +2/3 and -1/3 assigned to the u and d quarks respectively. The signs of the parton charges can in principle be measured in other experiments, via the asymmetry in e^+/e^- deep inelastic bremsstrahlung and via the $Z^0\gamma$ interference structure function in e or μ DIS, xG_3. The interference between the Bethe-Heitler process and virtual Compton scattering in DI bremsstrahlung on protons and deuterons yields information on the cubic charge of the partons according to the Brodsky, Gunion and Jaffe sum rule [3.12]. Its measurement was originally performed in an attempt to resolve some problems with inelastic Compton scattering results, which appeared in disagreement with the predictions of the QPM and to distinguish between the normal charge assignment, integral and third integral parton charges, and the large charges needed to explain those results [3.13]. The measurement on protons yields $2e_u^3 + e_d^3 = 0.89 \pm 0.34$ [3.13], compatible with the expectation of 0.56 and singling out a positive value for e_u, the error being too large to constrain e_d. xG_3 is a non-singlet structure function arising from the interference of the vector (weak and electromagnetic) and of the weak axial-vector amplitudes. It is similar to xF_3 measured in neutrino scattering and it can be expressed as

$$xG_3(x) = 2x \sum_f a_f e_f [f(x) - \bar{f}(x)]$$

with $a_u = 1/2$ and $a_d = -1/2$ the axial vector couplings of the quarks to the Z^0 in the standard model. Within the minimal model and neglecting the antiquarks, i.e. at large x, the ratio of the structure functions is $xG_3/F_2 = 9/5$. The result of the BCDMS collaboration for $x \geq 0.2$ is $1.87 \pm 0.25 \pm 0.24$, the first error being statistical and the second systematic, in agreement with the standard assignments for the weak and electric charges of the quarks and in particular with the sign of the difference $e_u - e_d$ [3.14].

Additional tests of the weak and electric charges of the partons have recently been published by the CHARM collaboration [3.15], which measures the x-distributions in neutral current (NC) ν and $\bar{\nu}$ scattering on marble

$$F_2^{NC}(x) = x \cdot [(u_L^2 + u_R^2 + d_L^2 + d_R^2) \cdot (q_V(x) + 2\bar{q}(x)) - 2(u_L^2 + u_R^2 - d_L^2 - d_R^2) \cdot (s(x) - c(x))]$$
$$xF_3^{NC}(x) = x \cdot (u_L^2 - u_R^2 + d_L^2 - d_R^2) \cdot q_V(x),$$

where $u_{L,R}^2 [d_{L,R}^2]$ are the left/right-handed weak couplings of the u[d] flavoured partons. The comparison with charged lepton data therefore measures a combination of weak and electric charges. The CHARM F_2 data are compared in Fig. 3.3a with the BCDMS F_2 muon-carbon data. For large x (x ≥ 0.3) the valence quark approximation may be assumed valid and antiquarks may be neglected, obtaining the ratio averaged over x

$$R_2 = 18/5 \cdot (g_L^2 + g_R^2) \cdot F_2^{lN}/F_2^{NC} \quad = \quad 1.13 \pm 0.10,$$

where $g_{L,R}^2 = u_{L,R}^2 + d_{L,R}^2$, in agreement with the expectations. The valence NC structure function is plotted in Fig. 3.3b together with the electroweak interference structure function xG_3 determined in DI μ scattering [3.14]. The ratio is

$$R_3 = 2(g_L^2 - g_R^2) \cdot xG_3/xF_3^{NC} \quad = \quad 0.98 \pm 0.19,$$

averaged over x, again in agreement with the quark model predictions. These data confirm the universality of the nucleon structure functions as probed by the weak NC neutrino and antineutrino interactions and by the e.m. current.

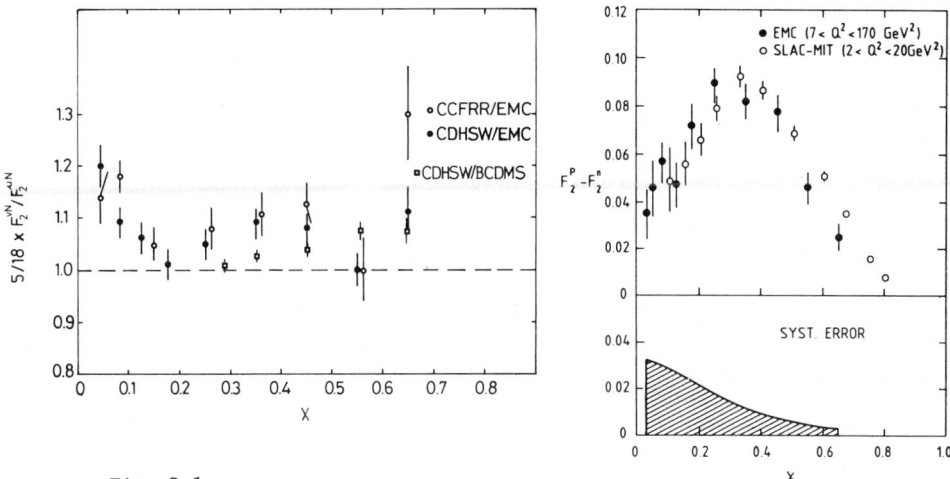

Fig. 3.1

Fig. 3.2

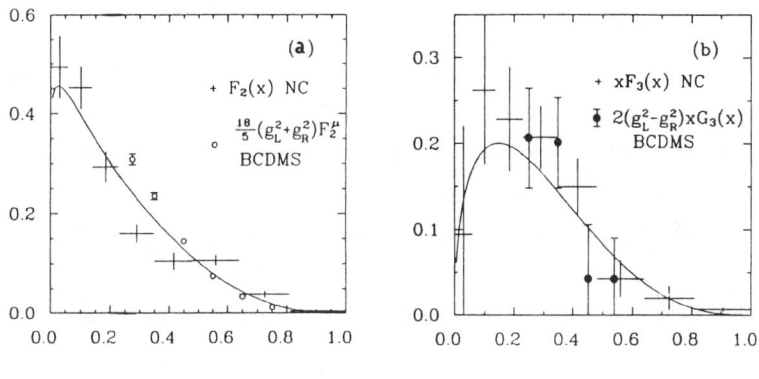

Fig. 3.3

3.1 The ratio between the structure function F_2 measured in neutrino and in muon scattering on isoscalar nuclei. The dotted line shows the effect of the strange and charm seas at small x.

3.2 $F_2^p - F_2^n$ averaged over Q^2 as a function of x. The systematic error for the EMC data is shown in the lower part of the figure.

3.3 The structure functions a) $F_2(x)$ and b) $xF_3(x)$ at $Q_0^2 = 10 GeV^2$ measured by the CHARM collaboration in NC neutrino interactions. The full curves show the results of a fit [3.15].

  a) In comparison with F_2^{NC} we show $18/5(g_L^2 + g_R^2)F_2^\mu$ with F_2^μ derived from DIS on carbon. An uncertainty of ±3% in the absolute normalization is not included in the error bars.

  b) In comparison with xF_3^{NC} we show $2(g_L^2 - g_R^2)xG_3$ with xG_3 electroweak interference strucure function measured from the charge asymmetry of DIS of muons.

3.3 The number of valence quarks in the proton

From eq. (3.4), dividing by x and integrating over all x, one has the Gross-Llewellyn-Smith sum rule [3.16] which yields the number of valence partons (apart from a small QCD correction [3.17])

$$\int_0^1 F_3^{\nu N} dx = \sum_f (N_f - N_{\bar{f}}) \cdot (1 - \alpha_s/\pi)$$

The data need to cover very low x at fixed Q^2, otherwise much of the uncertainty on the integral comes from extrapolation. As an example the data of the WA25 collaboration are shown in Fig. 3.4 [3.18]. The average of the available data [3.4,3.5,3.18,3.19] yields $\sum_f (N_f - N_{\bar{f}}) = 3.0 \pm 0.2$ [3.20] and is compatible with the QPM expectation of three valence quarks in the nucleon.

The separate number of u flavoured and d flavoured quarks in the proton can be obtained from neutrino data on hydrogen and deuterium targets

$$1/2 \int_0^1 dx (F_2^{\nu n} - F_2^{\nu p})/x = N_{u_V} - N_{d_V}$$

assuming that the sea is the same in the proton and in the neutron (Adler sum rule [3.21]). The data of the WA25 collaboration [3.18] give a mean value of 1.07 ± 0.20 averaged over Q^2, again in agreement with 1 expected in the QM. Solving for the sum and the difference, one has $N_{u_V} = 2.04 \pm 0.14$ and $N_{d_V} = 0.97 \pm 0.14$ in perfect agreement with the naive model. If the electric charges of the quarks are assumed from the naive model, one may also use charged lepton data and the value of the Gottfried sum rule is expected 1/3, which, as discussed in the previous section, is checked by the EMC data.

3.4 The momentum sum rule and the role of gluons

From eqs. (3.1) and (3.4) the integrals of the $F_2's$ yield at fixed Q^2 the charge weighted fraction of momentum of the nucleons carried by the partons, i.e.

$$\int_0^1 F_2^{lp} dx = \int_0^1 \sum_f e_f^2 x (f + \bar{f}) dx$$
$$\int_0^1 F_2^{\nu N} dx = \int_0^1 \sum_f x (f + \bar{f}) dx.$$

The data, both with neutrinos [3.4,3.5] and charged leptons [3.11,3.22], roughly saturate half of the sum rule, i.e. only half of the proton momentum is carried by quarks (see Fig. 3.5). The rest of the momentum must be carried by objects which do not interact weakly or electromagnetically, at least within the present resolution of the virtual boson probes, hinting towards the existence of neutral partons, the gluons. This conjecture is confirmed by the observation of scaling violations in nucleon structure functions, which follow the pattern predicted by QCD and allow a gluon structure function to be extracted from the data (see next section).

Additional striking evidence for gluons comes from three or more jet events in e^+e^- collisions [3.23] and from hadron jets at the CERN $p\bar{p}$ collider [2.22]. Further, the capability

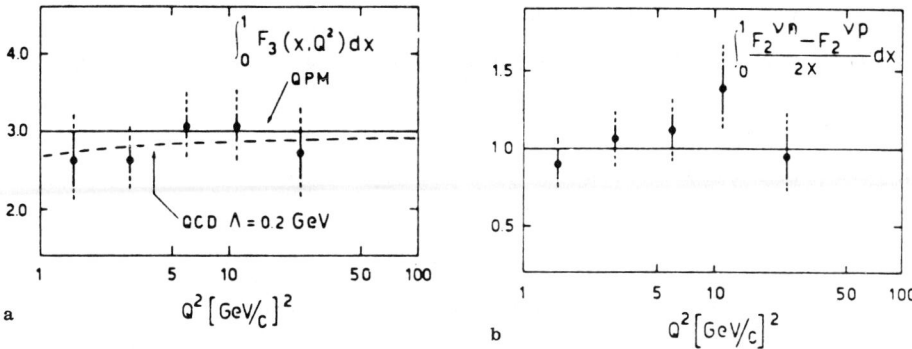

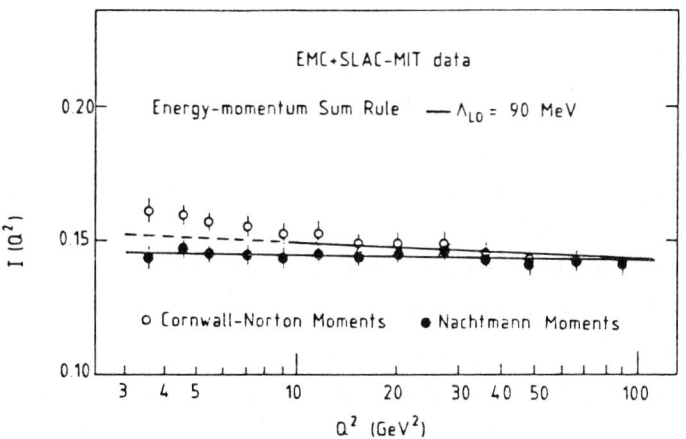

Fig. 3.4

Fig. 3.5

3.4 Evaluation of the Gross-Llewellyn-Smith(a) and Adler(b) sum rules as a function of Q^2 [3.18]

3.5 The energy momentum sum rule [3.1]

of QCD of explaining a wide variety of phenomena supports the evidence for the existence of gluons, which are assumed to be the quanta of the strong interaction binding the quarks into hadrons.

4 RECENT DATA ON STRUCTURE FUNCTIONS

> In nature's infinite book of secrecy a
> little I can read (William Shakespeare)

4.1 Precise determination of $R(x,Q^2) = \sigma_L/\sigma_T$

Several experiments have contributed recently precise data on R, both at low and high Q^2, between 1 and 100 GeV^2, for x between 0.05 and 0.7, changing substantially the scenario with respect to the old SLAC data, which yielded a value of R as large as 0.2 at large x (see Fig. 4.1). All the new results agree with each other and show clearly for the first time that R is definitely small and a smoothly decreasing function of both x and Q^2, in agreement with the QPM and QCD predictions, once target mass corrections are included at low Q^2. It is compatible with zero at $x \geq 0.3$ and $Q^2 \geq 50 GeV^2$ and $\simeq 0.1$ at $x \sim 0.2$ and $Q^2 \sim 10 GeV^2$. The data at large Q^2 are from neutrino-iron (CDHSW) [3.4,4.1], muon-iron and muon-proton (EMC) [3.6,3.22], muon-carbon and muon-proton (BCDMS) [3.7,4.2], neutrino-marble (CHARM) [3.19] scattering. Precise measurements at lower Q^2 come from electron scattering (E140) [4.3]. Fig. 4.2 illustrates the method used for the measurement of R at high energy. Rather than measuring the cross section in a given (x,Q^2) bin at fixed energy and different angles as it is done at low energy, the inelasticity and hence the virtual photon polarization, $\epsilon = 2(1-y)/[1+(1-y)^2]$, are varied by changing the beam energy. The data of the BCDMS collaboration at 120,200 and 280 GeV muon energy are shown in Fig. 4.2, where R = 0, consistent with the measurements where the data overlap, is used all over the (x,Q^2) region.

An empirical parametrization of $R(x,Q^2)$, comprising a term to account for the perturbative QCD calculation and a term in $1/Q^2$ for non perturbative contributions, gives a good representation of the data

$$R(x,Q^2) = [\alpha(1-x)^\beta/ln(Q^2/\Lambda^2) + \gamma(1-x)^\delta/Q^2]$$

where $\alpha \sim 1.1, \beta \sim 3.3, \gamma \sim 0.1, \delta \sim 1.9$ and $\Lambda = 0.2 GeV$. Notice also that the data show little dependence on target nuclei within errors. The conclusion on R is that it is consistent with QCD predictions once the target mass effects are included for a reasonable value of the QCD parameter, $\Lambda \sim 0.2 GeV$, thus confirming the spin 1/2 nature of the charged partons, without any need for invoking the scattering on spin 0 diquarks or large non perturbative and higher twist effects which were necessary to explain the old SLAC

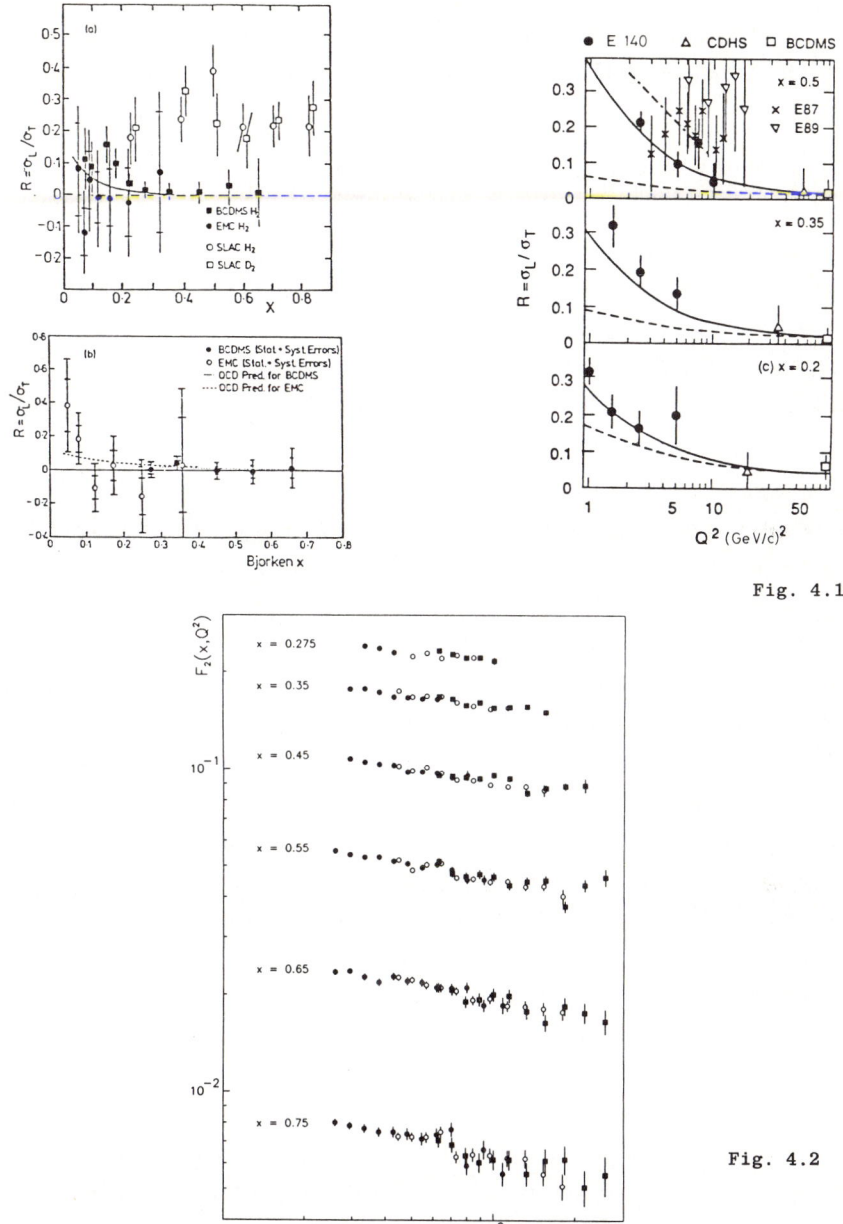

Fig. 4.1

Fig. 4.2

4.1 Measurements of R as a function of x from a) hydrogen targets, b) carbon and iron targets in DIS of muons. c) Measurements of R as a function of Q^2 from DIS of muons and neutrinos. The dashed line shows the predictions from perturbative QCD, the solid line shows the QCD predictions with target mass effects and the dash-dotted line is the prediction of a diquark model.

4.2 F_2 from BCDMS carbon data at different muon energies. Full dots: 120 GeV; open dots: 200 GeV; full squares: 280 GeV. R=0 has been at all (x,Q^2) values

data.

4.2 Scaling violations of the nucleon structure functions

All measurements indicate qualitatively the same behaviour: the nucleon structure functions at fixed x are roughly independent of Q^2 and show mild (log) scaling violation patterns. At small x (below 0.2) the structure functions increase with Q^2, whilst they fall with Q^2 at large x. The existence of gluons, the radiation of gluons by quarks and the pair creation of $q\bar{q}$ by gluons explain naturally this behaviour. Gluons are emitted from quarks in a way analogous to bremsstrahlung from accelerated electric charges. At fixed $x = Q^2/2M\nu$, if Q^2 grows so does ν, the energy of the virtual boson, and the radiation probability increases, leading to an overall shrinking of the nucleon structure functions.

The structure function F_2 measured by the BCDMS collaboration on carbon is shown in Fig. 4.3 [3.7]. Following the results of the previous section the value of F_2 is extracted assuming $R = R_{QCD}$ all over the (x, Q^2) range covered by the experiment, $0.25 \leq x \leq 0.8, 25 \leq Q^2 \leq 250 GeV^2$. Mild logaritmic scaling violations are observed in qualitative agreement with the trend seen in other experiments. Comparing the BCDMS data with other μ experiments on nuclei at large Q^2 some systematic difference is indeed observed. Within errors the disagreement with the BFP iron data [4.4] amounts to about 5 % difference in absolute normalization. The disagreement with the EMC iron data [3.6] (see Fig. 4.4) instead is both in magnitude and in shape. The BCDMS data decrease more rapidly with x than the EMC data and this effect is in any case in a direction opposite to the trend of nuclear effects, which however between C and Fe would be small. The same systematic difference is indeed observed between the preliminary BCDMS H_2 data [4.2] and the EMC ones [3.22](Fig. 4.5). They are 15 % lower at very low x and equal for x $\sim$ 0.6, i.e. the x dependence measured in different experiments is different. The origin of this discrepancy is not yet known. It should be remarked in this respect that small calibration errors may reflect in large systematic errors on the structure functions [4.5] and that in addition it is hard to determine the absolute normalization to better than a few percent. Recent preliminary data of the NEMC group in D_2 [4.6] are somewhat higher than the previous EMC measurements at x $\leq$ 0.10 and, since the discrepancy is essentially present for all targets, most of the discrepancy at low x would automatically disappear. However the same systematic trend with x is also observed when comparing BCDMS muon data with the CDHSW neutrino structure function, so not all problems would be solved.

4.3 Analysis of the scaling violations and measurement of α_s

The QCD analysis of the BCDMS muon-carbon data is complicated by the fact that F_2 has both a singlet and a non-singlet piece and that it is in general dependent on the

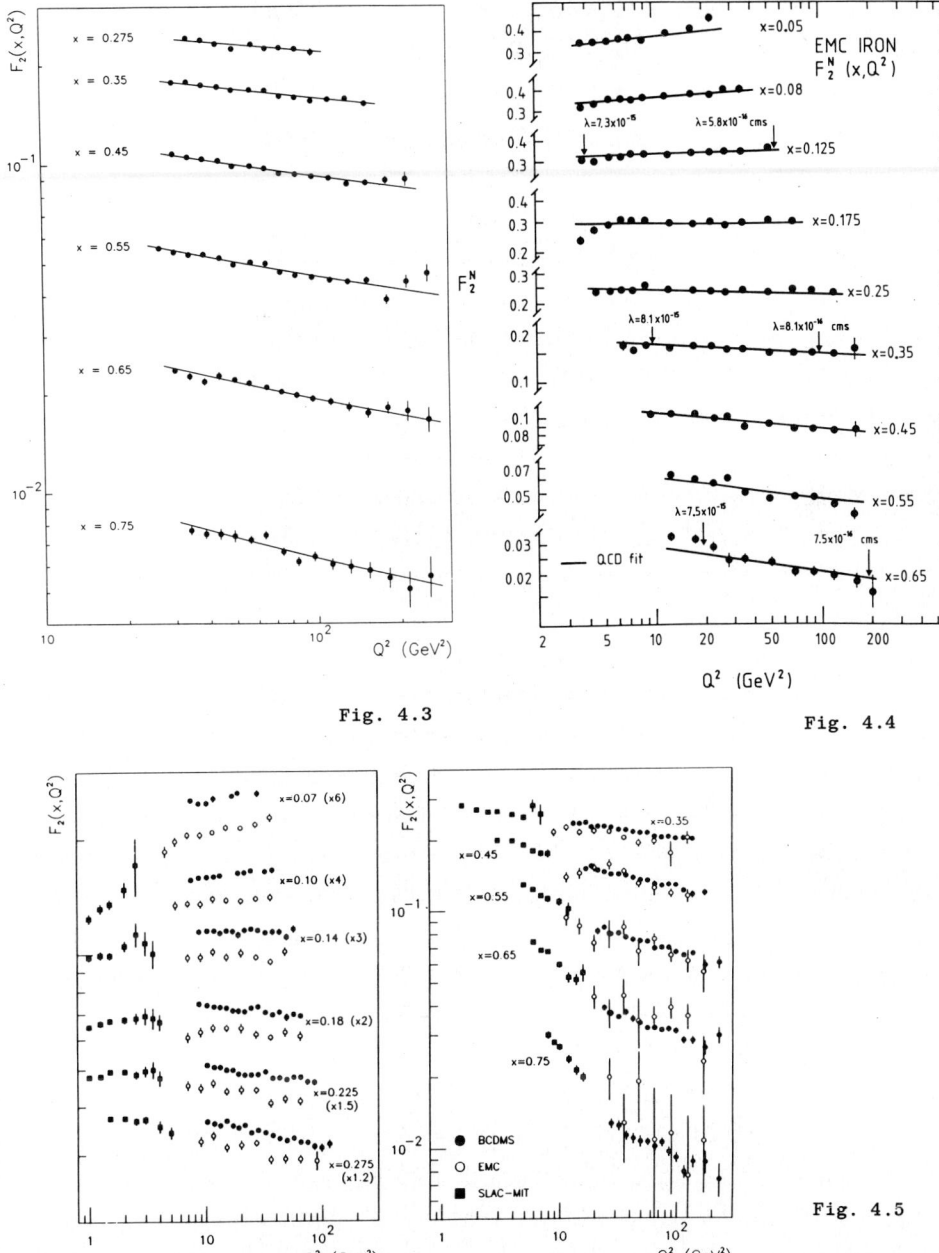

Fig. 4.3

Fig. 4.4

Fig. 4.5

4.3 F_2 from BCDMS carbon data obtained combining the measurements of Fig. 4.2. The solid line shows the results of a QCD fit discussed in the text
4.4 F_2 from EMC iron data, λ is the wavelength of the virtual photon. The solid line shows the results of a QCD fit.
4.5 The proton structure function. The preliminary BCDMS data are compared with the EMC and with the early SLAC-MIT data

gluon distribution, which is not directly observed. The Altarelli-Parisi (AP) equation for F_2 is

$$dF_2(x,Q^2)/dlnQ^2 = \alpha_s(Q^2)/2\pi \cdot \int_x^1 [P_{qq}(z) F_2(x/z,Q^2) + 2\sum_{i=1}^f e_i^2 P_{gq}(z) \cdot x/z \cdot G(x/z,Q^2)]dz$$

with

$$\alpha_s(Q^2)/4\pi = 1/\beta_0 ln(Q^2/\Lambda^2) \cdot [1 - \beta_1 lnln(Q^2/\Lambda^2)/\beta_0 ln(Q^2/\Lambda^2)],$$
$$\beta_0 = 11 - 2/3f, \qquad \beta_1 = 102 - 38/3f,$$

where α_s is given in next to leading order (NLO) and f is the number of active flavours. P_{qq} and P_{gq} are the splitting functions for the processes q→qg and g→qq respectively [3.1] and $G(x,Q^2)$ is the gluon structure function. However in the x-range covered by the BCDMS carbon data the gluon distribution is expected to die out, so that the non-singlet approximation can be used. The data have been fitted using the Altarelli-Parisi equation directly [4.7]. Fits have been made both in LO and NLO (in the modified minimal subtraction scheme,$\overline{MS}$) neglecting the gluon contribution in the previous eq. for x $\geq$ 0.3. Several fitting programs [4.8] all yield compatible results for Λ, both in LO and in the $\overline{MS}$ scheme

$$\Lambda_{\overline{MS}} = 230 \pm 20(stat) \pm 60(syst) MeV$$
$$\alpha_s(Q^2 = 100 GeV^2) = 0.160 \pm 0.003(stat) \pm 0.010(syst)$$

It has to be noticed that the BCDMS carbon data are all at large Q^2 ($Q^2 \geq 25$ GeV^2) and therefore the problems originating from non perturbative effects ($\sim 1/Q^2$) should be negligible. In addition the comparison between the predictions of QCD for the Q^2 dependence of F_2 at different x (the x-dependence of scaling violations) is in perfect agreement with the data, which is not always the case for other experiments (see Fig. 4.6).

Two different QCD analyses have been performed on the preliminary BCDMS H_2 data [4.2]. The first, for x$\geq$0.275 and $Q^2 \geq 20 GeV^2$, uses the non singlet approximation neglecting the contribution of the gluons. The same fitting programs were used in the analysis yielding

$$\Lambda_{\overline{MS}} = 205 \pm 22(stat) \pm 60(syst) MeV$$
$$\alpha_s(Q^2 = 100 GeV^2) = 0.156 \pm 0.004(stat) \pm 0.011(syst)$$

in agreement with the results of the fit to the carbon data. The second analysis uses data at all values of x and Q^2 (0.06$\leq x \leq$0.80, 8$\leq Q^2 \leq$260GeV^2) by decomposing the proton structure function into singlet and non singlet parts. They follow different evolution equations and the singlet part is coupled to the gluon distribution, which is parametrized as $xG(x,Q_0^2) = A(\eta + 1)(1 - x)^\eta$ at $Q_0^2 = 5 GeV^2$. The results of the fits yield values of $\Lambda_{\overline{MS}}$ close to the NS fit. The fitted value of A is checked to be compatible with the

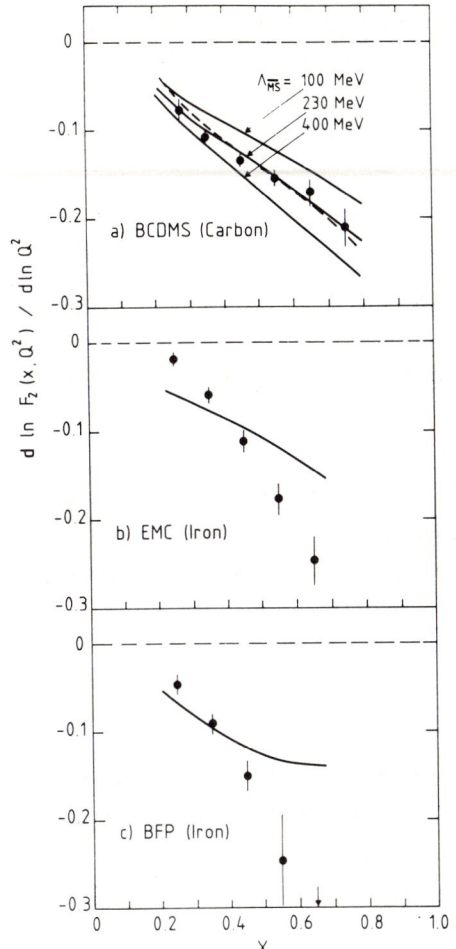

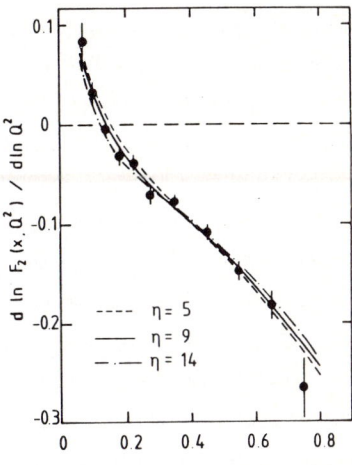

Fig. 4.7

Fig. 4.6

4.6 Scaling violations expressed as logarithmic derivatives, $dlnF_2(x,Q^2)/dlnQ^2$. The errors are statistical only. a) BCDMS carbon data. The solid lines are nonsinglets QCD predictions for different values of Λ. b) As in a) for the EMC iron data for $Q^2 \geq 10$ GeV^2. c) As in a) for the BFP data on iron for $Q^2 \geq 10$ GeV^2

4.7 Logarithmic derivatives of F_2 for the full kinematical range of the BCDMS hydrogen data. The singlet+nonsinglet prediction in NLO QCD is also shown for different exponents of the gluon distribution.

momentum sum rule within rather large errors. Two different minima are found in LO and NLO for the exponent η of the gluon distribution, the latter being much softer ($\eta \sim 9$). Again the observed logaritmic derivative of F_2 can be compared with the predictions of QCD and the agreement is remarkable (see Fig. 4.7).

Values of Λ and α_s have been obtained in other DIS experiments since about 10 years. All the experiments measure α_s at one point, which is between a few GeV^2 and 100 GeV^2; so in order to test the running of α_s, one has to combine the DIS results with the data from different processes. The values of α_s in LO are displayed vs Q^2 in Fig. 4.8. Measurements from DIS and other processes have been included . The muon DIS are the average of the measurements of the BCDMS collaboration while the other values are taken from the review paper of T. Sloan [4.9]. It is clear that at the moment, with the present value of Λ, it is not possible to see α_s run within a single experiment. Indeed whether it runs at all between 10 and 10,000 GeV^2 is far from clear, though its behaviour is compatible with the QCD prediction. Notice that the effect is small: over that large Q^2 interval covered by totally different reactions α_s is expected to vary by a mere factor of 2. When this business started with $\Lambda \sim 500$ MeV in 1978, there was reasonable hope to see α_s run. Now the best possibility seems to be within one single experiment on DIS at Q^2 moderately large, say 20 to 500 GeV^2, where the first limit is dictated by the need of keeping the target mass corrections and higher twist effects low, whilst the upper limit is dictated by statistics and existing machines. Needless to say, huge statistics and quite good control of systematics are a must: for once precision is at premium, rather than just an energy increase.

5 CONCLUSIONS

Predicting is difficult, expecially if it

concerns the future (R. Storm Petersen)

After 20 years of circumstantial evidence the quarks are still unobserved and moreover the theory which naturally explains the lack of observation of free quarks with the binding of coloured gluons has not been tested in its most peculiar prediction, the logarithmic decrease of α_s with Q^2. If the quarks really exist (and the indirect evidence is compelling), either they are seen and the theory must be changed or the theory which justifies why they are not seen must be verified quantitatively. After all, they could just be the ad hoc expression of some underlaying strong interaction dynamics and the answer is as usual left to experimentation. Perturbative QCD is probably best studied by very accurate DIS at fixed target machines. Large energy ep colliders such as HERA, with the potentiality of reaching the so far unexplored very small x region, should permit better measurements of

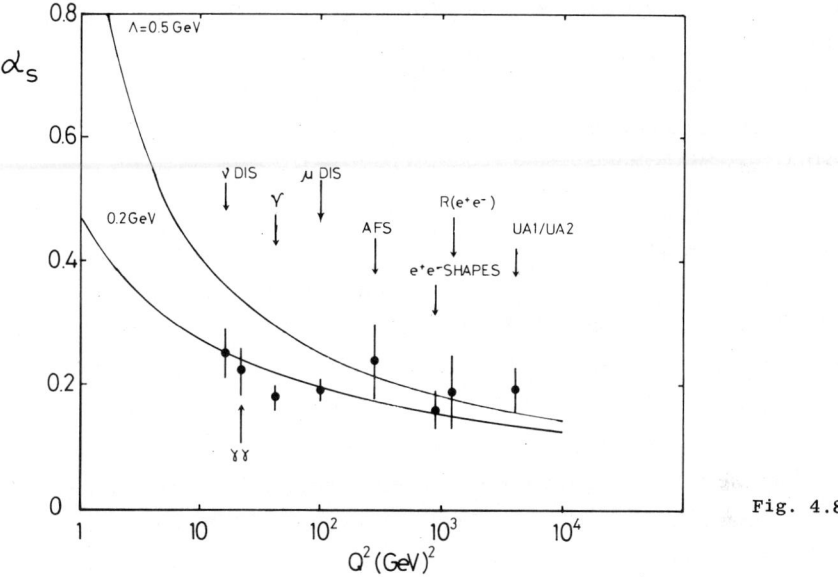

4.8 The measured values of α_s from different techniques, as a function of Q^2. The solid lines show the values of α_s computed in LO QCD.

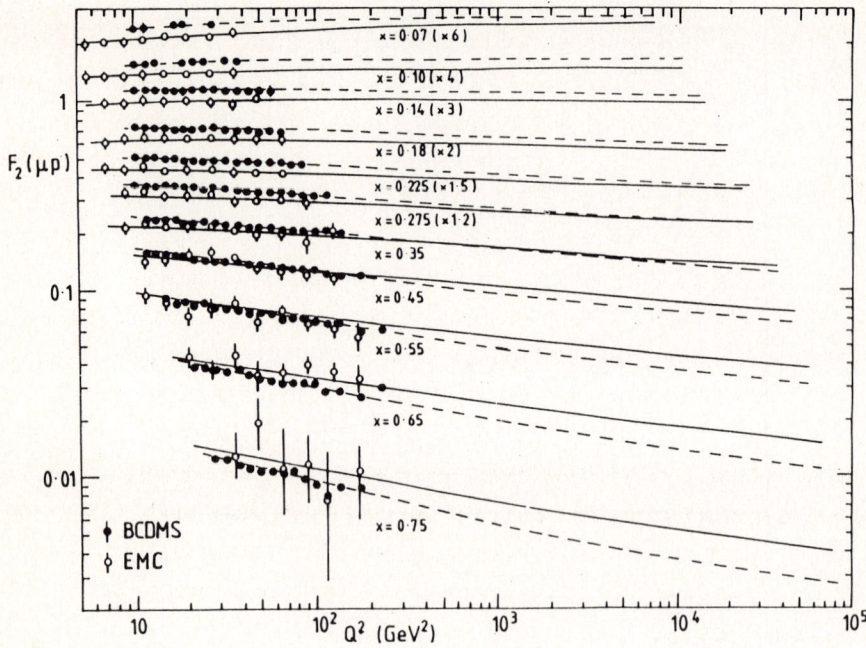

Fig. 5.1

5.1 Extrapolation of the fits to the structure function data to the HERA Q^2 domain [5.1]

the sum rules and in addition may just reveal a composite structure and/or a wealth of new particles. An extrapolation of fits of the present data on structure functions to the HERA Q^2 domain is shown in Fig. 5.1 [5.1]. The BCDMS and EMC hydrogen data were fitted separately in combination with high energy neutrino data (CDHSW): on a log-log scale the discrepancies do not appear now to be enormous and the data from HERA could perhaps just decide between them after taking into account statistical and systematic limitations. Non perturbative QCD may be tested by fixed target hadron spectroscopy (hadron factories), by large energy heavy ion colliders searching for quark-gluon plasmas and by lattice calculations of the hadron spectrum. Other questions concerning quarks are as well still unanswered. For example why, if they exist, they are so many - which again could point to some different explanation and/or to some further layer of elementary objects. What to me seems a conclusion is that, independently from the precision achieved in measuring sum rules and the accuracy in testing perturbative QCD, the most fundamental question is still unanswered and quarks do not exist in any ordinary sense. In this respect their 20 year old discovery could perhaps be compared to the revolution started with inelastic scattering off the nuclear atom which ended up in quantum mechanics.

REFERENCES
1.1 - W.K.H. Panofsky, 14th Int. Conf. on High Energy Physics, Vienna 1968, J. Prentki and J. Steinberger editors, CERN 1968, p. 23
 - E. Bloom et al., paper n.363 submitted to that conference
1.2 M. Gell-Mann, Phys. Lett. 8, 214 (1964)
 - G. Zweig, CERN Report 8419/Th412 (1964)
1.3 R.P. Feynman, Photon hadron interactions, W.A. Benjamin inc., 1972
1.4 see F.E. Close, An introduction to quarks and partons, Academic Press, New York (1979)
1.5 see I.J.R. Aitchison and A.J.G. Hey, Gauge theories in particle physics, Adam Hilger, Bristol (1982)
 - T. Muta, Foundations of quantum chromodynamics, World Scientific, Singapore (1987)
1.6 L. Lyons, Phys. Rep. 129, 255 (1985)
1.7 J.J. Aubert et al, EM coll., Phys. Lett. 123B, 275 (1983)
1.8 J. Ashman et al., EM coll., Phys. Lett. 206B, 364 (1988)
1.9 D.H. Perkins, Summary talk at the 24th Int. Conf. on High Energy Physics, Munich 1988
2.1 M. Aguilar-Benitez et al., Particle Data Group, Phys. Lett. 170B, 79 (1986)
2.2 see e.g. A. Le Yaouanc, L. Oliver, O. Pene and J.C. Raynal, Hadron transitions in the quark model, Gordon and Breach, 1988
2.3 M. Gell-Mann, Proc. of the XIII Int. Conf. on High Energy Physics, Berkeley (1967), p. 3
2.4 R.H. Dalitz, Proc. of the XIII Int. Conf. on High Energy Physics, Berkeley (1967), p. 236
2.5 W.O. Greenberg, Phys. Rev. Lett. 13, 598 (1964)
 - W.O. Greenberg and D. Zwanziger, Phys. Rev. 150, 1177 (1966)
 - J. Han and Y. Nambu, Phys. Rev. 139B, 1006 (1971)
 - M. Gell-Mann, Acta Physica Austriaca Suppl. 9, 733 (1972)
 - M. Gell-Mann and H. Fritzsch, Proc. XVI Int. Conf. on High energy Physics, eds. J.D. Jackson and A. Roberts, Fermilab (1972), vol. 2, p. 135
2.6 R. Hofstadter, Rev. Mod. Phys. 28, 214 (1956)

- R. Hofstadter, Ann. Rev. Nucl. Science 7, 231 (1957)
2.7 N.M. Kroll, 14th Int. Conf. on High Energy Physics, Vienna 1968, J. Prentki and J. Steinberger editors, CERN 1968, p. 75
- E. Lorhmann, Proc. Int. Conf. on Elem. Part. Phys., Lund (1969), p. 11
- A. Suri and D.R. Yennie, Ann. Phys. 72, 243 (1972)
- J.J. Sakurai, inProperties of the fundamental interactions, 1971 Int. School of Subnuclear Physics, ed. by A. Zichichi, Editrice Compositori, Bologna (1973), p. 242
2.8 J.D. Bjorken, Varenna, Enrico Fermi School, Course XLI, 1967, p. 55
2.9 J.D. Bjorken, Phys. Rev. 179, 1547 (1969)
2.10 R.P. Feynman, Phys. Rev. Lett. 23, 1415 (1969) and unpublished
2.11 J.D. Bjorken and E.A. Paschos, Phys. Rev. 185,1975 (1969)
2.12 C.G. Callan and D.J. Gross, Phys. Rev. Lett. 22, 156 (1969)
2.13 see C. Peroni, these proceedings
2.14 R.E. Taylor, Proc. of the 4th lepton-photon symposium, Liverpool 1969, p. 251
- E.D. Bloom et al., Phys. Rev. Lett. 23, 930 (1969)
- M.L. Breidenbach at al., Phys. Rev. Lett. 23, 935 (1969)
2.15 E.D. Bloom et al., SLAC-PUB-796 (1970), presented at the 15th Int. Conf. on High Energy Physics, Kiev 1970
2.16 R. Wilson, Proc. of the XV Int. Conf. on High Energy Physics, Kiev (1970) p. 219 and papers by G. Barbiellini et al., V. Alles-Borelli et al. and by D.Bartoli et al. presented at that conference
- V. Alles-Borelli et al., in Elementary Processes at High Energy, 1970 Int. School of Subnuclear Physics, ed. by A. Zichichi, Academic Press, New york (1971) p. 790
- S. Drell, Proc. of the Amsterdam Int. Conf. on Elementary Particles, Amsterdam (1971) p. 307 and papers by C. Bacci et al. and D. Bartoli et al. presented at that conference M. Bernardini et al., lectures at the Erice summer school, 1971
2.17 T. Eichten et al., Phys. Lett. 46B, 274 (1973)
2.18 H. Deden et al., Nucl. Phys. B85, 269 (1975)
2.19 e.g. J.G.H. de Groot et al., Zeit. Phys. C1, 143 (1979)
2.20 A.M. Boyarski et al.,Mark II coll., Phys. Rev. Lett. 35, 196 (1975)
2.21 J.J. Aubert et al., Phys. Rev. Lett. 33, 1404 (1974)
- J.E. Augustin et al., Phys. Rev. Lett. 33, 1406 (1974)
- S.W. Herb et al., Phys. Rev. Lett. 39, 252 (1977)
2.22 P. Bagnaia et al., UA2 coll., Phys. Lett 160B, 349 (1985)
- G. Arnison et al., UA1 coll., Phys. Lett. 172B, 461 (1986)
3.1 For a recent review of the quark structure of the nucleon see T. Sloan, G. Smadja and R. Voss, Phys. Rep. 162, 45 (1988)
3.2 H. Abramowicz et al., CDHSW coll., Z. Phys. C 25, 29 (1984)
3.3 J.J. Aubert et al., EM coll., Nucl. Phys. B213, 1 and 31 (1983)
3.4 H. Abramowicz et al., CDHSW coll., Z. Phys. C 17, 283 (1983)
3.5 D.B. Mac Farlane et al., CCFRR coll., Z. Phys. C 26, 1 (1984)
3.6 J.J. Aubert et al., EM coll., Nucl. Phys. B272, 158 (1987)
3.7 A.C. Benvenuti et al., BCDMS coll., Phys. Lett. 195B, 91 (1987)
3.8 J.J Aubert et al., EM coll., Nucl. Phys. B293, 740 (1987)
3.9 D. Allasia et al., WA25 coll., Z. Phys. C 28, 321 (1985)
3.10 K. Gottfried, Phys. Rev. Lett. 18, 1154 (1967)
3.11 A. Bodek et al., Phys. Rev. D20, 1471 (1979)
3.12 S.J. Brodsky, J.F. Gunion and R.L. Jaffe, Phys. Rev. D6, 2973 (1972)
3.13 D.L. Fancher et al., Phys. Rev. Lett. 38, 800 (1977)
3.14 A. Argento et al., BCDMS coll., Phys. Lett. 140B, 142 (1984)
3.15 J.V. Allaby et al., CHARM coll., CERN preprint EP/88-81
3.16 D.J. Gross and C.H. Llewellyn-Smith, Nucl. Phys. B14, 337 (1969)
3.17 G. Altarelli, Phys. Rep. 81, 1 (1981)

3.18 D. Allasia et al., WA25 coll., Z. Phys. C 28, 321 (1985) and Phys. Lett. 135B, 231 (1984)
3.19 F. Bergsma et al., CHARM coll., Phys. Lett. 123B, 269 (1983) and Phys. Lett. 141B, 129 (1984)
3.20 F.J. Sciulli, Proc. Int. Symp. on Lepton and Photon Interactions at High Energies, Kyoto (1985), eds. M. Kounma and K. Takahashi, [average = 2.81 ± 0.16]
3.21 S.L. Adler, Phys. Rev. 143, 1144 (1966)
 - G.C. Callan and D.J. Gross, Phys. Rev. Lett. 21, 311 (1968)
3.22 J.J. Aubert et al., EM coll., Nucl. Phys. B259, 189 (1985)
3.23 R. Brandelik et al., Tasso coll., Phys. Lett. 97B, 453 (1980)
 - H.J. Behrend et al., Cello coll., Phys. Lett. 110B, 329 (1982)
 - Ch. Berger et al., Pluto coll., Phys. Lett. 119B, 239 (1982)
4.1 P. Buchholz et al., CDHSW coll., Proc. of the Int. Europhysics Conf. on High Energy Physics, Bari 1985, L. Nitti and G. Preparata eds., (EPS 1986)
4.2 A.C. Benvenuti et al., BCDMS coll., CERN preprint EP/87-13, paper submitted to the Int. Europhysics Conf. on High Energy Physics, Uppsala, June 1987
4.3 S. Dasu et al., Phys. Rev. Lett. 61, 1061 (1988)
 - S. Dasu et al., University of Rochester report UR-1045 (1988)
4.4 P.D. Meyers et al., BFP coll., Phys. Rev. D34, 1265 (1986)
4.5 see e.g. F.L. Navarria, C. Zupancic and J. Feltesse, Nucl. Instr. Meth. 212, 195 (1983)
4.6 European Muon Coll., paper 0497 C presented at the Int. Conf. on High Energy Physics, Munich 1988
4.7 A.C. Benvenuti et al., BCDMS coll., Phys. Lett. 195B, 97 (1987)
4.8 L.F. Abbott and R.M.Barnett, Ann. Phys. (NY) 125, 276 (1980)
 - F.J. Yndurain, Phys. Lett. 74B, 68 (1978)
 - A. Gonzales-Arroyo, C. Lopez and F.J. Yndurain, Nucl. Phys. B174, 474 (1980); Nucl. Phys. B153, 161 (1979); Nucl. Phys. B159, 512 (1979)
 - A. Gonzales-Arroyo and C. Lopez, Nucl. Phys. B166, 429 (1980)
 - W. Furmanski and R. Petronzio, Nucl. Phys. B195, 237 (1982); Phys. Lett. 97B, 437 (1980)
 - M. Virchaux and A. Ouraou, Saclay preprint, DPhPe 87-15
 - V.G. Krivokhizhin et al., Z. Phys. C 36, 51 (1987)
4.9 T. Sloan, Proc. of the Int. Europhysics Conf. on High Energy Physics, Uppsala (1987)
 - T. Sloan, Nature 323, 405 (1986)
5.1 A.D. Martin, R.G. Roberts and W.J. Stirling, preprint DTP/88/2, January 1988

EXPERIMENTAL PROBLEMS IN THE MEASUREMENT
OF STRUCTURE FUNCTIONS FROM MUON DEEP-INELASTIC SCATTERING

Umberto Dosselli
I.N.F.N., Padova, Italy

1. INTRODUCTION

Since the early measurements at SLAC (ref. 1) on electron scattering off protons a great deal of experimental effort has been concentrated on the measurement of the nucleon structure functions using both charged leptons (electrons, muons) and neutral ones (neutrinos) as probes to study the innermost structure of the hadronic world. The onset of scaling, i.e. the independence of the measured cross-section with the momentum transfer Q^2, was first discovered and subsequently explained as almost incoherent elastic scattering onto hypothetical nuclear constituents called "partons" or "quarks".

Small, logarithmic deviations from the scaling pattern were then revealed by means of high precision experiments and, in the framework of one particular theoric scheme ("Quantum Chromo Dynamics") the measurement of the trend in such deviations as a function of both Q^2 and another variable called x_{Bj} allows the determination of the coupling constant of the strong interactions α_s.

I will focus my attention on two experiments that worked in muon deep-inelastic scattering with the aim to show in a didactic fashion which kind of problems have been takled in order to extract meaningful results.

A remarkable example of the quality finally achieved is shown in Fig. 1 by the structure functions measured by one of the two experiments (EMC) using an iron calorimeter as target; the scaling violation phenomenon, i.e. rise at low x_{Bj} and drop at high x_{Bj}, is clearly visible and the size of the error bars is impressive.

At the end of this paper I will also sketch the scenario for future experiments on structure-function measurements at even higher energies that will take place at the HERA e-p collider.

2. Kinematics of the reaction

The reaction:

lepton + nucleon → scattered lepton + X   (1)

is schematically described by the Feynman graph of Fig. 2 where an incident muon of four-momentum $k = E_i (1, 0, 0, 1)$ impinges on a target proton [four-momentum $p = M (1, 0, 0, 0)$] and is diffused with a four-vector $k' = E_f (1, \sin\vartheta, 0, \cos\vartheta)$. The scattering is mediated by a virtual photon γ_v whose mass squared is given by the momentum transfer squared:

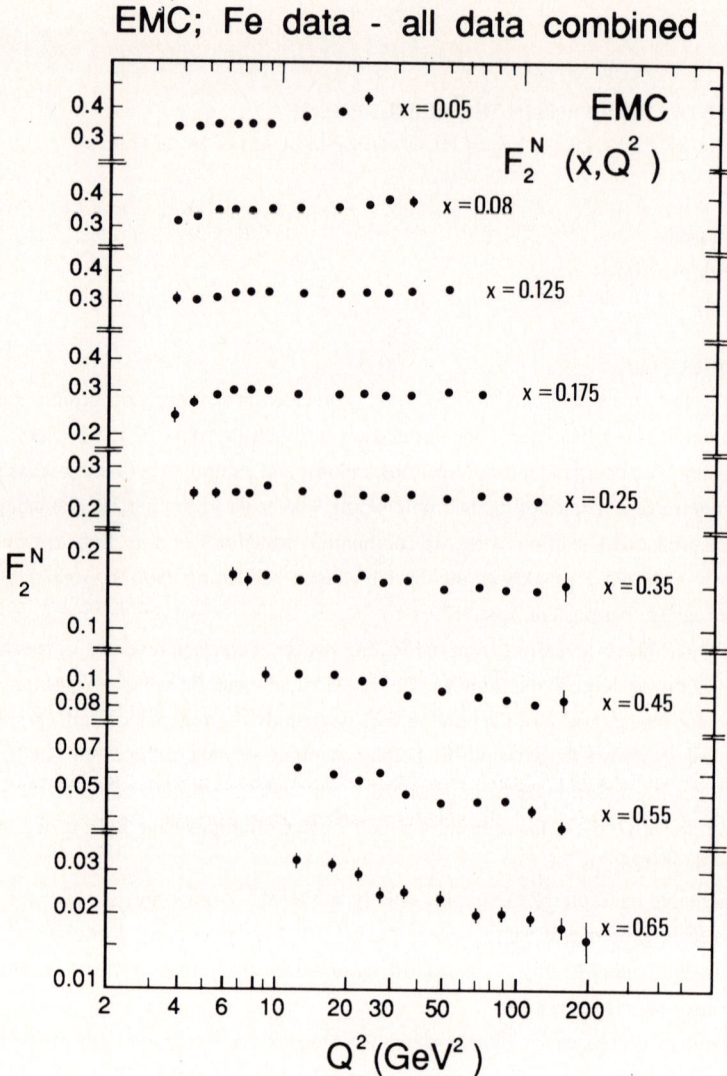

FIGURE 1

$$q^2 = -Q^2 = (k-k')^2 \approx -4E_i E_f \sin^2 \frac{\vartheta}{2} \qquad (2)$$

and its energy in the laboratory frame is:

$$\nu = \frac{pQ}{M} = E_i - E_f \qquad (3)$$

where M is the target proton mass. The fraction of the total energy carried by the current is expressed by the variable y as:

$$y = \frac{p \cdot Q}{p \cdot K} \approx \frac{\nu}{E_i} \qquad (4)$$

Another very widely used variable is the so-called Biorken x as:

$$x_{Bj} = \frac{Q^2}{2M\nu} \qquad (5)$$

This latter variable has a very simple and intuitive meaning in the framework of the QPM (Quark Parton Model) where it expresses the fraction of the proton momentum carried by the target parton onto which the interaction occurs.

It is important to notice that the complete kinematics is described by only two independent variables such as, for instance, (Q^2, ν), or (x_{Bj}, y) and so on; experimentally what one is actually measuring is just the incoming lepton energy E_i, the scattered lepton energy E_f and its scattering angle ϑ and we shall see in this paper which experimental procedures has been necessary in order to extract these 3 quantities with the smallest possible errors allowing the determination of the structure functions showed in Fig. 1.

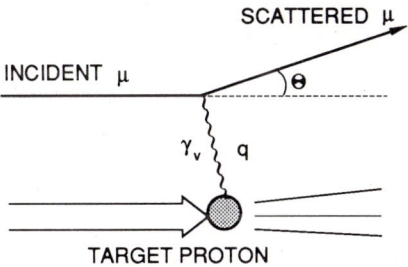

FIGURE 2

3. The CERN-SPS muon beam-line (M2)

A very important condition that has to be verified in order to have high quality measurements in muon scattering is that the beam has to be "first class" and the CERN μ beam at the Super Proton Synchrotron machine has been designed in order to reduce the halo.

For sake of comparison, the pre-existing μ beam at FNAL (U.S.A.) has as much as 50% of the total incoming flux present as halo and one can easily see how dramatic the situation can be with fluxes up to ~10^8 μ/s; halo means for instance false triggers and events with the incident muon of non defined momentum.

The techniques used at CERN in order to reduce the halo were:

- large aperture magnetic optics alternately focusing-defocusing (FODO);
- magnetic collimators;
- vertical beam deflection.

The FODO technique allows the beam transport without losses in a large momentum bite.

The magnet collimators (Fig. 3) are essentially toroids with zero magnetic field on the beam axis and allow to eliminate the halo near to the beam; these kind of collimators also produce as much halo as normal (non-magnetic) collimators but with a much more favourable space-distribution, i.e. much wider.

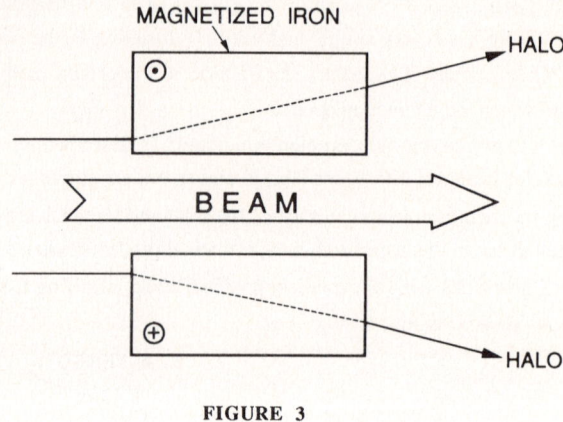

FIGURE 3

Finally a vertical deflection of the beam before the experimental area kills the far halo component.

The global scheme of the beam-line is reported in Fig. 4 and one can observe 4 principal

I. the hadron production
II. the decay channel
III. hadronic absorption
IV. momentum selection

Let us have a close look to each of this sectors.

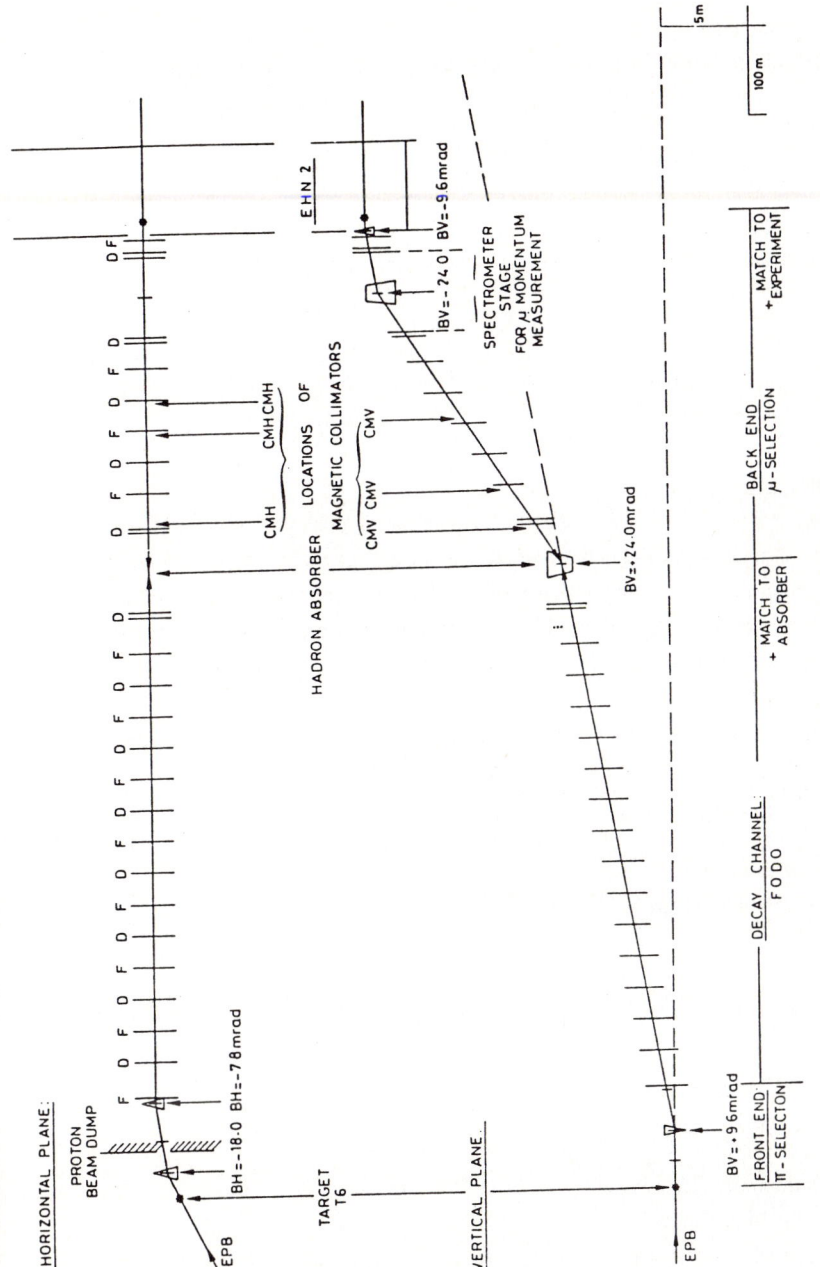

FIGURE 4

SCHEMATIC LAYOUT OF BEAM M2

I. - Hadron production

The 400 GeV primary protons interact on a tungsten target 50 cm long. The produced π and k beam is accepted under a solid angle of 20 µsr. The momentum is selected with a spread of $\frac{\delta p}{p} \leq 10\%$; the choice of the final muon charge is also performed here.

II. - The decay channel

It contains hadrons and the decay products in a momentum window:

$$0.57 p_\pi < p_\mu < p_\pi$$

This channel contains 16 quadrupoles and is 640 m long; at the end of it about 3% of the produced hadrons are decayed.

III - Hadron absorption

At this level the beam is still ~97% hadronic and, considering that the hadron cross-section is roughly 4 orders of magnitude higher than the electromagnetic one it is clear that a good hadron absorber is crucial for the beam quality downstream. Moreover, the chosen material has also to minimize Coulomb scattering experienced by the traversing muons. The two design goals has been met by an 11 m long Beryllium absorber, chosen by its excellent ratio L_{att}/L_{rad} between the attenuation and radiation lengths.

This absorber reduces by ~10^8 the hadronic component leaving a final contamination of

$$\frac{\text{hadrons}}{\mu} \leq 10^{-6}$$

IV - Momentum selection

A combined system of magnetic collimator selects a final momentum band of

$$\frac{\delta p_\mu}{p_\mu} \approx 5\%$$

We are here almost at the end beam-line; the muons are subsequently transported by a 2nd FODO channel on a spectrometer (BMOM) where the momentum of each individual muon is measured and hence focalized on the target.

3.1 - Beam parameters measurement

3.1.1 Momentum

With beam intensities as high as $1 \div 3 \cdot 10^7$ µ/s the average time distance between two µ is ~30 ns and hence high timing resolution hodoscopes are required in order to measure the position (and then the momentum) of the incoming particle.

The momentum measurement is performed using 3 dipoles magnets of the beam line, yielding a total of $\int B\, dl \approx 25\ \text{T}\cdot\text{m}$, and 4 hodoscopes planes, 2 before and 2 after the magnets. In order to minimize the flux per hodoscope element, each plane is composed by 64 scintillator with a resolution

$$\sigma(t) \approx 150\ \text{ps}$$

The intercalibration principle, sketched in Fig. 5

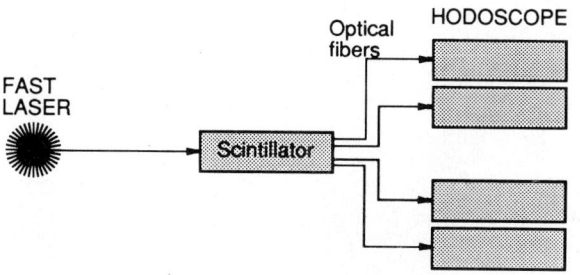

FIGURE 5

is based on a fast laser beam whose optical pulse is transported on a scintillator and the outgoing light is then fed, via a calibrated optical link, to the photomultipliers. With this system the trajectory of each incoming muon can be tracked yielding a final error on its momentum of

$$\sigma(p_\mu) \approx 1\ \text{GeV}$$

3.1.2 Position before target

The determination of the precise position and angle of the incoming muon upstream of the target is performed by means of 2 groups of scintillator, 2.2 mm wide; the $\sigma(t) \approx 700$ ps allowed an unambiguous assignment for each particle, with a space resolution of:

$$\sigma(y) \approx \sigma(z) = \frac{2.2\ \text{mm}}{\sqrt{12}} \approx 0.7\ \text{mm}$$

3.2 - Summary of the μ beam characteristic

Energy	50 → 280 GeV
Intensity	$10^4 \to 10^8$ μ/s every 7 s
Polarization	−80% → +80%
Halo	7% of the total flux
Average radius	15 mm
μ per proton delivered by the SPS	$1.6\ 10^{-6}$ at 280 GeV (μ^+)
	$0.7\ 10^{-6}$ at 280 GeV (μ^-)

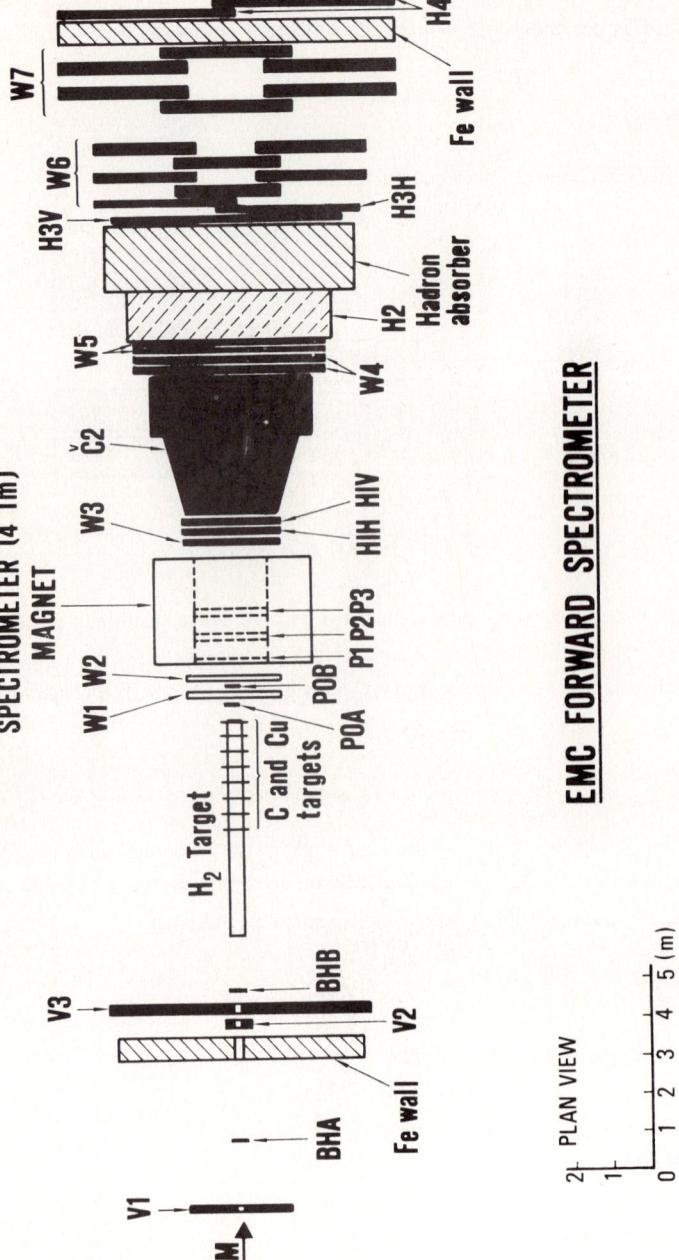

FIGURE 6

EMC FORWARD SPECTROMETER

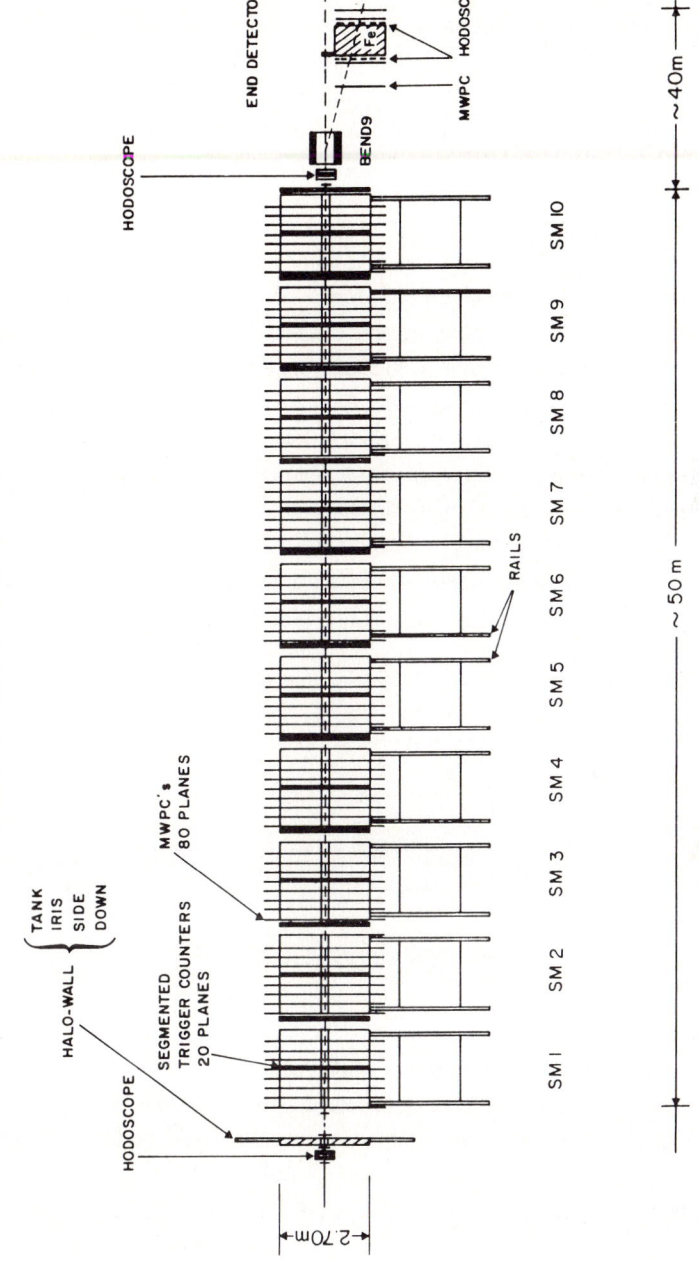

FIGURE 7

4. The experiments

On the muon beam-line two apparatus measured the deep-inelastic muon scattering: the European Muon Collaboration (EMC, Fig. 6) and the BCDMS (NA4, Fig. 7) experiments. The characteristics of these two detectors are vastly different and, after a brief illustration of the BCDMS experiment, I shall spend most of this paper describing in some detail the EMC spectrometer highlighting some relevant experimental features.

4.1 - The BCDMS experiment

The set-up of the NA4 experiment (Fig. 7) is based on 10 identical Supermodules placed downstream to a veto-wall of scintillators; each Supermodule consists in a toroid magnet with a ~5 m long target along its central axis and segmented with trigger counters (total of 20 planes) and Multi-Wire Proportional Chambers (total of 80 planes) for the trajectory reconstruction. The total length of the experiment is roughly 55 m, with ~40 m of target and hence this apparatus is very suited for very rare events (i.e. events at the highest Q^2) where high luminosity pays-off.

It is a very simple and ingenious experiment the principle being that (Fig. 8) an incident μ with 4-momentum $(E, \vec{k})$ scatters from the target and, trapped by the magnetic field, follows a sinusoidal trajectory with wavelength λ and amplitude Δ. Along this trajectory the scattered muon can be measured many times before it leaves the apparatus and the kinematics of the event is completely determined by the sagitta Δ, directly proportional to Q^2/Q^2_{max}, and by the wavelength λ.

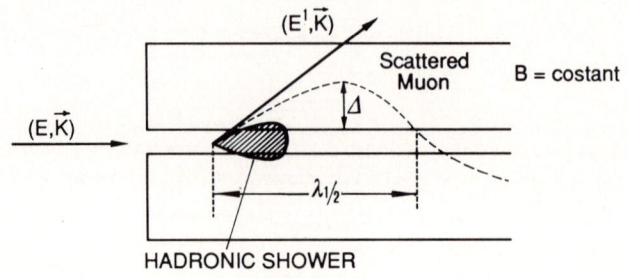

FIGURE 8

The measurement of the muon trajectory is also simplified because the hadronic shower accompanying each inelastic event is immediately absorbed by the magnet iron.

One delicate point of this kind of experiment is the perfect knowledge of the energy loss that the incident μ experiences before the interaction; one wrong estimation of this quantity will be transferred as an error on the initial momentum and hence on the quality of the final measurement.

4.2 - The European Muon Collaboration experiment

Designed as a more classical spectrometer (Fig. 6) the EMC apparatus starts with a set of veto wall to tag halo muons and, between and after, the two hodoscope stations for the position determination of each incoming muon. Following the beam-line, we find the target station; different kind of targets were used in different periods of data taking: 6 m long liquid hydrogen, same length of liquid deuterium, copper and carbon disks and an iron-scintillator calorimeter.

Downstream is located the spectrometer magnet (4 T · m bending power) and an hadron absorber. The trajectories of the scattered muon and secondary hadrons were tracked by 99 planes of drift and proportional chambers; scintillator hodoscopes complete the apparatus.

Hadronic identification is also possible by means of a gas Cerenkow; a full a detailed description of the apparatus can be found in ref. 2.

The first part of the data taking was performed using, as mentioned above, a target composed by slabs of iron interleaved with scintillator sheets; the experimental advantages of such a strategy are manyfold starting from the consideration that this heavy target increases the luminosity, for instance, by a factor 16 compared with the D_2 one; this enables the study of rare events such as multi-muon production.

Even more, if the target is a calorimeter one has a redundant measurement of the energy loss because

$$\nu = E_i - E_f = E_{hadronic}$$

and hence a cross-check about the performances of the off-line is possible.

More, the geometric reconstruction of tracks is easier because the target itself absorbs all hadrons and hence, downstream, only one track, namely the scattered muon, is present.

Last, but not at all least, it is possible to implement the readout of the target directly into the trigger enhancing the quality of the row data.

5. Cuts

The name "cut" hide most of the problems that an experimentalist has to face during the analysis of his experiment.

Each new accelerator brings with it usually a completely new kinematic domain that is shown, during "propaganda" seminars, to illustrate how interesting and exciting experiments there will be. What is *not* shown normally is what could be called the "problems diagram" (Fig. 9) that, by indicating areas of problems, tells the reader where the actual kinematic range is and where corrections, and hence cuts, are important.

In the following paragraphs I will review the more important corrections that we had to apply to the EMC data and, for each specific point, indicate which part of the kinematic domain had to be removed because there the necessary corrections were too large.

5.1 - Radiative corrections

The basic hypothesis is that deep inelastic scattering is described by only 2 structure functions in the framework of the one photon exchange approximation.

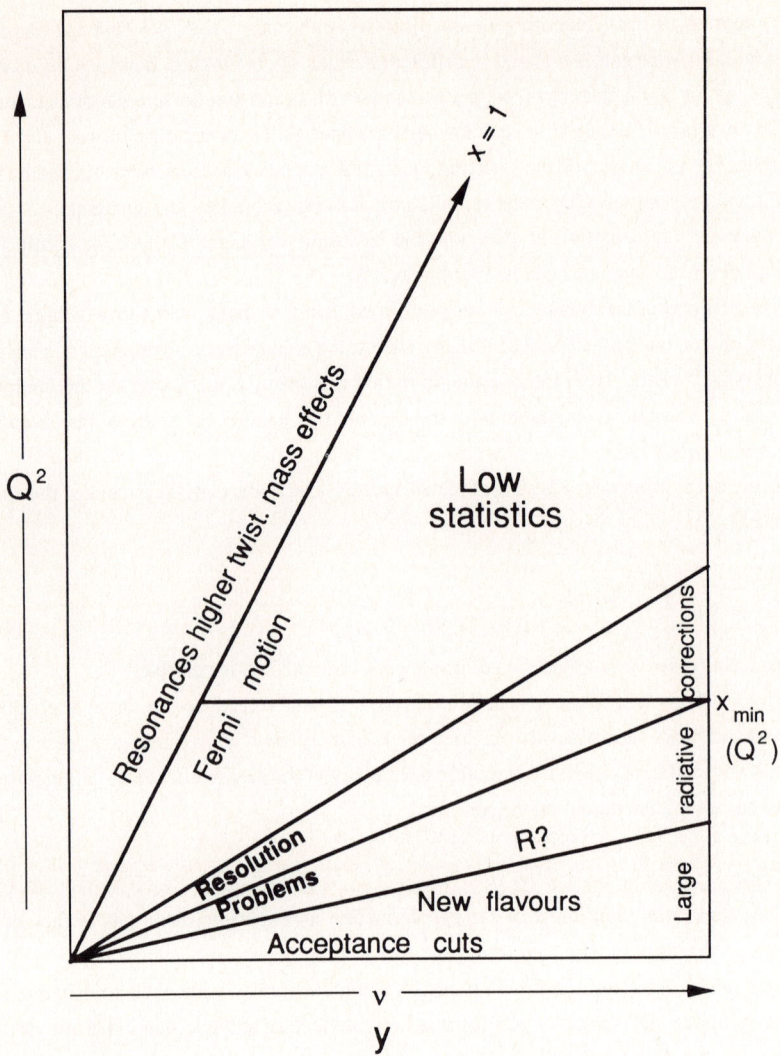

FIGURE 9

Structure Functions from Muon Deep-Inelastic Scattering

One has then to verify the hypothesis and obtain the cross section for the process

from experimental measurements.

The cross-section is, by definition, the sum over all term of order α^n:

In practice one limits his analysis at terms of $0\,(\alpha^3)$; from this limitation arises the first cut:

$$\nu < 0.9\,E_i$$

because in this area the corrections are smaller than 30%.

In this view the cross-section, i.e. the square of transition matrix, is given by:

$$|M^2| = \left|\;\right|^2 + 2\text{Re}\left[\;*(\;+\right.$$

$$\left.+\;+\;+\;\right)^*\right] +$$

$$+\left|\;+\;+\;+\;\right|^4 + 0(\alpha^2)$$

Diagrams 1, 2, 3, 6, 7, with just 1 photon exchanged, are precisely computable knowing W_1 and W_2 whereas for contributions 4, 5, 8, 9, when a photon interacts with the nucleon constituents, a model for the hadronic matter is necessary; their contribution being very small they are usually neglected. Elastic scattering events (diagram 6 and 7) with one photon irradiated simulate high Q^2 events; the phase-space of these contributions is well determined ($\nu \approx E\gamma$, high ν, low x_{Bj}) and hence by restricting at

$$x_{Bj} > 0.3 \quad \text{(or smaller for } H_2 \text{ and } D_2 \text{ targets)}$$

one avoids heavy corrections.

5.2 - Systematic errors

5.2.1. Background

Background means that in a given (x_{Bj}, Q^2) bin there are events that shouldn't be there coming for instance from halo muons or decays in flight of hadrons. To verify the contamination of such events one studies the distributions of opposite-sign μ's (muons with electric charge opposite to the beam polarity); these muons cannot be the scattered ones, but they come from π decays, charm decays, etc. and hence they represent a sample of the background muons (the same-sign ones). By studying this events one finds that they are concentrated at

$$y > 0.85$$

and then, by avoiding this region, the estimated residual background is less than 1%.

5.2.2 Event losses

This is a pedagogical example of how an error in the planning of the experiment could have turned in a major problem. In fact the maximum event dimension was fixed to 4 Kbytes (16 bits/byte) and, of course, if one event was longer than that it was simply cut by the Data Acquisition System and the event itself was non useable any more. But, with 99 planes of drift and proportional chambers, the probability of having a "ringing" wire in a plane is not zero and finally the number of truncated events was much higher than thought; the problem was to check whether they evenly populated the phase-space (GOOD) or if they peaked in some peculiar area of the (x, Q^2) plane (BAD). The chance was that during the data-taking periods with the calorimeter target the ADC's of the target itself were written at the beginning of the event record and hence it was possible to verify that the ν distribution of the truncated events was flat (I recall here that $\nu = E_i - E_f = E_{hadronic}$). As a result, the truncated events problem was considered as a flux loss (order of $1 \div 5\%$).

5.2.3 - Flux error

The absolute error per each taking period was $2 \div 3\%$ mainly coming from the TDC calibrations.

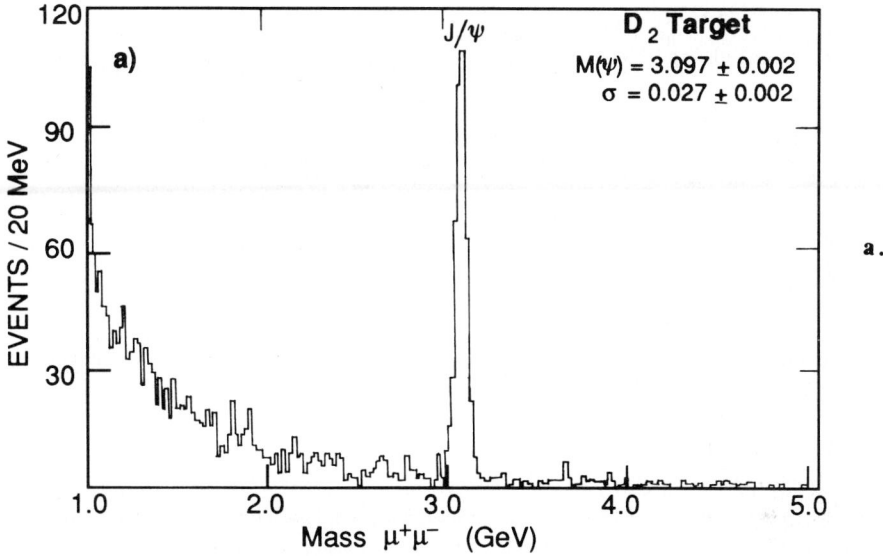

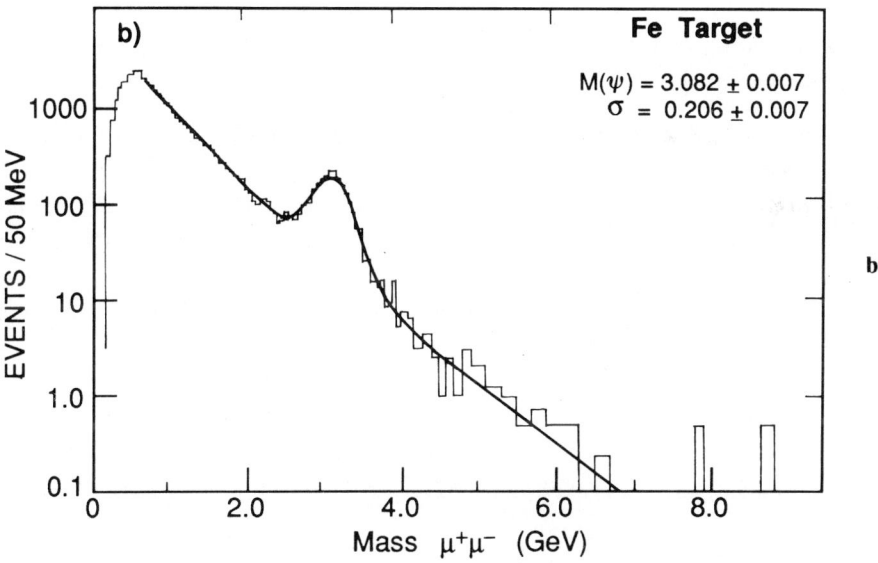

FIGURE 10

5.2.4 - Acceptance corrections errors

They were reduced by an extremely careful alignment of each plane of the detector. The final relative error on F_2 from this source is ~1%.

5.2.5 - Magnets calibrations errors

This point is a crucial one because small errors on the determination of E_i (incident μ) and E_f (scattered μ) reflect in a large errors on the number of events in $\Delta x \, \Delta Q^2$ bins and hence in the F_2 measurement. It is also very important to have an intercalibration between E_i and E_f.

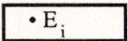

All the beam-line elements were verified with high precision; the residual error is estimated around δE_i ~0.4%.

- E_f

The J/ψ mass from D_2 target (Fig. 10a) ensures that the measurement is better than 0.3% and the width of the peak (20 MeV) is correct. From the J/ψ events obtained with F_e target (Fig. 10b) one is sure that the energy losses are correctly taken into account.

- E_i vs E_f

To intercalibrate E_i and E_f the beam was directly deflected into the spectrometer and hence its momentum is measured twice: in the beam line and in the apparatus. The results (Fig. 11)

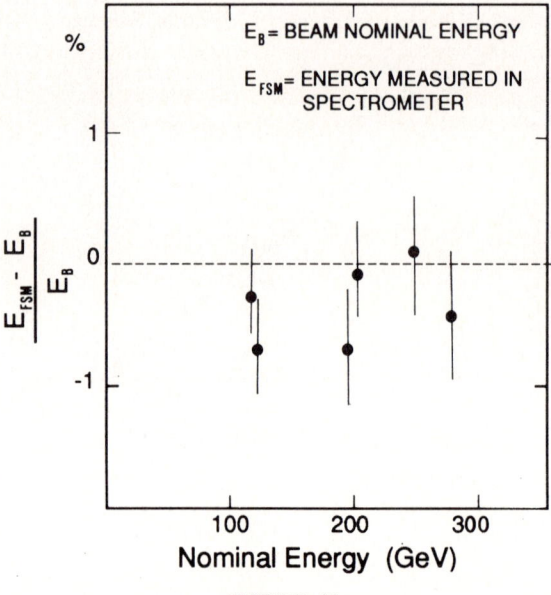

FIGURE 11

show that the observed difference of 0.3 ÷ 0.4% is consistent with what foreseen.
It was then assigned:

$$\frac{\delta E_i}{E_i} = 0.4\% \qquad \frac{\delta E_f}{E_f} = 0.2\%$$

6. Extraction of the structure functions

The last section of this paper will deal on how to extract the structure functions from the experimental data.

The starting point is that, in the framework of the already mentioned one photon exchange hypothesis, the cross-section for the process can be written as:

$$\frac{d^2\sigma}{dQ^2 dx} = \frac{4\pi\alpha^2}{Q^4}\left[\left(1 - y - \frac{Mxy}{2E}\right)\frac{F_2^N(x, Q^2)}{x} + y^2 F_1^N(x, Q^2)\right] \tag{6}$$

and the ratio between the longitudinal and transverse cross-sections is defined as:

$$R = \frac{\sigma_L}{\sigma_T} = \frac{4M^2 x^2}{Q^2}\frac{F_2(x, Q^2)}{2xF_1(x, Q^2)} + \frac{F_2(x, Q^2) - 2xF_1(x, Q^2)}{2xF_1(x, Q^2)} \tag{7}$$

Hence in each (x, Q²) bin the double-differential cross section $\frac{d^2\sigma}{dxdQ^2}$ is a function of both F_2 (x,Q²) and R (x, Q²). The first interest of the measurement is F_2 and hence, a priori, a plausible value for R has to be assumed. Since R is strictly zero in the Quark Parton Model and anyhow very small also in QDC, the most common choice is to assume R = 0.

In each surface element $\Delta x \, \Delta Q^2$ the physical events are related to the cross-section by the relationship:

$$N(\Delta x \Delta Q^2) = L \iint_{\Delta x \Delta Q^2} \frac{d^2\sigma}{dx\, dQ^2}\, dx\, dQ^2 \tag{8}$$

where $N (\Delta x \Delta Q^2)$ is the number of events in that bin;
        L = luminosity   is the number of scattering centers multiplied by the number of muons in the beam.

The simple expression (8) is unfortunately complicated by corrections on the acceptance, reconstruction efficiency, resolution effects and so on, and then it has to be re-written in a more complete fashion as:

$$N_{bin} = L\varepsilon_{beam} \iint_{\Delta x \Delta Q^2} dx dQ^2 A\varepsilon \iint dx' \, dQ^{2'} \frac{d^2\sigma}{dx' \, dQ^{2'}} \cdot \text{Radcor} \cdot \text{Fermi} \cdot \rho(x, x', Q^2, Q^{2'})$$
(9)

where

L = flux * number of scattering points

ε_{beam} = probability to measure a μ of the beam;

$A(x, Q^2)$ = average acceptance of the scattered muons;

$\varepsilon(x, Q^2)$ = average reconstruction programs efficiency;

$\dfrac{d^2\sigma}{dxdQ^2}$ = kinem (R) $F_2(x, Q^2)$ is the cross section where kinem (R) is the standard kinematic factor;

Radcor (x, Q^2, F_2, R) = correction factor for radiative effects;

$\rho(x, x', Q^2, Q^{2'})$ = resolution effects; it is the probability that an event migrate from its bin (x, Q^2) to another (x', Q^2) due to the finite resolution of the apparatus.

The goal now is to extract, from relation (9), $\dfrac{d^2\sigma}{dxdQ^2}$ starting from $N(\Delta x\Delta Q^2)$ that is what one actually measures. It is then necessary to invert the (9) and one has to iterate the process because R varies very much even within a give $(\Delta x\Delta Q^2)$ bin.

The method used consists in comparing the content of what has been measured in a given interval $[\Delta x\Delta Q^2]$, NDST, to the Monte Carlo value, in the same bin, NMC, calculate simulating the experiment and assuming a starting value for the cross-section. Each difference between NDST and NMC is attributed to the difference between the real cross-section and the hypothetical one, the latter is modified accordingly and the process starts again.

Hopefully the method reaches a convergence point and the correct value for the cross-section, and hence for F_2, <u>in that bin</u>, is found.

It is important to stress that, at the end, the experimental determination of $F_2(x, Q^2)$ is valid only in the measured (x, Q^2) bins and not outside.

7. The final results

After having described the beam-line, the apparatus, the various corrections and the method for the extraction I shall conclude showing a sample of the EMC and BCDMS results. Fig. 12 shows the F_2 of the nucleon in different x_{Bj} bins obtained with D_2 target and a 280 GeV beam by the EMC; the total number of events is 216.000 and the systematic error vary from 2% to 10 ÷ 20% at high x_{Bj} ($x_{Bj} = 0.75$).

Fig. 13 represents the proton F_2 obtained with H_2 target (EMC experiment) and in this data sample the systematic error is always below 6%.

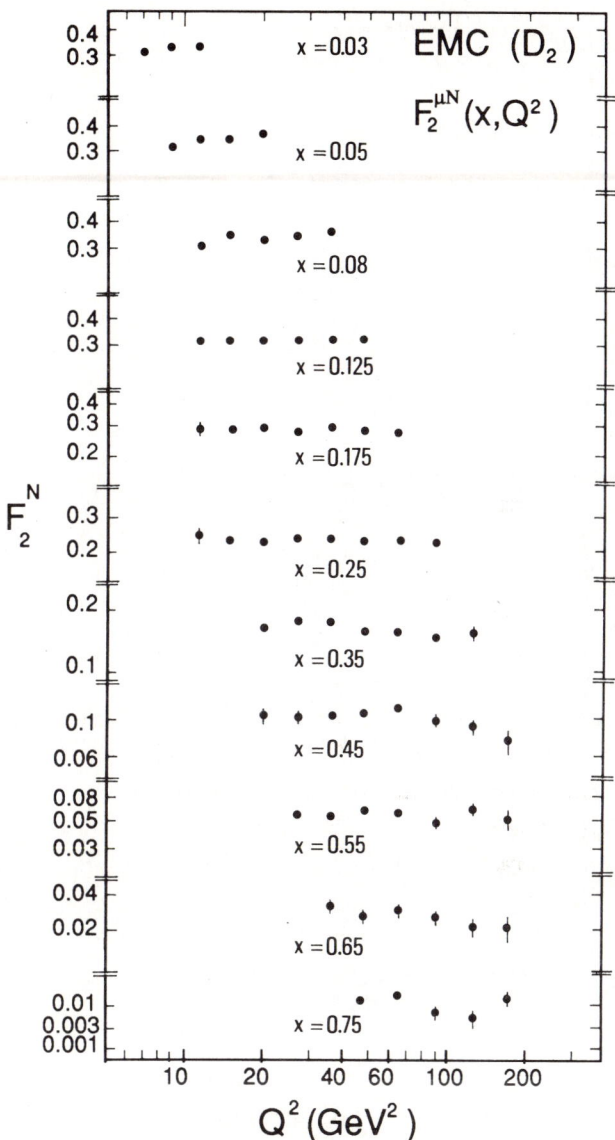

FIGURE 12

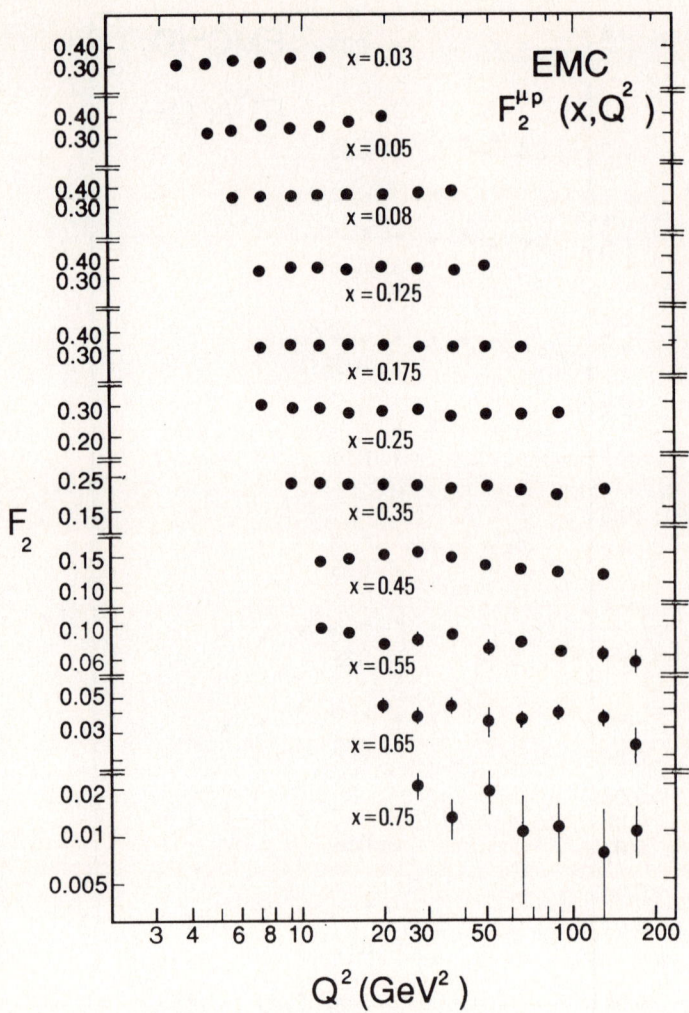

FIGURE 13

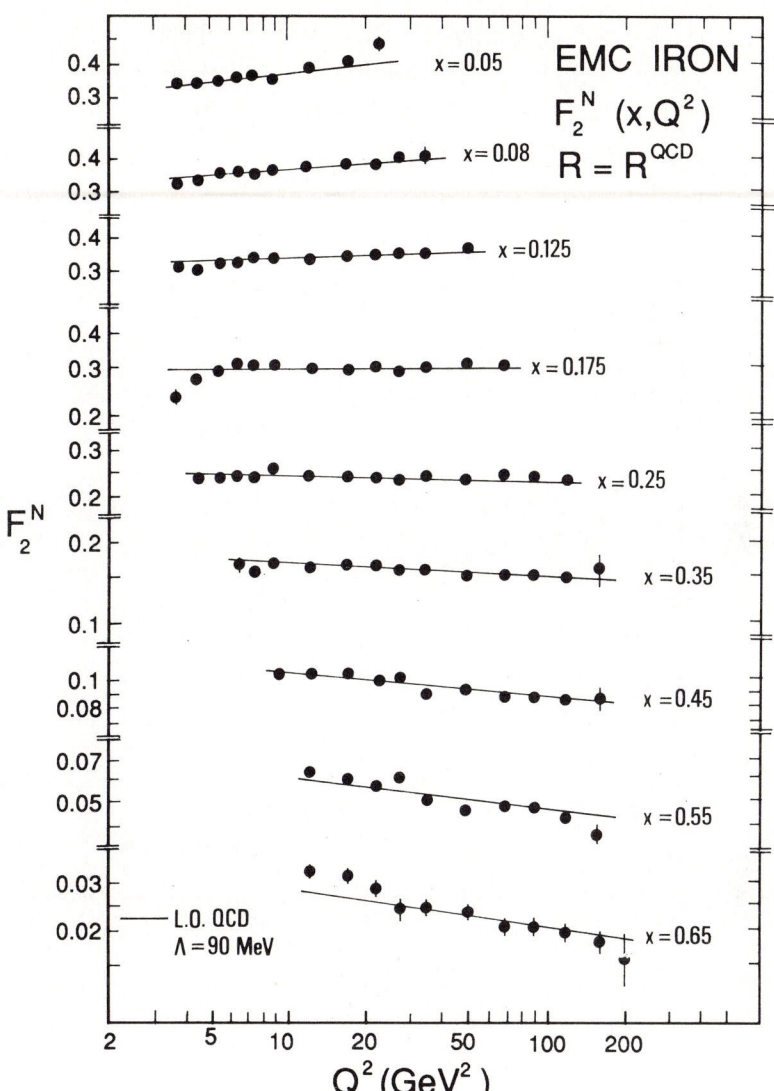

FIGURE 14

BCDMS (NA4) ; C data
x=0.275
x=0.35
x=0.45
x=0.55
x=0.65
x=0.75
10^{-1}
10^{-2}
$F_2(x_1 Q^2)$
10
10^2
Q^2 (GeV^2)

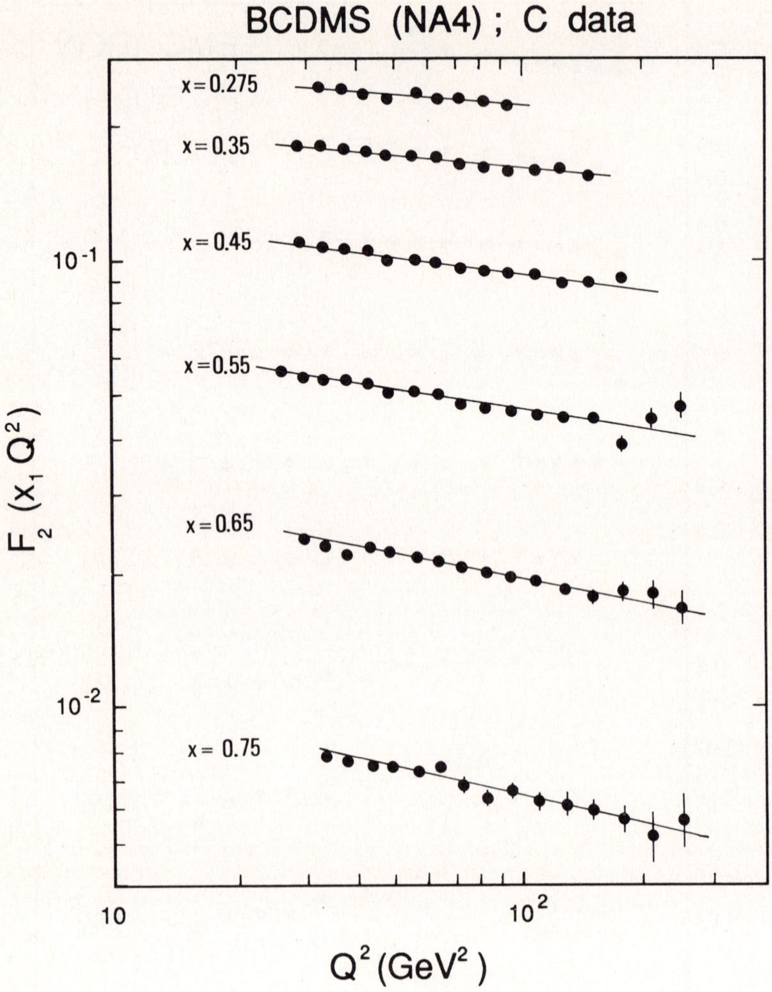

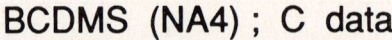

FIGURE 15

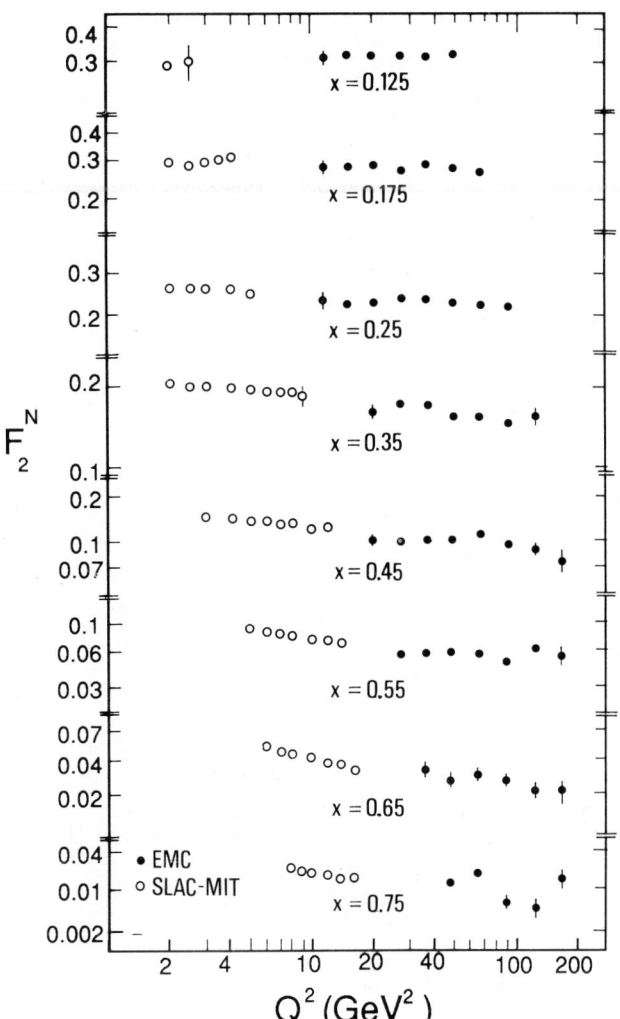

FIGURE 16

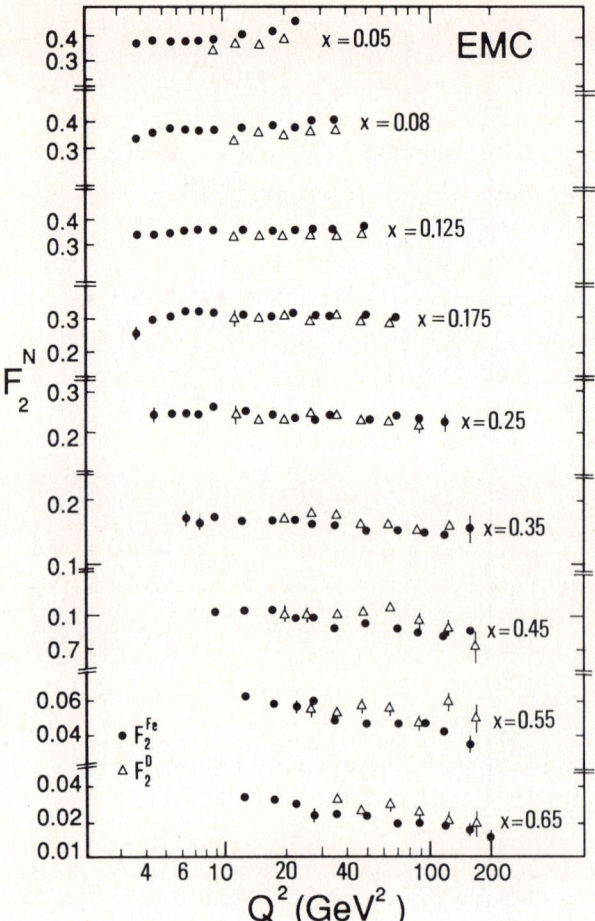

FIGURE 17

Also from the nucleon structure function from Fe target (Fig. 14) the pattern of scaling violation, namely rise of F_2 at low x_{Bj} and drop at high x_{Bj}, is clearly visible; superimposed is shown a leading order QCD fit with the scale parameter Λ fixed at 90 MeV.

From the BCDMS similar patterns are observable (Fig. 15); in this figure data were obtained from a carbon target and the quality of the measurements is remarkable.

A comparison between the EMC H_2 data and the old SLAC-MIT F_2 (Fig. 16) indicates a coherent transition from low Q^2 to the high Q^2 regime.

Finally Fig. 17 shows together the nucleon structure functions for Fe and D_2 targets; a closer look reveals a small but systematic trend of the D_2 data to lie below the F_2 ones at low x reversing the trend at high x; this effect is nowadays well known as "EMC effect".

8. Conclusion

We have explored the long, sometimes tedious, always difficult path from the design of an experiment to its final results, trying to underline the crucial points that are really relevant to the final quality of the data.

I would like to conclude with an outlook to the future of the structure functions

Large experiments are planning to measure F_2 at the HERA Collider at DESY, Hamburg (D), where a new challenge is open: 30 GeV electrons colliding against 800 GeV protons yielding unusual events topologies and with a bunch crossing every 96 ns! The potentialities of this machine are huge and, amongst others, the highest Q^2 available will be of the order of few 10^4 GeV^2; Fig. 18 shows the variation of the distance probed inside the nucleon as a function of Q^2. At HERA distances as short as 10^{-16} cm will be explored and the question

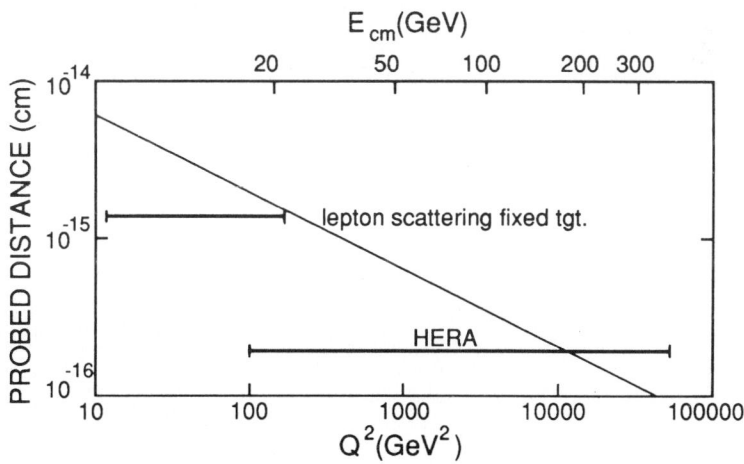

FIGURE 18

whether at these distances quarks are still point-like objects will find an answer.

REFERENCES

1. W.K.H. Panofsky, Proceedings of 14th International Conf. on H.E.P., Vienna, Austria, 1968, J. Prentki and J. Steinberger editors, p. 23.
   M. Breidenbach et al., Phys. Rev. Lett. 23 (1969) p. 935.
   E. Bloom et al., Phys. Rev. Lett. 23 (1969) p. 930.

2. EMC, O.C. Allkofer et al., Nucl. Instr. and Meth. 179 (1981) p. 445.

NUCLEAR EFFECTS IN DEEP INELASTIC MUON SCATTERING

Cristiana PERONI
I.N.F.N. Torino,
C. Massimo D'Azeglio, 46, I-10125 Torino, Italy

1. INTRODUCTION

Deep inelastic lepton scattering has long been used to probe the structure of the nucleons. Fundamental experiments at SLAC and later at FERMILAB and CERN have helped in discovering and clarifying the quark-parton picture of the nucleon.

In the following we will be concerned with the study of the effects that the nuclear medium produces onto the quark and gluon distributions.

Since the beginning of the experimental and theoretical investigations on the nucleon inelastic structure functions, and for a long time, nuclear effects have been considered and studied quite independently from the sub-nuclear effects, which were the subject of such investigations. This approach was based on the assumption that the corrections to be applied to the deep inelastic lepton nucleon cross section measured on a bound nucleon, as opposed to a free or quasi-free one, were either small and calculable, or confined to specific kinematic regions. Under this assumption deep inelastic experiments requiring high luminosity to compensate for the small cross section of the reaction under study, like neutrino experiments or charged lepton ones at high Q^2, were carried out on nuclear targets rather than hydrogen or deuterium. It was in one of these experiments, done by the European Muon Collaboration, using a beam of high energy muons on various targets, that the nucleon structure function F_2 measured using an iron target was observed to differ from the same F_2 measured on deuterium. Such difference was smaller than the relative systematic error between previous experiments measuring each the nucleon structure function either on iron or on deuterium; for the first time however, it was the same experiment to measure on the two targets, so that the systematic experimental errors on the two measurements could cancel to a large extent in the ratio between the two, leaving the deviation from one of the ratio to be interpreted as a real physics effect: what came to be called the EMC effect [1].

2. NUCLEAR TARGETS AND FERMI MOTION

In deep inelastic scattering the typical energy transfers are three to four orders of magnitude bigger than the average nuclear binding energy of 8 MeV/nucleon. It is then reasonable to assume that collective effects in scattering off nuclei can be neglected, which amounts to saying that each nucleon contributes incoherently to the total deep inelastic cross section. Any structure function measured on a nucleus of charge Z and mass A is then given by an incoherent superposition of the individual structure functions. In the case of iron and deuterium, as originally measured by EMC, once a small correction is applied to take into account the neutron excess in iron, the expected result is:

$$R(x) = \frac{F_2^{Fe}(x)}{F_2^{D}(x)} = 1$$

$$F_2^{Fe}(x) = \frac{1}{56}\left[26 F_2^{p}(x) + 30 F_2^{n}(x)\right]$$

$$F_2^{D}(x) = \frac{1}{2}\left[F_2^{p}(x) + F_2^{n}(x)\right]$$

where $x = Q^2/2M\nu$ is the well known Bjorken variable [2].

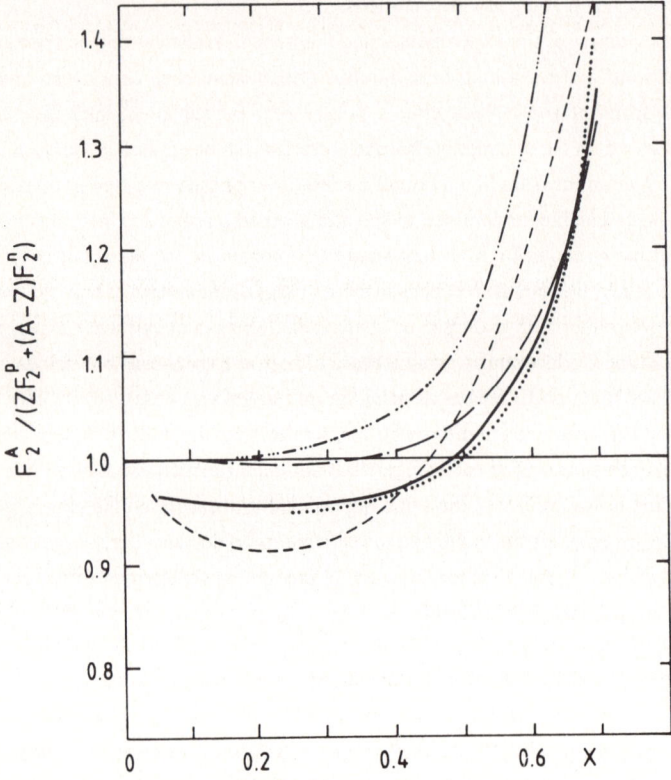

Fig. 1. Theoretical predictions for the Fermi motion correction of the nucleon structure function F_2^{Fe} from the models of Berlad et al., [3] (dashed line), Frankfurt and Strikman [4] (dotted line), and Bodek and Ritchie (solid line) [5]. The dash-dotted line and triple-dot-dashed line are variants of the Bodek and Ritchie calculations which indicate the sensitivity to different model assumptions.

In reality, a deviation from unity of this ratio is to be expected once one considers that the nucleons are not stationary in the nucleus, their collective motion being the Fermi motion, with average momentum:

$$P_F = \left(\frac{3}{2}\pi^2 \rho\right)^{1/2}$$

where $\rho = A/V$ is the nuclear density.

The variable x, in the quark-parton model, is the fraction of the nucleon momentum carried by the quark that has undergone deep inelastic scattering, and it is experimentally determined in the approximation that the nucleon is stationary.

The structure function describes the momentum distribution of the quarks inside the nucleus, weighted by the square of their charges; this is only true however, for a free or quasi-free nucleon, i.e. for hydrogen or deuterium targets. The structure function measured on nuclear targets, because of the Fermi motion, is the bare nucleon structure function convoluted with the momentum distribution of the nucleon in the nucleus. The effect on the ratio R has been calculated by several authors for various nuclear wave functions [3-5] and is shown in fig. 1. The strong enhancement at large x, is the consequence of the smearing of F_2^A towards higher values of x.

3. THE EARLY MEASUREMENTS AND SHADOWING

Already a the time of the original EMC results, considerable experimental evidence existed of the deviation from unity of the ratio R at very low x values [6-12]. This region, extensively studied in the 1970's mainly by low energy electron experiments, is known as "shadowing" region.

Intuitively, a photon can be viewed as a superposition of virtual hadrons (essentially due to fluctuations of the "bare" photon into quark-antiquark pairs, behaving like vector mesons) and a point-like component. When the photon energy is sufficiently high these hadrons may live long enough to traverse a whole nucleus. As hadrons are absorbed strongly in a nuclear medium, nucleons inside the nucleus are "shadowed" by the nucleons on the surface.

Such effects are well established for real photons [13-16]; they seem to be there also for virtual photons as long as Q^2 is small enough. It should be noted that, although appealing, this Vector Meson Dominance model is by no means the only one proposed to explain the data in this region (see for example [17] and references therein). A partial compilation of experimental results for Al/D is presented in fig. 2. The data are qualitatively in agreement with the VMD model sketched above, but it is immediately clear that no simple and consistent experimental picture emerges from them.

One of the problems with many of these early experiments is that they were carried out with low energy electron beams, in a kinematic region (high y = v/E beam) where radiative events are abundant (in a radiative event a real photon is emitted along the incoming or outgoing electron line; the photon is normally undetected and the apparent Q^2, v and consequently x are therefore wrong).

Since at high y the radiative corrections become sizable, systematic errors thereby introduced on the data are important.

It should also be noted that the medium and large x range was not covered by these experiments, which were typically limited to $x \leq 0.1$. Up to 1982 the only effect thought to influence F_2^A/F_2^D at larger x was supposed to be Fermi smearing, discussed above, important only for $x \geq 0.6$.

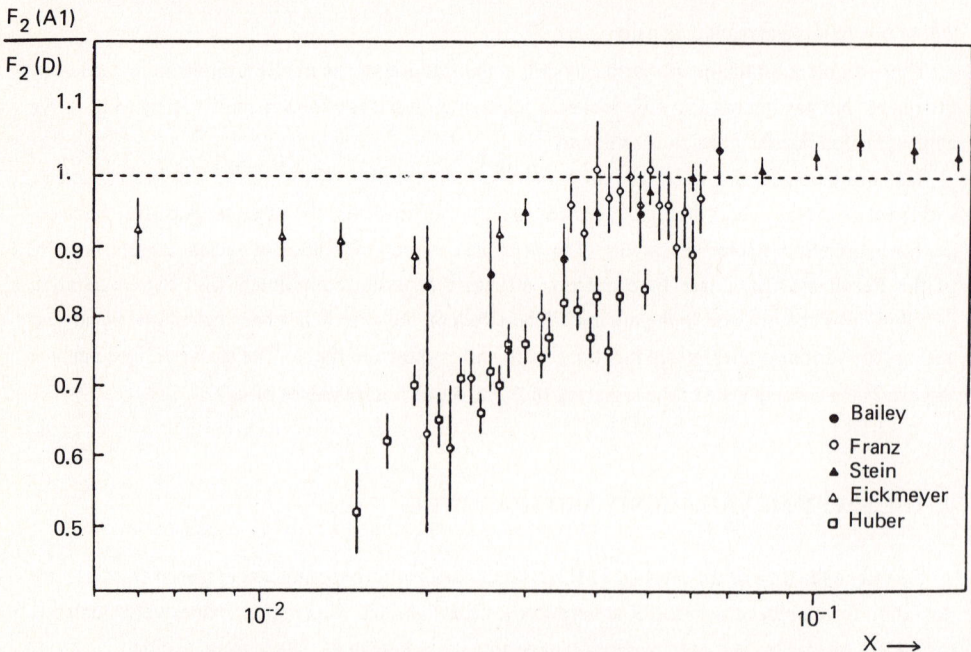

Fig.2. x-dependence of the shadowing effect in aluminium, from different leptoproduction experiments: Bailey [9], Franz [8], Stein [10], Eickmeyer [6], Huber [7].

4. THE SECOND GENERATION EXPERIMENTS AND THE "DEDICATED" EXPERIMENTS

In 1982 the European Muon Collaboration published data on the comparison of F_2 extracted from Fe and D targets [1].

Unexpectedly the Fermi motion prediction for the ratio F_2^{Fe}/F_2^D was strikingly contradicted by the data (fig. 3). The EMC result was soon reconfirmed by a reanalysis done at SLAC of old electroproduction data on aluminium and deuterium [18,19].

This led to a flood of theoretical papers attempting to interpret the findings; some of the proposed ideas will be discussed below; for now it will suffice to stress that it is by no means possible to consider the distribution of quarks inside a nucleon (hence F_2) independent of the nuclear environment in which the nucleon is immersed.

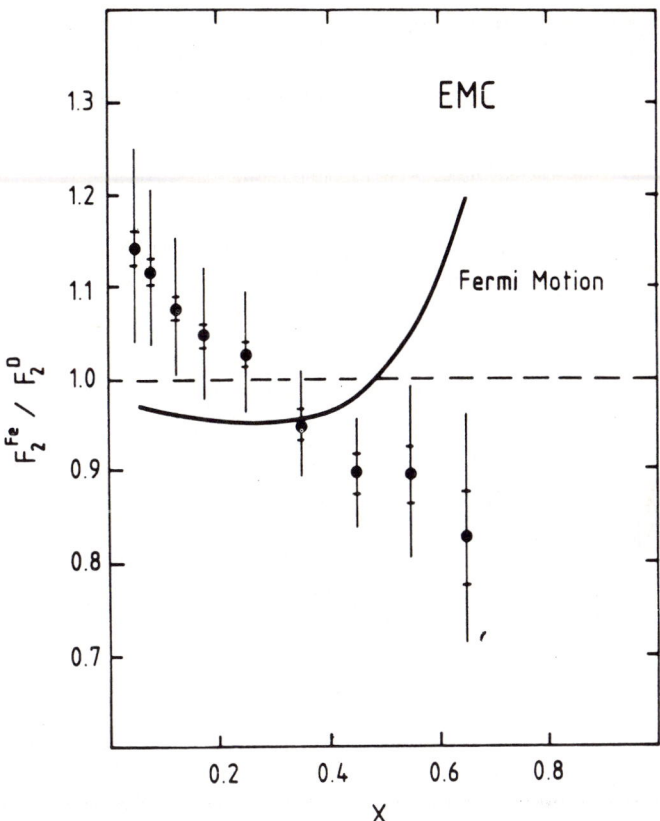

Fig. 3. The structure function ratio for Fe and D measured by the EMC [1]. The inner error bars are statistical, the outer error bars statistical and systematic errors combined in quadrature. A further normalisation uncertainty of ± 7% is not shown in the figure. The smooth curve shows the behaviour expected from Fermi motion calculations [5].

It should be emphasised that the 1982-83 experimental results, first from the EMC then from SLAC [20,21] and BCDMS [22,23] came from the comparison of data sets taken at different times and in different experimental conditions. None of those experiments were optimised a priori to determine ratios of structue functions. This should be borne in mind when comparing these data with later ones appearing mainly after 1985 [24,25,26]. The latter were produced by "dedicated" experiments, designed to minimise systematic errors on the ratio of structure functions; as an example, target arrangements have been designed to allow cancellation of geometrical acceptance, efficiency corrections and/or beam fluxes and to minimise the effects of time variations of apparatus acceptance and efficiencies.

5. THE EXPERIMENTAL SITUATION

The experimental situation in summary can be described as follows (fig. 4).

i) For $x \leq 0.1$ a downward trend is present suggesting that shadowing sets in regardless of Q^2 being quite large (> 10 GeV^2); this may suggest that the simplest version of the VMD models is not correct and that the VMD approach has to be refined; on the contrary totally different models in which shadowing arises due to interactions at the partonic level may be favoured (see [17] and references therein).

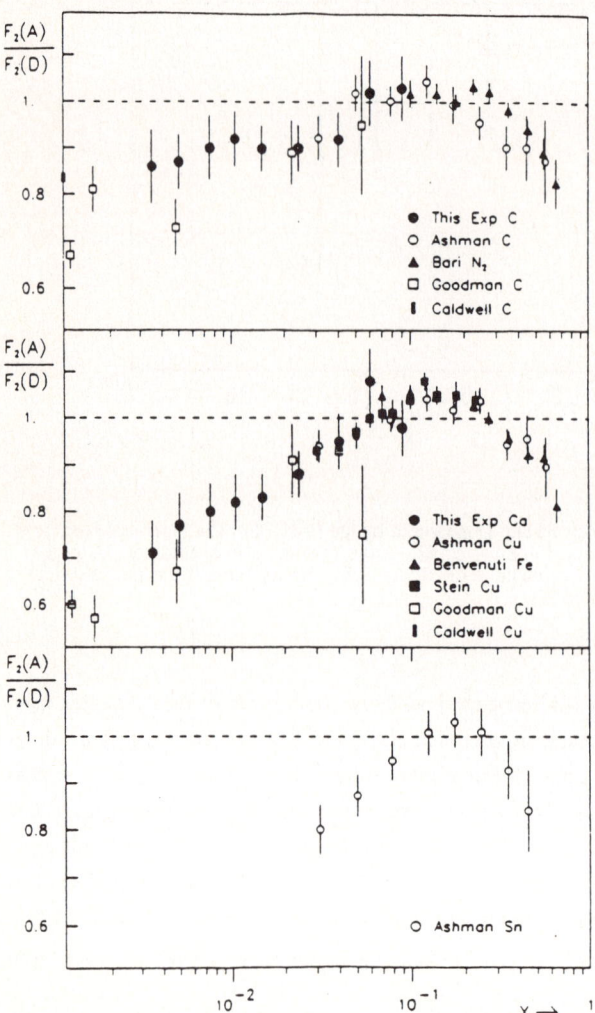

Fig.4. Structure function ratios obtained in leptoproduction experiments, for different target nuclei: Arneodo [17] (labelled "this exp"), Ashman [24], Bari [23], Benvenuti [22], Goodman [11]. The points on the vertical axis are ratios from a photo-production experiment, Caldwell [13], where the energy of the photon beam was 60 GeV. Error bars represent statistical and systematic errors summed in quadrature

ii) At low x (0.1 ≤ x ≤ 0.3) the effect is substantially reduced as compared with the original EMC data (but consistent with them within the quoted systematic error). Together the data indicate some antishadowing (enhancement of F_2^A with respect to F_2^D) in this region.

iii) R(x) decreases linearly between x ~ 0.3 (where it starts becoming smaller than one) and 0.6.

The dip around x ~ 0.6 becomes more pronounced as we move to higher values of A (approximately as logA); the position of the dip is however independent of A.

iv) For x ≥ 0.6 the ratio starts to rise again, as expected from Fermi motion effects.

v) There is so far no convincing experimental evidence of a significant Q^2 dependence of R.

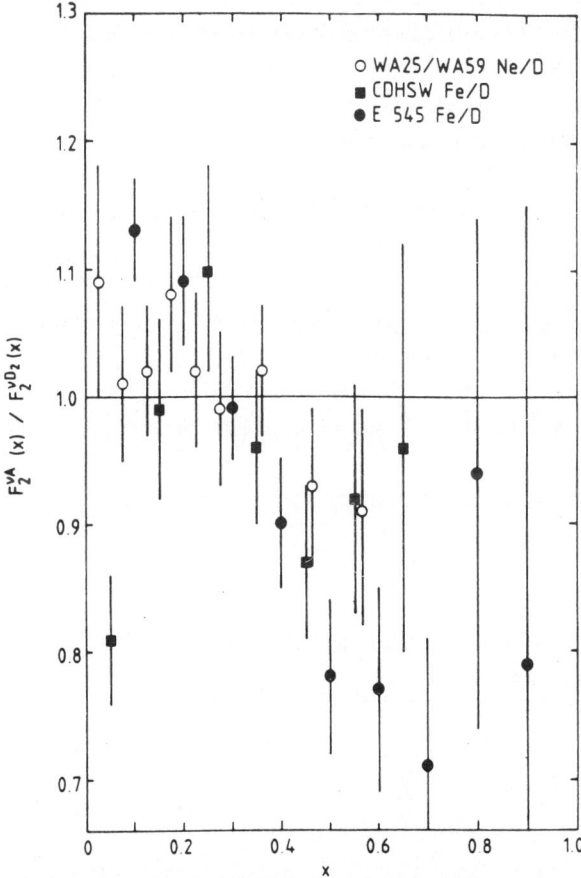

Fig. 5. Compilation of nuclear effects in deep inelastic scattering for recent neutrino data

In principle the EMC effect is better studied in neutrino than in electron or muon experiments, as they allow to probe sea and valence quarks separately. In practice however, the statistical signficance of neutrino data is rather poor; no neutrino experiment [27-33] is precise enough to confirm the experiment on its own; the overall trend of the data is nonetheless similar to the charged lepton experiments (fig. 5).

6. THEORETICAL IDEAS

As mentioned above, the EMC effect has aroused a strong interest in the theoretical community, testified by a massive production of literature on the subject. We shall not give any sort of comprehensive review of the theoretical ideas emerged in the last years (see for instance [2]). We shall only attempt to give a broad classification of the theoretical models proposed.

One can distinguish the following (not mutually exclusive) categories:

a) the nucleons in the nucleus have different properties than the free ones (e.g. different effective mass, different confinement radius, etc);

b) the nucleus is the incoherent sum of free nucleons and other elementary particles (e.g. pions, baryon-resonances);

c) the nucleus is an incoherent sum of "quark-clusters" (e.g. diquarks, deuterons = 6 quark clusters, α-clusters = 12 -quark clusters, etc.);

d) the nucleus is a plasma of quarks and gluons (colour conductivity of the nucleus);

e) the scattering mechanism in the nucleus is different than in free nucleons (shadowing).

From a less phenomenological point of view a different classification can be proposed:

1) the structure function of the nucleus is the incoherent sum of the structure functions of the components (nucleons, quarks, clusters):

$$F_2^A(x,Q^2) = \sum_{cl} n^{cl} \cdot F_2^{cl}(x,Q^2)$$

n^{cl} = number of the component cl objects in the nucleus

2) the structure function of the nucleus is the folding of the momentum distribution of the constituents with their structure functions:

$$F_2^A = \int_x^A f^{cl}(y) \cdot F_2^{cl}\left(\frac{x}{y}, Q^2\right) dy$$

$F^{cl}(y)$ = distribution function of a cluster in the nucleus

$F_2^{cl}(\xi, Q^2)$ = structure function of a constituent

3) the structure function of the nucleus can be obtained from the structure function of the nucleon by rescaling either of the two variables x, Q^2.

$$F_2^A(x, Q^2) = F_2^N(x, \xi \cdot Q^2)$$

or $\quad F_2^A(x, Q^2) = F_2^N(x, \xi \cdot Q^2)$ with ξ = rescaling parameter

The theoretical situation can be possibly summarised by the statement that unfortunately there is no single theoretical picture of the EMC effect so far describing all the features of the experimental data.

The compatibility of some of the different approaches lends however hope that the EMC effect may provide a key to translate between the language of conventional nuclear physics and the one describing the nucleus in terms of quark degrees of freedom, in addition to having promoted a new kind of collaboration between the communities of nuclear and particle physicists.

References

1. EMC, J.J. Aubert et al., Phys. Lett. B123 (1983) 275.
2. See for instance, T. Sloan, G. Smadja, R. Voss, Phys. Rep. 162 (1988) 45-167.
3. G. Berlad et al., Phys. Rev. D22 (1980) 1547.
4. L.L. Frankfurt and M.I. Strikman, Nucl. Phys. B181 (1981) 22.
5. A. Bodek and J.L. Ritchie, Phys. Rev. D23 (1981) 1070;
   A. Bodek and J.L. Ritchie, Phys. Rev. D24 (1981) 140.
6. J. Eickmeyer et al., Phys. Rev. Lett. 36 (1976) 289.
7. G. Huber et al., Z. Phys. C2 (1979) 279.
8. J. Franz et al., Z. Phys. C10 (1981) 105.
9. J. Bailey et al., Nucl. Phys. B151 (1979) 367.
10. S. Stein et al., (SLAC E71), Phys. Rev. D12 (1975) 1884.
11. M.S. Goodman et al., Phys. Rev. Lett. 47 (1981) 293.

12. M. Miller et al., Phys. Rev. D4 (1981) 1.
13. D.O. Caldwell et al., Phys. Rev. Lett. 42 (1979) 553.
14. V. Heynen et al., Phys. Lett. 34B (1971) 651.
15. S. Michalowski et al., Phys. Rev. Lett. 39 (1977) 737.
16. G.R. Brookes et al., Phys. Rev. D8 (1973) 2826.
17. M. Arneodo et al., CERN-EP/88-56, to appear in Phys. Lett.
18. A. Bodek et al., (SLAC E87) Phys. Rev. Lett. 50 (1983) 1431.
19. A. Bodek et al., (SLAC E49B) Phys. Rev. Lett. 51 (1983) 1431
20. R.G. Arnold et al., (SLAC E139) Phys. Rev. Lett. 52 (1984) 727;
    R.G. Arnold et al., (SLAC E139) SLAC-Report SLAC-PUB-3257 (1983).
21. S. Dasu et al., (SLAC E140) Phys. Rev. Lett. 60 (1988) 2591.
22. BDCMS, A.C. Benvenuti et al., Phys. Lett. 189B (1987) 483.
23. BCDMS, G. Bari et al., Phys. Lett. B163 (1985) 282.
24. EMC, J. Ashman et al., Phys. Lett 202B (1988) 603.
25. EMC, L. Ropelewski, Talk presented at the XIXth Symposium on Multiparticle Dynamics, Arles, France, June 13-17, 1988.
26. NMC, M. Arneodo, Talk presented at the XXIVth International Conference on High Energy Physics, Munich, August 4-10, 1988
27. M.A. Parker et al., Nucl. Phys. B232 (1984) 1.
28. A.M. Cooper et al., Phys. Lett. B141 (1984) 133;
    J. Guy et al., Z. Phys. C236 (1987) 337.
29. CDHS, H. Abramowicz et al., Z. Phys. C25 (1984) 29.
30. J. Hanlon et al., Phys. Rev. D32 (1985) 2441.
31. V.V. Ammonosov et al., Pis'ma Zh. Eksp. Teor. Fiz. 39 (1984) 327.
32. T. Kitagaki et al., Proceedings of the 12th International Conference on Neutrino Physicsand Astophysics (Sendai, Japan, 1986).
33. A.M. Cooper, Proceedings of the 11th International Conference on Neutrino Physics and Astrophysics (Nordkirchen near Dortmund, 1984). eds K. Kleinknecht and A. Paschos, p.381.

IV

UNDERGROUND AND COSMOLOGICAL PHYSICS

ELEMENTARY PHYSICS OF THE GRAVITATIONAL COLLAPSE

Giulio Auriemma

Dipartimento di Fisica e Sezione INFN, Università di Roma "La Sapienza", Roma, Italy

1. Introduction

Recently, the more then 30 years old idea that the huge amount of energy liberated in type II supernovae explosions is produced by the collapse of the core of a massive star, has received a dramatic experimental confirmation from the detection of neutrinos emitted few hours before the detection of the optical outburst of the Shelton 1987A supernova in the Large Magellanic Cloud. The observational aspects of this historical event are reviewed in the lecture of prof. Galeotti, therefore I will not discuss them here.

The collapse of a massive stellar core[1] is a physical event whose detailed description is awkwardly complicated by the underlying micro-physical processes occurring at the high density and temperature in the matter of the stellar core.

Nevertheless it is remarkable that the general outlines of the process and nearly all the measurable, such as the total energy of the neutrino burst, the spectrum of the neutrinos of each flavor, and the time scale of the burst can be easily derived from first principles, in a single lecture. In the following I will exhibit this "back of the envelope" theory of the stellar collapse, after a short digression upon the cosmological relevance of the supernovae explosion.

2. Some nuclear Cosmogony

It is accepted by almost everyone that the Hot Big Bang cosmological model[2] is at the moment the most satisfying approach to the problem of the origin of the Universe and its early evolution. However the so called Standard Cosmological Model, as reviewed in another lecture of this course , in which is also incorporated the standard model of electroweak interactions and the hadronic phenomenology, is highly unsatisfactory, when compared with the observable Universe. I will recall here only three major problems, which are, more or less, loosely related to the subject of the present lecture.

The first problem is the well known baryon-antibaryon asymmetry problem[3]. It is rather natural to assume that at the Big Bang the Universe was symmetric, or *i.e.* with zero net baryonic number. But in this case , soon after the temperature of the Universe drops below $\sim 2\text{GeV}$ most of the baryon are annihilated and the specific entropy of the Universe, measured by the ratio $\eta = n_B/n_\gamma$ should be today $\eta \simeq 10^{-18}$ instead of the observed $\eta_{obs} = 10^{-(9\pm1)}$. This is a very fundamental problem, indeed, because since the late '60 the russian physicist A. Sakharov has convincingly shown that we cannot avoid a nearly empty Universe, if we do not accept CP violation and baryon non-conservation at work in its very early phase[4]. This is considered to be at present, the only hint for physics beyond the Standard Model, from an experimental (observational) point of view.

However even if the Grand Unified Theories could give us a satisfactory explanation of the present baryon asymmetry, we have to face another difficulty, the extremely small anisotropy of the 3 K relic radiation[5]. As a matter of fact accurate measurements show that the fluctuation of the radiation is $\Delta\rho_{rad}/\rho_{rad} \simeq 10^{-5}$ when the matter is clusterized in structures which have $\Delta\rho_{mat}/\rho_{mat} \simeq 1$ up to a scale of the order of 50 Mpc. This is a problem because matter and radiation are strongly coupled by e.m. interactions, therefore we have to postulate that the anisotropy of the matter distribution grew up from the original 10^{-5} to 1 in a timescale short compared to the age of the Universe, due to a physical process that is quite unknown.

The third problem is that the Big Bang nucleosynthesis cannot produce nuclei heavier the the ^7Li. This is an effect of that "perversion" of the Nature called weak interaction, which makes the neutron unstable via the β-decay reaction

$$n \to p + e^- + \nu \quad , \tag{1}$$

which has a width

$$\Gamma \sim G_F^2 \Delta^5 \quad , \tag{2}$$

where $G_F \simeq 1.16 \times 10^{-5}$ GeV^{-2} is the Fermi constant and $\Delta = m_n - m_p$ is the isospin mass splitting. Free neutrons, which can more easily enter the Coulomb field of nuclei to build up the atomic number, have a lifetime of $\Gamma^{-1} = 898 \pm 16$ sec and no complex nucleus could be synthesized in the Big Bang.

The Universe which would be constructed according to the Standard model is extremely unappealing, though, because it would be nearly void, with no Galaxies or stars and with no complex elements. This catastrophic failure of the Standard model is the most fascinating aspect of Cosmology. The challenge is in fact that in order to understand the most common facts of everyday life, including our own existence, we are forced to look for new physics, perhaps using the Universe as the "poor man's accelerator" as pioneering suggested by the recently disappeared, russian physicist and cosmologist Zeldovich [6] .

The rest of my talk will be devoted to the fact that, once we have made the Universe baryon asymmetric and largely anisotropic, we could solve the problem of constructing heavy nuclei, via nuclear reactions in the core of massive stars. In fact when stars are formed from the initial H and He, plus the little Lithium contamination, in a time estimated to be of the order of 10^9 yrs, the nuclear reactions taking place in the hot core can build up all the elements, with abundances which will be dictated by the stability of the nuclei. But the new problem is that without a dramatic explosion which disrupts the core of the star, spreading out the elements, we could not have planets or stars like the Sun, which have a suitable fraction of heavy elements. Therefore in this process the explosion of supernovae have the fundamental cosmogonic role of spreading around the heavy nuclear species. As was recalled in the previous lecture on supernovae, the Type I supernovae are originated by detonation of intermediate mass stars, and liberate the fundamental CNO elements, but only Type II supernovae which actually disrupt a large fraction of the core of massive stars, can freed the really heavy elements, like Fe, which we find in our mines. So I think that is a cosmic justice that the weak interactions which prevented the formation of heavy element in the Big Bang, are responsible of the instability of massive star at the end of their evolution.

3. Instability of the stellar core

A very simple argument originally due to Landau (1932) can convince that the core of a massive star, formed by Fe should undergo to a dramatic implosion, due to the effect of the weak interactions. This implosion is called a gravitational collapse. Let us consider the Hamiltonian of a self gravitating ensemble of Iron nuclei and electrons

$$H = T + U \quad , \tag{3}$$

where $T = \sum p_i \, \dot{q}_i$ the total kinetic energy of the particles, and U_{grav} is the total gravitational potential of the ensemble. The latter will be dominated by the nuclei, or (if we neglect terms of the order of π)

$$U_{grav} = -\frac{G_N \, N_N^2 \, m_N^2}{R} \quad , \tag{4}$$

where $G_N \simeq 10^{-39}$ GeV^{-2} is the Newton constant, N_N the number of nucleons, $m_N \simeq 1$ GeV the nucleon mass and R the radius of the core[‡1]. At high density and low temperature the Pauli exclusion principle will force the electrons to have a Fermi momentum

$$p_F \sim (N_e/R^3)^{1/3}. \tag{5}$$

It is easy to see that in this situation almost all the kinetic energy of the system is the kinetic energy of the electrons. If the electrons are relativistic the hamiltonian is

$$H = N_e \left(\frac{N_e}{R^3}\right)^{1/3} - \frac{G_N N_N^2 m_N^2}{R}, \tag{6}$$

or introducing $Y_e = N_e/N_N$ the leptonic fraction, which is for ordinary matter $Y_e \sim 0.5$

$$H = \frac{N_N^{4/3} Y_e^{4/3}}{R} - \frac{G_N N_N^2 m_N^2}{R}. \tag{7}$$

From the canonical equation $\dot{q}_i = \partial H/\partial p_i$ we know that the sign of $\dot{q}_i$ is the same of the sign of H, then if $H < 0$ the system will contract. Therefore from Eq. (7) we see that the core will contract if the total number of nucleons is

$$N_N > N_{Ch} = \left(\frac{1}{G_N m_N^2}\right)^{3/2} Y_e^2 \simeq 10^{57}. \tag{8}$$

This corresponds to the Chandrasekar mass limit for the stability of a white dwarf star. How can this be applied to the core of a massive star? In fact when the core of a star is composed by Fe, only endothermic reactions can take place, therefore the core will contract, but its temperature will not grow. The reactions which convert lighter nuclei into iron can proceed, at the boundary of the core, therefore the mass of the core will increase until the Chandrasekar limit is crossed. At a certain point the Fermi momentum of the electrons will be larger then the threshold of the inverse β-decay reaction

$$^{56}Fe + e^- + (3.965\ MeV) \rightarrow\ ^{56}Mn + \nu_e. \tag{9}$$

The effect of this reaction is catastrophic, because it reduces Y_e driving the core further away from the equilibrium. The collapse cannot be arrested until all the protons are converted into neutrons. At this point the degeneracy pressure of the neutrons will grow enough to halt the collapse.

[‡1] All formulas are written with the convention $\hbar = c = k_B = 1$. Therefore the unit of length is GeV^{-1}.

4. Energy released in the collapse

The gravitational binding energy of a stellar core is given by Eq. (4), thus the energy released in the collapse is

$$\Delta E \simeq G_N \, N_{Ch}^2 \, m_N^2 \left(\frac{1}{R_2} - \frac{1}{R_1} \right) \quad , \tag{10}$$

where R_1 is the radius at the onset of the instability and R_2 the radius after the collapse. The initial radius, as we have discussed above, is the radius at the moment when the "neutronization" of the core starts, or when p_F is of the order of few MeV's. From Eq. (5) we have

$$R_{Ch} \simeq \frac{1}{p_F} \left(\frac{Y_e^2}{G_N \, m_N^2} \right)^{1/2} \sim 10^8 \text{ cm} \quad . \tag{11}$$

The collapsed will be stopped, and very likely a shock wave is bounced back into the external layer of the stars, when the degeneracy pressure of the relativistic neutron gas, "stiffens" the collapsed core, or when $p_F^{neut} \simeq m_n$. This happens when the radius is

$$R_{n.s.} \simeq \frac{1}{m_n} \left(\frac{1}{G_N \, m_n^2} \right)^{1/2} \sim 10 \text{ km} \quad , \tag{12}$$

then we have $R_{n.s.} \ll R_{Ch}$ and the total energy released in the collapse is

$$\Delta E \simeq \frac{G_N \, N_{Ch}^2 \, m_N^2}{R_{n.s.}} \sim 10^{53} \text{ ergs} \quad . \tag{13}$$

It is worth noticing that Eq. (13) is correct only in flat space-time. Relativistic correction should be applied when the curvature of the metric is not negligible. It is demonstrated on standard textbooks [7] that the relativistic corrections for a spherically symmetric field without angular moment are of the order of $G_N \, M/R$ or comparing with Eq. (13), $\Delta E/M \sim .1$.

This is the proof that relativistic corrections do not change the order of magnitude of the estimates, obtained with non relativistic formulae.

5. Collapse time scale

Once triggered the neutronization has the effect of reducing the lepton fraction Y_e, thus reducing the degeneracy pressure of the electrons. We can describe the dynamics of

be $\sim 10^{53}$ ergs, in a burst with a duration of ~ 10 sec and that the average energy of the neutrinos will be ~ 10 MeV.

Whether or not this theory apply to type II supernovae is to be decided on the observational bases. Likely the detection of neutrinos from the Shelton SN in the Magellanic Cloud is a leap in that direction.

REFERENCES

[1] For a detailed discussion of the physics of supernova explosions see G.E. Brown, H.A. Bethe and G. Baym; *Nuclear Physics, A375*, (1982), 481 and references therein.

[2] For a general introduction see S. Weimberg; *"Gravitation and Cosmology"*, J. Wiley & Sons, Inc, 1972, page 469ff.

[3] G. Steigman; *Ann. Rev. Nucl. Part. Scie.*, 29, (1979), 313.

[4] A.D. Sakharov; *Sov. Phys.-JETP*, 22, (1965), 241.

[5] R.A. Sunyaev and Ya.B. Zeldovich; *Ann. Rev. Astr. Astrophys.*, 18, (1980), 537.

[6] A.D. Dolgov and Ya.B. Zeldovich; *Rev. Mod. Phys.*, 53, (1981), 1.

[7] L. Landau and E. Lifchitz; *"Théorie des Champs"*, Éditions MIR, Moscou, 1970, page 391.

[8] D.Z. Freedman, D.N. Schramm and D.L. Tubbs; *Ann. Rev. Nucl. Scie.*, 27, (1977), 167.

4. Energy released in the collapse

The gravitational binding energy of a stellar core is given by Eq. (4), thus the energy released in the collapse is

$$\Delta E \simeq G_N \, N_{Ch}^2 \, m_N^2 \left(\frac{1}{R_2} - \frac{1}{R_1} \right) \, , \qquad (10)$$

where R_1 is the radius at the onset of the instability and R_2 the radius after the collapse. The initial radius, as we have discussed above, is the radius at the moment when the "neutronization" of the core starts, or when p_F is of the order of few MeV's. From Eq. (5) we have

$$R_{Ch} \simeq \frac{1}{p_F} \left(\frac{Y_e^2}{G_N \, m_N^2} \right)^{1/2} \sim 10^8 \text{ cm} \, . \qquad (11)$$

The collapsed will be stopped, and very likely a shock wave is bounced back into the external layer of the stars, when the degeneracy pressure of the relativistic neutron gas, "stiffens" the collapsed core, or when $p_F^{neut} \simeq m_n$. This happens when the radius is

$$R_{n.s.} \simeq \frac{1}{m_n} \left(\frac{1}{G_N \, m_n^2} \right)^{1/2} \sim 10 \text{ km} \, , \qquad (12)$$

then we have $R_{n.s.} \ll R_{Ch}$ and the total energy released in the collapse is

$$\Delta E \simeq \frac{G_N \, N_{Ch}^2 \, m_N^2}{R_{n.s.}} \sim 10^{53} \text{ ergs} \, . \qquad (13)$$

It is worth noticing that Eq. (13) is correct only in flat space-time. Relativistic correction should be applied when the curvature of the metric is not negligible. It is demonstrated on standard textbooks [7] that the relativistic corrections for a spherically symmetric field without angular moment are of the order of $G_N \, M/R$ or comparing with Eq. (13), $\Delta E/M \sim .1$.

This is the proof that relativistic corrections do not change the order of magnitude of the estimates, obtained with non relativistic formulae.

5. Collapse time scale

Once triggered the neutronization has the effect of reducing the lepton fraction Y_e, thus reducing the degeneracy pressure of the electrons. We can describe the dynamics of the collapsing core just assuming that the particles fall freely in the gravitational potential well. The time scale of the collapse will be

$$t_{coll} \simeq [(R_2 - R_1)/g]^{1/2} \tag{14}$$

where $g \simeq G_N\, M/R^2$ is the acceleration. From this simple estimate we have

$$t_{coll} \simeq \left(\frac{R_{Ch}^3}{G_N\, M_{Ch}}\right)^{1/2} \sim 1 \text{ ms}. \tag{15}$$

6. Neutrino burst from neutronization

It is straightforward to compute the energy taken away from the core by the neutrinos produced in the neutronization, in fact the total number of neutrinos is equal to the number of protons or

$$N_\nu^{neutr} \simeq Y_e\, N_{Ch} \tag{16}$$

and the energy of the neutrino is $E_\nu \sim p_F$. However we consider that during the collapse the Fermi energy changes fast. On the other side the neutrinos can escape only when the radius is $R \leq t_{coll} \sim 10^7$ cm, at this point the Fermi momentum of the electrons is $p_F \sim 20\ MeV$. The energy carried away by the neutrino burst is

$$L_\nu^{neutr} \times t_{coll} \simeq E_\nu\, Y_e\, N_{Ch} \sim 10^{52}\ ergs \tag{17}$$

which is only 10% of the total energy liberated by the collapse. We must conclude that about 90% of the energy of the collapse is still contained inside the neutron star.

7. Neutrino diffusion

The cooling of the neutron star formed in the collapse proceed via pair neutrino radiation, because the opacity of the matter for γ-rays is too large.

We can apply here a result of the theory of random walk, namely that in the diffusion limit a particle can diffuse out of a sphere of radius R after a number of scatterings N_{scatt}, given by

$$\lambda\, N_{scatt}^{1/2} \simeq R \quad . \tag{18}$$

From this formula we have that the diffusion time is for a relativistic particle

$$t_{diff} \simeq N_{scatt}\, \lambda = \frac{R^2}{\lambda} \quad , \tag{19}$$

where λ is the mean free path. The diffusion time of neutrinos is short because their mean free path is much longer.

The neutrino opacity of the neutron star is dominated by Neutral Current elastic scattering, because for a neutrino energy $E_\nu \lesssim 300\ A^{1/3}\ MeV$ the scattering is coherent[8], and the cross section is

$$\sigma_{coh} \simeq G_F^2\ \sin^4\theta_W\ A^2\ E_\nu^2\ , \qquad (20)$$

where $\theta_W \simeq 30°$ is the weak angle. From Eq. (19) we could estimate the diffusion time, once that we have determined the neutrino energy.

8. Neutrino spectrum

The spectrum of emitted neutrino will be a Fermi-Dirac distribution with zero chemical potential since the neutrinos are produced by the reaction

$$\gamma + \gamma \rightleftharpoons \nu + \bar\nu\ . \qquad (21)$$

The total luminosity of a black body neutrino emitter is

$$L_\nu = \frac{\pi^2}{30}\frac{7}{8} N_f\ T^4\ 4\pi\ R^2\ , \qquad (22)$$

where N_f is the number of neutrino flavors. The temperature of neutrinos will be fixed by the condition

$$\Delta E \simeq t_{diff} \times L_\nu\ . \qquad (23)$$

We can estimate the diffusion time from Eq. (19), assuming $E_\nu \sim T$.

The average mean free path for neutrinos is

$$\lambda_\nu^{-1} \simeq \frac{N_{Ch}\ G_F^2\ \sin^4\theta_W\ A^2\ T^2}{R^3} \qquad (24)$$

Inserting this value in Eq. (23), we can derive the temperature of the neutrino spectrum, which turns out to be $T \sim 10$ MeV and from Eq. (19) we have $t_{diff} \sim 10$ sec.

9. Conclusions

We have seen as an elementary treatment of the physics of stellar collapse leeds to the determination of almost all the observable quantities of the neutrino burst. In fact on these solid ground we could expect that most of the energy emitted in neutrinos will be $\sim 10^{53}$ ergs, in a burst with a duration of $\sim$10 sec and that the average energy of the neutrinos will be $\sim$10 MeV.

Whether or not this theory apply to type II supernovae is to be decided on the observational bases. Likely the detection of neutrinos from the Shelton SN in the Magellanic Cloud is a leap in that direction.

REFERENCES

[1] For a detailed discussion of the physics of supernova explosions see G.E. Brown, H.A. Bethe and G. Baym; *Nuclear Physics, A375,* (1982), 481 and references therein.

[2] For a general introduction see S. Weimberg; *"Gravitation and Cosmology"*, J. Wiley & Sons, Inc, 1972, page 469ff.

[3] G. Steigman; *Ann. Rev. Nucl. Part. Scie.,* 29, (1979), 313.

[4] A.D. Sakharov; *Sov. Phys.-JETP,* 22, (1965), 241.

[5] R.A. Sunyaev and Ya.B. Zeldovich; *Ann. Rev. Astr. Astrophys.,* 18, (1980), 537.

[6] A.D. Dolgov and Ya.B. Zeldovich; *Rev. Mod. Phys.,* 53, (1981), 1.

[7] L. Landau and E. Lifchitz; *"Théorie des Champs"*, Éditions MIR, Moscou, 1970, page 391.

[8] D.Z. Freedman, D.N. Schramm and D.L. Tubbs; *Ann. Rev. Nucl. Scie.,* 27, (1977), 167.

ON THE STANDARD COSMOLOGICAL MODEL

Pierluigi FORTINI

Università di Ferrara - Dep. of Physics - INFN - Sezione di Bologna

The main properties of the standard cosmological model are reviewed. Some problems connected with the physics of elementary particles are particularly discussed; suggestions for further readings and research are given.

1. GENERAL RELATIVITY IN OUTLINE

Like all equations of physics also Einstein's equations

$$G_{ik} \equiv R_{ik} - \frac{1}{2} g_{ik} R = \frac{8 \pi G}{c^4} T_{ik} \tag{1}$$

derive from a variational principle. Gravitational action S_g is given by Hilbert variational principle

$$S_g = \frac{c^3}{16 \pi G} \int_V R \sqrt{-g} \, d^4 x \tag{2}$$

where:

a) V is the volume of curved four dimensional space and $d^4 x$ is the (four) volume element $dxdydzd(x^0 = ct)$,

b) g is the determinant of the metric g_{ik} which is linked to the distance ds of two neighbouring points (with coordinaes x^i and $x^i + dx^i$) by:

$$ds^2 = g_{ik} \, dx^i \, dx^k$$

c) by means of differential combinations of g_{ik} one gets:

the Riemann tensor $\quad R^i{}_{klm}$

the Ricci tensor $\quad R_{ik} = R^j{}_{ijk}$

the Ricci scalar $\quad R = R^i{}_i$

Dimensionally is:

$$[R] = [R_{ik}] = [R^i{}_{klm}] = [\text{length}^{-2}]$$

so that R is the inverse square of the curvature radius of space-time.

d) T_{ik} is the energy - momentum tensor of matter so that

e) Einstein equations mean that space-time curvature, expressed through the combination G_{ik} (Einstein tensor), i.e. geometrical properties of the space are determined by physical properties of matter through the tensor T_{ik}.

f) Solving Einstein equations means: given the properties of matter (i.e. given the tensor T_{ik}) determine a metric tensor such that $G_{ik}(g)$ is equal to $\frac{8\pi G}{c^4} T_{ik}$.

Generally speaking $G_{ik}(g)$ is a second order expression (i.e. it contains the second derivatives of g_{ik}) and is not linear.

2. THE COSMOLOGICAL PROBLEM

Basically a cosmological model is: i) a simplifying assumption on the structure and distribution of matter in the universe as seen through astronomical measurements and; ii) a simplifying assumption on the geometry (i.e. on the metric tensor) of the universe. Of course one assumption must be a consequence of the other through Einstein equations.

Let us start from ii).

Hubble's law $v = Hd$ suggests the introduction of the following postulate: the world lines of the particles of the cosmological fluid (i.e. the galaxies) form in four dimensional space-time a bundle of non intersecting geodesics orthogonal to the three dimensional space.

As a second postulate one assumes the so called cosmological principle: at every instant of time the Universe is homogeneous and isotropic that is the hypersurface t = const. (i.e. ordinary three-space) is of constant curvature.

It has been known since last century that constant curvature three-dimensional hypersurfaces can be of three types: flat hypersurfaces (euclidean spaces); positive curvature hypersurfaces (like a sphere) which are closed (i.e. of finite extension) and negative curvature hypersurfaces (like a Beltrami pseudo-sphere) which are open (i.e. of infinite extension).

Starting from these properties one can show that there is only one line element which satisfies the above two postulates and which is the so called Robertson-Walker line element:

$$ds^2 = c^2 dt^2 - S^2(t)\left[\frac{dr^2}{1-kr^2} + r^2(d\theta^2 + \sin^2\theta d\varphi^2)\right] \qquad (k = 0, +1, -1) \qquad (3)$$

where (r, θ, φ) are dimensionless coordinates of three-space, linear dimensions are given through the scale-factor $S(t)$ which gives at every instant of time the linear dimensions of three-space. In other words the distance between any two bodies is a functions of time: physical space therefore develops according to a law of similitude. The number k has the following meaning: k = 0 means a flat three space (universe of Einstein-de Sitter); k = 1 means a positive curvature three-space i.e. closed and k = -1 means an open, that is negative curvature, three space.

Let us now consider i) i.e. the physical state of matter. From homogeneity and isotropy of matter distribution (as it seems to follow from astronomical observations, though recent observations seem to cast serious doubts on this) one is lead to imagine cosmic matter as a perfect fluid with pressure p and energy density ε. Now total energy density ε is equal to kinetic-energy density (which is the random motion energy of the fluid) plus rest energy density ϱc^2 (where ϱ is matter density). According to the kinetic theory of gases the density of kinetic energy is equal to three times the pressure. So we have:

$$\varepsilon = \varrho c^2 + 3p \qquad (4)$$

The non vanishing components of the energy-momentum tensor are:

$$T^1_1 = T^2_2 = T^3_3 = -p \qquad T^0_0 = \varepsilon$$

In present day universe the random motion of the galaxies ($v \sim 10^3$ km/sec) is negligible respect to the rest energy. In fact:

$$p \sim \varrho v^2 \sim c^2 \varrho \frac{v^2}{c^2} = (\frac{10^3}{3.10^5}) \varrho c^2 \sim 10^{-5} \varrho c^2$$

and therefore the cosmological fluid is well described in its present state by a classical fluid for which one has:

$$p = 0 \qquad\qquad \varepsilon = \varrho c^2 \qquad\qquad (5)$$

An expanding universe means that in the past its dimensions were smaller that those measured to day and therefore that pressure was greater.

In other words the more one approaches the big-bang the more the cosmological fluid becomes relativistic i.e. from (4):

$$p = \frac{1}{3} \varepsilon \qquad\qquad (6)$$

Obviously in this situation is meaningless to speak of stars and galaxies: developing from big-bang the universe went through various phase transformations which brought it from a hot state, in which matter and radiation transformed actively one into the other, to a cold state such as the present one in which matter and radiation are completely decoupled. Then if we take Robertson-Walker metrics (3) and (5) as an equation of state then Einstein equations (1) become

$$2\frac{\ddot{S}}{S} + \frac{\dot{S}^2 + kc^2}{S^2} = 0 = \frac{8\pi G\, T_1^1}{c^2} = \frac{8\pi G\, T_2^2}{c^2} = \frac{8\pi G\, T_3^3}{c^2} \qquad (7)$$

$$\frac{\dot{S}^2 + kc^2}{S^2} = \frac{8\pi G\, T_0^0}{3c^2} = \frac{8\pi G\, c^2}{3c^2} \varrho = \frac{8\pi G}{3} \varrho$$

which are the equations for a matter dominated universe.

For the relativistic case (radiation dominated universe) we have from (6):

$$2\frac{\ddot{S}}{S} + \frac{\dot{S}^2 + kc^2}{S^2} = -\frac{8\pi G}{3c^2}\varepsilon$$

$$\frac{\dot{S}^2 + kc^2}{S^2} = \frac{8\pi G}{3c^2}\varepsilon \qquad (8)$$

The quantity

$$H = \frac{\dot{S}}{S}$$

is called "Hubble constant" at time t; its present value is:

$$H_o = 100\ h_o\ \text{km/sec Mpc} \qquad (0.5 \leq h_o \leq 1)$$

According to the values $k = 0, +1, -1$ we have three solutions which, as can be seen from the second equation (7), correspond to $\rho < \rho_c, \rho < \rho_c, \rho = \rho_c$ where

$$\rho_c = \frac{3H_o^2}{8\pi G} = 1.2\ 10^{-5}\ h_o^2\ \text{nucleons/cm}^3$$

If we put

$$\rho = \rho_c \Omega$$

we get

$$\varrho = 1.2 \ 10^{-5} \ h_o^2 \ \Omega \ \text{nucleous/cm}^3 \qquad (9)$$

3. THERMODYNAMICS OF THE UNIVERSE:

For every particle of mass m_i one must distinguish between:

a) a high temperature regime when $T > T_i = \frac{m_i c^2}{kB}$ where particles are relativistic and b) a low temperature regime when $T < T_i$ where particles are non relativistic.

a) $T > T_i$. In this case one must distinguish between bosons and fermions. Standard thermodynamics gives:

$$n_B = \frac{g_B \xi(3)}{\pi^2} \left(\frac{k_B T}{\hbar c}\right)^3 \qquad n_F = \frac{3}{4} \frac{g_F}{2} \xi(3) \left(\frac{k_B T}{\hbar c}\right)^3$$

$$\varepsilon_B = \frac{g_B}{2} a T^4 \qquad \varepsilon_F = \frac{7}{8} \left(\frac{g_F}{2}\right) a T^4 \qquad (10)$$

$$s_B = \frac{4}{3} \frac{\varepsilon_B}{T} = \frac{2}{3} a T^4 \qquad s_F = \frac{4}{3} \frac{\varepsilon_F}{T} = \frac{7}{12} g_F a T^4$$

where:

$$a = \frac{\pi^2 k_B^4}{15 \hbar^3 c^3} = 7.56 \ 10^{-15} \ \frac{\text{erg}}{\text{cm}^3 \ {}^o k^4}$$

is the black body radiation constant related to Stefan-Boltzmann constant by $a = \frac{4\sigma}{c}$ and k_B is the Boltzmann constant; n, ε and s are the particle number density, energy density and entropy density (for bosons and fermions) respectively; g_B, g_F is the spin multiplicity for bosons and fermions (i.e. $g_{B,F}$ = 2s+1 were s is the spin).

b) $T < T_i$. In this case bosons and fermions behave in the same way and one has:

$$n_i = \frac{g_i}{\hbar^3} \left(\frac{m_i k_B T}{2h} \right)^{3/2} e^{-T_i/T} \qquad (11)$$

$$\varepsilon_i = m_i c^2 n_i$$
$$s_i = m_i n_i c^2/T$$

Besides the above formulae one needs also the following thermodynamical identities:

$$s = \frac{p + \varepsilon}{T} \qquad (12)$$

$$\frac{dp}{dT} = \frac{p + \varepsilon}{T}$$

From Friedmann equations (in both regimes (7) and (8)) one gets the following compatibility relation:

$$\frac{d}{ds}(\varepsilon s^2) + 3 p s^2 = 0 \qquad (13)$$

which, combined with (12), gives $s^3 s$ = const i.e.: the total entropy of the universe keeps constant while specific entropy decreases with expansion. Then:

$$s^3(t) = \frac{p + \varepsilon}{T} = \text{const.}$$

during expansion.

It is now very important to state general conditions under which thermodynamical equilibrium holds. Given a gas which undergoes thermodynamical transformations, thermal equilibrium holds if thermalization processes take place in a time which is shorter than the characteristic time of the transformation. Therefore given a certain reaction in the cosmic fluid with cross-section σ it can go on in a condition of equilibrium if the rate of the process is greater than that of the expansion of the gas. Now the rate is given by $\sigma n v$ (where n,v are the number density and mean velocity of the particles involved) while the expansion rate is $H = \frac{\dot{S}}{S}$

If the reaction is to take place at equilibrium one must have

$$\sigma \, nc > H \qquad (14)$$

while if

$$\sigma \, nc < H \qquad (15)$$

the reaction is out of equilibrium.

If (14) holds one has a continuous creation and annihilation of particles in equilibrium with radiation while if (15) holds the reaction cannot produce particles any longer and the existing one propagate independently with a mean free path which is the same as the rudius of the universe; in fact, in the latter case, the reaction time is longer than the age of the universe at the epoch considered.

So, for instance, consider the reaction $e^+ + e^- \rightleftarrows \gamma + \gamma$: before a certain time t_* when (14) held true, radiation interacted continuously with the electrons; after this time, (15) was satisfied and radiation acquired a mean free path equal to the radius of the Universe.

From this time on, radiation, upon adiabatic expansion, cooled down to the present 3°K temperature.

4. THE RADIATION DOMINATED UNIVERSE:

Henceforth we shall consider only the regime represented by equations (8) i.e. a radiation dominated universe: this was the state of matter before the recombination era envisaged at the end of Section 3. It is very easily seen from (8) that, in this regime, the behaviour of the Universe is quite insensitive to the actual value of k so that we can take, without loss of generality, k = 0.

From the second of (8) we get:

$$\frac{\dot{S}^2}{S^2} = \frac{8\pi G}{3c^2} \varepsilon$$

From (10) we see that

$$\varepsilon = x\, a\, T^4 \tag{16}$$

where the number x is obtained summing on every kind of particles. It turns out therefore that:

$$\frac{\dot{S}^2}{S^2} = \frac{8\pi G x\, a}{3c^2} T^4$$

but from (13), with $p = \frac{\varepsilon}{3}$ one gets:

$$\varepsilon = \frac{B}{S^4} \qquad (B = \text{const.})$$

and there fore

$$T = \frac{A}{S} \qquad (A = \text{const.})$$

Integrating the resulting differential equations we get:

$$T(°k) = \left(\frac{3c^2}{32\pi Ga\chi}\right)^{1/4} t^{-1/2} = 1.52 \ 10^{10} \ \chi^{-1/4} \ t^{-1/2} \ (\text{sec})$$

which gives the time scale used in this paper. Other useful expressions are:

$$t \ (\text{sec}) = \frac{2.3 \ 10^{20}}{T(°k)^2 \ \chi^{1/2}}$$

$$T \ (\text{MeV}) = 1.3 \ \chi^{-1/4} \ t^{-1/2} \ (\text{sec})$$

$$t(\text{sec}) = \frac{1.7}{T(\text{MeV})^2 \ \chi^{1/2}}$$

The mean number of particles of every kind is simply given by:

$$n(\text{cm}^{-3}) = \frac{\varepsilon}{3k_B T} = \frac{\chi aT^3}{3k_B} \simeq 6 \cdot 10^{31} \ \chi^{1/4} \ t^{-3/2} \ (\text{sec})$$

Using this time scale we can conventionally divide the period, which goes from the big-bang to the end of the radiation era, into the following stages:

	T (MeV)	t(sec) after big bang	
1)	$T > 10^{22}$	$t < 10^{-43}$	Quantum gravity era
2)	$10^{22} > T > 5 \cdot 10^3$	$10^{-43} < t < 10^{-6}$	Grond unified theory era
3)	$5 \cdot 10^3 > T > 10^2$	$10^{-6} < t < 10^{-4}$	Hadron era
4)	$10^2 > T > 10^{-1}$	$10^{-4} < t < 10^2$	Lepton era
5)	$10^{-1} > T > 4 \cdot 10^{-7}$	$10^2 < t < 10^{13}$	Nucleosynthesis takes place
6)	$4 \cdot 10^{-7} > T > 10^{-10}$	$10^{13} < t < 10^{18}$	Radiation decouples; stars and galaxies; present Universe

We shall now take into account some (out of the many) problems connected with the above periods.

1) <u>Quantum gravity</u>: it is obvious that classical theory of gravity (i.e. general relativity) breaks down as long as the gravitational action is of the order of the Planck constant. So, starting from (2) we can give an order of magnitude estimate of S_g taking as a fundamental length the radius of the Universe at time t, i.e. ct. We can therefore consider $R \sim \frac{1}{c^2 t^2}$ and $V \sim (ct)^A$. In this way we get:

$$S_g \approx \frac{c^3}{G} \frac{1}{c^2 t^2} (ct)^A \approx \frac{c^5 t^2}{G}$$

so for S_g ℏ we see that quantum gravity holds for times less than

$$t_p = \sqrt{\frac{\hbar G}{c^5}} \sim 5.39 \ 10^{-44} \text{ sec (Planck's time)}.$$

So for $t < t_p$ gravity is quantized; the big-bang must therefore be thought of as a purely quantum effect. Most probably this is the reason why classically the big bang appears as a singularity of space-time. The radius of the universe at Planck time is

$$l_p = ct_p = \sqrt{\frac{\hbar G}{c^4}} \sim 1.62 \ 10^{-33} \text{ cm (Planck's length)}$$

The quantity Gm^2 has dimensions of an energy times a length so that $\frac{Gm^2}{\hbar c}$ is dimensionless. This quantity can be taken as a measure of the strenght of gravity just as $\frac{l^2}{\hbar c}$ is a measure of the electromagnetic field strength. This quantity becomes of the order unity for masses of the order of

$$m_p = \sqrt{\frac{\hbar c}{G}} \sim 2.18 \ 10^{-5} \text{ gr (Planck's mass)}$$

This means that when gravity is quantised it is also very strong. Finally the temperature at this epoch is of the order of

$$T_p = m_p c^2 = \sqrt{\frac{hc^5}{G}} \sim 1.96 \; 10^{16} \text{ erg } (1.23 \; 10^{22} \text{ MeV}) \text{ (Planck's temperature)}$$

No surprise if under such extreme conditions physics should appear quite different from its present day form.

2) <u>Evaluation of decouplings</u>: in what follows we shall often refer to decoupling time and temperature for various processes. It is useful to sketch here the procedure followed to evaluate these conditions for a general reaction of a particle N coupled, through its antiparticle, to radiation

$$N + \bar{N} \rightleftarrows \gamma + \gamma$$

In order to evaluate the decoupling time for a reaction of this kind one must take the following standard steps:

a) start from classical distribution (11) to get:

$$n_N = n_{\bar{N}} = \frac{g_N}{\hbar^3} \left(\frac{m_N k_B T}{2\pi}\right)^{3/2} \exp\left(-\frac{m_N c^2}{kT}\right)$$

b) use cross section from laboratory experiment;

c) evaluate the reaction time:

$$\tau \sim \frac{1}{\sigma n v}$$

d) equate this to cosmic time in order to get the decoupling time t_* and therefore the decoupling temperature T_*;

e) evaluate n_N and $n_{\bar{N}}$ at $t = t_*$ and $T = T_*$.

3) <u>Quark confinement</u>:

a very important clue one can have from GUT era is about quark confinement. In fact if one assumes that during this period quarks (q) were free and therefore

$$q + \bar{q} \rightleftarrows \text{hadrons} \rightleftarrows \text{photons}$$

if one takes the steps from a) to e) of Section 2, using as a cross section $\sigma \sim (\frac{h}{m_q c})$ one gets a density of $\sim 10^{-10}$ quarks/barions. These "fossil" quarks have been looked for unsuccessfully in cosmic rays, in accelerators, in ordinary matter (Millikan like experiments) and even is solar photosphere.

A way out of this difficulty could be the fact that quarks can exist only confined inside hadrons but cannot be found free (quark confinement). It would be very interesting to repeat these naive calculations in the light of QCD.

4) <u>Hadron era</u>:

by applying steps from a) to c), Section 2 one gets that at $T_* \approx 10^2$ MeV the reaction

$$N + \bar{N} \rightleftarrows \gamma + \gamma$$

(where N is a hadron) decouples giving

$$\frac{n_N}{n_\gamma} = 2 \cdot 10^{-18} \text{ barions/photons}$$

Now the actual ratio is 10 orders of magnitude greater than this! In fact from (10) one gets the value of the photon number density

$$n_\gamma = \frac{\xi(3) K_B^3}{\pi^2 \hbar^3 c^3} \left(\frac{T}{3} \cdot 3\right)^3 = 550 \left(\frac{T}{3}\right)^3 \text{ photons/cm}^3$$

and from (9)

$$\frac{n_N}{n} = \frac{1.2 \cdot 10^{-5}}{550} h_0^2 \Omega \left(\frac{T_0}{3}\right)^{-3} = 2.2 \cdot 10^{-8} h_0^2 \Omega \left(\frac{T_0}{3}\right)^{-3}$$

which, for reasonable values of h_o and Ω (i.e. of order unity), is just ten orders of magnitude greater than that of relic barions. These considerations lead to two problems which are still unresolved:

a) Why is the present-day number of barions greater that that of antibarions?

b) Why don't we see $10^{-18}/10^{-10} = 10^{-10}$ antibarions/barions as remnant of the big-bang? One should notice that this abundance is by no means negligible: in fact for gold and radium we have respectively

$$Au/H \sim 10^{-12} \qquad Ra/H \sim 10^{-18}$$

5) Production of elements:

one of the great success of the theory of big-bang, besides the explanation of the 3°K black-body radiation, is to give a satisfactory account of the abundance (0.20-0.30 by weight) of helium.

At the end of leptonic era and for about 300 sec. were realised the conditions for the formation of heavy elements. At this epoch the neutrinos, though decoupled from other leptons were still interacting with proton to give:

$$n + e^+ \leftrightarrows p + \bar{\nu}_e$$
$$p + e^- \leftrightarrows n + \nu_e$$

Following the usual standard steps of Section 2 we get that this happened when the age of the universe was about 100 sec. and the temperature about 2.10^9 °K. From this time on the ratio n/p is determined only by the expansion of the Universe and denterium can be produced for about 300 sec through the reaction:

$$n + p \rightarrow d + \gamma$$

and He^4 through the main process

$$d + d \rightarrow He^4 + \gamma$$

or also:

$$d + p \to He^3 + \gamma$$
$$He^3 + n \to He^4 + \gamma$$

The abundance is determined by the "decoupling" time and the results of calculations (in this case by no means trivial due to the many intermediate steps which are possible) can be summarized as follows:

i) He^3 and He^4 abundances observed to day are, within the rather big experimental errors, in agreement with the ones calculated in this way.

ii) Elements heavier than He (except Li^7) cannot be produced during cosmic expansions and stellar evolution must be invoked.

6) <u>Lepton families</u>:

another limit imposed by cosmology to the physics of elementary particles is on the number of existing lepton families. In fact, as already observed, the decoupling times of reactions depend critically on the number of species which are present in cosmic plasma through (see formula (16)).

From a very simple calculation it turns out that:

$$\chi = \frac{7}{8}(2+L) + 1$$

where L is the number of lepton families. Because $T^2 \alpha \chi^{1/2}$ one easily sees that the fraction of elements produced increases as L increases and therefore that the abundance of elements is a function of the number of lepton families which are present. In particular it has been shown (see for instance[2]) that for an increase of L from 2 (ν_e, ν_μ) to $3(\nu_e, \nu_\mu, \nu_\tau)$ the abundance increases by 0.02 so that, being the uncertainty of the order of 0.1, one reasonably expects that lepton families are less than 5 (3 of them already known). Of course if one can determine the abundance with a smaller error it is possible to put more stringent limits on this important number.

REFERENCES

1) J.V.Narlikar, Introduction to Cosmology (Jones and Bartlett 1983)

2) J.V. Narlikar and T. Padmanabhan, Gravity, gauge theories and Quantum Cosmology (Reidel 1986).

3) Ya.B. Zeldovich and I.D. Novikov, Relativistic Astrophysics. (University of Chicago Press, 1971)

4) Ya.B. Zeldovich and I.D. Novikov, Stroienie i evolutsia vselennoj (Nauka-Moscow 1982)
(italian Translation: Struttura ed Evoluzione dell'Universo - Editori Riuniti, 1985)

SN 1987a AND EXTRA SOLAR NEUTRINO ASTROPHYSICS

Piero GALEOTTI

Istituto di Fisica Generale e Sez. INFN, Università di Torino
Istituto di Cosmogeofisica del CNR, Corso Fiume 4, Torino, Italy

Abstract

In this paper we summarize our understanding of supernova 1987a, one year after the explosion occured in the Large Magellanic Cloud. The problem of two neutrino bursts, recorded in different underground experiments and separated in time by 4.7 hours, is also discussed. It is shown that there is no contradiction from the experimental point of view among the different observations, and some refinements on the current predictions from theoretical models are needed to fit all the observations. The combined analysis of the data recorded in neutrino and gravitational wave detectors at the onset of the collapse clearly indicates a long duration of the phenomenon. Thus, any serious (even if difficult) tentative to explain how a star ends its life as a supernova should be based on all these new experimental data.

1. Introduction

Certainly, the most exciting astronomical event recorded in 1987 was the Large Magellanic Cloud supernova. This event was rather surprising for several reasons:
- the supernova exploded in a small irregular galaxy where, according to supernova statistics[1], the rate of explosions should be much lower than in our Galaxy;
- the progenitor star was the blue supergiant Sanduleak -69.202, while all previuos models predicted a red supergiant progenitor;
- the maximum luminosity was much smaller than that of all other supernovae of the same type;
- finally, the search for collapse neutrinos was too successfull, since two bursts were reported.

One year later, our understanding of SN 1987a has improved, but there are still problems in interpreting the initial phases of the development of the collapse. The light curve is explained as powered by the decay of radioactive nuclei, and doesn't show evidence of a remnant neutron star.

The total mass expelled in about one year, and passed through the photosphere, is of order 10 $M_\odot$, confirming that a rather massive star was the progenitor of the supernova. Still unexplained is the nature of the mistery spot[2,3], observed for some times as a bright companion or a light echo a few months after the explosion.

However, the main puzzle related to SN 1987a is that two bursts have been detected in underground detectors, separated in time by 4.7 hours, which, assuming they are due to neutrino interactions, impose serious difficulties in understanding the stellar collapse from which this supernova originated. Indeed, according to the "standard" collapse models of non-rotating (but neutron stars are the fastest rotating objects in the sky), non-magnetic (but neutron stars have the strongest magnetic field in the sky) stellar cores, only one neutrino burst is predicted to be emitted over time scales of the order of a few seconds.

Nevertheless, it has been shown that the two events are not contradictory from the experimental point of view[4], and that a degenerate neutrino gas with low temperature (T ≈ 0.5 MeV), and chemical potential μ ≈ (12-15) for the first burst, can fit[5] the data recorded in the Mont Blanc and in other detectors at the same time, since it predicts a relatively low total energy outflow $W_\nu \approx (2-6)\ 10^{54}$ ergs, and a small number of expected interactions in Kamiokande.

Finally, a correlation analysis[6,7] of the data recorded in underground detectors and in gravitational wave antennas shows a statistically significant excess of coincidences at a time close to the Mont Blanc event, for a duration of order of 1-2 hours. Certainly, this is the most unexpected (and thus exciting) result obtained observationally from SN 1987a.

2. Neutrino observations

Neutrino observations from SN 1987a were reported[8,9,10,11] by the Underground Neutrino Observatory (UNO) of the Mont Blanc Laboratory, by the Baksan Scintillation Telescope (BST), by the Kamioka and IMB water Cerenkov detectors. The UNO, described[12,13] in detail elsewhere, was designed to be a detector of low-energy neutrinos from collapsing stars; Kamioka and IMB were designed as detectors of higher energy processes, mainly to search for proton decay candidates. For this reason, the light yield in UNO is much higher (15 phe's per MeV) than in Kamioka (3.4 phe's per MeV) and IMB (1.2 phe's per MeV), and also the energy threshold in these detectors is different: 5-6 MeV in UNO, 9-10 MeV in Baksan and Kamioka, and about 20 MeV in IMB.

On February 23.12, 1987 ($2^h 52^m 36.8^s$ UT) the Mont Blanc computer printed out at the occurence a burst of 5 pulses with a duration of 7 s and an imitation rate from the background of 1.78 10^{-3}/day. A second burst of 3 pulses with a duration of 0.5 s and an imitation rate of 1.74 10^{-2}/day was printed out at $5^h 2^m 0.7^s$ UT, the same day. All the experimental data recorded by the UNO during a 2-days period encompassing SN 1987a are reported in ref.13.

Later on Kamioka, IMB and Baksan reported observations of a second burst, delayed by 4.7 hours in comparison with the Mont Blanc one; within the universal time uncertainty of Kamioka ($\pm$ 1 min), these observations coincide. The need for a time correlation with a msec UT clock in such kind of experiments was very soon recognized, because only the UNO and IMB, among the 4 experiments, have a good UT timing.

In the Mont Blanc data acquisition system, on-line software identifies on real time and prints on the computer output any burst candidate, i.e. any burst of pulses recorded in any interval of time Δt between 1 ms and 600 s, with a low imitation rate from the background. In this on-line analysis, all events involving more than 1 counter within the resolution time (150 ns) of the experiment, or with an energy release in excess of 60 MeV, are rejected because due to cosmic ray muons crossing the detector. Off-line analysis is further made in a similar way, but including a more detailed analysis of the single pulses in the burst. In the other experiments, the burst identification is performed off-line, and different cuts are made to subtract high energy events, for example cosmic ray muons crossing the detector. In water Cerenkov detectors, however, there is the advantage to measure also the direction of the observed particle so, when studying point sources such as SN 1987a, an additional information is available in these experiments.

Cosmic ray muons are the main source of background in underground detectors: it is always easy to identify througoing muons, but it is rather difficult to identify the low energy particle (mainly neutrons and gammas) produced by cosmic ray muon interactions in the rock surrounding the apparatus. A very large coverage of rock is needed to reduce the muon flux and the connected contribution from their interactions in the rock; in the UNO (5,200 hg/cm^2 underground) only 3.5 muons per hour are recorded on the average, while this contribution is $\approx$ 1300 muons per hour in Kamioka and $\approx 10^4$ muons per hour in IMB, which are located at shallower depths.

Several times, see for example ref.[14,15], the Mont Blanc and Kamioka results have been compared each other. Indeed, the energy threshold of these detectors is similar while the sensitive mass of Kamioka is much higher, and nevertheless only a small signal has been recorded[16] in Kamioka at the time of the Mont Blanc event. Thus, a brief review of the basic properties of the Mont Blanc and Kamioka experiments is needed to compare the data recorded in these detectors.

In any experiment, the main signal induced by supernova neutrinos is given by positron pulses, produced through the capture reaction ($\bar{\nu}_e p, n e^+$) with free protons, with a cross section:

$$\sigma = 2.43 \frac{E_e}{m_e c^2} \left[\left(\frac{E_e}{m_e c^2}\right)^2 - 1 \right]^{1/2} 10^{-44} \text{ cm}^2 \tag{1}$$

and an energy threshold $E_{th} \approx 1.8$ MeV.

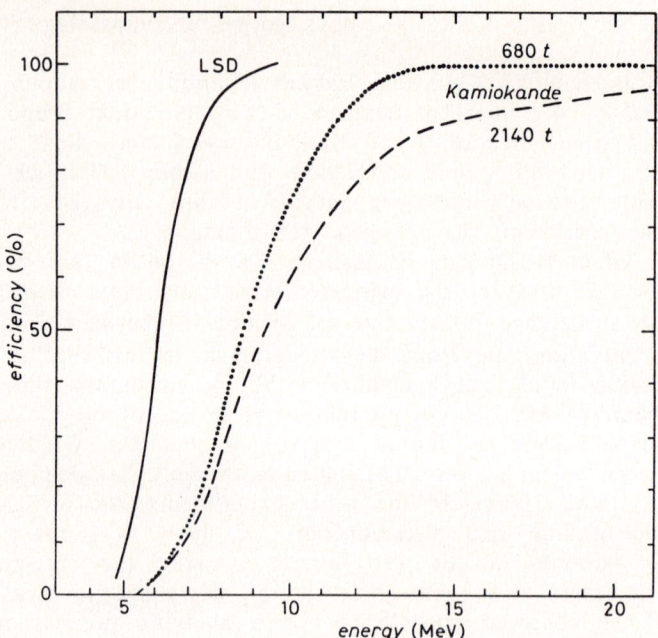

Fig. 1. – Efficiencies of LSD and Kamiokande detectors.

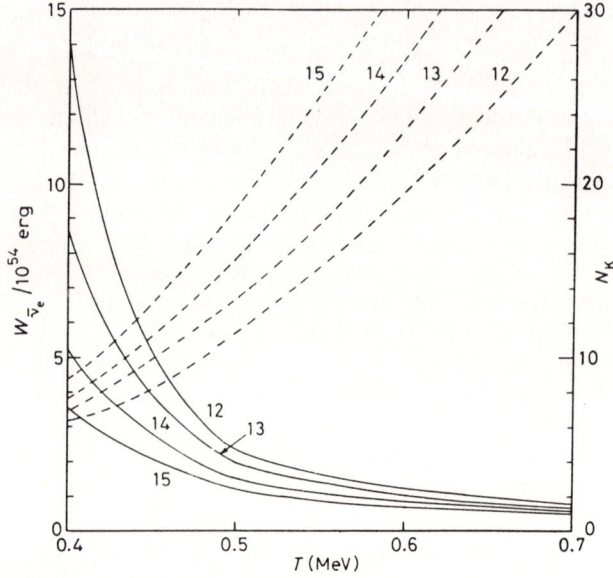

Fig. 2. – Total energy output released in $\bar{\nu}_e$-neutrinos (full curves) and expected number of events (dashed curves) in Kamiokande for a degenerate neutrino spectrum with dimensionless chemical potentials $\mu = \psi/T$, indicated on each curve.

A comparison of the signals expected in the Mont Blanc and Kamioka experiments, based only on the mass ratio $N_H^K/N_H^{MB} = 17$, is misleading because positrons are detected in a different way in the two experiments. In water (Kamiokande) the positron is detected through the emission of Cerenkov light, and the total visible energy released by a positron with kinetic energy T_e is $E_{vis}^C = T_e - T_c$, where $T_c \approx 0.25$ MeV is the threshold for Cerenkov light production by positrons in water. Therefore, in water the visible energy is $E_{vis}^C = E_v - 2.1$ MeV. In liquid scintillator (Mont Blanc) the visible energy is given by the total positron energy (kinetic and rest-mass): $E_{vis}^S = T_e + 2m_e = E_v - 0.8$ MeV, because the positron annihilates into 2 gammas and produces a 1 MeV extra energy release, detectable in the scintillator but not in water. Therefore, for the same neutrino energy, the visible energy in scintillator is higher than in water by the quantity $\Delta E = E_{vis}^S - E_{vis}^C \approx 1.3$ MeV for positron annihilation at rest, and a little bit more for positron annihilation on flight.

In addition, the detection efficiency of electrons with total energy E_e is also higher in scintillator than in water at the energies of interest, as can be seen from fig.1 where the efficiency curves of the Mont Blanc and Kamiokande experiments are reported. For example, from fig.1 one deduces a 50% efficiency at ≈ 6 MeV in Mont Blanc, and at ≈ 9 MeV in Kamioka.

Finally, we notice that the two bursts (detected at the Mont Blanc time and 4.7 hours later) are both difficult to explain within the framework of the current models of stellar collapses. Indeed, the first one involves a high luminosity outflow ($W_v > 10^{54}$ erg) of low energy neutrinos ($E_v \approx 7-9$ MeV); the second one shows a considerable anisotropy of the recorded pulses (high energy pulses are observed within a small solid angle from the supernova), which cannot be explained as due to (v_e-e) scattering, since it would require an even larger energy output because of the smaller cross section of this reaction.

3. Model predictions for the first neutrino burst

In the "standard" calculations[17,18,19,20] of neutrino emission, a non-rotating stellar core is considered, and the effect of a strong magnetic field is also neglected. The basic conclusion is the prediction of a Kelvin cooling stage during the collapse (after the core becomes non-transparent to neutrinos), where the volume emissivity of neutrinos is replaced by the surface emissivity. The internal thermal pressure halts the collapse even in the case of very massive cores, untill the neutrino emission from the core surface cools the matter, a phase which lasts for about 10 seconds. Subsequently the collapse enters into the relativistic stage of contraction, with the final formation of a black hole. For smaller core masses, a hot neutron star is formed, and also in this case the cooling from the neutrino-sphere lasts for about 10-20 seconds; thereafter the core becomes an usual cold neutron star.

During this stage, matter consists mainly of neutrons, with a small admixture of protons and electrons. The equilibrium is maintained by the usual reactions: ($e^+e^- \leftrightarrow 2\gamma$), ($e^+n \leftrightarrow \bar{\nu}_e p$), ($e^-p \leftrightarrow \nu_e n$). Also muon and tau neutrinos, generated in the reaction ($e^+e^- \rightarrow \nu\bar{\nu}$), are expected to be in thermal equilibrium with matter inside the core, and trapped because of elastic scattering with nucleons. In the scattering reactions neutrinos are thermalized, and they diffuse out of the core by electron scattering. As a consequence, in the standard models all the 3 neutrino species are in thermal equilibrium inside the neutrino-sphere.

This picture, however, must be considered as a first aproximation of the real collapse. In particular, opposite results have been obtained on the high energy tail of the neutrino spectrum: there could be an enhancement because of accretion in the process of collapse, or a cut-off because of re-absorption of the high energy neutrinos in the stellar envelope outside the neutrino-sphere. In any case, whether or not high energy neutrinos are produced by accretion, their absorption outside the neutrino-sphere is an unavoidable physical process, since the neutrino-sphere is defined as the layer where neutrinos with the average energy are absorbed, and, since the neutrino cross section increases as E_ν^2, those with higher energy are stronger absorbed outside the neutrino-sphere.

The "standard" models of collapsing stellar cores are unable to fit the experimental data. However, the present predictions on the neutrino outflow from a gravitational collapse will probably change drastically in the future, when magnetic field and expecially rotation will be taken into account. Two basic differences with non-rotating models were found in some preliminary calculations[21]: a lowering of the temperature of the neutrino-sphere, and of the neutrino luminosity.

A non-standard possibility has been suggested[5] for which rotation plays a crucial role, and implies a low energy spectrum of degenerate neutrinos, from which both the total energy output and the number of expected events in Kamiokande are reduced, as shown in fig.2. Indeed, the low temperature (T $\approx$ 0.5 MeV) implies a small number of high energy neutrinos, for which the Kamiokande detection efficiency is high, and gives a low density of e^+e^- pairs in equilibrium and, consequently, a very small flux of muon and tau neutrinos, since they are produced only through the ($e^+e^- \rightarrow \nu\bar{\nu}$) channel. Thus, the total energy outflow is drastically reduced, because of the suppression in the production of these two flavours.

According to this model, a massive rotating star passes through the stage of a flattened spheroid and then brakes into two fragments with masses M $\approx$ 20 $M_\odot$ and m $\approx$ (1-2) $M_\odot$. The massive component continues to collapse and produces the first neutrino burst during the black hole formation. The low mass star approaches the massive component, orbiting around it, because of gravitational radiation losses, untill it is disrupted and its matter is accreted by the massive black hole, thus producing the second neutrino burst.

4. Conclusions

We have discussed here what we presently know on extra solar neutrino astrophysics after the first experimental data are available. The unsolved problems are probably still more numerous than the solved ones, and further work is needed before reaching a conclusion on what we really observed from the Large Magellanic Clouds the 23rd of february 1987.

Because of the different techniques and energy thresholds, from the experimental point of view there is no contradiction among the data recorded in several underground detectors, and a correlation analysis has indeed shown a significant excess of coincidences, over time-scales of 1-2 hours, between underground experiments and gravitational wave antennas. Hence, the experimental data are clear and the information they give is wide, probably too wide because a collapse on two stages or with a long activity was never predicted by the past theoretical models.

The exact time of the collapse could, in principle, be obtained from the evolution of the optical light-curve of the supernova during its early stages, extrapolated back to the pre-supernova magnitude. However, these extrapolations have given[22,23] opposite results because of the uncertainty in the photospheric parameters.

Contrarily to the data, there is still the need for a model able to give an interpretation of all the experimental data, and to suggest a possible scenario for SN 1987a. Unless throwing away some of the experimental evidences (and somebody did it, even if this event is the first optical supernova observed in a technological era) one should try to have an interpretation based on a gravitational stellar collapse over long time-scales, or at least on two stages. This is a very difficult job, but surely more interesting than neglecting some data in order to fit what we already know. A suggestion has indeed been proposed[16,24] very soon after the explosion: a collapse into a neutron star and subsequently its transition into a black hole. More recently, another possibility, based on the rotation of the collapsing core, has been also suggested[5,25]. Anyway, the final answer to this problem will probably be reached only when the supernova shell becomes transparent.

References

1- P.Galeotti, C.Raiteri, Astron Astrophys., 190, 69, 1988
2- C.Papaliolios et al., Proc.IV G.Mason Astrophys.Workshop, Fairfax (USA), M.Kafatos, A.Michalitsianos Eds., Cambridge Univ.Press, 1988, p.225
3- W.P.S.Meikle et al., Nature, 329, 608, 1987
4- M.Aglietta et al., Europhys. Lett.,3,1321,1987
5- V.S.Berezinsky et al., Nuovo Cim., 11C, 287, 1988

6- E.Amaldi et al., Europhys. Lett., 3, 1325, 1987
7- E.Amaldi et al., Proc. 2nd Rencontres de Physique de la Vallèe d'Aoste, 1988, Ed. M.Greco, La Thuille, Ed. Frontières, 1988
8- M.Aglietta et al., Europhys. Lett.,3,1315,1987
9- K.Hirata et al. Phys. Rev. Lett., 58,1490,1987
10- R.M.Bionta et al. Phys. Rev. Lett., 58,1494,1987
11- E.M.Alexeyev et al., Sov. Jetp Lett., 45,461,1987
12- M.Aglietta et al., Nuovo Cim., 9C,185,1986
13- M.Aglietta et al., Proc. 2nd Rencontres de Physique de la Vallèe d'Aoste, 1988, Ed. M.Greco, La Thuille, Ed. Frontières, 1988
14- J.N.Bahcall, Proc.IV G.Mason Astrophys.Workshop, Fairfax (USA), M.Kafatos, A.Michalitsianos Eds., Cambridge Univ.Press, 1988, p.172
15- A.Wolfendale, Proc. XX Int. Cosmic Ray Conf., Moscow, 1987
16- A.De Rujula, Phys. Lett., B193,514,1987
17- D.K.Nadjozhin, Astrophys. Space Sci., 53, 131, 1978
18- D.K.Nadjozhin and I.V.Ostroscenko, Sov. Astron., 24, 47, 1980
19- J.R.Wilson, Astrophys. J., 163, 209, 1971
20- R.Mayle, J.R.Wilson, and D.N.Schramm, Astrophys. J., 318, 288,1987
21- V.S.Imshennik and D.K.Nadjozhin, Astrophys. Space Phys. Rev., (R.A.Syunyaev Ed), Harwood Acad. Publ., 2, 75, 1983
22- D.N.Arnett, Astrophys. J., 319, 136, 1987
23- E.J.Wampler et al., Astron. Astrophys., 182, L51, 1987
24- W.Hillebrandtet al., Astron. Astrophys., 180,L20, 1987
25- L.Stella, A.Treves, Astron. Astrophys., 185,L5, 1987

SUPERNOVAE: THEORY VERSUS OBSERVATIONS. (THE CASE OF SN 1987A AND THE CASE OF TYPE Ib SUPERNOVAE)

Amedeo TORNAMBE'

Istituto di Astrofisica Spaziale, CNR, CP 67, I-00044 Frascati, Italy

This talk is devoted to discuss some observational facts concerning the supernova 1987a in the Large Magellanic Clouds (LMC) and their theoretical implications on the evolution of the progenitor stellar system and of the supernova itself.
I will successively briefly discuss the observational properties of a new class of supernovae now addressed as type Ib supernovae.
Two theoretical models which have been proposed for these supernovae progenitor systems and exploding mechanisms will be shortly analysed since these present implications that are strictly connected with the arguments discussed in this section devoted to the underground particle physics.
However, since this audience is mostly of physicists not strictly involved in astrophysics, I want to recall firstly some main observational and theoretical features concerning supernovae; this will be done in the introduction.

1. INTRODUCTION

Supernovae (SNe) are the sites where nuclearly processed heavy elements are returned to the interstellar medium. Depending on the supernova type, heavy elements (A>4) are mainly synthetized during the explosion or during the previous nuclearly supported phases, of the evolution of the stellar progenitor.
Supernovae are hence at the basis of the nuclear evolution of the matter in the universe.
This is not, however, the only relevant aspect of supernovae in (astro)physics. The interaction between supernovae driven shocks and the interstellar medium can induce star formation; some are standard candles and can be used as distance indicators; some others are fully efficient in producing relevant neutrino fluxes and gravitational waves; and so on.
There are, however, many still open points concerning the nature of the stellar progenitors, the explosion mechanisms, the nature of the compact remnants (when formed), the nucleosynthesis and the like.
I want, first of all, give some general information on the classification of supernovae, on the observational properties of the various classes and on the current theoretical beliefs on such subclasses.

1.1 On the observational aspect

Up till now (with the only exception of SN 1987A) all the observational informations on SNe events have been fully derived by means of the electromagnetic radiation, mostly in the visual portion of the spectrum and only recently in other wave bands (as UV, IR, X and radio). As a consequence spectroscopic and photometric observations in the optical wavelenghts have historically rapresented the means to classify the various observed SN events in external galaxies (the last SN event in our galaxy having occurred in 1604, well before the developement of modern detection and analysis techniques).

It is possible to divide supernovae in two main classes: type I SNe and type II SNe (see table 1). Such a first classification has been performed on the basis of spectroscopical properties. Type I supernova show no Hydrogen lines at all in their spectra while H lines are fully dominant in the spectra of type II supernovae.

Up to few years ago the subdivision of supernovae in two main groups, was rather sufficient to give an idea of the observational situation once the following informations were added: the light curves (luminosity in a given pass-band versus time) of type I supernovae are extremely homogeneous while those of type II SNe are very non homogeneous; type I SNe can be found both in elliptical galaxies (where star formation stopped presumably several Gyrs ago and only old, low mass, stars exist) and in spyral galaxies (where star formation is an ongoing phenomenon and young massive stars are present), while type II supernovae are found only in spyral and irregular galaxies (in both of which young massive stars are present).

More recently, due to the identification of a certain number of type I supernovae showing observational constraints well distinct from those of other type I SNe, the subdivision of type I SNe in the subclasses type Ia (the classical ones) and type Ib has been adopted. Type Ib supernovae are found only in star forming systems (i.e. spyral galaxies) nearby to the star formation zones. An attempt has been made also to subclassify the family of type II SNe from their light curves but with some difficulty which derives from the large variety that such light curves show. However two main groups can be evidentiated being the so called type II-L and type II-P. L and P stand for linear and plateau, depending from the shape of the light curve in its descending phase after the maximum (see, e.g., 1). While it is still unclear what is the physical reason which induces the difference on the type II light curves, it is quite certain that type Ia and type Ib SNe represent a quite different physical phenomenon.

The energetics that can be derived by means of the e.m. radiation (i.e. radiated and kinetic energy with the exclusion of neutrino and gravitational wave energy) are quite comparable for any kind of event and are of the order of 1.E 51 ergs (1 foe, foe standing for ten to Fifty One Ergs).

TABLE 1

	Type Ia	Type Ib	Type II
spectrum	no H lines	no H lines	H lines dominate
population	all the types	young	young
galaxy type	ellipticals and spyrals	spyrals	spyrals and irregulars
rate in our Galaxy (SNe per 100 yrs)	0.3	0.5	1.1
max luminosity (Mv)	-20	-18.5	-18
expected neutrino energy (ergs)	10^{49}	?	3×10^{53}
expected GWR	no	?	yes
expected explosion mechanism	C-deflagration	?	core collapse and bounce

1.2. on the explosion mechanisms

There are two main physical processes that can give rise to an energy release of the order, or in excess, to 1 foe; these are:

a) a collapse of a normal star (or of a part of it) to a neutron star in which more than 100 foes are released at the expence of the gravitational energy field

b) the degenerate carbon flash during which the nuclear conversion of carbon and oxygen to nuclear statistical equilibrium (NSE) matter (or "incineration") gives rise to 1 foe of energy release for every solar mass (2.E 33 grms) of incinerated matter at the expence of the nuclear energy.

Other forms of nuclear energy releases of the order of 1 foe per event can be identified (as discussed in the last section).

The current belief is that massive stars (M>8-10 solar masses) end their evolution suffering a core collapse after all the nuclear burning phases up to Si burning have occurred at their center. Only an extremely small fraction of the energy released during the core collapse to a neutron star (or to a more compact object) should be sufficient to eject the envelope giving rise to the observed supernova phenomenon. There are, however, several difficulties to reproduce theoretically such an occurrence. Most of the attempts made till now, to get rid of the shock wave damping down in the inner part of the star, have been unsuccessful.

Even if it is quite evident that the explosion mechanism is to be related to an ongoing shock wave generated by the core collapse and bounce, there must be some still unclear physical reason inducing the rejuvenation of the shock wave or the formation of a more energetic one (current research points toward the neutrino physics or to nuclear matter properties; for a review on the argument see 2).

There is, however, a longstanding general agreement, brightly confirmed by SN 1987a, that such a massive stars are to be related to type II supernovae. Such an occurrence is supported by several constraints like the presence of H in the spectra, the occurrence of type II SNe only in systems where young massive stars exist and their vicinity to the star forming regions, the mass of the remnants, the good coincidence between the theoretically expected rates and the observed ones and that on the estimated pollution degree and the observed one (also in the composition details, see, e.g., 3) and others.

The situation is more embarassing for the case of type I SNe. While it seems quite evident, for several reasons, that the explosion mechanism is the degenerate carbon ignition in a core exceeding the Chandrasekhar mass limit, completely unknown are the nature and the evolutionary properties of the progenitor stellar system and the mechanisms that induce the CO core mass to get over the Chandrasekhar critical mass.

Every model till now proposed has its weak points.

The most accredited model is presently the one by Iben and Tutukov (see, e.g., 4) in which two degenerate CO stars in a binary system, previously deprived of their envelopes do to common interactions, merge due to gravitational wave energy loss by the system, exceeding, in such a way, the Chandrasekhar mass limit.

This model presents several properties that are accounted by the observations like the absence of Hydrogen, the possibility to obtain an explosion at practically any time after the star formation, the good agreement between the theoretically expected rates and the observed ones, the possibility to explain the uniformity of the light curves and their shapes, the composition of the ejecta and others. The weakest point is that to obtain a close double degenerate star system, evolutionary behaviours are required that are extremely difficult to compute theoretically and to identify observationally. As a result many uncertainties are still affecting the model.

The case of type Ib supernovae will be more extensively discussed in the last section.

2. THE CASE OF SN 1987A

This is the first supernova observable to the naked eye since 1604.

It is a type II supernova since H lines are dominating the optical spectrum.

It is the first supernova for which the properties of the progenitor star can be easily derived by the previously recorded plates. These are:

$L = 1.3 \times 10^5 \, L_o$ (L_o=solar luminosity=4×10^{33} erg/sec)

$T_{eff} = 15850$ K

It is possible, as a consequence, to derive the original mass of the star, the one it had at the beginning of its evolution, but only to put constraints on the mass at the moment of the explosion. The reason for this is rather simple. In the last evolutionary phases the luminosity of the star is determined by the physical conditions at its inside (which, in turn, are linked to the size of the He core), while is fairly independent on the mass amount of the external H-rich envelope. In turn the internal conditions are determined by the original mass of the star. Hence, if mass loss from external layers occurred during the advanced evolutionary phases, such an occurrence will not sensitively affect the total luminosity.

The measured luminosity of the supernova progenitor star suggests, as a consequence, the following physical conditions: the mass of the He-core (the H-exhausted inner layers) is about 6 solar masses corresponding to an initial mass star of about 20 solar masses, the actual mass of the exploding star lying between these two values and the portion exceeding the inner 6 solar masses being mostly componed by H (75% by mass) and He (with a small amount of heavy elements). It is also possibel to derive the radius of the progenitor star at the moment of the explosion. It turns out to be about 30 solar radii (the radius of the Sun being 7.E 10 cms).

We are hence facing a pre-exploding massive star of an unusually small size: not a red supergiant but a blue supergiant.

An analysis of the stars surrounding Sk 69-202, the progenitor star to the supernova 1987A, allows us to derive some further constraints on the evolutionary history of the exploded star. The presence, in the field, of red supergiants at the same and larger luminosity, firmly suggests that the progenitor star passed through a supergiant phase (about 300 solar radii) before shrinking to a smaller size at which it successively suffered the explosion. This occurrence has been further on confirmed by other observational facts allowing to determine also the time at which this transition occurred.

Stars evolving through the red (super)giant phase (where extended outer convective layers develop) suffer, during this phase, strong mass loss and the interaction between the supernova and its previously lost mass can be revealed in several ways.

All these properties of the progenitor star are to be taken in mind when discussing the properties of the 1987A supernova which has been, improperly, some times, addressed as a peculiar one.

In fact, even being a nearby supernova, SN 1987A is not a galactic supernova; the Large Magellanic Cloud (LMC) is a nearby small irregular galaxy whose young stars have physical properties different from the ones of young stars in our galaxy and, presumably, from young stars in other spyral galaxies.

A fundamental difference is in the chemical composition of these stars, which, in the LMC, is sensitively less abundant in heavy elements (A>4, generally addressed as "metals"). It turns out that metals are, in the LMC, a factor of four less abundant than in the Sun and more than in young stars in our galaxy. (This is easily understood in terms of nuclear processing of the original matter that is more advanced in our more massive galaxy where a larger specific supernova rate has occurred).

The abundance of heavy elements affects sensitively, through the nuclear burning and opacity sources, star's evolutionary behaviours.

This is, most likely, the reason why the explosion occurred in a massive blue supergiant star.

In fact, evolutionary excurses from the red (extended envelopes) to the blue (compact envelopes), or vice versa, are a well known behaviours which depend on the balance between the energy generated in the inner layers of the star and the efficiency of the external layers of the star to transfer such energy by means of radiative transport (see 5 for a more precise discussion of the topic). Bearing in mind that only the external layers of the star are involved in such expansions or contractions, one can easily understand that, when the inner conditions (i.e. the energy production) change, the external layers can change accordingly.

When and how the star becomes a red giant or turns back to a blue giant depends on the mass of the star, on its chemical composition and on the mass of the H-rich envelope (since it is possible that, even if the required conditions to change the structure are fulfilled, the thermal time scales of the envelope are longer than the nuclear timescales which would trigger the changes with the net effect that the star remains in the previous structural shape).

In any instance, massive stars in our galaxy and in other spyrals, due to their metal rich envelopes (2-3% and more, by mass, of heavy elements) cannot avoid to become red supergiants, a condition from which they can hardly recede due to the large surface opacity, once the energy flux from the inner side of the star becomes sensitively large during the evolutionary phases following the central H or He-burning ones.

On the contrary, in the case of a less metal rich envelope (i.e. lower opacity source contribution), it can occur that, after the starting of a central burning successive to that of Hydrogen (e.g. central He-burning or even central carbon burning), the star can recede from its giant structure (attained at the end of the previous burning) due to a decreased central flux. Mass loss or chemical mixing to the surface layers can play also a relevant role in the quoted behaviour.

The progenitor star of SN 1987A, presumably become a red giant at the end of the central H-burning phase, successively (presumably during the central He burning phase), due to its low opacity envelope, become a blue giant changing no more its structural configuration up to the SN explosion for the reason that from that point on the thermal timescales had become longer than the nuclear time scales.

This picture, strongly supported by the observations, has not been, however fully confirmed, in the details, by theoretical computations and few open problems still exist; the discussion of these details should go beyond of the purposes of this lecture.

The fact that a type II supernova explosion occurred in a rather compact star has determined its unusual, but not anomalous, behaviour.

Fig. 1 shows the light curve of SN 1987A compared to that of a more "common" type II supernova respect to which it reveals a pronounced underluminosity. Such a behaviour is well understood and, in its main features, had been theoretically predicted well before the explosion of SN 1987A (see, e.g., 6). It simply depends from the physical and geometrical configuration of the matter in which the energy of the shock wave is deposited which, in the case of a compact object, is in favour of the kinetic energy at the expence of the thermal one (see, e.g., 21).

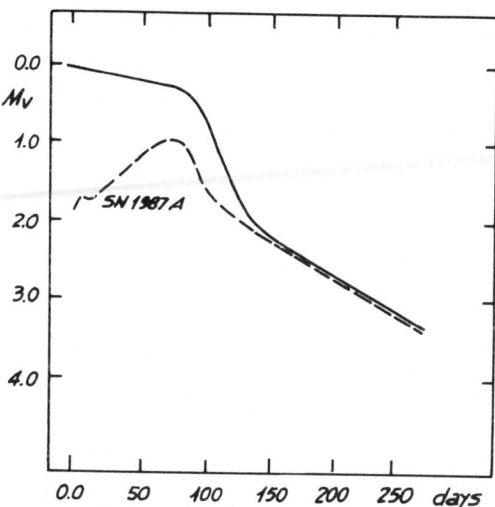

fig 1. A schematical behaviour of the light curve of SN 1987A as compared to that of a more "normal" type II-P supernova (full line). The abscissa reports the magnitudes below the maximum of the type II-P SN.

The problem now arises of how many type II supernovae have we lost due to their underluminosity. Of course the question is not of easy solution and depends on how much matter in the universe is in form of small, poorly evolved, irregular galaxies. There is, however, not such a problem for our galaxy and for spyrals where such a subluminous type II SNe are not expected and for which the current rates should be rather well established both from the observational point of view (see, e.g., 7 and 8) and on the basis of theoretical considerations (see 9 and 10).

3. THE EVOLUTIONARY BEHAVIOUR OF SN 1987A

As discussed in the previous section, the properties of the progenitor star of SN 1987a are rather well understood in their main lines; so it is for the evolution of the supernova.

The evolutionary behaviour of a supernova can be crudely understood bearing in mind few basic considerations. The shock wave passing through the stellar structure determines a strong temperature increase and the formation of an outward velocity field, the velocities being larger for larger distances from the center of the star.

As such a structure expands some simple physical phenomena

occur: external layers start to become transparent to the optical radiation as a consequence the last scattering surfaces while incresing in size (due to the general expansion) moves inward the structure toward originally hotter regions; the expansion induces a general adiabatic cooling; energy continues to be deposited in the structure by radioactive decays; at a certain moment the lagrangian inward motion of the last scattering surface becomes larger than the general expansion velocities, as a consequence, also its geometrical size starts to recede after having reached a maximum.

Since (according to the Stefan-Boltzmann law) the luminosity is proportional to the fourth power of the effective temperature and to the second power of the photospheric radius one can in principle understand, from the quoted occurrences, the general behaviour of the light curve.

Of course, in facts, the situation is, for several reasons, more complicated and the correct simulation of a SN light curve requires the use of a detailed hydrodynamical code.

Bestfitting theoretical light curves with the observed ones it is possible to derive several informations on the supernova phenomenon, like the velocity field of the expanding matter, the amount of radioactive material formed during the explosion, the mass of the shocked envelope, the total amount of energy deposited by the shock wave and others. Of course, observational constraints help to reduce the number of free parameters.

According to Nomoto and coworkers (11) the bestfit to the observed curve of light is obtained for an H-rich envelope mass of about 7 solar masses, an explosion energy of about 1 foe and an amount of 0.07 solar masses of radioctive Ni56 produced during the explosion and mixed through the internal layers. The just now quoted energy refers to the one deposited in the envelope and not to the much more energetic one of the neutrino pulse. If part of the neutrino pulse energy has been used or not to rejuvenate the shock wave (the so called delayed explosion mechanism) it remains a still open problem (see, e.g., 12).

For the first time, however, a neutrino energy pulse of the order, or in excess to, 3.E 53 ergs has been observed during a type II supernova event. This is the bright confirm of the old idea that a collapse to a neutron star (or to a more compact object) is at the basis of a SN explosion (of the second type).

For a discussion on the more complicated debate of the two neutrino pulses (see, e.g., 18, 19 and 20) I leave the reader to the lecture of Piero Galeotti in this volume.

I want to conclude this section shortly addressing to the non optical satellite observations in the UV and X-ray bands and to what we have learnt from them.

As soon as the shock wave gets out the surface of the exploding star this brights up in the UV. In the case of SN 1987a the UV went down quite fastly reaching a plateau that was at a first time interpreted as the evidence that the presumed progenitor star didn't explode. A carefull analysis of the observational constraints allowed (13) to show that Sk-69 202 had indeed disappeared of sight. This still when the optical brightness of the supernova was masking, at the optical wavelenghts, a large number of stars of the field.

Another important contribution from space UV observations came some 100 days after the explosion, when narrow emission lines from highly ionized gas appeared in the UV spectrum.

According to 14 these are due to the interaction between the UV flash (at the shock wave breakout into the star's surface) and a preexisting circumstellar matter lost by the progenitor star during a previous evolutionary phase.

The carefull analysis of these emission lines allowed 14 to give constraints on some important properties of the progenitor star. First of all it has been possible to confirm that Sk-69 202 passed through a red supergiant phase, where strong wind was experienced by the star, before returning to the blue some 1.E 4 yrs before the explosion.

A second important point, that has to be accounted by stellar evolution theory, is that the previously lost material shows a composition of the C, N and O elements that clearly indicate that the material has been nuclearly processed by the CNO p-burning cycle. As a consequence mixing occurred between the most external layers and the H-burning shell, previously to, or during the, mass loss event. Even He should have been mixed to the surface presumably at the same time increasing, in such a way, the already large surface He abundance. Such an occurrence, not predicted before for large mass stars, could help to understand the evolutionary behaviour of Sk-69 202 before the explosion.

As for the interpretation of satellite and balloon borne X-ray observations, the situation is still a bit confusing.

As a matter of the fact there are several physical mechanisms that can give rise to X-ray emission, each one following a precise expected behaviour at the various energy ranges. The most probable of these mechanisms are: the compton degradation of gamma emission from radioactive nuclei or from the pulsar; the interaction between the ejecta and the interstellar medium; the emission from the central sincro-nebula (if formed); the black body emission from the surface of the neutron star.

The observations showed, up till now, a constant X-ray flux at larger energies (say from 30 Kev up to 200 Kev), but a variable x-ray flux at lower energies (say from 2 Kev to 20 Kev) starting from some five months after the explosion.

Unfortunately there is not yet a straightforward interpretation on the origin of such a radiation since the expected behaviours are not fulfilling the observed one at a sufficient degree; it is however very likely that gamma rays from radioactive Co decay are playing the relevant role.

4. CONCLUSIONS ON SN 1987A

Even if is still premature to draw conclusions on what we have learnt from the unique supernova SN 1987A some first considerations can be done.

First of all there is the bright confirm that a type II SN event takes origin from the explosion of a massive star.

The core collapse, and the formation of a neutron star, are at the basis of the phenomenon; the predicted energy release, from such an occurrence, has been for the first time observed in the neutrino pulse (18, 19, see also 20); it is, however, still unclear if part of this neutrino energy has been used or not for the explosion of the envelope or if it has been simply triggered by the shock wave generated at the core bounce (i.e. delayed or prompt explosion).

There is no anomaly or strangeness in the evolutionary behaviour of the supernova: its uncommon behaviour depends only on the fact that the explosion occurred in a blue compact star.

The evolutionary story of the progenitor star is rather whell understood in its main lines, we still lack, however, of a detailed computed evolution, accounting for all the properties and behaviours observationally deduced from the occurrence of the supernova explosion.

The low metal abundance, typical of the Large Magellanic Clouds, is, most presumably, the main responsible for the compact structure at the moment of the explosion.

It is unlikely that supernovae of the type II af such a kind are common in the universe and have been lost due to their underluminosity, this at least in normal spyral galaxies, as our one, where young stars are much more metal rich and explode as red supergiants.

5. ON TYPE Ib SUPERNOVAE

I want to conclude my contribution addressing to a current problem on the identification of the progenitor stellar systems of type Ib supernovae; this argument can be, in fact, of interest to an audience interested in problems like the gravitational wave or the neutrino underground detection from core collapses.

A certain number of type I supernovae have been recently found to show observational properties well distinct from those of classical type I supernovae. The new class has been defined as type Ib SNe while the classical ones are now addressed as type Ia SNe (see, e.g., 17).

Type Ib supernovae are a couple of magnitudes subluminous respect to the type Ia ones; they are radioemitting, radioemission being promptly showed after the explosion and followed by a fast decay; type Ib supernovae are found only in the vicinity of star forming regions (contrary to the type Ias that are found practically anywhere); their spectra resemble those of type Ia ones some month after the explosion, this feature allowed Panagia (17) to define them as born old type I SNe; some other peculiarities are present in the spectrum of these SNe.

A theoretical model for the evolution of the progenitor and for its explosion has then to fulfill a number of requirements: an exploding mechanism able to release about 1 foe; the absence of H in the pre-exploding structure; fast evolutionary time scales of the progenitor stellar system; presence of circumstellar matter (to account for the prompt radio emission); composition of the ejecta in agreement with the observations.

Two models have been proposed, but, even being substantially different, none of the two can be presently preferred to the other given the paucity of the observational constraints (obtained from far extragalactic SNe) and because of theoretical difficulties.

The first model consists of a massive He star deprived of the H-rich envelope due or to a strong wind, or to an interaction with a companion star. In such a case the explosion is driven by the core collapse and bounce as in classical type II supernovae (see, e.g., 15).

A second model consists of a close binary system componed of a degenerate CO star and of a low mass He-star. These stars can merge, due to gravitational wave radiation energy loss, and He can be accreted on the surface of the CO star. He is accreted at a rate for which an He detonation is expected after about 0.4 solar masses of He have been accreted on the top of the CO dwarf. The degenerate ignition gives rise to a detonation with a release of about 1 foe or less (see, e.g., 16). Since in this case only 0.4-0.5 solar masses of material should be incinerated (i.e. one half of what expected in type Ia SNe) this explosion should generate a supernova 2-4 times less luminous than a type Ia.

It is clear, from the quoted properties of the two models, that in the first case, being the explosion determined by a core collapse and bounce, an energetic neutrino pulse and the emission of detectable gravitational waves are expected, while no of the two occurrences is expected in the second case.

However it seems that supernovae explosions belonging to such a class should be quite frequent. One can obtain from the data given by (8) the following rates for our Galaxy

    Type Ia SNe   0.3   every 100 yrs
    Type Ib SNe   0.5   every 100 yrs
    Type II SNe   1.1   every 100 yrs

It is of course not necessary to stress that such a data are to taken with the due caution since they are derived from a five years observations of external galaxies and the translation to our own galaxy needs some assumptions to be done; such assumptions can induce an error as large as a factor of three.

These considerations induce a twofold conclusion, from one side we have the risk that roughly one half of the expected supernovae in our galaxy could not be useful for gravitational wave antennas and neutrino underground detectors, from the other side I see the possibility that the next supernova event could be of the type Ib and, in such a case, the detection of neutrino and GWR signals, or its failure, could be crucial to improve our knowledge on type Ib Supernovae.

REFERENCES

1) J.B. Dogget and D. Branch, Astron. J. 90 (1985), 2303
2) S.E. Woosley and T.A. Weawer, Ann. Rev. Astron. Astrophys. 24 (1986), 205
3) F. Matteucci and A. Tornambe', Astron. Astrophys.142 (1985),13
4) I. Iben and A.V. Tutukov, Ap.J. Suppl. 54 (1984), 335
5) A. Renzini, acta of the IAU Symp. 105 (1984), 21
6) I. Yu. Litvinova and D.K. Nadezhin, Sov. Astron. Lett. 11 (1985), 145
7) E. Cappellaro and M. Turatto, Astron. Astrophys. (1988), in press.
8) S. van den Bergh, R.D. McClure and R. Evans, Dominion Astrophys. Obs. Preprint Jan. 1987
9) A. Tornambe' and F. Matteucci, Ap. J. Lett. 318(1987), L28

10) A. Tornambe', F. Matteucci, I. Iben and K. Nomoto, Lecture notes in Physics 287 (1987), 284
11) K. Nomoto, T. Shigeyama and M. Hashimoto, Proc. IAU Colloquium 108 (1988), in press
12) S.H. Kahana, Proc. IAU Colloquium. 108 (1988), in press
13) R. Gilmozzi et Al, Nature 328 (1987), 318
14) C. Fransson et Al, Ap.J.(1988), in press
15) J. C. Wheeler et Al., Ap. J. Lett. 313 (1987), L69
16) I. Iben, K. Nomoto, A. Tornambe' and A.V. Tutukov, Ap.J. 317 (1987),717
17) N. Panagia, Proc. 4th European IUE Conference, ESA SP-218 (July 1984), 21
18) Hirata et Al., Phys. Rev Lett. 58 (1987), 1490
19) R.M. Bionta et Al., Phys. Rev. Lett 58(1987), 1494
20) M. Aglietta et Al., Europhys. Lett. 3 (1987), 1315
21) I. Yu. Litvinove and D.K. Nadezhin, Astrophys. Space Sci. 89 (983), 89

A VERY LARGE TELESCOPE FOR NEUTRINO AND GAMMA ASTRONOMY

P. Pistilli
Physics Department
Lecce University
INFN Lecce

Abstract*

We discuss the feasibility of a telescope consisting in a sampling array for extensive air showers measure combined with a muon tracking device. The sampling array will extend over a surface of $\sim 10^6 \ m^2$ while the muon tracking device will cover $\sim 10^4 \ m^2$.

The telescope should be performed with a combination of streamer tubes and resistive plates counters and would became the most powerfull device to study high energy neutrinos and gamma rays astronomy as well as cosmic ray physics in the highest energy $(\sim 10^{18} \ eV)$ region.

---

* The content of this talk has been elaborated togheter with: G. Iaselli, F. Nuzzo, F. Romano (BARI); A. Rossi, G. Susinno (COSENZA); A. Grillo, A. Marini, F. Ronga, V. Valente (LABORATORI NAZIONALI DI FRASCATI); P. Bernardini, P. Pistilli (LECCE); M. Ambrosio, G. Barbarino, B. Bartoli, V. Silvestrini (NAPOLI); G. Bressi, A. Lanza, M. Cambiaghi, S. Ratti (PAVIA); M. Bonori, M. Conversi, G. D'Agostini; M. De Vincenzi, P. Lipari, F. Massa, M. Mattioli, A. Nigro, O. Palamara, S. Petrera, A. Sciubba (ROMA I); R. Cardarelli, R. Santonico (ROMA II); L. De Cesare, G. Grella, M. Guida, G. Marini, G. Romano, G. Vitiello (SALERNO).

Introduction

The ambitious project of exploring astrophysical point sources of very high energy ($\geq 10^{12}\ eV$) neutrinos may require detectors of such large area ($\geq 10^4\ m^2$), that it seems at present extremely difficult to have them placed in an underground laboratory.

On the other hand a surface experiment to search for such point sources must distinguish between upgoing muons produced in the interaction of the neutrinos with the terrestrial crust and downgoing muons of much more intense flux.

In this document we discuss the feasibility of a very large area telescope dedicated to high energy neutrino and gamma rays astronomy performed with an array of streamer tubes and resistive plate counters (RPC). The RPC technique allows to detect charged particles with good time ($\sim 1 \div 2\ ns$) and space ($\sim 3\ cm$) resolution covering large areas at low price; it is therefore expecially suitable to built an array of counters that samples the extensive air showers (EAS). The combination of streamer tubes and RPC allows the construction of a muon tracking device capable to measure the muon molteplicity and to distinguish between upgoing and downgoing muons, measuring the time of flight (TOF) between RPC's crossed by the muons.

With such a detector built in Italy will be possible a simultaneous observation of the very high energy γ-rays point sources located in the northern celestial emisphere and the very high energy ν-sources of the southern emisphere.

This document is organized as follows:

The main features of the apparatus are described in section I.

A short review of the physical items that can be studied with such a telescope is presented in section II.

The problems related to the trigger system, to the background rejection and to the DAQ system will be discussed in section III.

In section IV a critical comparison with other existing or proposed telescopes is presented.

In the appendix A we discuss the possibility of producing a large quantity of RPC's in reasonable time and we give an estimate of the construction time and of the cost of the telescope.

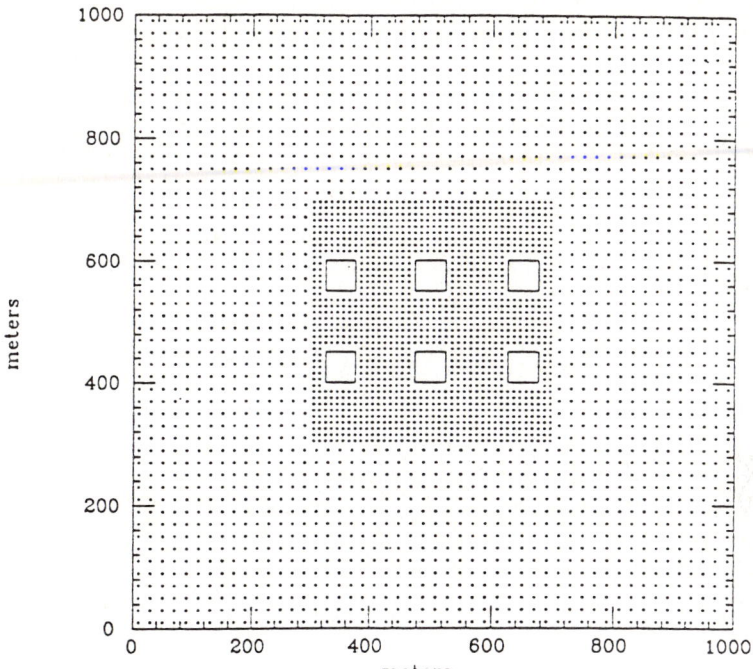

Fig. 1: Top view of the array; the dots rapresenting the sampling units, the shaked area the moun detectors.

§1. The apparatus

A detailed description of the experimental set up is behind the purpose of the present document. The final optimation of several parameters of the apparatus depends by the height at which it will be placed and by the experimental results obtained with a prototype. We present here a possible scheme of the telescope and we discuss its main characteristics.

The telescope consists in two parts: an array of counters that samples the EAS within an effective area of $\sim 10^6 \ m^2$ and a series of μ-tracking modules covering an area of $\sim 10^4 \ m^2$.

§§1.a The EAS detector

The top view of the array is presented in Fig. 1. The array surface of $1000 \times 1000 \ m^2$ is sampled with $\sim 3 \cdot 10^3$ units.

Each unit consists in two RPC's $(1 \times 2 \ m^2)$ placed in two planes and interspaced by a thin ($\sim 0.5 \ r.l.$) concrete layer. On the top of the unit a layer ($\sim 1 r.l.$) of lead can be placed. The lead acts as a converter of the γ-rays in the EAS, increases the number of the electrons crossing the detector with an improvement of the angular resolution of the shower direction as discussed in ref. [1].

The units form a grid $\sim 20 \ m$ spacing over the surface of $\sim 10^6 \ m^2$. In the central part $300 \times 300 \ m^2$ the grid becomes $\sim 10 \ m$ spacing. Such array is similar to several existing or proposed EAS detectors with the RPC's substituting the more popular scintillator counters. A detailed discussion on an EAS detector performed with scintillator counters can be found in ref. [1].

Most of the conclusions are identical using such RPC's array whose characteristcs we summarize in the following points:

a) Such a detector can search for point sources of EAS measuring the difference in the arrival times of the shower front on the grid.

b) With a resolution time of the detector of $1 \div 2 \ ns$ (easily achieveable by RPC tecnique) the direction of the primary particle that induces the shower will be measured with a angular resolution of $\sim 10 \ mrad$ for shower of $\sim 10^{14} \ eV$.

c) With an effective area of $\sim 10^6 \ m^2$ we expect $\sim 10^5 \ (10^3)$ *events/year* from a source like Cygnus–X3 with $E_\gamma \geq 10^{14} \ (10^{16}) \ eV$.

d) With an effective area of $10^6 \ m^2$ we expect $\sim 100 \ ev/year$ of ultra high energy $(E \geq 10^{18} \ eV)$ cosmic rays.

§§1.b The muon detector

The second part of the detector consists of 24 modules grouped in 6 supermodules dedicated to μ-tracking and discriminating between upgoing and downgoing muons by the measure of the muon TOF between RPC's.

A schematic view of a module is shown in Fig. 2. It consists of 9 planes of streamer tubes $20 \times 20 \ m^2$ area interspaced with concrete absorber layers each $\sim 50 \ cm$ thick. Three double planes of RPC's are respectively placed at the

Telescope for Neutrino and Gamma Astronomy

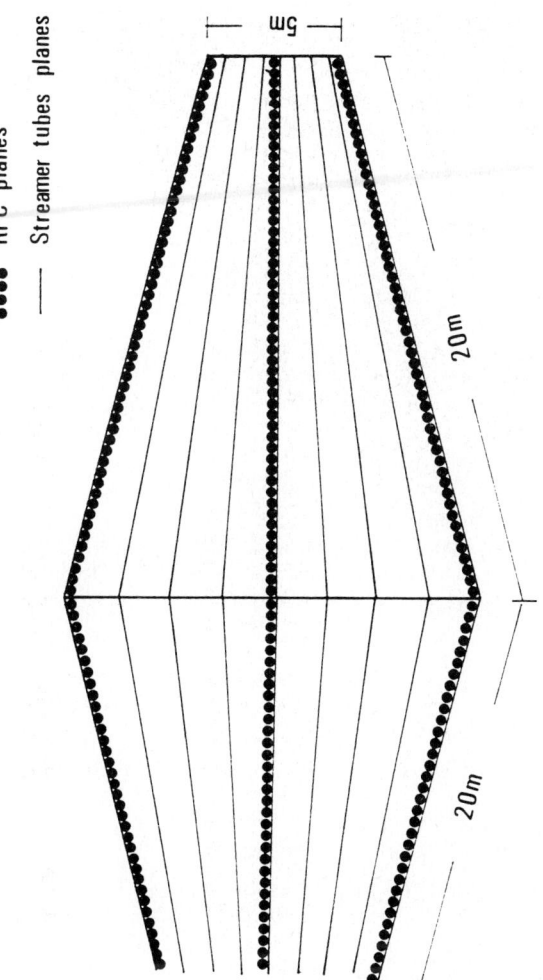

Fig. 2: Schematic view of a μ-tracking module.

basis, in the middle and on the top of the module, each 20×20 m^2 plane being composed of 20 RPC's 1×20 m^2. The streamer tubes have 3×3 cm^2 cross section and measure one coordinate on the wire; the other coordinate is given by an array of 3 cm wide pick-up strips orthogonal to the wire.

For the RPC's the measure of one coordinate is also performed with the pick-up strips system, while the other coordinate is known by the difference of two time measurements taken at the edges of the counter within an accuracy of ~ 10 cm.

The streamer tubes give the direction of the muons crossing at least 5 meters of concrete ($E_\mu \geq 1.5$ GeV) with an angular resolution $\Delta\theta_\mu \leq 10^{-2}$ rad, while the time information given by the RPC system allows to identify the upgoing muons out of the much more intense downgoing muon flux.

With this very large muon tracking system extending over an area of $\sim 10^4$ m^2, this device becomes by far the most powerful telescope for ν-astronomy looking at the upgoing muons; associated with the EAS grid permits to measure with excellent accuracy the muon molteplicity associated with a shower.

The result is a capability to distinguish between electromagnetic and hadronic showers with a strong enhancement of the signal-to-background ratio in the search for high energy γ-rays source.

Moreover the simultaneous measure of the electron and muon density of a shower coming from an identified point source is a very powerful tool to study eventually new exotic physics in the interaction of very energetic γ-rays with the atmosphere.

§2. Gamma and neutrino astronomy, cosmic rays physics

§§2.a Gamma astronomy

Recently the observation of extensive air showers ($E \geq 10^{14}$ eV) from point sources has been reported by independent observers [2]. The first reported point source of EAS was the now well established Cygnus–X3, other reported sources are Vela X–1, LMC X–4 and Hercules X–1. The detection of several other objects has been claimed in the TeV range with Cerenkov detectors. All these sources are also periodic emitters of X-ray radiation, and are interpreted as a binary systems composed of a compact object (a neutron star) in orbit with a normal star.

Given our present knowledge of particle physics these air showers must be initiated by γ rays. In fact charged particles are disordered by galactic magnetic field and even extremely energetic neutrons will decay before reaching the earth.

A possible physical mechanism for the UHE photon emission has been described by Vestrand and Eichler [3], and is sketched in Fig. 3. The source consists of a compact star in a periodic orbit with a star that has not yet collapsed. Protons

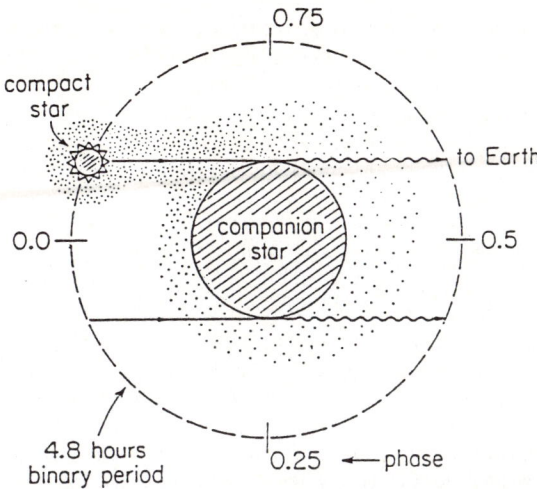

Fig. 3: Possible physical mechanism for the UHE photon emission.

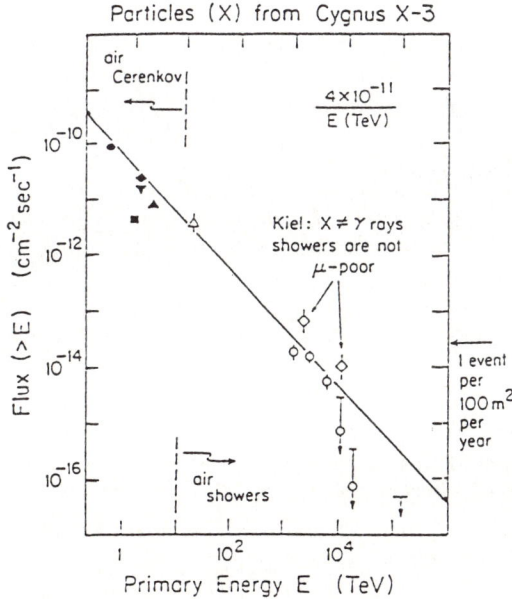

Fig. 4: Measured flux from Cygnus X-3.

accelerated by the compact star by means of a still unclear mechanism interact with the companion or the surrounding gas to produce a cascade of secondaries as in any beam dump. Gamma rays will be produced from π^0 decay and neutrinos from $\pi \to \mu\nu$ decay. Therefore it is possible to derive from a measured flux of high energy photons, the expected flux of neutrinos. These neutrinos will be detected through the charged current interactions occurring in the rock below the telescope giving rise to a detected muon.

A summary of the average measured flux from the best measured source Cygnus X–3 is reported in Fig. 4. An average of the data, can be represented[4] for $(E_\gamma \geq 0.1\ TeV)$ with the simple formula

$$F(\geq E) \simeq \frac{4 \times 10^{-11}}{E(TeV)}\ cm^{-2}s^{-1} \qquad (1)$$

corresponding to $10^4\ events/(year\ 10^5\ m^2)$ above a threshold energy of $100\ TeV$. This is an average flux. The signal has a rich time structure. The most important feature is that the signal has the same periodicity of the X–ray flux with a $4.8\ h$ period that is associated with the binary system orbital motion. The emission of UHE photon is however present only during a small fraction of order $\sim 10^{-2}$ of a period. Additional periodicities (for example of 19 days) have also been reported, as well as the presence of periods of increased activity with fluxes higher than the average by more than one order of magnitude.

Usually for background rejection, the experiments have made use of "phase analysis" exploiting periodicity and small duty cycle of the UHE gamma emission. On the other hand it would be very significant to be able to identify a source without making recourse to its periodic nature. This can be done with use of the muon detector.

The showers initiated by cosmic ray hadrons are not spectacularly different from those initiated by γ-rays. The prominent feature that can be used to identify γ-rays is the low muon content in the electromagnetic showers. In Fig. 5 (taken from ref. [4]) the number of muons in excess of $1\ GeV$ is plotted as a function of the shower size (number of electrons N_e) for γ and nucleon initiated showers. The result is qualitatively easy to understand: muons are the progeny of hadrons and the photon is hadronic at order 10^{-2}.

Therefore the measure of the muon molteplicity strongly enhances the sensitivity for the search of γ-rays source.

To estimate the background on the signal from a point like source we note that the diffuse background of atmospheric cosmic rays is:

$$\phi(\geq E) \simeq 1 \times 10^{-8} \left(\frac{E}{100\ TeV}\right)^{-1.75}\ cm^{-2}\ s^{-1}\ sr^{-1} \qquad (2)$$

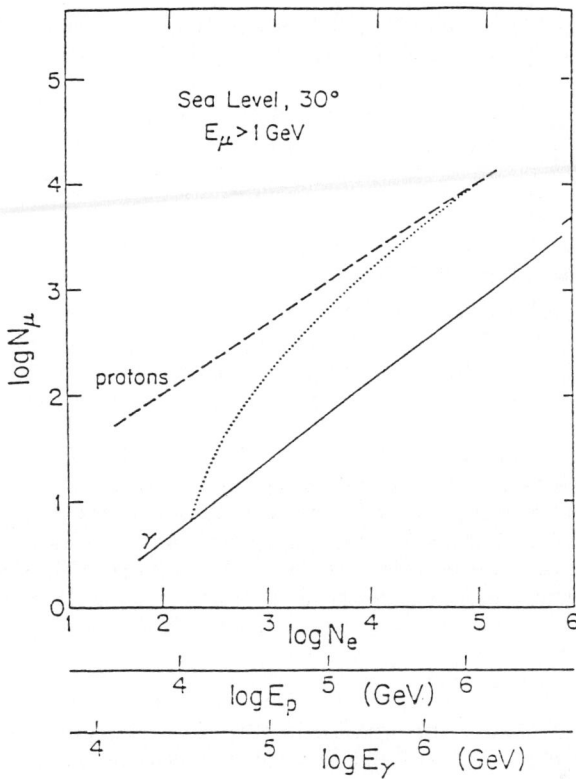

Fig. 5 Muon content in γ and μ initiated showres vs the shower size (number of electrons Ne).

With the angular resolution of 0.5° on the shower direction obtained from the sampling array, one has for a source like Cygnus a signal to background ratio $\frac{S}{B} \simeq 10^{-2}$ for $E_\gamma \geq 10^{14}$ eV.

With the measure of the muon multiciplity the hadronic background is reduced of about a factor 100 giving raise to $\frac{S}{B} \simeq 1$ and allowing identification of the source without recourse to its periodic nature.

The angular resolution for the shower could be also substantially improved using the measurements of the direction of the muons that are tracked with a resolution $\Delta\theta_\mu \simeq 10^{-2}$ rad with a further background rejection. The $\frac{S}{B}$ can also be considerably better during periods of high activity of the source.

On has to remember that up to now the observation of UHE photon showers has been accompanied by the still unexplained phenomenon of a muon multeplicity higher than what is normally expected for a gamma induced shower. This could be interpreted as exotic interaction of high energy photons, or with the exotic nature of the shower initiating neutral particle.

Making use of the periodic structure of the signal one could arrive without making use of the information from the muon tracking system to a $\frac{S}{B}$ of order unity or better, and then study in detail the muon content of the signal shower searching for exotic behaviour, as discussed in detail in ref.[4]. As an example in Fig. 5 is presented the muon signature in a model[4] where the photons become strongly interacting at very high energies.

§§2.b Neutrino Astronomy

As we have discussed in the previous section, we expect a neutrino flux from UHE photon sources. Such fluxes have not been detected up to now, but are expected to be near the level of sensitivity of the largest proposed neutrino detectors. Positive detection of these neutrino sources and study of the correlation between photon and neutrino fluxes will be one of the major area of research in high energy astrophysics.

To estimate the expected signal from a source, we will consider a source located at distance $d = 10$ kpc (1 $kpc = 3.085 \times 10^{21}$ cm, d is of the order of the galaxy radius), and emitting a power in neutrinos $P_\nu = 10^{38}$ erg/sec between 1 GeV and 10^7 GeV, with a power law spectrum $\propto E^{-\alpha}$ *. Using the muon yield calculated in reference [5] we obtain an upward muon flux:

$$F_{\alpha=2}(\geq 2\ GeV) \simeq 2.4 \times 10^{-15} \left(\frac{P_\nu}{10^{38}\ erg/sec}\right) \left(\frac{10\ kpc}{d}\right)^2\ cm^{-2}\ s^{-1} \qquad (3)$$

---

* Cygnus X-3 is located at a distance $d \geq 11$ kpc and emits in high energy photons a power $P_\gamma \sim 10^{37}$ erg/sec. The neutrino flux is expected to be about 10 times stronger because of the longer fraction of the orbit period in which neutrino production is effective

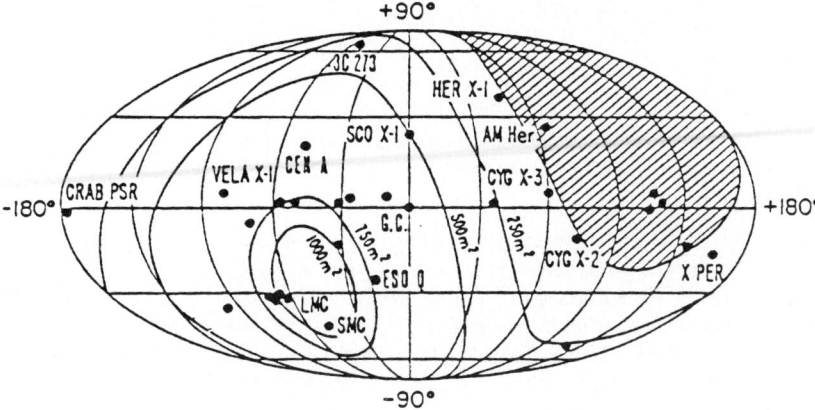

Fig. 6: Detector field of view in Galactic coordinates. Countors represent the time-averaged exposed areas. The dashed region in the Earth's northen hemisphere is inaccessible.

or $F \simeq 8$ $events/(year$ 10^4 $m^2)$

For steeper spectra, the same power in neutrino will result in a smaller upward muon flux. For a spectra $\propto E^{-2.2}$ the flux is reduced (with the same conditions) to 4 $events/(year$ $10^4 m^2)$ for a spectra $\propto E^{-2.4}$ to 2 $events/(year$ $10^4 m^2)$

Detailed estimates of the upward muon flux expected from Cygnus X-3 exist in the literature [6], and area of the order of $\simeq 10$ $events/(year$ 10^4 $m^2)$ for a fully efficient detector.

A detector located in Italy would be mainly sensitive to neutrino sources located in the southern celestial emisphere. In fact a point located at celestial coordinates (α, δ), will have in the laboratory system, variable polar coordinates (see Fig. 6, where we have assumed a detector of 10^3 m^2 at latitude 42°). For neutrino astronomy the observation of a point like source is possible only if it is below the horizon. In Fig. 7 we show the average acceptance in a sidereal day for a neutrino source located at celestial declination δ with a detector located at latitude of 42°. In the same figure we show the declination angle of various expected neutrino sources. As one can see Cygnus X-3 is at the edge of the observation window for neutrino emission, while the LMC X-4, Vela X-1, and Cen X-3 are all well located. Taking into account the measured UHE photon fluxes from these sources, we expect a signal of $\simeq 20$ $events/year$ from LMC X-4 and ~ 5 $events/year$ from Vela X-1 for a detector of area 10^4 m^2. It should be stressed that these estimates have an uncertainty of at least a factor of 5. It must be also remarked that the angular uncertainty in the direction of the neutrino sources is dominated by the resolution of the muon tracking device $\Delta\theta_\mu \sim 10^{-2}$ rad; the angle of the muon with respect to the neutrino line and the multiple scattering effects turn out to be negligible.

The most important physical background is due to the upward muon flux induced by the neutrinos produced in the earth atmosphere in the showers caused by primary cosmic rays. Detailed calculation of the expected fluxes have been performed [7] that are in good agreement with existing data. The upward muon flux for $(E_\mu \geq 2$ $GeV)$ is shown in Fig. 8. The average value is:

$$\overline{\Phi}(\geq 2\ GeV) \simeq 2.5 \times 10^{-13}\ cm^{-2}\ s^{-1}\ sr^{-1} \tag{4}$$

The expected rate in a flat horizontal detector is:

$$R_{hor}(\geq 2\ GeV) \simeq 2100\ \left(\frac{events}{year\ 10^4\ m^2}\right) \tag{5}$$

Lowering the energy threshold the rate increases (same units) to 2400 events for $E_\mu \geq 1$ GeV and to 2600 for $E_\mu \geq 0.5$ GeV.

The background B coming from celestial coordinates (δ, α) is weakly dependent on the declination angle δ, and is linearly proportional to the solid

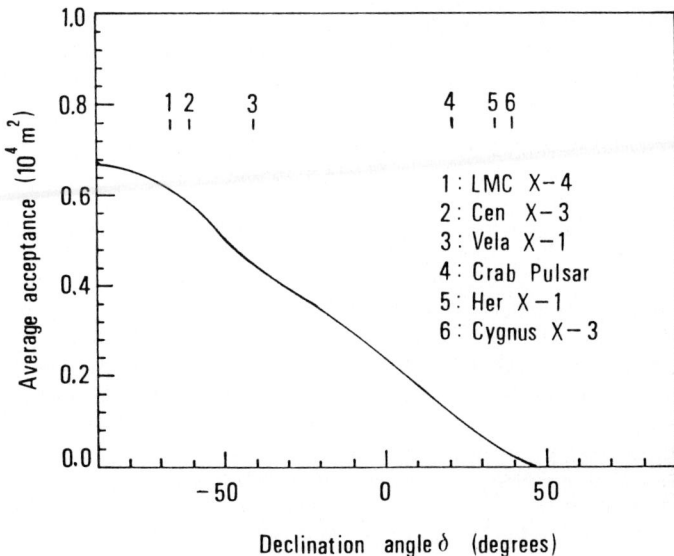

Fig. 7: Average acceptance of the apparatus as a function af the declination angle. The position of the most important sources is also shown.

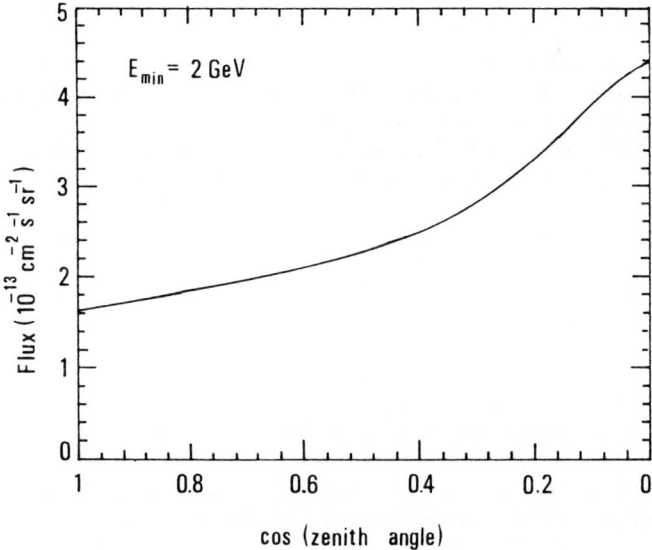

Fig. 8: Angular dependence of the upaward muons background.

angle resolution $\Delta\Omega \simeq \pi\Delta\theta_\mu^2$, Typically we have $B \simeq R/(2\pi)\Delta\Omega$. For an angular resolution of 0.5° we have:

$$B(\Delta\theta_\mu = 0.5°) = 0.08 \ (events/(year \times 10^4 \ m^2)) \tag{6}$$

If the angular resolution becomes $\Delta\theta_\mu = 1, 2$ and 5 degrees the background raises respectively to 0.3, 1.3 and 8 $(events/(year \times 10^4 \ m^2))$

We conclude that for a resolution $\Delta\theta_\mu \simeq 10^{-2} \ rad \simeq 0.5°$, the background is negligible, therefore the sensitivity of the experiment increase linearly with time.

§§2.c Cosmic rays at ultra high energy

The very large area of the EAS detector (that can be further enlarged with smaller sampling density) allows to detect with reasonable statistics $(100 \ ev/(year \times 10^6 \ m^2))$ ultra high energy cosmic rays $(E \geq 10^{18} \ eV)$. The study of the highest energy tail of the spectrum, and the identification of a possible cut-off turns out to be very important in the understanding of the cosmic rays origin[8].

§3. Trigger system, rejection power and DAQ system

§§3.a Trigger system

The trigger for an extensive air shower can be simply performed by demanding a definite multeplicity of the sampling units.

We discuss here more in detail the trigger system to select single upgoing muons that presents more problems.

A possible logical scheme can be summarized in the following steps (Fig. 9):
a) Out of the three double RPC planes the two on the top (T) and at the basis (B) of the module are used in the trigger.
b) We consider the RPC plane $20\times 20 \ m^2$ divided in 20 parts one meter wide each. We use OR's among the 32 pick-up strips from both sides.
c) We send the two outputs left and right to a meantimer circuit in order to have the time information independently on the hit coordinate along the RPC array.
d) For each plane the 20 meantimer outputs are sent to a OR which represents the plane signal and to a majority circuit which vetoes the previous signal if a spurious hit is present on the plane (2 % dead time).
e) We perform independently for B and T the coincidence within 5 ns between the output from the two contiguous RPC planes. The output of the coincidences will be about 25 ns wide.

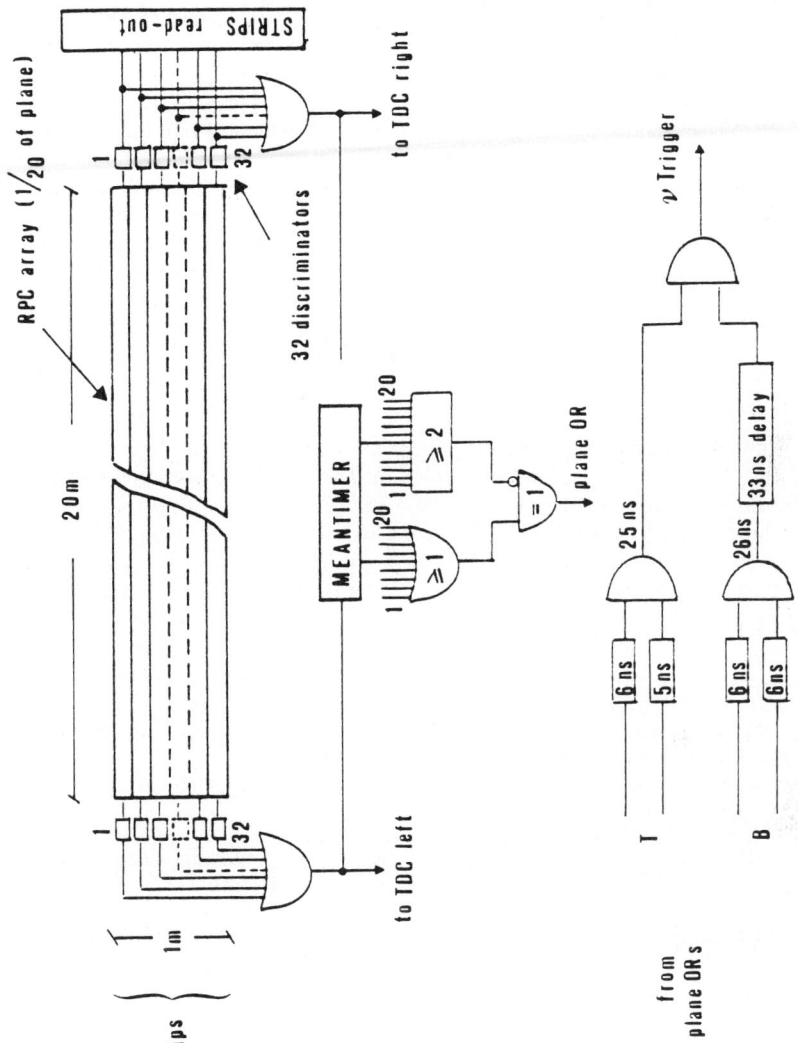

Fig. 9: Logical scheme for the trigger system.

f) We insert a delay at the output of the coincidence B corresponding to twice the TOF between B and T for a vertical trajectory. In such a way the upgoing (downgoing) muons will be in (out of) time (Fig. 10).

g) The coincidence between B and T allows a large angular acceptance with respect to the vertical direction. With the discussed geometry ($20 \times 20 \times 5\ m^3$) a time of $\sim 25\ ns$ correspond to an angular acceptance of $\sim 1.3\ rad$.

The accidental rate of the described trigger will be given by cosmic rays ($\sim 200\ Hz/m^2$), radioactivity ($\sim 300\ Hz/m^2$) and spurious RPC noise ($\sim 500\ Hz/m^2$). In this condition the accidental rate is less then 10 Hz per module. This rate can be strongly reduced with an online filter by using a smaller resolution time of the B-T coincidence correlated with a rough information of the track direction.

The overall electronics efficiency of this trigger is ~ 90 %.

§§3.b Offline rejection power

For the identification of the upgoing muons all the three double RPC planes will be used. The analysis will be performed on the events with only 1 muon track without spurious hits in the $1 \times 20\ m^2$ RPC array. By using the time ($\sim 2\ ns$) and space ($\sim 5 \times 10\ cm^2$) resolution we demand in at least 5 RPC planes space and time alignement of the track reconstructed by the streamer system. With the discussed configuration the rejection power will be dominated by the non gaussian response of the time measurements of the RPC. It has been measured [9] a tail at 20 ns smaller then 10^{-4}. With this number we obtain a rejection power of $\sim 10^{-12}$. Taking into account the downgoing muon flux of $3 \cdot 10^{12}\ muons/year/module$ we obtain an accidental background of few $event/year/module$ much smaller than the physical background due to atmospheric neutrinos (about 100 $event/year/module$).

§§3.c DAQ system

The acquisition system suitable for the proposed detector can be designed in such a way to take advantage of the modularity of the apparatus.

Therefore it will consist of a network of microcomputers, each one dedicated to the acquisition of the single module: this organization comes natural since the experimental apparatus consists of 6 equal supermodules, each almost independent from the others.

A central computer will perform all the functions of collection and organization of data coming from the network: event building, output to disk or tape, run control and monitoring.

This network can be ralised using Ethernet, at a rate of 10 $Mbaud$, simply using hardware and software commercial products available for network up to a maximum size of few Kms. If a higher speed or a longer network will be required one can think to use the optical cable link under development for the Gran Sasso Laboratory (up to 120 $Mbaud$ at a distance $\leq 50 Km$).

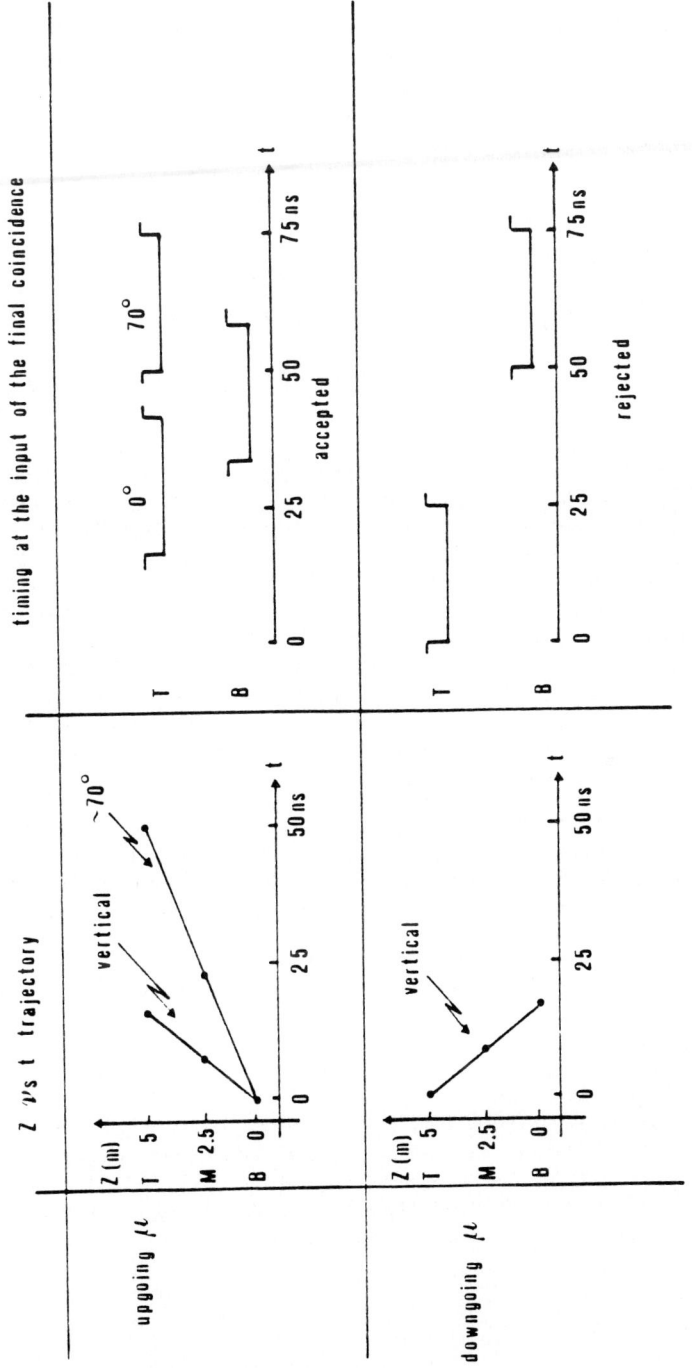

Fig. 10: Time delay for the trigger.

An acquisition system using this architecture is being used for MACRO[10] where the dedicated microcomputers are microVaxes running a VAXELN built-up operating system and the central computer is a VAX8200 running under VMS.

§4. Comparison with other experiments

The presently discussed telescope is simultaneously suitable for high energy neutrino and gamma astronomy.

The most powerful detector searching for point like high energy neutrino sources is MACRO placed in the underground Gran Sasso laboratory. The MACRO detector presently under construction will cover an area ($\sim 10^3\ m^2$) about ten times smaller than that of the discussed telescope, tracking muons practically with the same angular resolution.

As we have discussed in section 2 the background from atmospheric neutrinos turn out to be negligible with an angular resolution $\Delta\theta_\mu \leq 1°$, therefore the larger surface of the proposed detector would have a sensitivity to point like neutrino sources ten times larger than MACRO. This gain turns out to be crucial for a solid identification of neutrino sources.

It is important to remark that the excavation of an underground laboratory ten times larger than the Gran Sasso one is practically hopeless; therefore a surface muon tracking device exploiting the good time resolution given by the RPC to reject the background is perhaps the only way to study high energy neutrino astronomy.

A competing technique could be to use very large water pools and to detect the muons looking at their Cerenkov light with a PM array. In this case [11] to achieve an angular resolution for the muon tracking system $\Delta\theta_\mu \leq 1°$ (crucial for the rejection of the atmospheric neutrino background) implies the use of an extremely complex PM's array with an effort and a price of the same order of magnitude than the proposed detector, but with less reliability for the muon identificaion.

Regarding cosmic rays physics and ultra high energy gamma astronomy, the sampling system we are proposing is very similar in effective area and angular resolution to the EAS array of the Fly Eye II recently proposed by the Chicago group [1]. The muon device here proposed turns out to be more than one order of magnitude better than any existing or proposed telescope. In fact the existing or proposed telescopes either haver an effective muon tracking device over a much smaller area (typically few hundred m^2) or a much simpler muon identification performed with scintillation counters placed a few meters underground.

The proposed telescope will track the muon over $\sim 10^4\ m^2$ and will correlate the angular direction obtained independently by the EAS and on the muon bundles.

We conclude that the proposed telescope would become by far the most powerful device dedicated to high energy gamma and neutrino astronomy.

Appendix A
Construction time and cost exstimate

In ref. [12]-[18] one may found the main results on the RPC technique from the first study to the most recent ones.

The proposed telescope consists in three parts :
a) The sampling array performed with $\sim 4\ 10^3$ RPC units of $0.5 \times 1\ m^2$.
b) The streamer tubes system for the muon tracking covering a surface of $\sim 10^5 m^2$ with $\sim 1.5 \cdot 10^5$ channels.
c) The RPC planes used for the time information in the muon tracking modules for a total surface of $\sim 6 \cdot 10^4\ m^2$. The first two parts can be savely performed with the existing technology. A similar number of streamer tube channels has been put in operation in several experiments. The assembly problems can be scaled by the similar structure of the MACRO experiment with the advantage not operating in an underground laboratory.

We extimate that with a manpower of about 15 persons dedicated to the tubes constructions and to the general assembly the entire muon tracking device can be savely completed in about 4-5 years.

Also the construction of the EAS sampling array does not present particular problems with perhaps the only exception of the distribution of the gas over the very large $1000 \times 1000\ m^2$ surface. This latter problem is under investigation.

The construction and the assembly of the RPC planes for the muon tracking device is a major technical challenge. It implies the capability to built, to test and to assembly about 60 m^2 of RPC per day. Moreover the use of RPC in so large scale needs a realibility, a stability and a easy handling of this technique that up to now has been established only at the level of $\sim 200\ m^2$ area detector[18].

We want to point out however that an effort in the direction of large scale RPC production have been recently done in the Physics Departement of Rome in view of their use in the FENICE[19] and ICARUS[20] experiments.

A semiautomatic device capable to produce about 12 m^2 of RPC per day is already operating. The production and the operation of several hundred square meters is forseen before the end 1988 . We belive that this will be an excellent text to understand the problems related to the use of RPC over a very large scale and to judge the feasibility of the telescope we have discussed as well as the effort it needs.

With respect to the cost we have done a rough estimate based on the commercial prices of the various parts of the equipment. The cost of tubes and RPC's mechanics will be $\sim 20\ GL$, that of electronics (including DAQ system) $\sim 10\ GL$. The mechanical support of the muon tracking and the general facilities (cranes, hangars, electric power system, etc.) should cost $\sim 20\ GL$. It has to be remarked however that due to the very large size of the discussed apparatus, the cost can be substantially reduced with respect to these estimates.

REFERENCES

[1] M.K. Campbell et al. "A Proposal for a large sufrace array at Fly's eye II" Chicago 1987.

[2] M. Samorsky and W. Stamm, Astr.J. 268 L17, (1983). For a complete review see A.A. Watson, in Proceedings of 19th ICRC La Jollla 1985.

[3] W.T. Vestrand and D. Eichler, Astr.J. 261, 251, (1982).

[4] F. Halzen, K. Hikasa and T. Stanev, Phys. Rev. D34, 2061, (1985).

[5] T.K Gaisser and T. Stanev, Phys. Rev. D31, 2270, (1985).

[6] T.K Gaisser and T. Stanev, Phys. Rev. Lett. 54, 2265, (1985).
V.S. Berezinsky, C. Castagnoli, and P. Galeotti, Nuovo Cimento 8C, 185, (1985).
E.W. Kolb, M.S. Turner and T.P. Walker, Phys. Rev. D32, 1145, (1985); D33, 859 (E) (1986).

[7] T.K Gaisser and T. Stanev, Phys. Rev. D30, 985, (1984).

[8] Reference on High energy cosmic rays.

[9] R. Santonico and F. Massa, private communication.

[10] MACRO detector. Nucl.Instr.Meth. A264, 18-23, (1988).

[11] Koshiba, Cerenkov detector.

[12] J. W. Keuffel, Phys.Rev. 73, 531, (1948), and Rev.Sci.Instr. 20, 202, (1949).

[13] L. Madansky and R.W. Pidd, Phys.Rev. 73, 1215, (1948), and Rev.Sci.Instr. 21, 407, (1950).
F.Bella and C. Franzinetti, Nuovo Cimento 10, 1353 and 1461, (1953).
F.Bella, C.Franzinetti and D.W.Lee, Nuovo Cimento 10, 1338, (1953).

[14] Yu.N.Pestov and G.V.Fedotovich, Preprint IYAF 77-78, Slac Translation 184, (1978).
Yu.N.Pestov, Nucl.Instr.Meth. 196, 45, (1982).
W.R.Atwood et al., Nucl.Instr.Meth. 206, 99, (1983).

[15] R.Santonico and R.Cardarelli, Nucl.Instr.Meth. 187, 377, (1981).

[16] R. Cardarelli et al. Nucl.Instr.Meth. A263, 20, (1988).

[17] G. Battistoni et al. LNF 87-80, subm. to NIM.

[18] G. Bressi et al. Nucl. Instr. Meth. A261, 449, (1987).

[19] FENICE Proposal LNF 87/18(R).

[20] ICARUS Proposal 85/01 INFN/AE-85/7.

X-RAY ASTRONOMICAL RESEARCH AND THE ITALIAN SATELLITE SAX

Filippo FRONTERA

Dipartimento di Fisica, Università di Ferrara, Ferrara, Italy; Istituto Tecnologie e Studio Radiazioni Extraterrestri, CNR, Bologna, Italy

An outline of the status of X-ray Astronomy research, with particular emphasis on celestial sources of X-ray emission, is reported. In this context the Italian X-ray astronomy satellite SAX is presented and some of its scientific objectives are summarized.

1. INTRODUCTION

The SAX (acronym of Satellite Astronomia X) mission has already passed the study phases A and B and is now entering in its final phase which will terminate, at the end of the year 1992, with the launch of the first Italian scientific satellite completely devoted to X-ray astronomical observations in the 0.1-200 keV energy band. In this lecture I will present the scientific goals of the SAX mission in the context of current research in X-ray astronomy.

The study of X-ray emission from the sky has been shown to be a powerful probe of very high temperature ($>10^6$ K) plasmas, high energy particles, very strong (up to 10^{13} Gauss) magnetic fields, intense gravitational fields, and dark matter in the Universe.

X-ray astronomy is a very young science born in 1962 when a strong X-ray source in the constellation of Scorpio, named SCO X-1, was discovered by a group of four physicists working at MIT and American Science and Engineering[1]. Strong X-ray emission was not expected on the basis of theoretical considerations and actually the experiment was performed to investigate the X-ray fluorescence from the Moon.

Since then, numerous celestial X-ray sources have been discovered. Figure 1 shows a view of the X-ray sky in the 0.5-25 keV energy band obtained with the A-1 experiment on board the American satellite HEAO-1[2]. The picture of the X-ray sky which emerges depends not only on the sensitivity of the instrument used but also on the energy band explored. We find in fact many more sources below 1 keV than above. This is a consequence of the energy spectrum of X-ray photons which is, for most X-ray sources, a steeply decreasing function of energy. Figure 2 shows the energy spectrum of the Crab Nebula, one of the most powerful sources in the sky, and the result of a supernova explosion which occurred in the year 1054. A satellite which explored with a very high sensitivity the X-ray sky below about 1 keV was the "Einstein" Observatory[3]. An almost complete list of the

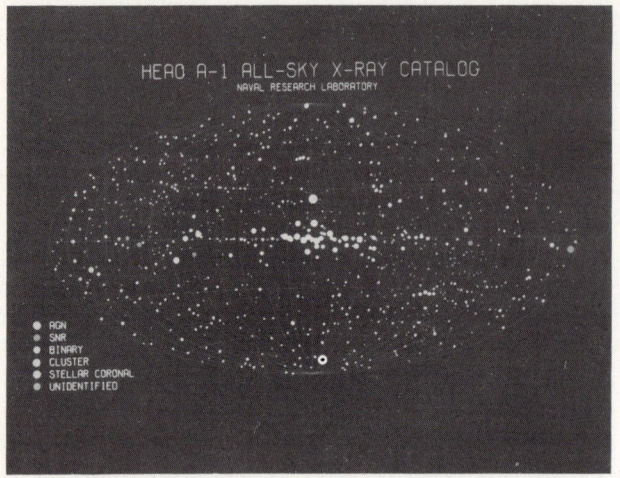

FIGURE 1

The X-ray sky at a sensitivity level of 0.2 mCrab in the 0.5-25 keV energy band (reprinted from Wood et al.[2])

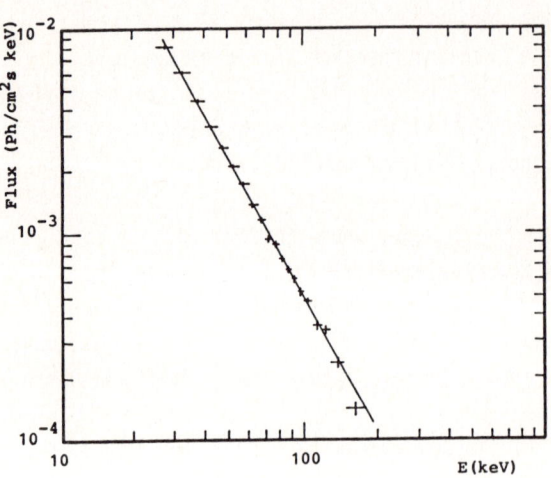

FIGURE 2

Hard X-ray spectrum of the Crab Nebula (reprinted from Frontera et al.[11])

X-ray satellite missions launched so far together with information on their operative energy band, energy resolution of the instrumentation and their sensitivity is given in Table 1. The sensitivity is given in units of 10^{-3} times the flux from the Crab Nebula in the operative energy band. One Crab unit in a given energy interval can be obtained from the Crab X-ray photon spectrum which is well described by a power law $N(E) = 9.7 \, E^{-2.1} \exp[-(E_a/E)^{8/3}]$ ph/cm^{2}s keV where $E_a = 0.97$ keV and E is measured in keV. In the 12-180 keV energy band 1

TABLE 1. List of X-ray astronomy missions

Satellite Name	Country	Launch Date	Energy band (keV)	Energy resolution ΔE/E	Field of view (FWHM)	Limit sensitivity (m Crab) (*)
UHURU	USA	1970	2-20	20% at 6 keV	1°x10°	~1
OSO 7	USA	1971	1.5-40	20% at 6 keV	1°/3°	~1
			7-500		6°.5	~20 at 100 keV
COPERNICUS	USA/UK	1972	0.5-4.5	20% at 6 keV	3°	~5
			2.5-7.5	20% at 6 keV	3°x10'	~5
ANS	NL/USA	1974	2-20			~5
			0.2-0.3			
ARIEL V	UK	1974	2-20	20% at 6 keV	0°.75x10°.6	~1
			26-1000	30% at 662 keV	4°	~100
SAS3	USA	1975	0.15-15		12° (2'.4 fringe)	~1
			1.5-15	20% at 6 keV	1°x32°	~1
OSO 8	USA	1975	0.13-3.5		5°x5°	~1
			2-60	18% at 6 keV	5°x5°	~1
			20-5000	21% at 662 keV	5°x5°	~20 at 100 keV
HEAO 1	USA	1977	0.15-60	16% at 6 keV	1°x4°/1°x1°	~0.2
			0.5-25		6°x6°/3°x3°	~1
			1.5-15		4° (0'.5 fringe)	~1
			12-180	25% at 60 keV	1°x20°	~10
HEAO 2 (Einstein)	USA	1978	0.15-4	35% at 1.5 keV	1°.2 (2" images)	~3x10^{-3}
HAKUCHO	JAPAN	1979	2-20	18% at 6 keV	1°.3x1°.3	~10
TEMNA	JAPAN	1983	0.15-30	18% at 6 keV	Wide (0°.1 fringe)	0.3
EXOSAT	ESA	1983	2-60	9.5% at 6 keV	~3°	3x10^{-2}
			0.04-2	poor	2°.2 (20" images)	0.3
			1-20	18% at 6 keV	0°.75	~10
MIR	USSR/NL/FRG	1987	2-15	9.5% at 6 keV	0°.75	
				18% at 6 keV	7°.5x7°.5 (3' images)	~1
			15-200	25% at 60 keV	1°.6x1°.6	~5
GINGA	JAPAN/UK/USA	1987	1.5-30	18% at 60 keV	0°.8x1°.7	0.1

(*) See text for definition

Crab unit corresponds to a flux of 2.9×10^{-8} erg/cm^2 s, while in the 2-10 keV band it corresponds to 2.1×10^{-8} erg/cm^{2}s. Of these missions only the last two are working today. As can be seen from Table 1 most of these missions achieved their maximum sensitivity in the energy band below ~20 keV. The hard X-ray energy band (15-200 keV) has been poorly explored. The most sensitive scanning of the sky in the 12-180 keV range was performed at a flux sensitivity of about 10^{-2} Crab units with the A-4 experiment on board the HEAO-1 satellite. About 70 galactic and extragalactic hard X-ray sources were detected. For comparison about 850 X-ray sources were detected with the HEAO 1-A1 experiment in the band 0.5-25 keV, while below ~1 keV with the *Einstein* Observatory about 4000 X-ray sources were detected in only 1% of the sky. Hard X-ray observations have also been performed with balloon-borne experiments from altitudes of 40 km, where the transparency of the atmosphere to hard X-rays is high (>20%).

2. STATUS OF THE X-RAY OBSERVATIONS

Given the numerous X-ray observations of celestial objects, it is very hard to give in a lecture a complete review of all the results obtained. I shall limit the discussion here to only two relevant aspects of these results: a) sites of X-ray emission; b) energy sources that power X-ray emission.

2.1. Sites of X-ray emission

Identification of many X-ray sources with objects already known at other wavelengths (mainly in the optical band) has contributed strongly to determining the sites of X-ray emission. Almost all classes of celestial objects known in the visible band emit X-rays. In Tables 2a and 2b I report some features of the X-ray emission from the best known classes of optical objects. As can be seen, among galactic X-ray objects the most luminous ones are those associated with collapsed objects (white dwarfs, neutron stars, black holes). The temporal and spectral features of their X-ray emission (Table 2a) has proved a powerful tool in the study of the physics of the environment near super-strong gravitational fields, with and without intense magnetic fields (X-ray pulsars, black holes). The high X-ray luminosity of most classes of objects, in particular of Active Galactic Nuclei (AGN) (Table 2b) has lead to the study of highly efficient mechanisms of energy production. The implications of diffuse X-ray emission measurements from clusters of galaxies are particularly noteworthy. From the values of spectral parameters of the X-ray emission from separated cluster regions it has been possible to derive the distribution of gas density and temperature as a function of distance from the cluster centre and therefore, assuming hydrostatic equilibrium, the gravitational mass of the cluster. This analysis, performed up to now for only a few clusters, has shown that the cluster mass which emits X-rays is much less than the gravitational mass and therefore that dark matter is present in the clusters of galaxies.

In addition to the classes of X-ray sources reported in Tables 2a and 2b there are other X-ray emitters that have not been identified yet with known objects. They include:

a. *X-ray transient sources*. These sources have been observed to brighten for short time periods from a few minutes to months. They could be nearby galactic stars (e.g. RS Canum Venaticorum-like objects) or extragalactic objects.

b. *Gamma-Ray Bursts*. These are events of short duration (less than ~ 2 minutes), with high energy fluence (up to ~10^{-4} erg/cm^2 seen at the Earth) and very hard spectra (bremsstrahlung-like spectra with kT $\geq$100 keV). Their celestial distribution is almost isotropic. Their origin is not clear yet. Several missions are planned for the next ten years that could solve the mystery of Gamma Ray Bursts.

c. *Other X-ray sources*. There are many X-ray sources not identified yet, due to their poorly known position in the sky. Observation of these sources with higher angular resolution telescopes will help to identify them and therefore to determine their nature.

d. *Diffuse X-ray background*. In addition to discrete X-ray sources, a diffuse X-ray emission has been detected. Its spectral shape is still relatively imprecise below 3 keV, while it is well known between 3 and 60 keV, where it is well fitted by a thermal bremsstrahlung law with kT $\simeq$ 40 keV (T $\simeq$ 4.64x10^8 K). The origin of this diffuse X-ray background is still unknown. Below a few keV, probably the major contribution to the X-ray background comes from evolved galaxies (such as quasars). However in the band above 3 keV, the question of its origin is completely open. At least two possible hypotheses are under discussion: one assumes that the X-ray background is due to the summed contribution of discrete X-ray sources (class or classes unknown)[5], and the other one assumes that the X-ray background is due to diffuse intergalactic emission[6]. Much more sensitive experiments in the hard X-ray band (>10 keV) are necessary to discriminate between these two possible hypotheses.

2.2. Energy sources of celestial X-ray emission

One of the major problems of high energy astrophysics is the justification of the strong X-ray luminosity of celestial objects (Tables 2a and 2b) which ranges from about 10^{27} erg/s from Sun-like stars to 10^{47} erg/s from quasars.

On the basis of the observational properties of the X-ray fluxes (mainly energy spectra and time variability) there are very strong arguments in favor of the conversion of gravitational energy into electromagnetic radiation as the basic mechanism that powers most of the X-ray sources in the sky.

A comparison of the energy released in the case of nuclear fusion with that released in the case of accretion of mass onto a stellar object can clarify this

TABLE 2a. Classes of optically identified galactic X-ray sources

Class (§)	Energy band (§§) (keV)	X-ray luminosity (X10^{38} erg/s)	X-ray features
Non-degenerate stars			
a. Early-type stars (O,B spectra)	<1	10^{-9}–10^{-5}	Spectra such as those expected from a single or multi-temperature plasma with $10^6 \lesssim T \lesssim 10^8$ °K. Many partially ionized element lines (Mg, Si, S, Fe). X-ray luminosity strongly related to total luminosity for O, B, A types.
b. Other stars (A-M spectra)	<1	10^{-11}–10^{-8}	
Isolated white dwarfs	<1	10^{-10}–10^{-7}	Spectra corresponding to emission from a hot ($T \lesssim 10^5$ °K) photosphere.
X-ray binaries with a degenerate companion			
a. Cataclismic variables (white dwarf companion)	0.1–100	10^{-9}–10^{-3}	Two spectral components: one hard with thermal bremsstrahlung spectrum ($kT \gtrsim 20$ keV) powered by accretion; one soft with blackbody spectrum ($kT \lesssim 10$ keV) from the heated white dwarf photosphere. Magnetized white dwarfs with $B \sim 10^8$ Gauss. X-ray periodicities.
b. High Mass X-ray binaries (neutron star companion)	1–100	10^{-5}–5	X-ray pulsation (69 ms $\lesssim P \lesssim 850$ s) from 30 objects. Chaotic time variability down to second time scales. Power law spectra with exponential cut-offs for $E \gtrsim 10$ keV. Cyclotron lines implying $B > 10^{12}$ Gauss. Model spectra based on scattering of up high energy electrons on low energy photons in superstrong B. Helium--like iron lines (6.7 keV).

Table 2a (Cont.d)

Class (§)	Energy band (§§) (keV)	X-ray luminosity ($\times 10^{38}$ erg/s)	X-ray features
c. Low Mass X-ray binaries	1-30	10^{-3}-4	Quasi periodic oscillations (5-65 Hz) with frequency depending on luminosity. Complex X-ray spectra with more than one component: e.g. thermal bremsstrahlung + blackbody. Helium-like iron lines. X-ray bursts (Durations $\lesssim$40 s) with blackbody spectra (T$\lesssim 10^6$ °k).
d. Black hole candidates	1-500	10^{-2}-4	Four-five candidates (the best known being Cyg X-1). At least two intensity states for Cyg X-1. Chaotic flux variations down to ms time scale. Power-law spectra with Wien high energy tails. Comptonization of low energy photons by high energy electron clouds consistent with observed spectra.

Supernova Remnants

Class (§)	Energy band (§§) (keV)	X-ray luminosity ($\times 10^{38}$ erg/s)	X-ray features
a. Shell-like	0.5-25 keV	<10	Emission from the external rim of the remnant. Thermal energy spectra with at least two components (one with kT$\lesssim$1 keV, the other with kT$\gtrsim$1 keV). Superimposed on continuum ion-emission lines (S, Au, Ca, Fe). Spectra such as those expected from non-ionization equilibrium plasmas (electron temperature $T_e > T_{ion}$) with some inconsistencies.
b. Crab-like (plerions)	0.1-300 keV	10^{-6}-3	The emission fills the whole remnant, including the centre. Non thermal spectra (power law) up to highest X-ray energies consistent with syncrotron emission.

(§) with reference to optical properties
(§§) where is the bulk of X-ray flux

TABLE 2b. Classes of optically identified extragalactic X-ray sources

Class (§)	Energy band (§§) (keV)	X-ray luminosity ($\times 10^{38}$ erg/s)	X-ray features
Normal Galaxies	0.1–10 keV	$1-10^2$	X-ray emission mainly studied below few keV with the HEAO-2 satellite. X-ray spectra roughly known. X-ray luminosity related to radio/optical luminosity.
Active Galactic Nuclei			
a. Seyfert Galaxies	0.5–200	10^4-10^7	Power law energy spectra with mean spectral index ~ 0.7 up to $\gtrsim 100$ keV. Poorly known at higher energies. Flux variability down to hundred seconds. Flare-like activity, isolated outbursts, different intensity states.
b. BL Lac Objects	0.5–200	10^5-10^8	Power law spectra with different slopes depending on energy band. Strong time variability such as Seyfert galaxies.
c. Quasi Stellar Objects	0.5–200	10^5-10^{11}	Power law energy spectra with slope 0.4–0.5. Flux time variable.
Clusters of Galaxies	<30 keV	10^3-10^7	Thermal bremsstrahlung spectra with $2 \lesssim kT \lesssim 10$ keV due to hot gas among galaxies of the cluster. Evidence of variable kT with distance from centre of the cluster. Visible $K\alpha$ and $K\beta$ Iron lines.

(§) with reference to optical classes
(§§) where is the bulk of X-ray flux

point. The fusion energy that can be extracted for conversion of a quantity m of hydrogen (the most abundant element in the Universe) to helium is given by:

$$\Delta E_F = 6.3 \times 10^{18} \left(\frac{m}{1g}\right) \text{ erg}$$

Instead the gravitational potential energy released by accretion of a mass m on to the surface of a body of mass M_x and radius R_x si given by

$$\Delta E_{acc} = 1.3 \times 10^{20} \left(\frac{M_x}{M_\odot}\right) \left(\frac{R_x}{R_6}\right)^{-1} \left(\frac{m}{1g}\right) \text{ erg}$$

where: $M_\odot = 1.989 \times 10^{33}$ g is the solar mass and $R_6 = 10^6$ cm (typical radius of a neutron star). The ratio $\Delta E_{acc}/\Delta E_F$ depends on M_x/R_x (compactness of the object). $\Delta E_{acc}/\Delta E_F$ is equal to about 70 in the case of mass accretion onto a black hole ($R_x = 2GM_x/c^2$ where G = is the gravitation constant and c is the velocity of light), it is about 30 in the case of a typical neutron star (R_x = 10 Km, $M_x = 1.4 M_\odot$), it is about 0.02 in the case of a typical white dwarf ($R_x = 10^3$ km, $M_x = 1 M_\odot$), and it is about 3×10^{-4} in the case of a Sun-like star ($R_\odot = 6.96 \times 10^5$ km is the solar radius). This means that the mass accretion process is a more powerful energy source than nuclear fusion when the source compactness is high. A high compactness of most of the X-ray sources is in agreement with their short time variability (see Tables 2a and 2b). Note that the source size $\Delta \ell \simeq c \Delta t$ where c = vacuum light velocity and Δt = variation time scale.

The conversion of gravitational energy into electromagnetic radiation is in agreement with the X-ray luminosities observed from compact objects:

$$L_x \simeq 7 \times 10^{36} \left(\frac{M_x}{M_\odot}\right) \left(\frac{\dot{M}}{10^{-9} M_\odot/yr}\right) \left(\frac{R_x}{R_6}\right)^{-1} \text{ erg/s}$$

where $\dot{M}$ is mass accretion rate per year. It is also consistent with the temperature of the emission region:

$$T_{bb} \simeq 10^7 \left(\frac{L_x}{L_{37}}\right)^{1/4} g^{-1/4} \left(\frac{R_x}{R_6}\right)^{-1/2} {}^\circ K$$

under the hypothesis that the X-ray radiation has a blackbody spectrum, where $L_{37} = 10^{37}$ erg/s and f = fraction of stellar surface responsible of X-ray emission (in the case of an X-ray pulsar f ≪ 1). Finally the observations are consistent with the Eddington luminosity $L_E \simeq 1.3 \times 10^{38} (M_x/M_\odot)$ erg/s which gives the maximum luminosity of objects accreting mass when gravitational attraction which drives the mass accretion is balanced by outward radiation pressure[7].

However not all the celestial objects emit X-rays via mass accretion. On the basis of the observational features (Tables 2a and 2b), we can state that this mechanism is operative in X-ray binaries, in normal galaxies and in active galactic nuclei.

Thermonuclear fusion is expected to work in the X-ray burst sources and when there is a nova outburst. Nuclear fusion is also the mechanism that precedes the supernova explosion with the synthesis of heavy elements such as Ni, Fe, etc. The recent explosion of the supernova SN1987a in the Large Magellanic Cloud, an extraordinary event at such a small distance from us (55 kpc), will allow the better understanding of supernova phenomenon. X-rays have already been observed from SN1987a[8,9]. A probable origin is downwards Compton scattering of Co^{56} γ-ray photons produced in the explosion with the expanding ejecta. However supernova explosions are also the stellar events in which compact objects (neutron stars, degenerate dwarfs, black holes) can be formed. This means that X-rays can also be produced at the birth of such objects (in the case of SN1987a a neutron star or possibly a black hole). In the next few years from the time behaviour of its X-ray emission we can establish if a collapsed object was born in the SN1987a explosion.

Other energy sources of X-ray emission are conversion of rotational energy in the case of X-ray isolated pulsars inside supernova remnants (e.g. the pulsar NP0532 in the Crab Nebula), conversion of kinetic energy of the material expelled in a supernova explosion for its interaction with the interstellar medium in the case of supernova remnants, high temperature gas (10^8 °K) that could be produced by galactic explosions in the case of clusters of galaxies.

3. THE SAX MISSION

The SAX mission is a scientific program for the most part supported by the Piano Spaziale Nazionale managed by CNR. A consortium of scientific institutes has the scientific responsibility for the mission. The Italian hardware Institutes are: Istituto di Fisica Cosmica e Tecnologie Relative, CNR, Milano; Istituto di Tecnologie e Studio Radiazioni Extraterrestri, CNR, Bologna; Istituto di Astrofisica Spaziale, CNR, Frascati; Istituto di Fisica Cosmica e Applicazioni Informatiche, CNR, Palermo. There are also contributions from Dipartimento di Fisica, Università, Milano; Istituto Astronomico Università, Roma I; Istituto di Fisica, Università, Palermo; Dipartimento di Fisica, Università, Ferrara. Also involved in the SAX program are the Laboratory for Space Research, Utrecht, Holland, and the Space Science Department of the European Space Agency (ESA). The main industrial contractor for the production of SAX is Aeritalia, Torino.

3.1. Observational objectives and Payload Configuration

The observational goals of the SAX mission can be summarized as follows:

a. to achieve the same limiting sensitivity as the *Einstein* Observatory (Table 1) but in a broader energy range (0.1-10 keV). This will be accomplished by using 4 X-ray concentrators (C/S) with 1' imaging capabilities;

b. to improve the energy resolution by a factor~ 2 in the 1-10 keV energy band and by a factor from 2 to 5 in the 10 to 80 keV band with respect to the instruments flown in previous missions. This goal will be accomplished by using gas scintillator proportional counters (GSPC) instead of traditional proportional counters as detectors in the focal plane of C/S, and a direct-viewing high pressure GSPC (HP-GSPG) detector in the 3-120 keV energy band;

c. to improve the limiting sensitivity in the 15-200 keV energy band by a factor >10 with respect to the UCSD/MIT high energy instrument (A4) on board HEAO 1 (see Table 1). This will be accomplished by using a low background, high efficiency *PHOS*phor sand*WICH* X-ray detector PDS (= *P*hoswich *D*etection *S*ystem) and by choosing an almost equatorial orbit for the SAX satellite;

d. to achieve more balanced sensitivities in the low (<10 keV) and in the high (>10 keV) energy bands with respect to the previous missions;

e. to transmit the data with high time resolution (<50 μs);

f. to improve the limiting sensitivity to source lines superimposed on the X-ray continuous spectra of the celestial sources by a factor of about 10 with respect to the average sensitivities obtained to date, with the additional advantage of a better energy resolution (see b. above);

g. to continuously monitor wide sky fields at a sensitivity level of ~1 mCrab in order to discover new X-ray sources and to study transient phenomena. This will be accomplished by using 2 Wide Field Cameras (WFC). Each consists of a position sensitive proportional counter coupled with a pseudo-random mask, to achieve a few arc minutes imaging capability.

We expect the following limiting sensitivities in X-ray observations of 10^5 s duration: i. ~4 × 10^{-15} erg/cm^2s keV (~1 μCrab) in 1-10 keV; ii. ~2 × 10^{-14} erg/cm^2s keV (~0.1 mCrab) in 5-80 keV; iii. ~1 × 10^{-13} erg/cm^2s keV (~1 mCrab) in 30-200 keV.

These figures are referred to Crab Nebula-type spectra. The main features of each instrument are summarized in Tables 3, 4, 5 and 6. The narrow field instruments, C/S, HP-GSPC and PDS, will have their axes coaligned, while the two WFCs will have their axes in opposite directions and at 90° from the narrow field instruments.

3.2. Satellite characteristics

A complete sketch of SAX is shown in Fig. 3. Its main characteristics are: dimensions: 2.85 m diameter × 2.8 m height; total mass: 1200 kg; payload mass:

TABLE 3. Characteristics of Concentrators (C/S)

Mirror System (4 identical units)

Double cone approximation to the Wolter-I Configuration
- Mirror material   Ni, Au coated
- Technique for mirror production   Au evap. over Ni electroformed
- Length of a single cone section   15 cm
- Number of nested mirror/unit   30
- Aperture of the outermost mirror   16.2 cm
- Aperture of the innermost mirror   6.0 cm
- Grazing incidence angle (external)   0.62°
- Grazing incidence angle (internal)   0.23°
- Geometrical collecting area/unit   120 cm^2/unit
- Focal length   185 cm
- Optics on axis resolution (half power radius)   1 arcminute
- Effective coll. area (on axis)/unit
    1 keV   94 cm^2
    8 keV   46 cm^2
- Field of view (FWHM) at 1 keV   30 arcminute

Focal plane detectors GSPC with 25 μm Be window (3 units)
- Effective area/unit:
    6 keV   60 cm^2
    8 keV   41 cm^2
- Energy resolution   0.08/ $\sqrt{(E/6\ keV)}$ keV
- Position resolution   0.5 / $\sqrt{(E/6\ keV)}$ mm
                                         equivalent to 0.9 / $\sqrt{(E/6\ keV)}$ arcmin
  Background rejection   99%
  Residual background   0.002 counts/cm^2 s keV

Focal plane detector GSPC with 1 μm Polyprop.window (1 unit)
- Effective area:
    0.25 keV   60 cm^2
    0.61 keV   32 cm^2
    1   keV   66 cm^2
    > 5 keV   as for Be window

TABLE 4. SAX-HPGSPC Main Characteristics

Exposed geometrical area		450 cm^2
Pressure of Xenon		5 atm
Depth of drift region		10 cm
Acceptance window		Be/900 μm
Field of view		1° (collimated)
Energy range		3-120 keV
Effective area:	3 keV	60 cm^2
	6 keV	320 cm^2
	60 keV	280 cm^2
	100 keV	160 cm^2
Energy resolution:	$\Delta E/E$	0.25 $\sqrt{E(keV)}$
		(0.03 at 60 keV)
Background rejection		99%
Residual background:	E < 35 keV	10^{-3} counts/cm^2 s keV
	E > 35 keV	10^{-4} counts/cm^2 s keV
	(via detection of the K-fluorescence photon)	

TABLE 5. Characteristics of the Phoswich Detector System (PDS)

Exposed geometrical area		800 cm^2
Energy range		15-200 keV
Field of view		1.4° (collimated)
Effective area:	20 keV	680 cm^2
	60 keV	680 cm^2
	100 keV	500 cm^2
	200 keV	140 cm^2
Energy resolution ($\Delta E/E$)		1.40 $\sqrt{E(keV)}$
Residual background:	30-40 keV	2.3×10^{-4} counts/cm^2 s keV
	40-80 keV	1.7×10^{-4} counts/cm^2 s keV
	80-200 keV	1.1×10^{-4} counts/cm^2 s keV

TABLE 6. WFC Main characteristics

Number of units		2
Effective area per unit		250 cm^2 (through mask)
Energy range	Full field	2-10 keV
	Center of field	2-30 keV
	Timing (no image)	2-35 keV
Field of view		20°x20°
Angular resolution		5'
Energy resolution:		20% at 6 (keV)
Sensitivity per unit in 10^4 sec		1 mCrab

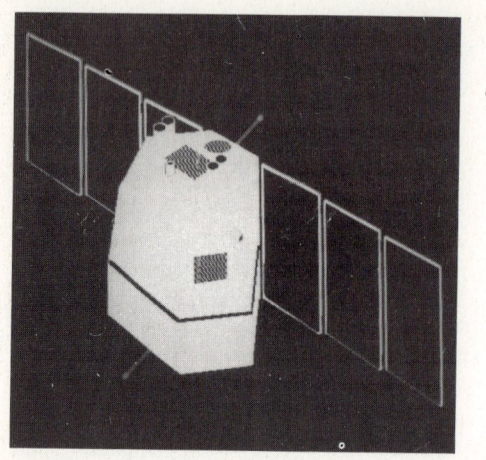 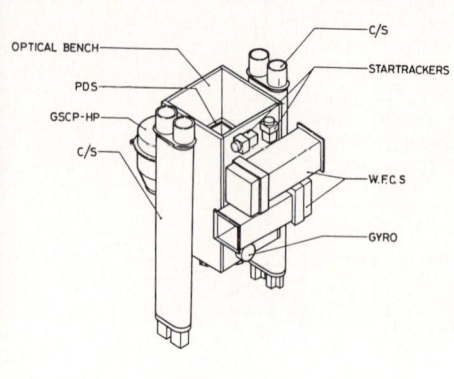

FIGURE 3

Overall view of the SAX satellite (left) and payload configuration (right)

385 kg; orbital characteristics: 2° inclination, circular orbit, 600 km altitude; life time: two years minimum; power: 1,600 watts at end of life; on board memory: 300 Mbit; attitude: three-axis-stabilized within 1', sun pointing within $\pm 30°$.

3.3. Scientific objectives

Some of the scientific objectives of the SAX mission, subdivided according to the classes of celestial objects shown in Tables 2a and 2b can be summarized as follows.

a. *High Mass X-Ray Binaries*

Most of these sources are X-ray pulsars (~30). The main goals will be that of an accurate measurement of their energy spectra in the broad (0.1-200 keV) energy range of SAX, the study of the dependence of their features on the pulse phase and on the phase of the neutron star along the binary orbit, the study of the chaotic noise (non poissonian) in the pulsar period and in the flux with time. We expect to discover new cyclotron lines.

b. *Black hole candidates*

The main goals are the study of the chaotic short time variability (0.1 ms time scale) of the flux in different energy bands with their mutual correlations, and the accurate measurement of the energy spectra of black hole candidates up to 200 keV. We expect to discover new features in the spectra thanks to our better energy resolution. The PDS instrument will prove very important in the study of the environment close to the hole.

c. *Low Mass X-ray binaries*

One important goal will be that of studying the quasi periodic oscillations (frequency, amplitude), simultaneously at low (<10 keV) and high (>10 keV) energies, in particular the time delay of the hard X-rays with respect to the soft X-rays. Another goal will be that of extending the measurements of the energy spectra to hard X-rays, in order to better understand the origin of the high energy X-ray spectra from these sources.

d. *Transient Sources*

One important goal will be that of monitoring sky fields with transient sources in order to localize them when they brighten.

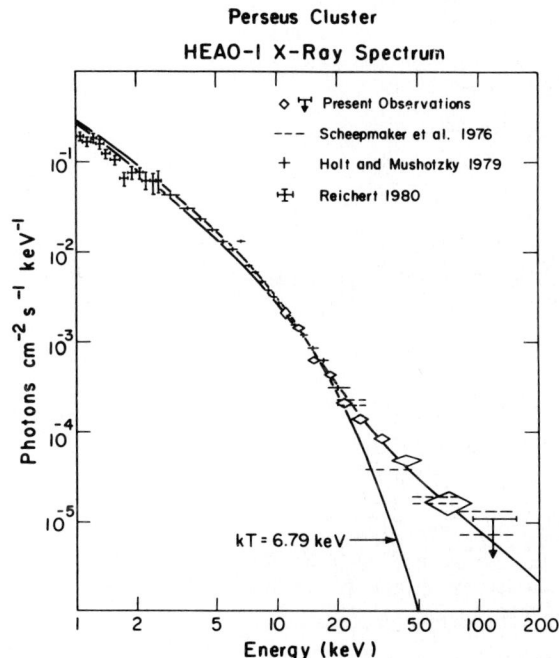

FIGURE 4

X-ray spectrum of the Perseus cluster of galaxies (reprinted from Primini et al.[12]).

e. *Supernova Remnants (SNR)*

The most important goal will be that of accurate spectroscopy (lines, continuum) up to 10 keV of nebular regions spatially resolved, in order to discriminate among different models now under discussion[10]. Other goals are the measurement of spectral features of SNR of different ages for evolution studies, and the study of SNR known at radio wavelengths to derive SNR distances.

f. *Active Galactic Nuclei*

The most important goals are:

i. Accurate measurements of their spectra up to 200 keV for a conspicuous sample of sources. This can help to solve the open question on the origin of the X-ray diffuse background.

ii. Cross-correlation between intensity variations below 1 keV and slope variation of the energy spectra above 10 keV. This will help in understanding the origin of the variations (absorption effects, X-ray production mechanism variation).

g. *Clusters of Galaxies*

The scientific goals will be the following:

i. Study of the X-ray spectrum of cluster regions spatially resolved. This can be accomplished for about 40 nearby clusters and will give further hints on dark matter.

ii. Study of the evolution of the cluster energy spectra with distance (up to redshifts of ~1).

iii. Study of hard (>20 keV) X-ray tails observed in some clusters in order to understand their origin (diffuse emission or AGNs in the cluster?). One example of the energy spectrum of a cluster of galaxies, Perseus X-1, with hard tail is shown in fig. 4.

REFERENCES

1) R. Giacconi, H. Gursky, F. Paolini, and B. Rossi, Phys. Rev. Lett. 9 (1962) 439.
2) K.S. Wood et al., The Astrophys. J. Suppl. 56 (1984) 507.
3) R. Giacconi et al., The Astrophys J. 230 (1979) 540.
4) A.M. Levine et al., The Astrophys. J. Suppl. 54 (1984) 581.
5) T.T. Hamilton and D.J. Helfand, The Astrophys. J., in print.
6) E.A. Boldt, NASA Technical Memorandum No. 78106 (1978).
7) J. Frank, A.R. King and D.J. Raine, Accretion Power in Astrophysics (Cambridge University Press, Cambridge, 1985).
8) R. Sunyaev et al., Nature 330 (1987) 227.
9) T. Dotani et al., Nature 330 (1987) 230.
10) B. Aschenbach, Space Science Reviews 40 (19857 447.
11) F. Frontera et al., Balloon observation of hard X-ray sources, in: Proc. of the 3rd Cosmic-Physics Nat. Conf., ed. C. Castagnoli (SIF, Bologna, 1987) pp. 159-164.
12) F.A. Primini et al., The Astrophys J. Lett. 243 (1981) L13.

GALLEX: AN EXPERIMENT TO MEASURE THE SOLAR NEUTRINO FLUX
WITH RADIOCHEMICAL TECHNIQUES

A Brookhaven, Heidelberg, Karlsruhe, Milano, Munich, Nice, Paris, Rehovot, Roma, Saclay collaboration.

Presented by R. Santonico
Dipartimento di Fisica, Universitá di Tor Vergata, Roma,
and INFN Sezione di Roma, Italy

1. INTRODUCTION

According to the standard solar model[1] about 90% of the energy produced by the sun is originated in the so called hydrogen or p-p cycle consisting of the following nuclear reactions :

$$p\,p \rightarrow D\,e^+\,\nu_e \quad \text{(2 times)}$$

$$D\,p \rightarrow {}^3He\,\gamma \quad \text{(2 times)}$$

$$^3He\,{}^3He \rightarrow {}^4He\,p\,p$$

wich can be summarized in the process

$$4p \rightarrow {}^4He\,2e^+\,2\nu_e\,2\gamma \qquad (1)$$

delivering a total energy

$$4M_p - M_{^4He} = 28.1 \text{ MeV}$$

and a pair of neutrinos whose energy distribution extends up to a maximum value

$$E_{max} = 420 \text{ KeV}$$

with an average value

$$<E_\nu> = 250 \text{ KeV}$$

It turns out from process (1) that for any produced neutrino, a "visible" energy

$$1/2(4M_p - M_{^4He} - 2<E_\nu>) = 13.8 \text{ MeV}$$

is irradiated. As a result, the solar neutrino flux Φ_ν is *strictly correlated* to the visible power W_v irradiated by the sun:

$$W_v = 3.9 \times 10^{33} \text{erg/sec} = 4\pi R^2 \Phi_\nu \times 13.8 \text{ MeV}$$

For $R = 1.5 \times 10^8$ Km, the earth-sun distance, it turns out

$$\Phi_\nu = 6.1 \times 10^{10} \text{neutrinos/cm}^2 \text{ sec}$$

In addition to the pp neutrinos, ν_{pp}, produced in the above process, there is futher neutrino emission from the following processes.

a) The berillium produced in the process

$$^4\text{He } ^3\text{He} \rightarrow {}^7\text{Be } \gamma$$

emits by electron capture

$$^7\text{Be e}^- \rightarrow {}^7\text{Li } \nu_e$$

monochromatic neutrinos of two different energies: the 384 KeV line (B.R.=10%) which merges in the ν_{pp} band and the 862 KeV line (B.R.=90%). The flux of the berillium neutrinos ν_{Be}, $\sim 8\%$ of the solar flux, is not so strictly correlated to the sun luminosity and is predicted with a $\sim 10\%$ error due to the uncertainty in nuclear cross sections and helium abundance.

b) The boron ^{8}B produced in the reaction

$$^7\text{Be p} \rightarrow {}^8\text{B } \gamma$$

decays with a lifetime $T_{1/2} = 0.77$ sec according to the process

$$^8\text{B} \rightarrow {}^8\text{Be } e^+ \nu_e$$

The boron channel produces the most energetic neutrinos (ν_B) irradiated by the sun, their energy distribution extending up to 14 MeV, but the corresponding flux is a negligible fraction of the ν_{pp} one, $\Phi_{\nu_B} \simeq 10^{-4} \Phi_{\nu_{pp}}$, and is predicted with a $\pm 17\%$ uncertainty.

c) Other neutrino sources (p-e-p neutrinos and C-N-O cycle) that will not be described in detail contribute only 2% of the total solar flux.

The solar neutrino spectrum according to the standard solar model is shown in fig. 1.

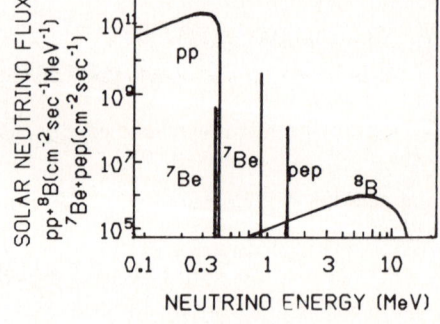

Fig. 1: Solar neutrino spectrum predicted by the Standard Solar Model.

2. THE CHLORINE EXPERIMENT[2]

It is the first experimental test of the standard solar model, based on the solar neutrino flux measurement with radiochemical techniques.

The transition

$$^{37}\text{Cl } \nu_e \rightarrow {}^{37}\text{Ar } e^- \qquad (2)$$

is induced by neutrinos of energy $E_\nu > 814$ KeV in a mass of 600t of C_2Cl_4 placed in the gold mine of Homestake (South Dakota). The produced argon is extracted and concentrated inside a proportional counter where it is identified from the ^{37}A decay.

The neutrino capture rate measured in this way is 2.0 ± 0.3 Solar Neutrino Unit (SNU) instead of 5.8 SNU as predicted by the standard solar model. Here the rate unity

$$1 \text{ SNU} = 10^{-36} \text{captures/target nucleus} \times \text{sec}$$

has been utilized.

This discrepancy, if the experiment is correct, can be explained by two hypotesis:

1) The standar model is incorrect. However, the discrepancy concerns mainly the boron neutrinos, due to the high energy threshold of the process (2). The model is readjustable.

2) The neutrino oscillation phenomenon

$$\nu_e \to \nu_\mu \quad \nu_e \to \nu_\tau$$

reduces the flux of electron neutrinos, the only ones which can induce the observed transition.

3. THE GALLIUM EXPERIMENT[3]

It is based on the observation of the process

$$\nu_e \, ^{71}\text{Ga} \to \, ^{71}\text{Ge} \, e^- \tag{3}$$

induced by the solar neutrinos and has the following advantages with respect to the chlorine experiment.

1) Due to the low energy threshold $E_\nu > 236$ KeV most of the observed processes are induced by the pp neutrinos.

2) The detector can be calibrated with neutrinos produced by a ^{51}Cr source. In this way, in particular, the cross-sections of the ν_{Be} neutrinos can be exactly measured.

A comparison of the ^{37}Ar and ^{71}Ge production rates is given in Table 1.

The experiment will be carried out in the Gran Sasso underground laboratory and is based on the following principle.

Table 1 — Neutrino fluxes and production rates according to the standard solar model[1b]

Reaction	B.R. (%)	Max E_ν (MeV)	Flux (10^{10}cm^{-2} sec^{-1})	^{37}Ar production rate (SNU)	^{71}Ge production rate (SNU)
$p + p \rightarrow D + e^+ + \nu_e$	99.75	0.42	6.1	0	70.2
$p + e^- + p \rightarrow D + \nu_e$	0.25	1.44	1.5×10^{-2}	0.24	2.5
$D + p \rightarrow {}^3He + \gamma$	100				
$^3He + {}^3He \rightarrow {}^4He + 2p$	86				
or					
$^3He + {}^3He \rightarrow {}^7Be + \gamma$	14				
$^7Be + e^- \rightarrow {}^7Li + \nu_e$	99.89	0.86 (90%) 0.38 (10%)	4.3×10^{-1}	0.95	27.0
or					
$^7Be + p \rightarrow {}^8B + \gamma$	0.11				
$^7Li + p \rightarrow 2\, {}^4He$	100				
$^8Be + e^+ + \nu_e$	100	14.06	4×10^{-4}	4.3	1.2
$^{12}C + p \rightarrow {}^{13}N + \gamma$					
$^{13}N \rightarrow {}^{13}C + e^+ + \nu_e$		1.20	5×10^{-2}	0.08	2.6
$^{13}C + p \rightarrow {}^{14}N + \gamma$					
$^{14}N + p \rightarrow {}^{15}O + \gamma$					
$^{15}O \rightarrow {}^{15}N + e^+ + \nu_e$		1.75	4×10^{-2}	0.24	3.5
$^{15}N + p \rightarrow {}^{12}C + {}^{14}He$					
Total: $4p + 2e^- \rightarrow {}^4He + 2\nu_e + 26.73$ MeV		14.06	6.63	5.8	107.0

A mass of 30t of natural gallium (39.6% ^{71}Ga and 60.4% ^{69}Ga) is accumulated in a tank as liquid GaCl$_3$. Solar neutrinos induce the transition (3) on ^{71}Ga producing ^{71}Ge nuclei at a rate of about 1 event/day. The ^{69}Ga isotope, due to its high energy threshold $E_\nu > 2.2$ MeV is almost insensitive to solar neutrinos.

The ^{71}Ge nucleus is unstable and decays in ^{71}Ga via K or L electron capture with a lifetime $T_{1/2}=11.4$ days so that after e.g. 2 weeks there are about 9 atoms of ^{71}Ge still accumulated in the tank. These produce germanium chloride GeCl$_4$ which is highly volatile and can be swept out from the solution by a circulating stream of a gas such as helium. Subsequently the germanium chloride is transformed in germane which is introduced in a gaseous proportional counter of 0.5 cm^3 active volume where the ^{71}Ge atoms are identified from the signal induced in the counter by the above mentioned decay. The decay modes of ^{71}Ge are listed in Table 2. A simplified sketch of the experimental set-up showing the gallium target and the ^{51}Cr source is given in Fig. 2.

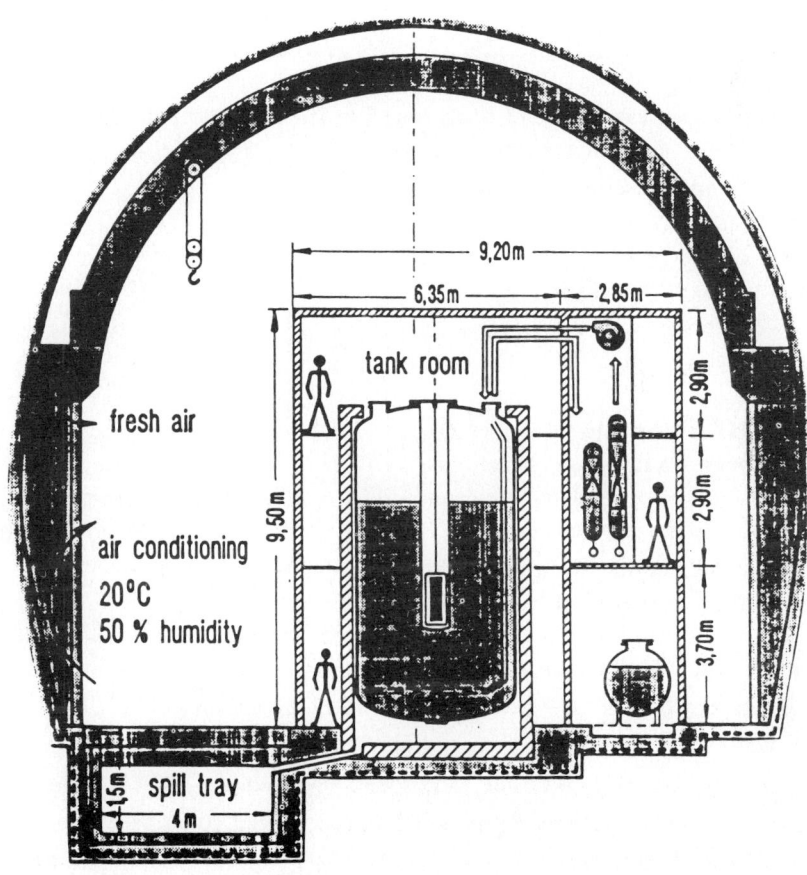

Fig. 2: Sketch of the experimental set-up showing the gallium target and the ^{51}Cr source

Table 2 — ^{71}Ge decay

Branching Ratio	Kind of capture	Auger electrons Energy (KeV)	X rays energy (KeV)
41.5	K	10.37	-
5.3	K	0.11	10.3
27.2	K	1.12	9.3
14.0	K	1.15	9.2
10.3	L	1.30	-
1.7	M	.16	-

The most important source of background in the proportional counter is due to β and γ emission of the materials of the counter.

The background discrimination is based on the following criteria:

1) The energy distribution shows two peaks at 1.2 and 10.4 KeV (L and K-capture) for the genuine events (fig. 3), and is flat or peaked at E=0 for the background.

2) The pulse rise time is a few nsec for the genuine events (the Auger electrons are freeded in a small volume) and ~100 nsec for the background (fig. 4).

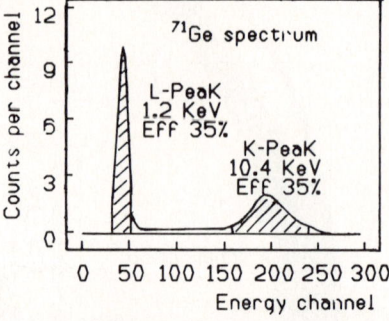

Fig. 3: Measured energy distribution at the occurrence of a ^{71}Ge decay in the proportional counter.

A test of background rate has been carried out during several months on three counters shielded with low activity lead and placed in the bypass n 12, about 0.5 Km far from the Gran Sasso laboratory. The measured background rate turns out:

0.7 events/day for the L − Peak (0.95KeV < E < 1.45KeV)

0.09 events/day for the K − peak (8.8KeV < E < 12KeV)

Based on this criterium a futher background rejection of ~ 90% is expected.

In addition to the spurious signals of the proportional counters another important kind of background is due to the ^{71}Ge production from other sources than solar neutrinos such as: 1) cosmic muons, 2) α particles generated by natural emitters contained in the $GaCl_3$ solution, 3) fast neutrons emitted by the rocks.

1) The cross section for the muon-induced transition $^{71}Ga \rightarrow {}^{71}Ge$ has been measured at Fermilab with muons of 8.5 and 225 GeV energy. The cosmic muon flux in the

Gran Sasso laboratory has also been measured and turns out to be ~ 1 muon/m^{2}h. The resulting background rate has been estimated to be $\sim 10^{-2}$ events/day.

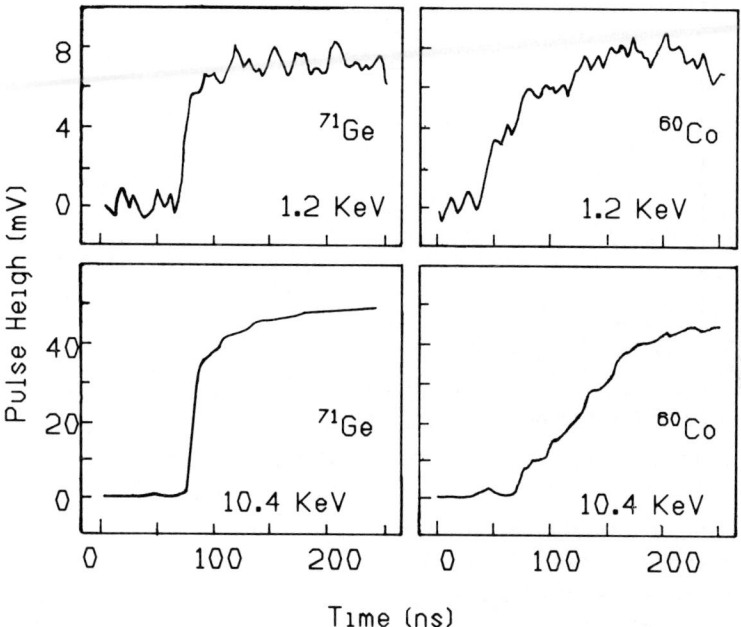

Fig. 4: Pulse shape of proportional counter pulses

2) A 10^{-9} thorium or 10^{-7} uranium concentration inside a 50t Gallium target produce ^{71}Ge atoms at rate of 10^{-2}/day. The same effect is caused by a ^{226}Ra activity of 0.2 pCurie/Kg. It has been shown that the target activity can be kept under the above values.

3) The fast neutron background in the Gran Sasso laboratory has been measured by several groups and turns out to be of the order of 10^{-6}n/cm^2 sec. The corresponding ^{71}Ge production rate is $\sim 10^{-3}$ atoms/day.

4. CALIBRATION OF THE DETECTOR

A ^{51}Cr source will be used to calibrate the detector[3c]. This calibration allows in particular to measure the cross-sections of the berillium neutrinos whose values are not exactly known.

The ν_{Be} neutrinos are monoenergetic ($E_\nu = 862$ KeV) and can produce, in addition to the ground state, also the 170 and 500 KeV excited states of the ^{71}Ge nucleus. The same states can be also produced by the neutrinos emitted in the ^{51}Cr decay (see fig.5) via K electron capture

$$^{51}\text{Cr } e^- \to {}^{51}\text{V } \nu_e$$

whose energy is 764 KeV.

If the absorption cross section of ^{51}Cr neutrinos in the gallium target is measured, the ν_{Be} neutrino cross sections can be obtained from the ratios

$\sigma_{\nu_{Be}}/\sigma_{\nu_{Cr}}=1.25$ for the ground state

$\sigma_{\nu_{Be}}/\sigma_{\nu_{Cr}}=1.30$ for the 170 KeV excited state

$\sigma_{\nu_{Be}}/\sigma_{\nu_{Cr}}=1.57$ for the 500 KeV excited state

which have been exactly calculated.

A $0.6 \div 0.8$ MCurie neutrino source will be obtained through the process ^{50}Cr$(n,\gamma)^{51}$Cr irradiating a 125 Kg cromium mass at the neutron beam of the Siloe Grenoble reactor. With the purpose of reducing the irradiation time the cromium will be enriched with ^{51}Cr isotope which constitutes only the 4.4% of the natural cromium.

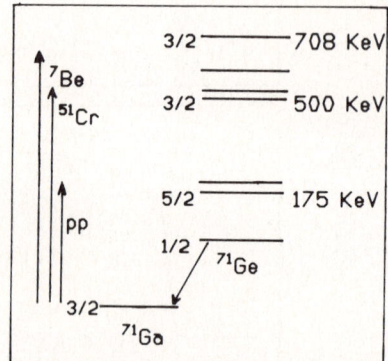

Fig. 5: Energy levels of the ^{71}Ge nucleus and scheme of its decay to ^{71}Ga.

The detector will be calibrated by submerging the neutrino source in the GaCl$_3$ target and counting the produced ^{71}Ge atoms. As the lifetime of the ^{51}Cr isotope is $T_{1/2} = 28$ day a reasonable strategy would be to extract the ^{71}Ge atoms every e.g. 10 day. The number of ^{71}Ge atoms at any extraction is listed in the following, assuming a 0.6 MCurie source activity:

1st extraction 14.5 ^{71}Ge atoms   2nd extraction 11.3 ^{71}Ge atoms

3rd extraction 8.8 ^{71}Ge atoms   4th extraction 6.8 ^{71}Ge atoms

5th extraction 5.3 ^{71}Ge atoms   6th extraction 4.1 ^{71}Ge atoms

Note that due to its γ-activity of ~ 60 KCurie (E_γ=320 KeV) the source has to be screened with a 10 cm thick lead shield.

5. INTERPRETATION OF THE RESULTS

The gallium experiment result will be particularly clear in the two following cases.

A measured rate of ~ 110 SNU would be a convincing demonstration that the standard solar model is correct and the anomalous result from the chlorine experiment concerns only the boron channel which is responsible for a negligeable fraction of the solar neutrino flux as specified above.

On the other hand, if the measured rate is < 70 SNU, which is the value expected from the contribution of the ν_{pp} neutrinos alone (see Table 1) this would strongly suggest that the oscillation phenomenon is responsible for the missing neutrinos.

6. NEUTRINO OSCILLATIONS

According to a Pontecorvo[4] suggestion, if the neutrinos are massive and the lepton family number is not conserved, the ν_e ν_μ neutrinos may differ from the mass eigenstates ν_1 ν_2 of mass m_1 and m_2 respectively. In this case the flavour and mass eigenstates are correlated by the following relationships:

$$|\nu_e> = \cos\theta \, |\nu_1> + \sin\theta \, |\nu_2>$$

$$|\nu_\mu> = -\sin\theta \, |\nu_1> + \cos\theta \, |\nu_2>$$

where θ is the mixing angle. If $0 \leq \theta \leq 45°$, as it is normally assumed, ν_e is "mostly" ν_1. Here, for reasons of simplicity, only two neutrinos have been considered.

The $|\nu_e>$ $|\nu_\mu>$ states are not stationary. The oscillation phenomenon is due to the time dependent phase difference ϕ between the ν_1 and ν_2 components:

$$d\phi/dt = E_1 - E_2 = \sqrt{p^2 + m_1^2} - \sqrt{p^2 - m_2^2}$$

in units $\hbar = c = 1$

If the masses m_1, m_2 are small with respect to the momentum p

$$d\phi/dt \simeq (m_1^2 - m_2^2)/2p$$

The probability that a ν_e neutrino emitted from a given source will be detected still as ν_e at a distance L is:

$$P(\nu_e \to \nu_e) = 1 - \sin^2 2\theta \cdot \sin^2(\pi L/l_v) \qquad (4)$$

where l_v is the (vacuum) oscillation length

$$l_v = 4\pi p/\Delta m^2 \quad \Delta m^2 = m_1^2 - m_2^2$$

The radiochemical experiments exposed above have high sensitivity in Δm^2, due to the large sun-earth distance but low sensitivity in $\sin^2 2\theta$ due to the limited statistics. Fig. 6 shows the probability $P(\nu_e \to \nu_e)$ vs. Δm^2 for solar neutrinos detected in the Gallium(a) and Chlorine(b) experiments[5].

When neutrinos propagate in matter (instead of in vacuum as supposed until now) a different oscillation mechanism may appear due to the different elastic scattering of ν_e and ν_μ on matter electrons (Wolfenstein [6]). For the scattering $\nu_e\ e^- \to \nu_e\ e^-$ indeed, both, W and Z^0 exchange graphs are operative (Fig. 7a and b); for the scattering $\nu_\mu\ e^- \to \nu_\mu\ e^-$ only the latter is possible (Fig. 7c).

The $\nu_e \to \nu_e$ transition probability in matter is given by

$$P(\nu_e \to \nu_e) = 1 - \sin^2 2\theta_m \cdot \sin^2(\pi L/l_m)$$

with

$$\tan\theta_m = \sin 2\theta/(\cos 2\theta - l_v/l_0)$$

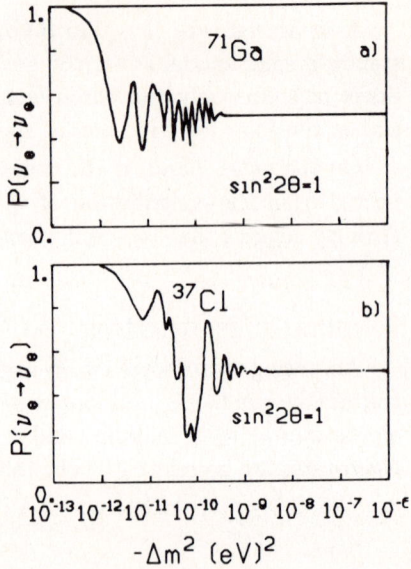

Fig. 6: Probability $P(\nu_e \to \nu_e)$ vs Δm^2 for solar neutrinos in vacuum detected in the Gallium and Chlorine experiments.

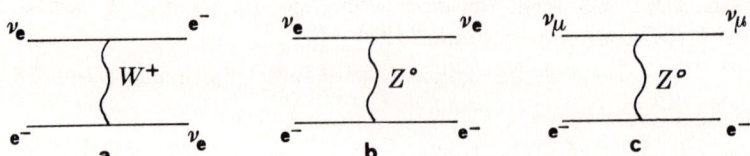

Fig. 7: Diagrams for $\nu_e\ \nu_\mu$ neutrinos elastic scattering

The oscillation length in matter is

$$l_m = l_v/[1 - 2(l_v/l_0)\cos 2\theta + (l_v/l_0)^2]$$

with

$$l_0 = 2\pi/\sqrt{2}\ G\ N$$

where N is the electron density and G the Fermi coupling constant.

If $m_1 < m_2$ the ν_e neutrinos propagating in the solar matter can be dramatically suppressed[7] for some values of Δm^2, even for very small values of the mixing angle ($\sin 2\theta \ll 1$). This absorption effect, for the gallium experiment, is shown in fig. 8.

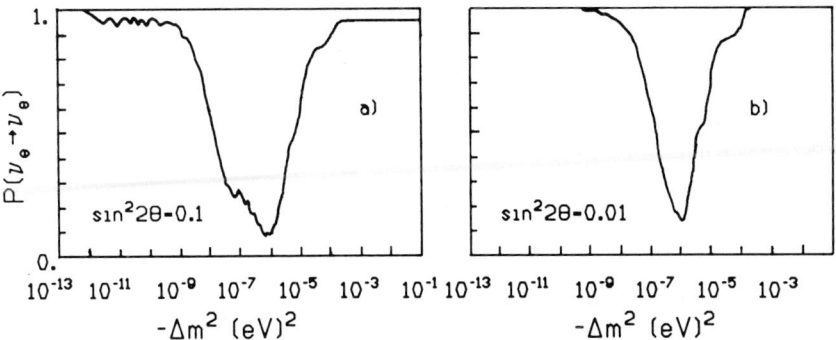

Fig. 8: Probability $P(\nu_e \to \nu_e)$ vs. Δm^2 for Gallium experiment[5b] taking into account matter effects.

References

[1] (a) J.N.Bahcall et al., Rev.Mod.Phys.**54**(1982)767

(b) J.N.Bahcall et al., Astrophys.J.**292**,L79(1985)

[2] (a) R.Davis et al., Phys. Rev. Letters **20**(1968)1205

(b) R.Davis et al., "Underground science at Homestake", in Science Underground, AIP Conference Proc. n.**96** (1982)p.2

(c) J.K.Rowley, B.T.Cleveland, R.Davis, in: "Solar neutrinos and neutrino astronomy. Homestake(1984), AIP Conf. Proc. n**126**, p.1

[3] (a) W.Hampel, "The Gallium Solar Neutrino Detector", in Science Underground, AIP Conference Proc. n**96**(1982)p.88

(b) T.Kirsten, "The Gallium Neutrino Experiment", paper presented at RIS 1984, Knoxville, Tenn.,16-20 April 1984

(c) M.Cribier et al. Internal report 13/Dec/1984. Département de Physique des Particules Elémentaires. Centre d'Etudes Nucléaires de Saclay.

[4] B.Pontecorvo: Sov.Phys.JETP **26**, 984(1968); V.Gribov, B.Pontecorvo: Phys. Lett. **28B**, 493(1969); For a comprehensive review of the formalism see for example S.M.Bilenky, B.Pontecorvo: Phys.Rep. **41**, 225(1978)

[5] (a) W.Hampel: The gallium solar neutrino detector and neutrino oscillation. In:Proc. of the Int. Conf. on Neutrino Physics and Astrophysics, Nordkirchen, p.530 (1984) ed. K.Kleinknecht, E.A.Paschos.

(b) J.Boucher et al., Z.Phys.C-Particles and Fields **32**, 499-511(1986)

[6] L.Wolfenstein: Phys. Rev. **D17**,2369(1978)

[7] S.P.Mikheyev, A.Yu Smirnov: Yad. Fiz. **42**, 1441(1985); Sov.J.Nucl.Phys. **42**, 913(1985); Nuovo Cimento **9C**,17(1986); A.Yu Smirnov: Proc. of the Moriond Workshop on Massive Neutrinos in Particle Physics and Astrophysics, Tignes,(1986)

THE LVD EXPERIMENT

G. Bari, M. Basile, G. Bruni, G. Cara Romeo, A. Castelvetri, L. Cifarelli, A. Contin, C. Del Papa, P. Giusti, G. Iacobucci, G. Maccarrone, T. Massam, R. Nania, V. O'Shea, F. Palmonari, E. Perotto, G. Sartorelli, M. Willutzky

University of Bologna and INFN Bologna, Italy.

M. Aryal, K. De, A.M. Shapiro, M. Widgoff

Brown University, Providence, Rhode Island 02912, USA

J.A. Chincellato, C. Dobrigkeit Chincellato, A.C. Fauth, A. Turtelli

Department of Rajos Cosmicos, University of Campinas, Brazil.

F. Rohrbach, A. Zichichi,

CERN, 1211 Geneva 23, Switzerland

L. Caputi, G. Susinno

Calabria University, Cosenza and INFN-LNF, Italy

G. Barbagli, G. Conforto, G. Landi, P. Pelfer

University of Florence and INFN Florence, Italy

G. Anzivino, S. Bianco, R. Casaccia, F. Cindolo, M. Defelice, Y. Dong, M. Enorini, F.L. Fabbri, C. Jing, I. Laakso, S. Qian, A. Rindi, Z. Shi, A. Spallone, Y. Sun, L. Votano, A. Zallo

INFN National Laboratories of Frascati, Rome, Italy

K. Lau, F. Lipps, B. Mayes, G.H. Mo, L. Pinsky, J. Pyrlik, D. Sanders, W.R. Sheldon, R. Weinstein

University of Houston, Houston, TX 77004, USA

Y. Dai, L. Din, G. Jing, Z. Lu, P. Shen, Q. Zhu

IHEP, Beijing, China

D. Alyea

Indiana University, Bloomington, IN 47401, USA

T. Kitamura

Institute for Science and Engineering, Kinki University, Higashi-Osaka, Japan

Y. Minorikawa

Department of Physics, Kinki University, Higashi-Osaka, Japan

G. Di Sciascio, R. Scrimaglio

University of L'Aquila, Italy

P. Rotelli

University of Lecce, Italy

G.E. Kocharov, V. Vasileyev

Ioffe Physical Technical Institute, Leningrad, USSR

M. Deutsch, E.S. Hafen, P. Haridas, B. Jeckelmann, G. Ji, H.H. Huang, C.S. Mao, A. Pitas, I.A. Pless, S.W. Wang, Y.R. Wu, Y.R. Yuan, C.Z. Zhao

Massachusetts Institute of Technology, Cambridge, Massachusetts 02139, USA

V.S. Berezinsky, V.L. Dadykin, F.F. Khaichukov, E.V. Korolkova, P.V. Kortchaguin, V.B. Kortchaguin, V.A. Kudryavtsev, A.S. Markov, V.G. Ryassny, O.G. Ryazhskaya, V.P. Talochkin, V.F. Yakushev, G.T. Zatsepin

Institute for Nuclear Research, Moscow, USSR

J. Moromisato, E. Saletan, D. Shambroom, E. von Goeler

Northeastern University, Boston, MA 02115, USA

N. Takahashi, I. Yamamoto

Department of Electrical Science, Okayama University of Science, Japan

T. Wada

Department of Physics, Okayama University, Okayama, Japan

G. D'Ali, S. De Pasquale

University of Palermo and INFN Bologna, Italy

B. Alpat, F. Artemi, C. Cappelletti, P. Diodati, P. Salvadori,

University of Perugia and INFN Perugia, Italy

A. Misaki, N. Inoue

Department of Physics, Saitama University, Japan

T. Hara

Institute for Cosmic Ray Research, University of Tokyo, Tanashi-shi, Japan

C. Aglietta, G. Badino, L. Bergamasco, C. Castagnoli, A. Castellina, G. Cini, M. Dardo, W. Fulgione, P. Galeotti, P. Ghia, C. Morello, G. Navarra, L. Periale, P. Picchi, O. Saavedra, G.C. Trinchero, P. Vallania, S. Vernetto

Institute of Cosmo-Geophysics, CNR, University of Turin, and INFN Turin, Italy

F. Grianti, F. Vetrano

University of Urbino, Italy

1. THE LVD EXPERIMENT

The large volume detector (LVD) can be defined as an underground observatory devoted to neutrino astronomy and to the measurement of the penetrating components of cosmic rays. The experiment has been optimized for major physics items such as the detection of the neutrino bursts of collapsing stars and the search for astrophysical point sources of ultra-high energy gammas and neutrinos. It can also be defined as a multipurpose experiment where many important issues, especially neutrino oscillations, standard cosmic ray physics, proton decay in the channel ($p \rightarrow K^+ + \bar{\nu}$), solar neutrinos and monopoles can be studied with varying degrees of sensitivity and on a competitive basis.

The experiment will be located in Hall A of the Gran Sasso Laboratory and will benefit from the very favourable conditions of the site: it is next to the Gran Sasso Rome-Adriatic motorway tunnel, permitting ease of heavy transport; it is only 130 km from Rome; it is at a depth of ~3600 mwe, which is an excellent compromise for the measurement of the penetrating components of cosmic rays; and, particularly important, geologically speaking, the Gran Sasso is low in radioactive elements (for instance, the gamma and neutron fluxes are from 5 to 10 lower than those measured in Mont Blanc).

2. THE LVD DETECTOR

The LVD detector[1] consists of a large volume of liquid scintillator interlayered with streamer chambers and is specifically designed to permit a multipurpose experiment. One of the major experimental objectives is the detection of mu mesons and the measurement of their trajectories and direction-of-flight; the muons can be either cosmic, produced in an atmospheric shower and with an energy of the order of a magnitude of TeV to be able to reach the apparatus, or muons induced by neutrino interactions in the rock surrounding the apparatus. Another important objective is the detection of neutrino interactions inside the detector, either for low energy neutrino interactions and the measurement of their energy, or the pattern identification of neutrino-induced events of higher energy.

Figure 1 is a general view of the LVD which basically consists of 190, 6.6m×2.1m×1.1m or 6.6m×2.7m×1.1m, identical modules inserted into an iron support structure[2]. They are arranged in 8(7) layers in each of five main towers which are divided by corridors transversal to the laboratory axis. Each module comprises

- ~9.6 tons of liquid scintillator divided into 8 modular counters (tanks) housed in Fe module "porta tanks";
- an L-shaped chamber module containing 80 limited streamer tubes on the bottom and on one side of the module, forming the tracking system.

The dimensions of the detector are 40m×12m×13m, and its total weight including the support structure and tracking system is ~3600 tons.

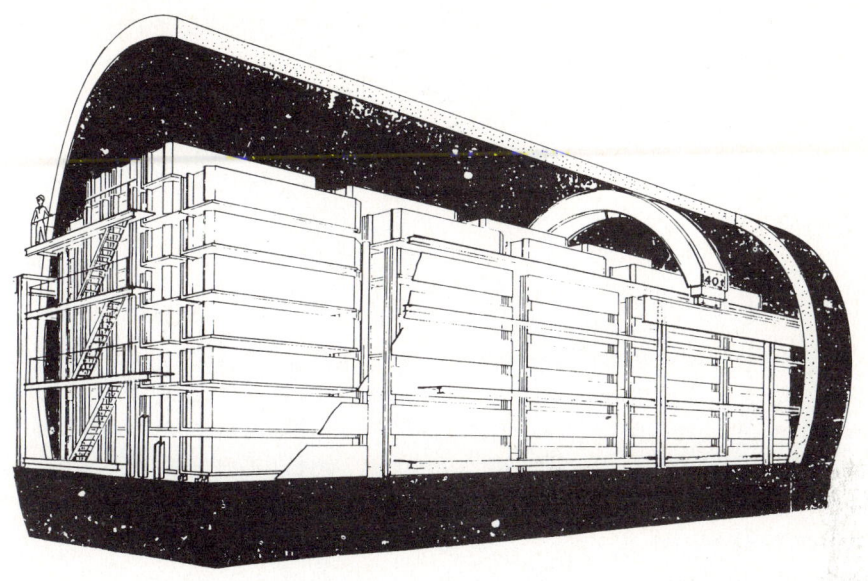

FIGURE 1
General view of the LVD

2.1 Scintillator system

The scintillator system consists of 1520, 1m×1m×1.5m stainless steel boxes filled with well-tested scintillator liquid with the following characteristics: structure C_nH_{2n+2} with n=10; density 0.8 g cm^{-3}; attenuation length 20m (λ=420nm); output light ~5 photoelectrons each PM for 1 MeV of energy loss; energy resolution 20%/$\sqrt{E}$.

Each tank is viewed by three phototubes for energy determination and time-of-flight measurements. The pulses of the three photomultipliers of one counter are amplified and discriminated at two different thresholds: a high level threshold of ~6 MeV and a low level one of 0.8 MeV. The high-level output of each discriminator is fed into a threefold coincidence which is the general trigger from the scintillators for the whole apparatus. The general trigger also opens a 500-μsec-gate where the second-level trigger for the low energy antineutrino detection operates. Steel structures house groups of eight tanks, as shown in Fig.2, and form part of the basic module of the experiment. The "porta tanks" also support the horizontal and vertical segments of the tracking chamber. The total mass of the liquid scintillator is 1800 tons.

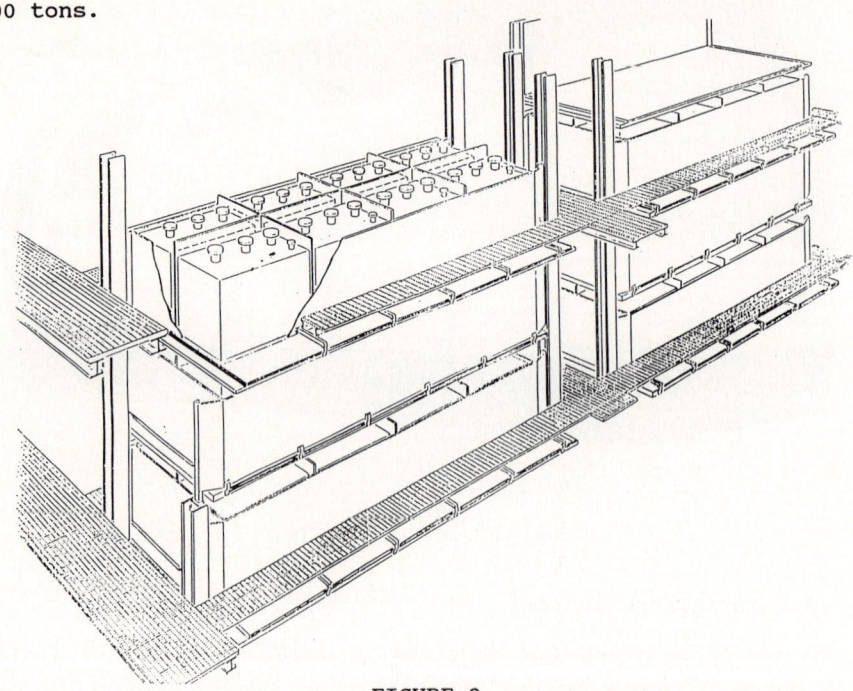

FIGURE 2
Detail of the LVD assembly system

2.2 Tracking system

The tracking system consists of 190 L-shaped chambers containing a double layer of limited streamer tubes. Each chamber is formed of two iron panels - a horizontal one and a vertical one - linked by hinges and fastened to each "porta tank"; thus, each basic module is surrounded by a double layer of streamer tubes (see Fig.3). The streamer tubes total 15000, arranged in 8(7) horizontal planes and 5 vertical planes, each one consisting of 8-wire PVC chambers, 6.3m long and 1cm×1cm in cross section. Each tube layer is equipped with a digital read-out in two coordinates and the signal is picked up by 4-cm-wide pick-up strips. The double layers are staggered by 2 cm. The resulting precision is ~0.5 degrees, while the efficiency of the double layer is ~100%.

Each chamber has 470 read-out channels and the complete tracking system contains ~80,000 channels which are each discriminated and fed into a shift register. A serial read-out chain is then performed by a Camac module. The tubes are filled with a standard (30/70) (Argon, Isobutane) mixture. The geometrical acceptance of the apparatus for an isotropic flux is 7000 m^{2}sr.

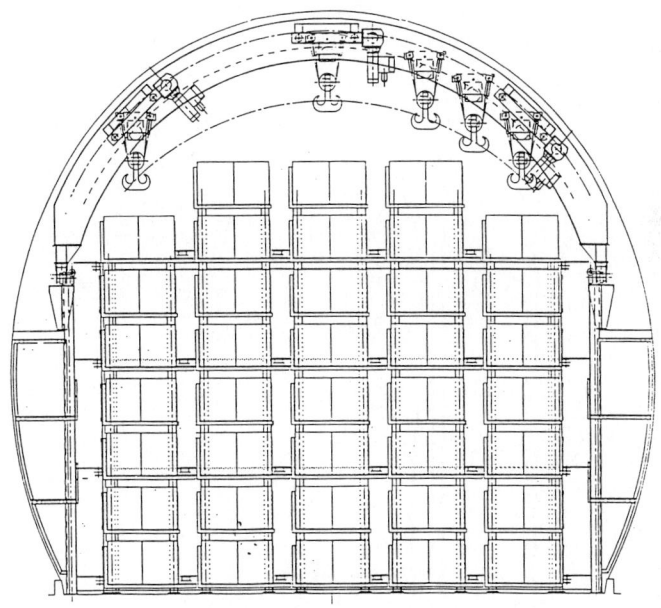

FIGURE 3
Cutaway view of the LVD

3. NEUTRINOS FROM GRAVITATIONAL COLLAPSE

Type II supernovae, one of the most spectacular events in the sky, are generally believed to occur at the end of the evolution of massive stars (M>8M$_\odot$). When the nuclear fuel is exhausted, the star cannot support itself against gravity and its core becomes dynamically instable and begins to implode. The final evolution of the core collapse is a neutron star, or, eventually, a black hole. If the outer parts are ejected, the brilliant supernova event can be seen as the external sign of a catastrophe marking the end of the star evolution. However, as regards our underground experiment, the most peculiar characteristic of this class of supernovae is the copious emission of neutrinos. In fact, the most important form of energy transport comes from neutrino interaction.

Type II supernovae are characterized on the basis of the optical spectrum by the presence of hydrogen lines and are seen only in the spiral arms of spiral galaxies, confirming that they are associated with massive stars[3]. Their evolution is so rapid that they cannot appear a long way from the sites of star formation.

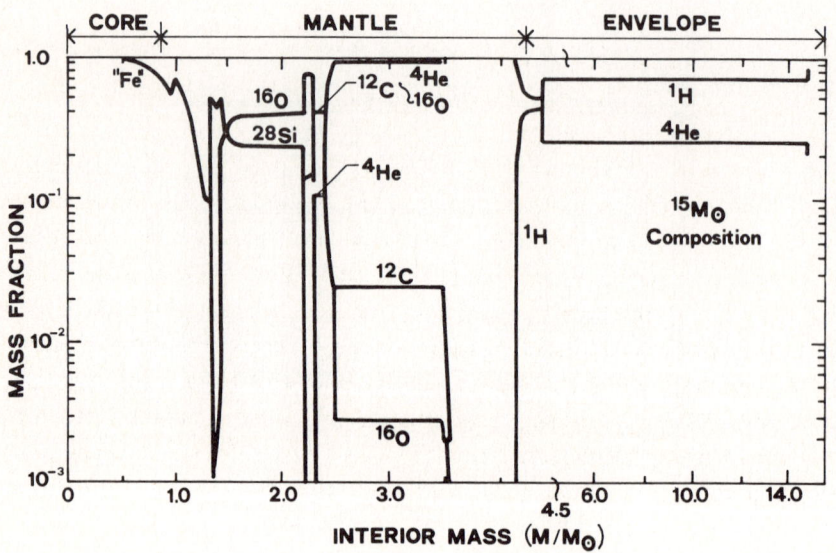

FIGURE 4
Composition of a 15 $M_\odot$ presupernova star when the edge of its iron core begins to collapse at 1000 km/s.
(Woosley and Weaver)

We know that the final stage in the evolution of a massive star is an "onion skin" with a central iron core surrounded by burning layers of SI, O, Ne, C, He, H. The advanced burning stages proceed through the ignition of heavier nuclei to the nuclear statistical equilibrium (A~Fe), with the star contracting at each stage as the burning of heavier fuel requires a higher temperature due to the increasing Coulomb barriers. The properties of a 15-$M_\odot$ presupernova star are shown in Fig.4.

After the final esoenergetic nuclear reaction (Si ignition), the iron core becomes a mass in excess of what the electron pressure can support and begins to collapse. Two processes then lead to pressure decrement accelerating the collapse of the core:

1. Photodisintegration of Fe nuclei on alpha-particles. This reaction ($^{56}Fe \rightarrow 13\alpha + 4n$) subtracts energy (124.4 MeV) as the electron pressure does not rise as rapidly as it would in the absence of breakup and so it cannot support the star against gravity.

2. Electron capture on "Fe" nuclei ($e^- + p \rightarrow n + \nu_e$) which removes the electrons required to support the star.

When either of these dynamical instabilities is encountered, the iron core begins to collapse very rapidly.

One of the most debated problems in the past was how the large[4] gravitational energy released in the core infall is transfered to the mantle and envelope of the star leading to their ejection and explaining the energies and luminosity of type II supernovae. As a major part of the heat generated by the gravitational collapse is converted into neutrinos through electron capture, it was long believed that their interaction with the mantle would transmit their outward momentum causing the explosions. With the discovery of weak neutral currents in the mid '70's, this mechanism was inadequate to explain explosion because a great number of neutrinos remain trapped in the core due to these currents.

Since then, theoreticians have been studying a model where the outward ejection of the mantle and envelope is due to the formation of a hydrodynamic shock wave formed after the "bouncing" of the central collapsing core as supranuclear density is reached.

The solution of the equation-of-state shows that, during the collapse, the iron core breaks up into an inner part (0.6, 0.8 $M_\odot$) which collapses homologously ($u \propto r$) with its surface falling at about the sound of speed, and an outer part falling at supersonic speed ($u \propto r^{-1/2}$), roughly half free-fall speed.

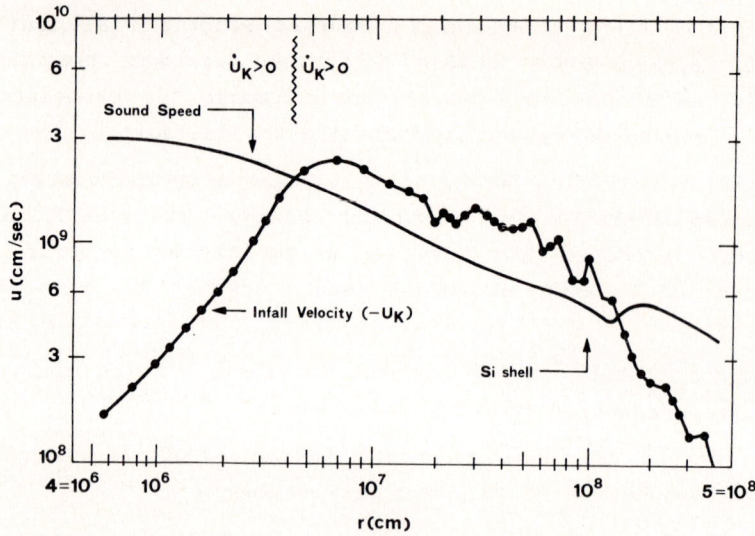

FIGURE 5
Material velocity u and sound speed as a function of r, about 1 ms before bounce. (Arnett)

Figure .5 shows the velocity of the infalling material, u, as a function of its distance r from the center at 1 ms before bounce[5]. Up to r~40 km, the velocity is proportional to r; it then reaches a maximum at ~75 km and is supersonic until r~1000 km. In the sonic point r_{sp}, the infall speed $|u|$ is equal to the sound of speed a, so that u+a = 0. This means that a signal generated in a point r<r_{rs} cannot move outwards from the sonic point, and that the material outside r_{sp} is unable to receive signals from inside r_{sp}. The material can, therefore, preserve homologys only inside r_{sp} through the rearrangement of pressure and density, while the exterior matter continues to move in a quasi-free fall.

As soon as the density of the inner region of the core exceeds the nuclear density, the collapse of the homologous core is brought to a sudden halt. Peak of pressure reaches $3 \cdot \rho_{nucl} = 8 \cdot 10^{14}$ g cm^{-3} with a temperature of ~10-15 MeV.

A pressure wave accumulates in the sonic point from where a shock wave begins to move out into the outer layer of the star. However, it is important to underline that the shock does not start at the center of the star, but around (0.8, 0.9 $M_\odot$). The energy available to the shock can be evaluated by calculating the gravitational binding energy of the homologous unshocked core. This is why the inner core quickly reaches hydrostatic equilibrium and it is possible to calculate the energy of the hydrostatic configuration, which is approximately $4-7 \cdot 10^{51}$ erg.

Neutrino emission and photodisintegration of iron nuclei cause strong energy losses to the wave during its passage through the remainder of the infalling core. If the shock wave reaches the envelope of the star with enough energy, the resulting explosion and ejection of the envelope is the spectacular event of type II supernova. This favorable condition occurs only with mass range $8_{M_\odot} < M \leq 16_{M_\odot}$ when it is possible to have a "prompt explosion".

Thus, an important criterium for the success of the core bounce mechanism is that the total mass of the collapsing core must not be too large and, as a consequence, the mass difference between the homologous and total core mass not too great to consume the total energy shock in stripping iron nuclei in free nucleons.

It has been confirmed that, before the bounce, the core is no longer transparent to neutrinos for a density greater than $\sim 2 \cdot 10^{11}$ g cm^{-3}. At this density, the drift velocity V_d of neutrinos relative to the nuclear material which causes the scattering is such that the net velocity of neutrinos relative to the star center is $u + V_d \leq 0$ with u being the infall velocity.

Once neutrino trapping has occurred, the total number of leptons remains constant and the neutrinos are in equilibrium with the matter. When the shock wave is formed, the inner core is never touched by the shock, so the nuclear matter is in the form of complex nuclei which makes neutrino diffusion very slow. It has been estimated that it takes ~100 ms for a neutrino to diffuse out of the core, which is very long compared to dynamic time.

On the other hand, the nuclei are completely dissociated into nucleons inside the shock wave; outside, the material is still in its original status of heavy nuclei. The neutrino mean free path is, therefore, smaller outside than inside, and it is the external matter that determines the ability of the neutrino to flow away.

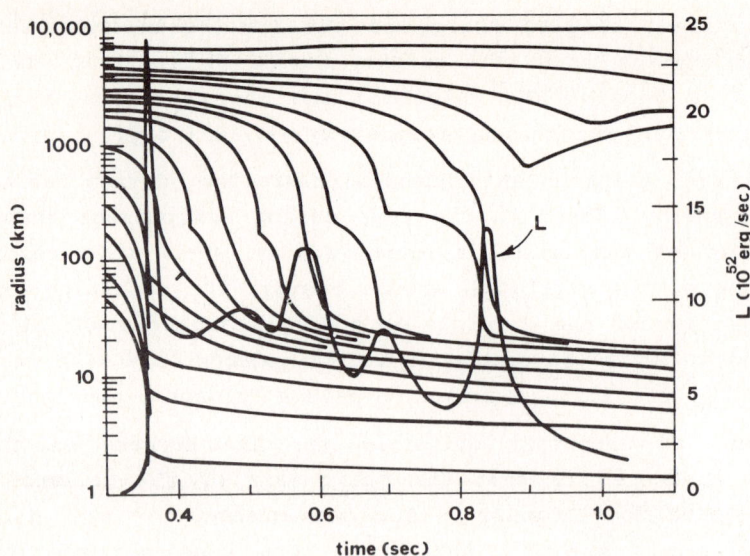

FIGURE 6
Electron neutrino luminosity vs time for a $25 M_\odot$ star superposed on the constant mass point trajectories.
(Mayle, Wilson and Schramm)

We can define[6] a "neutrinosphere" r_ν by the condition that there is one neutrino mean free path between r_ν and infinity. When the shock radius r_S is less than r_ν, the neutrinos remain confined in the shock because if the neutrinos diffuse out of the shock, their drift velocity due to the high density is less than the velocity of the shock, and the neutrinos are reassorbed by the shock. When $r_S \geq r_\nu$, the neutrinos can move freely away.

However, the prompt explosion described previously cannot be invoked for stars in the range $16_{M_\odot} \leq M \leq 80_{M_\odot}$. During their evolution, the shock wave stalls and becomes accretionary. Following the failure of the shock, a stationary "neutrinosphere" develops at about 40 km, while the stalled or accretion shock remains external to the neutrinosphere at ~100-300 km. Wilson et al. have proposed a model where the envelope explosion can also occur in this case. This is possible because the neutrinosphere heats the region just below the shock wave causing it to resume its outward course. As a result, the envelope explosion occurs, but is delayed by the amount of time characteristic of neutrino diffusion. However, Wilson's "delayed mechanism" has not yet been verified by other theoreticians.

Figure 6 shows the electron neutrino emission superposed on the mass point trajectories of $25M_\odot$ collapsing star.

3.1. Energy and time considerations

It has been calculated that $\sim 3 \cdot 10^{53}$ ergs (binding energy)[7] must be released to form a neutron star. During the supernova explosion, $\sim 10^{49}$ ergs of photons and $\sim 10^{51}$ ergs of kinetic energy are released. The difference in the calculated and observed energies is emitted in the form of neutrinos or gravitational waves. As the gravitational radiation can only be 1%, most of the binding energy is released in the form of neutrinos.

The collapsing core has $\sim 10^{57}$ protons which are converted to neutrons through electron capture

$$p + e^- \rightarrow n + \nu_e$$

As the average energy of ν_e emitted from the core is about 10 MeV, 10^{52} ergs are emitted by neutronization during the initial collapse, which is less than 10% of all the neutrinos radiated. The remaining neutrinos are given in pair processes (deleptonization)

$$e^+ + e^- \rightarrow \nu_i + \bar{\nu}_i$$

with $i = e, \mu, \tau$.

Muon and taon neutrinos are produced via neutral currents, while electron neutrinos are produced through both charged and neutral currents. The electron capture (neutronization) occurs in the initial collapse: initial ν_e burst takes place in times of the order of a few milliseconds during which the lepton rich core settles into hydrostatic equilibrium. The neutrino pairs are thermally radiated on a time scale of the order of the diffusion process ($\sim$ seconds). Core implosion, bounce and shock wave take about 1 s during which the first half of neutrino emission is released. The second half is emitted over the next few tens of seconds as the hot newborn neutron star cools down to become a standard cold neutron star (cooling phase). In any case, nearly all the neutrinos are emitted in about 20 s.

The luminosity of the mu and tau neutrinos during the deleptonization phase is less than the electron neutrino luminosity because they are emitted only via neutral current processes. It should also be pointed out that as the first burst ν_e comes from the homologous core region, whose size is largely independent of the original iron core mass, the first signal is model independent, while differences can be seen in the other phases.

In the "delayed" Wilson model for $M>16M_\odot$ stars, apart from the first burst, the neutrino luminosity has an oscillatory behavior superimposed on an exponentially decaying signal.

However, if we want to summarize the situation, we can say that average neutrino luminosity, mean neutrino energy and total emitted energy depend only on the initial iron core mass and are independent of the explosion mechanism, while the time structure of the neutrino luminosity is strictly related to these mechanisms.

On the other hand, the general scenario of the collapse is well known, while the mechanism for the ejection of the envelope in a supernova requires further study.

We can conclude that more information on the time scale structure of the neutrino emission is necessary to clarify the hotly-debated explosion mechanism of type II supernovae.

As we have seen, supernova events are involved in several important phenomena such as the production of neutron stars and most of the known X-ray sources of high energy cosmic ray emission. They also play an important role in the synthesis of almost all the elements which are heavier than helium, especially of the intermediate mass elements ($16 \leq A \leq 60$). Thus, the ejecta abundance well fits the observed cosmic abundance pattern.

Despite the great importance of the type II supernova event, our understanding of the phenomenon is incomplete for many reasons: Studies have, so far, been based on the observation of only electromagnetic radiation from the explosion or its remnants, but, due to the association of type II supernovae with massive stars, they are formed near the spiral arms of spiral galaxy regions that are often optically obscured. Furthermore, the expected rate of supernovae is also very low, approximated for galaxies and clusters of galaxies at (3-4) events in one hundred years. However, the fact that most of the energy is released through neutrinos indicates them as being the most important probe for studying stellar collapse.

As already stated, the LVD experiment is optimized for the detection of low energy neutrinos and antineutrinos from collapsing stars. Neutrinos, of course, are not obscured by matter, so their detection will resolve many of the difficulties in the current studies of collapsing stars, either by statistics so that the type II supernova event is not obscured by optical effects, or in model identification by measuring the time structure of neutrino luminosity.

4. DETECTION OF LOW ENERGY NEUTRINOS IN THE LVD

The most convenient way to study stellar collapse in the LVD is to detect the antineutrinos from the collapsing object through the inverse β decay reaction on free protons[8]

$$\bar{\nu}_e + p \rightarrow n + e^+$$
$$\downarrow$$
$$n + p \rightarrow D + \gamma \qquad (E_\gamma = 2.2 \text{ MeV})$$

If the scintillator modular counters are operated in a low background environment, they are suitable for detecting both the pulses which give the signature for an antineutrino:

- a prompt pulse from the positron with the energy above the high energy threshold;
- a delayed pulse from a neutron during a gate width $\Delta t = 500$ μs and an energy threshold of about 0.8 MeV.

This method has already been checked in the LSD at Mont Blanc.

The neutron moderation time plus deuterium fusion have been measured experimentally using a ^{252}Cf source in the Mont Blanc scintillator counter:

$$^{252}Cf \to n+\gamma+X$$
$$n+p \to D+\gamma \quad (E_\gamma=2.2 \text{ MeV})$$

The equipment is triggered by detecting the decay of ^{252}Cf. By comparing the number of triggers with the number of fusion events, it is possible to measure the efficiency for the detection of the neutron. The measured efficiency is 70%. Due to the lower background expected at the Gran Sasso and to the presence of an inner well-shielded fiducial volume in the LVD, the energy threshold will be lower.

From a collapse in the galactic center (10 Kpc), we can expect about 1000 interactions in the 1800 tons of liquid scintillator of LVD at a threshold of (5-6 MeV). With the expected noise level of 0.1 counts/sec in the 20 s of collapse time, the signal/noise ratio will be optimal.

More information about the dynamics of collapse can also be obtained observing the ν_e's through the elastic scattering reaction which produces, however, a lower number of interactions in the detector. As the ν_e also occurs in the first neutronization phase, the signal from the ν_e can provide information immediately at the very beginning of the collapse process.

Other reactions can also be detected in the LVD:

1. $\nu_e + ^{12}C \to \nu_e + ^{12}C^*$ is detected through the ^{12}C de-excitation with emission of one gamma of 15.1 MeV (96%), and two gammas of 10.7 and 4.4 MeV (4%).

2. $\nu_e + ^{12}C \to ^{12}N+e^-$ with $E_{th}=16.4$ MeV whose signature is a prompt electron of energy $(E-E_{th})$ followed after 11 ms by a positron from the β decay of the N with $E_{max}=15.4$ MeV and $\bar{E}=3$ MeV.

3. $\bar{\nu}_e + ^{12}C \to ^{12}B+e^+$ with the same signature as 2, but with a prompt positron pulse followed by an electron with $E_{max}=13.4$ MeV and $\bar{E}=4.5$ MeV.

For a supernova collapse occurring in the Magellanic clouds at 60 Kpc, the signals will be weaker but still detectable.

The LVD will no doubt be superior for detailed studies of stellar collapse in our galaxy.

5. THE PROBLEM OF THE ORIGIN OF COSMIC RAYS

Another of the most debated problems in the field of astrophysics concerns the origin of cosmic radiation. It is mostly formed of atomic nuclei which move with nearly the speed of light and is composed of

- protons - 92%
- helium nuclei - 6%
- heavier nuclei - 1%
- electrons - 1%
- gamma rays - 0.1%

The power radiated as cosmic rays in our galaxy is much stronger than that radiated in the form of radio waves and X-rays. The energy distribution of cosmic rays has a tail extending to energies higher than thousands of TeV.

The main characteristic of the composition of cosmic rays is the fact that its chemical composition exactly reproduces that of the whole galaxy, which excludes some of the hypotheses on the origin of cosmic radiation. For example, we cannot say that the rays form part of what remained after the big bang which produced nearly exclusively hydrogen. The most credible models today hold that cosmic rays must be generated either by an object of particular composition, or by many exotic sources. Several different sources have been proposed which would be able to act as extremely powerful particle accelerators and thus explain the energy of the cosmic rays, such as pulsars, binary systems, supernova explosions, quasars, etc. It is also believed that cosmic radiation comes mostly from outside our galaxy. A few discrete sources of cosmic radiation have been discovered in recent years, in particular of ultra-high gamma energy such as Cygnus-X3[9], Vela-X1, LMC-X4 with the gammas having a spectral power of E^{-2} and a characteristic modulation time of the signal emitted.

In order to study these point-form sources of cosmic rays, it is necessary to use neutral particles only, as charged particles have completely lost the memory of their initial direction due to the magnetic field of the galaxy through which they have travelled for at least 20 million years.

The obvious candidates are the gamma-rays and neutrinos; in fact, the neutrons should have too high an energy for their average life to be long enough to survive a long journey.

There are essentially two kinds of earth surface apparatus which can detect this radiation. They both exploit the property of the gamma rays of being able to produce electromagnetic showers in their interaction with the higher atmospheric layers. The apparatus can either be of the Cerenkov-light type and in this case covers an interval in energy between $3 \cdot 10^{11}$ and 10^{13} eV, or an extensive air shower array which measures directly the density of the particles in the shower using scintillation counters and streamer tubes and, in this case, the threshold at sea level is 10^{15} eV.

In order to explain how a cosmic object can produce particles with an energy >TeV as well as the emission periodicity, Verstand and Eichler[10] have proposed a model where it is held that the sources are binary systems formed from a massive star (neutron star, pulsar) orbiting around another standard star. In this model, charged particles (mainly protons) are accelerated in the strong electromagnetic field of the massive star and thus interact in the atmosphere of the neighboring star producing π^0 which then decay into gammas. The atmosphere must be thick enough to create π but not too thick to reabsorb the gamma rays (see Fig.7).

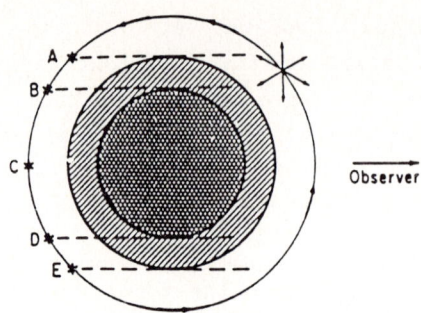

FIGURE 7
Verstand and Eichler model

These models also claim the emission of charged pions and thus of neutrinos from the decay of muons. As the ν_μ are not obscured by the neighboring stars, their intensity should, therefore, be at least a factor 10 greater than that of the gammas. Research regarding discrete sources can be successfully carried out using underground apparatus[11] by examining the muons produced either by gammas in their interaction with the atmosphere, or by neutrinos in their interaction with the rock surrounding the apparatus. The studies regarding muons produced by gammas are limited to sources such as Cygnus-X3 which are low over the horizon so that the signal expected is not covered by the normal flow of cosmic muons. Research on muons produced by neutrinos is restricted to sources below the horizon like VELA-X1 or LMC-X4 and, in this case, the target is the whole earth[12].

As the LVD has both horizontal and vertical tracking surfaces distributed across the entire apparatus, it is an excellent omnidirectional detector for the study and research of discrete sources of cosmic rays with its geometrical factor of 7000 m^{2}sr.

Discussions with Dr. A. Tornambè concerning stellar collapse physics are gratefully acknowledged.

REFERENCES

1) C. Alberini et al., Nuovo Cimento C, 9 (1986) 237.
2) G. Bari et al., Nucl. Instrum. Methods A264 (1988) 5.
3) R.P. Kirshner, Supernovae: A survey of current research, eds. Rees and Stoneham (Reidel 1982) 1.
4) S.E. Woosley and T.A. Weaver, Ann. Rev. Astron. Astrophys. 1986, "The Physics of Supernovae Explosions".
5) G.E. Brown, Supernovae: A survey of current research, eds. Rees and Stoneham (Reidel 1982) 13.
6) H.A. Bethe, Supernovae: A survey of current research, eds. Rees and Stoneham (Reidel 1982) 35.
7) R. Mayle et al., Astrophys. J. 318 (1987) 288.
8) G. Badino et al., Nuovo Cimento C, 7 (1984) 573.
9) M. Samorski and W. Stamm, Astrophys. J. (1983) 268; R.J. Protheroe et al., Astrophys. J. (1984) 280; J. Boone et al., Astrophys. J. (1984) 285.
10) W.T. Vestrand and D. Eichler, Astrophys. J. 261 (1982), 251.
11) E. Kolb et al., Phys. Rev. D32, (1985) 1145; A. Dar, Phys. Lett. 159 (1986) 205; T. Stanev et al., Phys. Rev. D32 (1985) 1244.

12) M.L. Marshak et al., Phys. Rev. Lett. 54 (1985) 2079; G. Battistoni et al., Phys. Lett. B155 (1985) 465; Oyama et al., Phys. Rev. Lett. 56 (1986) 991.

THE MACRO DETECTOR AT THE GRAN SASSO LABORATORY

Sergio PETRERA

I.N.F.N. and Dipartimento di Fisica, Università di Roma "La Sapienza", Rome, Italy

The MACRO Collaboration *

The MACRO detector is presently under construction at Gran Sasso. It is a large area detector, the acceptance for isotropic particle fluxes being around 10000 m^2 sr, dedicated to the search of rare phenomena in the cosmic radiation. It makes use of three detection techniques: liquid scintillator counters, plastic streamer tubes, and track-etch detectors. It will perform a search for supermassive particles (like GUT monopoles), a survey of astrophysical sources of H.E. gammas and neutrinos, a systematic study of the penetrating cosmic ray muons, and will be sensitive to neutrino bursts from gravitational collapses in the Galaxy.

1. INTRODUCTION

The MACRO (Monopole, Astrophysics and Cosmic Ray Observatory) detector[1] is presently under construction in the Hall B of the Gran Sasso Laboratory at a depth of 3600 mwe (meter water equivalent).

The general purpose of the experiment is to search for new particles and rare phenomena in the penetrating cosmic radiation. In order to achieve this goal the MACRO detector was designed to have a large acceptance for isotropic particle fluxes (above 10000 m$^2 \times$ sr). This feature will allow to push the search for GUT magnetic monopoles (as well as any other supermassive charged particle) well beyond the astrophysical bounds for the first time.

---

* BARI: M.Calicchio, C. De Marzo, O. Erriquez, C. Favuzzi, N. Giglietto, E. Nappi, F. Posa, P. Spinelli; BOLOGNA: S. Cecchini, G. Giacomelli, F. Grianti, G. Mandrioli, A. Margiotta, L. Patrizii, G. Sanzani, P. Serra, M. Spurio; BOSTON: S. Ahlen, M. Felcini, D. Ficenec, E. Hazen, J. Incandela, D. Levin, D. Magand, A. Marin, J. Stone, L. Sulak, W. Worstell; CALTECH: B. Barish, C. Gustavino, G. Liu, C. Peck; CERN: G. Poulard, H. Sletten; DREXEL: N. Ide, C. Lane, R. Steinberg; FRASCATI: G. Battistoni, H. Bilokon, C. Bloise, P. Campana, V. Chiarella, A. Grillo, E. Iarocci, A. Marini, J. Reynoldson, F. Ronga, L. Satta, M. Spinetti, V. Valente; INDIANA: C. Bower, R. Heinz, L. Miller, S. Mufson, J. Petrakis, G. Spizak; L'AQUILA: P. Monacelli, A. Reale; LECCE: G. Mancarella, P. Pistilli; MICHIGAN: M. Longo, J. Musser, C. Smith, G. Tarlé; NAPOLI: M. Ambrosio, B.C. Barbarino, F. Grancagnolo, V. Palladino; PISA: A. Baldini, C. Bemporad, V. Flaminio, G. Giannini, G. Grassi, R. Pazzi; ROMA: G. Auriemma, M. De Vincenzi, M. Iori, E. Lamanna, P. Lipari, G. Martellotti, O. Palamara, S. Petrera, L. Petrillo, G. Rosa, A. Sciubba, M. Severi; TEXAS A&M: P. Green, R. Webb; TORINO: V. Bisi, P. Giubellino, M. Masera, A. Marzari-Chiesa, L. Ramello; VIRGINIA TECH: D. Solie, P. Trower

The capabilities of the apparatus (muon tracking with good angle accuracy, velocity and energy deposition measurements) will allow the investigation of other important physics objectives, e.g. the detection of the astrophysical point sources of H.E. gammas and neutrinos and of neutrino bursts from gravitational collapses within our galaxy.

In this paper the detection of magnetic monopoles by the MACRO detector will be discussed in detail; the other physics items interesting for MACRO are more widely developed in other contributions of these Proceedings. Section 2 will give a brief review of the evolution of the concept of monopole as well as the current expectations in the cosmic radiation. The interaction of supermassive monopoles with matter and their detectability in MACRO will be also discussed. In section 3 the other physics objectives of MACRO will be briefly outlined. Section 4 will be dedicated to the description of the MACRO detector in all its constituents whereas section 5 will more explicitly deal with the detection of supermassive particles. Finally section 6 will give a short outlook of the future time schedule of the experiment.

2. MAGNETIC MONOPOLES

2.1. Evolution of the concept of monopole

The possible existence of magnetic monopoles has been postulated several times.

Firstly its existence was merely considered in order to allow the symmetrization of the Maxwell equations, through the introduction of magnetic charge and current densities.

It was in 1931 that Dirac[2] introduced for the first time the magnetic monopole in the framework of modern physics, in order to explain the quantization of the electric charge. In particular, he considered the system electron-monopole and showed that, imposing the quantization condition to the radial component of the angular momentum, it follows that

$$\frac{eg}{c} = \frac{1}{2}\hbar n, \qquad n = 1, 2, 3, ...,$$

(where g is the magnetic charge) which implies the quantization of the electric charge. Conversely, all magnetic charges should be integral multiples of an elementary magnetic charge

$$g_D = \frac{\hbar c}{2e} \simeq \frac{137e}{2}. \tag{1}$$

g_D is usually referred to as the Dirac magnetic charge.

In this formulation there was no prediction for the monopole mass. Naive arguments about the "classical monopole radius" led to a mass of the order of 2.5 GeV/c^2. This fact gave rise to many accelerator experiments searching for the direct production of monopole-antimonopole pairs in $pp, p\bar{p}, e^+e^-$ collisions[3].

A new era began in 1974 when 't Hooft[4] and Polyakov[5] showed that Grand Unified Theories (GUT) naturally imply the existence of magnetic monopoles, whose most relevant characteristics become predictable.

For the simplest possible GUT the symmetry breaking occurs in two different stages

$$SU_5 \stackrel{10^{15}GeV}{\longrightarrow} SU_{3\ color} \times (SU_2 \times U_1)_{e.w.} \stackrel{10^2 GeV}{\longrightarrow} SU_{3\ color} \times U_{1\ e.m.},$$

corresponding to transitions from grand unification to electro-weak unification and then to separate interactions. In such framework the magnetic monopole appears at the first transition and its mass is strictly related to the mass of the vector boson which acts as the carrier of the grand unification. In SU_5 one expects a mass of the order of 10^{16} GeV/c^2 or more. Other gauge groups generally lead to higher masses; unified pictures including also gravity, like Kaluza-Klein theories[6], predict even higher values ($m_M \geq 10^{19}$ GeV/c^2).

It is on account of such enormous mass that the approach to monopole detection changed thoroughly after GUT prediction. GUT monopoles cannot be produced at any conceivable accelerator; they could only be produced in the "Early Universe" and therefore can only be searched for in the cosmic radiation.

2.2. Astrophysical bounds on flux and velocity

Several upper bounds for the monopole flux can be obtained on the basis of astrophysical considerations. Most of them are deduced from the observation of measured parameters of the universe, like its mass density or galactic magnetic field. The bound obtained from the existence of the galactic field leads to the most commonly accepted constraint on the monopole flux:

$$F \leq 10^{-15} \quad cm^{-2}s^{-1}sr^{-1} \quad \text{for } \beta \leq \beta_c,$$
$$F \leq 10^{-15}(\beta/\beta_c)^2 \, cm^{-2}s^{-1}sr^{-1} \quad \text{for } \beta > \beta_c,$$

where $\beta_c \approx 3 \times 10^{-3}$. This limit is usually referred to as the Parker bound[7] and is considered as a reference value which until now has not been reached by experiment.

Astrophysics imposes also some bounds to monopole velocity. If they are gravitationally bound to our galaxy, monopoles would have a velocity of the same order of the escape velocity from it, i.e. $\beta \sim 10^{-3}$. If they are bound to our solar system, $\beta \sim 10^{-4}$ is expected. Even lower values are expected in case they are bound to the Earth. On the other hand also the flux of poles is expected to decrease with increasing velocity.

2.3. Interaction of magnetic monopoles with matter

At high velocity, $\beta > 10^{-2}$, the interaction of magnetic monopoles with matter is well understood. A pole with magnetic charge g behaves like a particle carrying an electric charge $q = g\beta$:

$$(dE/dx)_M = (dE/dx)_e \times (\beta g/e)^2$$

This means that a highly relativistic pole (i.e. with $\beta \approx 1$) with magnetic charge equal to g_D (1) is $\sim$ 4700 times more ionising than a proton of same β.

Much more difficult is the treatment of monopole energy loss at lower velocities. When considering the penetration of a slow particle in matter it is important to take into account the adiabatic excitation of atoms, arising from the interaction of the atomic electrons with the monopole magnetic field. In this case Zeeman crossing of the atomic levels may occur and, as a result, the passage of a monopole may leave atoms in an excited state. This is known as the "Drell effect", being calculated for the first time by Drell et al.[8] for atomic hydrogen and helium. A calculation for neon is currently in progress[9]. Using this effect it may be possible to detect monopoles at velocities $10^{-3} < \beta < 10^{-4}$ by observing the ionization caused by energy transfer from excited atoms to complex molecules with low ionization potential (Penning effect). This detection technique is well exploitable in gaseous detectors (like streamer tubes) where both He (or Ne) and a complex component (like n-pentane) are used in the gas mixture.

For more complex materials the calculation of the energy loss at low β depend much more on the model used to describe the medium. These calculations have been performed by several authors[10] using different techniques. Furthermore Ahlen and Tarlé[11] have pointed out that when using scintillators one cannot limit the considerations to the ionization energy loss, but one should consider the overall scintillation photon yield obtained taking into account the specific features of the detector. In this way they obtained a sharp threshold at $\beta \approx 6 \times 10^{-4}$.

Recent measurements[12] of scintillation light emitted by slow recoil protons from neutron exposure of scintillator at the Brookhaven High Flux Beam Reactor suggest for monopoles a threshold β as low as 3×10^{-4}.

2.4. Monopole detection in MACRO

The MACRO detector expected performance in the search for monopoles is shown in fig. 1, together with the most significant existing results as obtained with the various detection techniques. With an acceptance of 10000 m²× sr, in a few years of operation MACRO will push the search well beyond the Parker bound (for monopole mass around 10^{16} GeV/c²) covering the velocity range where monopoles are effectively expected to exist. More in general MACRO will search for any charged supermassive penetrating constituent of dark matter, with a sensitivity, at the same mass scale, of a few percent of the critical density of the universe.

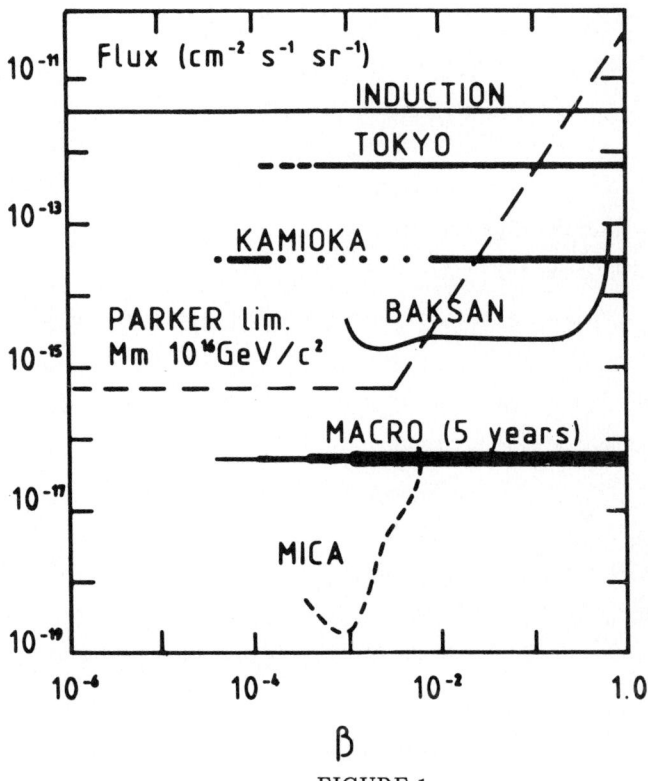

FIGURE 1

MACRO 90% C.L. limit on monopole flux after 5 years of operation. The different thickness of the line is to indicate the redundancy of information from the different detectors. The most significant existing results are also shown. The mica track-etch experiment is very sensitive, but relies on a number of restrictive assumptions.

3. OTHER PHYSICS OBJECTIVES

3.1. Gamma and neutrino astronomy

In recent years, multi-TeV γ-rays have been detected from a number of astrophysical sources such as Cygnus X3, Vela X1 and LMC X4. If it is true that the high energy γ-rays are a result of π^0 decay in hadronic cascades, then also neutrinos from the decay of the accompanying charged pions must be produced and their flux can be estimated from the observed γ-ray fluxes. Therefore deep underground observation of muons induced by these γ's and ν's can be viewed as a new tool to study astrophysical sources.

For sources above the horizon at Gran Sasso, like Cygnus X3, MACRO will track the penetrating muons produced in the γ-induced atmospheric showers. Assuming a standard extrapolation of photoproduction mechanisms to higher energies one expects about 10 events/yr from Cygnus X3 . The number of muons associated with UHE γ-showers could however be larger than theoretical predictions, as suggested from the anomalous μ-excess observed by the Kiel experiment[13], possibly increasing the muon signal by more than one order of magnitude. The associated background from atmospheric muons within a $\pm\,0.5°$ window is of the order of 10 events/yr, but can be significantly reduced taking into account the duty cycle ($\sim 1\%$) in the time modulated signal.

Neutrino sources below the horizon at Gran Sasso can be visible looking for directional signals in upward-going muons crossing the detector. Those muons are produced by neutrino interactions in the rock below the detector. Monte Carlo calculations show that the detected muons have an average energy of $\simeq 1$ TeV and preserve well the direction of the parent neutrinos. In fact, taking into account the $\nu - \mu$ angle at production and multiple scattering in the rock, it is found that 90% of the muon signal is in a cone of semiangle $\simeq 1°$. The background from atmospheric neutrinos in this angular window is negligible (0.1 events/yr). Extrapolating from the measured γ-ray fluxes, some of the sources could have measurable (few events/yr) fluxes of upward-going muons.

3.2. Neutrino bursts from stellar collapses

The large volume of liquid scintillator (~ 1000 m^3) contained in MACRO makes it possible to detect gravitational collapse events through the neutrinos that are emitted. The total energy release associated to neutrino emission is estimated to be few times 10^{53} ergs and is due to two distinct phases: the core neutronization resulting in a ν_e-burst over a time scale of milliseconds and the subsequent core cooling, which produces the large majority of energy emission and gives rise to $\nu\bar{\nu}$ pairs of all flavors with an approximate Fermi-Dirac spectrum with $< E_\nu > \simeq 10$ MeV emitted over a time scale of many seconds.

MACRO will be able to detect the $\bar{\nu}_e$ component of the cooling burst through charged current interactions with the protons in the liquid scintillator. In a typical gravitational collapse event located at the center of our galaxy, we expect to see of the order of 100 secondary positron induced by $\bar{\nu}_e$'s. Background measurements with the full scale prototypes at Gran Sasso, equipped with specially designed fast energy-reconstructing electronics, show that it will be possible to operate the detector with 5 and 10 MeV thresholds, respectively at the trigger and final event selection level.

3.3. Cosmic rays

MACRO with its large acceptance and tracking capability will also allow an accurate and systematic study of the penetrating component of the cosmic rays with very large statistics (10^7 events/yr). Furthermore it will be possible to obtain unbiased multiplicity distributions of multiple muon events, since the minimum detector dimension is more than twice the multimuon average size.

These samples of events will allow to investigate both in the field of cosmic ray physics (the composition of the primary spectrum, anisotropies, etc.) and of particle physics (anomalously high p_t's, "delayed" tracks, etc.).

4. THE DETECTOR

The MACRO detector is a modular array of liquid scintillation counters, plastic streamer tubes and track-etch detectors, which fill a box-shaped volume of 12×78 m^2 base and 9 m height. The three different types of detectors are used to measure particle trajectories, velocities and energy losses. This redundancy of the information was designed in order to attain unequivocal understanding of the data on the basis of only a few events.

Fig. 2 shows a vertical cut of the detector volume. It consists of a horizontal sandwich of three scintillation counter layers, 18 streamer tube layers, and one track-etch multilayer. Passive absorbers (iron and $CaCO_3$) are in between the sensitive layers, in order to identify penetrating particles, setting a threshold for through-going muons at 1 GeV. The four vertical faces of the detector are closed by one layer of scintillation counters and five layers of streamer tubes. Fig. 3 shows a general view of one of the 12 MACRO "supermodules", each one of dimensions $12 \times 12 \times 4.5$ m^3.

4.1. The liquid scintillation system

The liquid scintillation counters consist of long, FEP teflon-lined polyvinyl-chloride (PVC) tanks. The three horizontal layers use 12 m × 75 cm × 25 cm modules, viewed by a pair of 20-cm-diameter hemispherical photomultiplier tubes (PMT's) at each end as

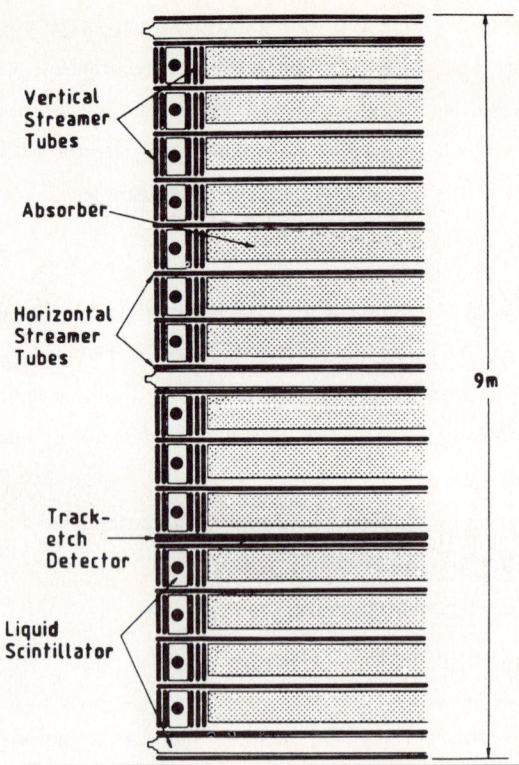

FIGURE 2   Detector cross section.

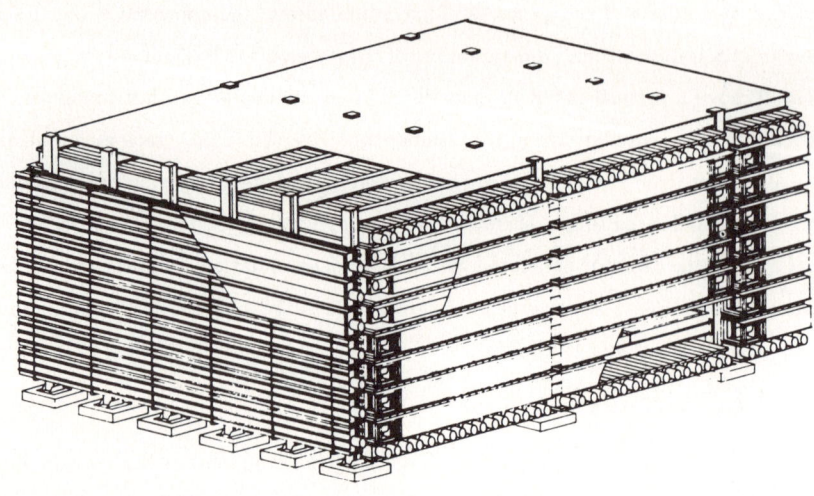

FIGURE 3
General view of one of the 12 MACRO supermodules ($12 \times 12 \times 4.5$ m^3).

shown in fig. 4. The vertical layers use 12 m × 50 cm × 25 cm modules, viewed by one 20 cm PMT at each side.

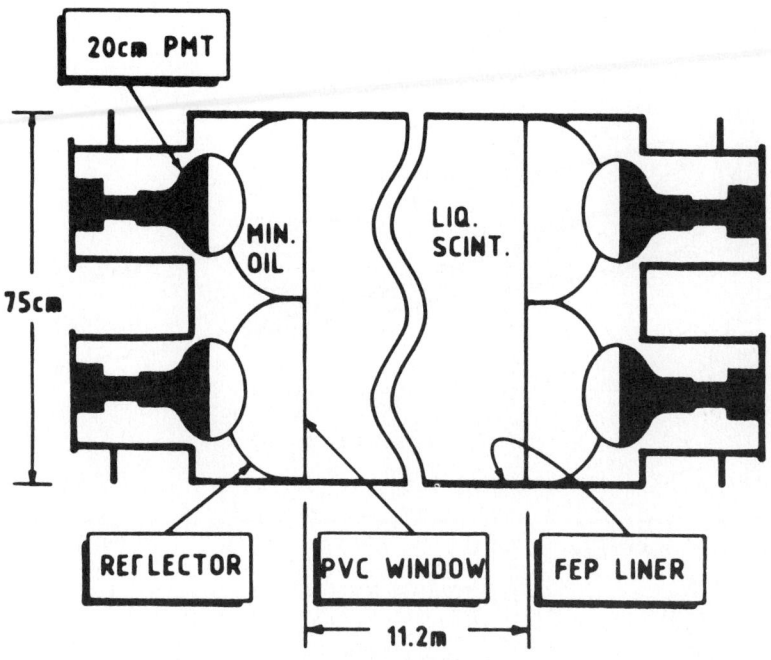

FIGURE 4
A schematic drawing showing the design of an horizontal scintillation counter.

PMT's are surrounded by parabolic reflectors and immersed in non scintillating mineral oil, to get rid of the relatively large light pulses due to very near soft electrons produced by radioactivity gammas. A clear PVC window separates the sensitive region containing the liquid scintillator from the PMT region.

The total number of counters is 484, the total mass of liquid is around 1 kton. Their main features, as measured on full scale prototypes, are: light attenuation length above 7 m (see fig. 5); trigger energy threshold around 5 Mev (10% of muon energy loss); time accuracy $\sim$ 1 ns (necessary for upward-going muon selection).

4.2. The streamer tube system

The streamer tube layers have a modular structure consisting of 8-tube PVC chambers with dimensions 25 cm × 3 cm × 12 m. The individual cell cross section is 3 × 3 cm^2 with 100 μm anode wire and graphite cathode (see fig. 6).

The tubes are equipped with digital readout in two coordinates. One coordinate is

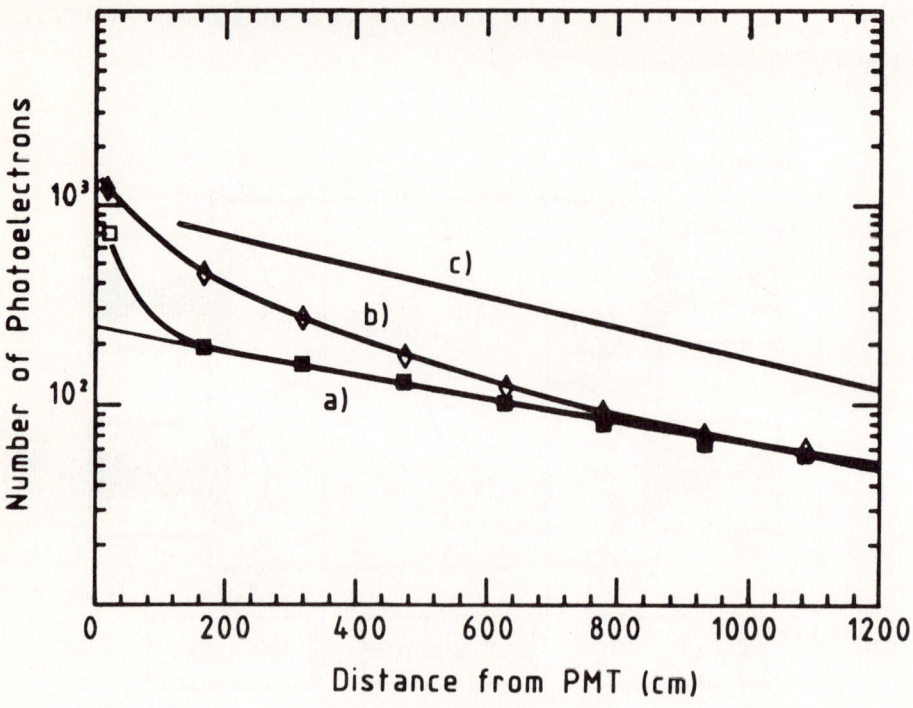

FIGURE 5
Scintillation counter response. a) and b): single PMT response of full scale prototype with different light collection arrangements; c): inferred response for the final counter design.

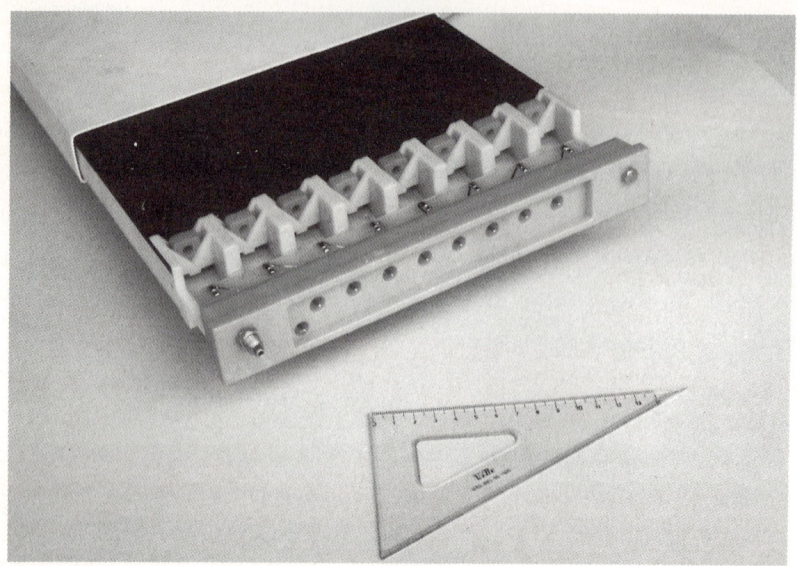

FIGURE 6   Streamer tube chamber shown open at one hand.

provided by the wires, while the other is provided by a layer of pick-up strips (D-strips) below the tube layer, placed at $\sim 30°$ with respect to the wire direction. Such strips are made out of an aluminium ribbon (3 cm wide, 40 μm thick) attached to a PVC foil (1 mm thick)

The total number of 8-tube chambers is about 7000, corresponding to ~ 125000 readout channels (55000 wires + 70000 D-strips).

The gas is a mixture of He, CO_2 and n-pentane. Space accuracy is about 1 cm, time accuracy is 50 ns for a through-going muon. The overall streamer charge response to ionization losses is summarized in fig. 7. It is substantially flat from the threshold (1% min. ion.) up to the minimum ionization loss, then is logarithmic above it.

4.3. The track-etch detector

The track-etch layer is a sandwich of Lexan and CR39 plastic sheets, with an aluminium absorber in between. The different thresholds of the two plastic materials allow the identification of different particles as shown in fig. 8. In particular, the passage of a monopole should result, after chemical etching, in collinear etch-pits of equal size on both faces of CR39 sheets.

5. THE DETECTION OF SUPERMASSIVE PARTICLES

The search for supermassive GUT monopoles (as well as any other charged slow penetrating particle) will be performed independently by the three detector subsystems described in the previous section, in order to exploit different excitation-ionization mechanisms of energy loss.

As shown in sect. 2, a crucial detector parameter is the monopole velocity threshold, above which it is detectable. Liquid scintillation counters should respond to monopoles, independent of charge state, with velocity $\geq 3 \times 10^{-4}$. The velocity threshold for the streamer tubes, for the conventional ionization mechanism, is at 10^{-3}. However the Drell excitation mechanism on He atoms, accompanied by ionization via the Penning effect, lowers the threshold down to 10^{-4}, for bare monopoles and for negative dyons. In the track-etch detector the CR39 covers the widest range, with the threshold at 2×10^{-5}, except for bare monopoles which are expected to be detectable above 10^{-3}.

Let us now consider separately the monopole triggers developed for the liquid scintillator and for the streamer tube systems. These two triggers will act quite independently one from each other. The track-etch detector is thought to be used in

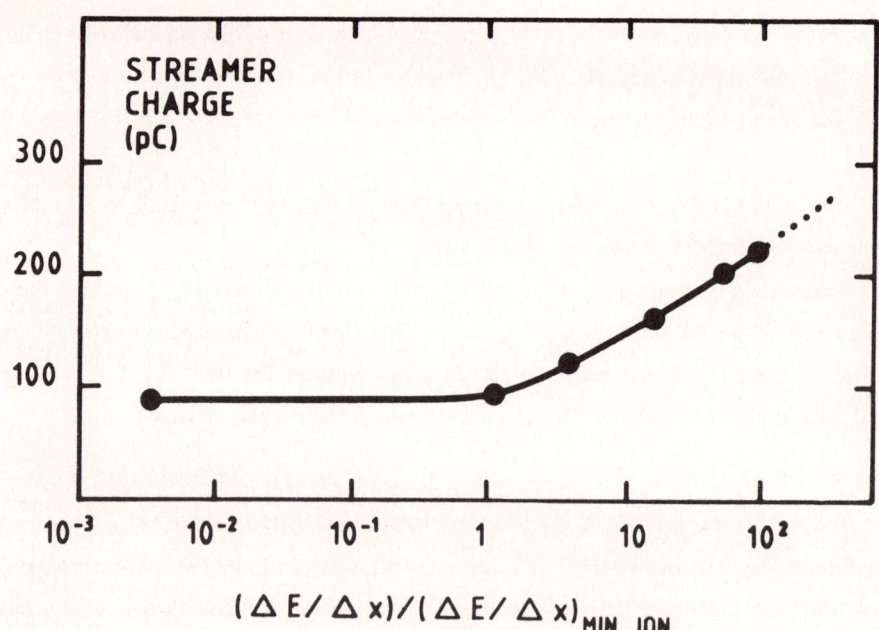

FIGURE 7
Streamer charge response as a function of ionization loss in the 3×3 cm^2 tube. The experimental points have been obtained with (from the left) single photoelectrons, muons, relativistic ions.

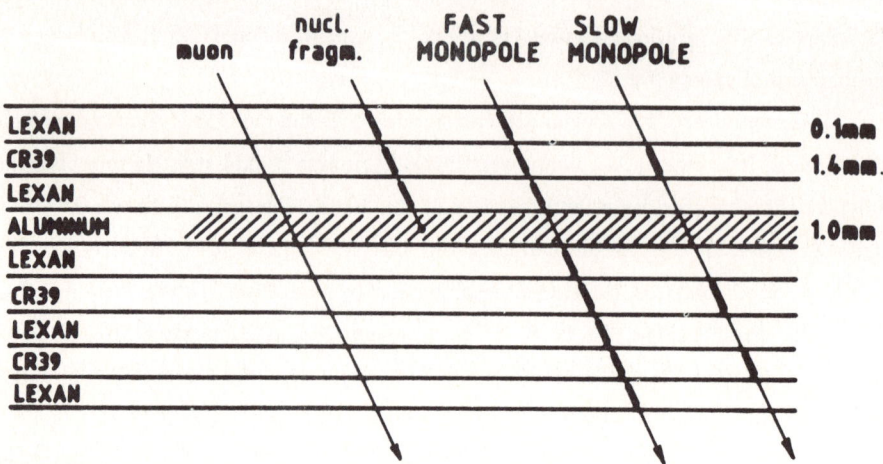

FIGURE 8
The track-etch detector sandwich, with schematic description of its response to different particles.

a "triggered" mode, that is removing, etching, and looking for a track in the module crossed by a candidate track in the active detectors.

5.1. Liquid scintillation counters

A characteristic signature of a slow monopole is its long transit time through the detector, resulting in a long light pulse. Even if the light level of a low β monopole is rather uncertain, for a $\beta = 8 \times 10^{-3}$ monopole we might expect to see something like 1600 photoelectrons (PE) produced during the 1 μs transit at the far end of the counter. However, the trigger has been designed in such a way to be sensitive to integrated pulses of as few as ≈ 20 PE's.

The PMT signal is input to a discriminator with a threshold at less than 1 PE, the output of which is a train of logic pulses. A trigger is generated if a pulse train of length T_0 is observed containing no gap larger than t_0, where T_0 and t_0 are adjustable parameters of the circuit. The PMT pulses are also recorded on waveform digitizers (6 and 22 μs ranges), with a time granularity to 1% of the total pulse duration. This will allow a detailed offline analysis of the triggered events including the correlation of the signals coming out from different scintillator counters.

5.2. The streamer tubes

The slow monopole trigger for the streamer tube system is based on the fact that such a particle crossing the apparatus with uniform speed appears in a space vs. time diagram as a straight line, whose slope corresponds to a distinct value of β, while the background is randomly distributed.

The trigger circuit[14] is driven by the OR signals of the tube planes of two adjacent modules (12×12 m² area). Each OR signal is sent to a shift register chain (serial in - parallel out). These shift register chains store the arrival time of each event (within a 1 μs time slot) for an overall depth of 480 μs. Such a time nearly corresponds to the maximum crossing time within a MACRO supermodule for a particle with $\beta = 10^{-4}$.

The shift register matrix is partitioned into 160 slices (β-slices), each one corresponding to a different β of the slow particle. The trigger is generated when for one of these β-slices the number of signals exceeds a preselected threshold. Requiring, for instance, 7 out of 10 horizontal planes, and at least 3 out of 4 lateral planes, one expects a satisfactory trigger efficiency ($\sim 80\%$) at a trigger rate compatible with the acquisition system. A second level selection of events, requiring space alignment, will drastically suppress the remaining background contribution.

6. OUTLOOK

Construction of the detector at hall B of the Gran Sasso Laboratory started on October 1987 and presently the first supermodule is completely finished. Fig. 9 shows a photograph of this supermodule after the mechanical mounting of the 7th streamer tube plane (February 1988). After some time dedicated to assessing backgrounds and fixing triggers, installation will continue, with the objective of having the full detector operational by 1990.

FIGURE 9
The first supermodule of MACRO in the Hall B at Gran Sasso, after the mechanical mounting of the 7th streamer tube layer in Feb. '88 (Photo Diotallevi).

REFERENCES

1) The MACRO Collaboration, Nucl. Instr. and Meth. A264 (1988) 18.
2) P. A. M. Dirac, Proc. Roy. Soc. , A133 (1931) 60.
3) For a review on magnetic monopole searches, see: Monopole '83, edited by J. L. Stone, (Plenum, New York, 1984); G. Giacomelli, Magnetic Monopoles, Rivista Nuovo Cimento, Vol. 7, N. 12 (1984).
4) G. 't Hooft, Nucl. Phys., B79 (1974) 276.
5) A. M. Polyakov, JETP Lett., 20 (1974) 194.
6) M. J. Perry, Monopole '83, pag. 29; Q. Shafi, Monopole '83, pag. 47.
7) M. S. Turner, E. N. Parker and T. J. Bogdan, Phys. Rev., D26 (1982) 1296.
8) S. D. Drell et al., Phys. Rev. Lett. 50 (1983) 644.
9) V. Patera, private communication.
10) D. M. Ritson, Magnetic monopole energy losses, SLAC-PUB-2950 (1982); S. P. Ahlen and K. Kinoshita, Phys. Rev. D26 (1982) 2347; S. P. Ahlen and G. Tarlé, Phys. Rev. D27 (1983) 688.
11) S. P. Ahlen and G. Tarlé, Phys. Rev. D27 (1983) 688.
12) D. J. Ficenec et al., UM-HE-87-0001, submitted to Phys. Rev. Lett.
13) M. Samorski and W. Stamm, Astrophys. J. Lett. 268 (1983) L17.
14) G. Auriemma et al., Nucl. Instr. and Meth., A263 (1988) 249.

ULTRA-HIGH SENSITIVITY MASS SPECTROMETRY: APPLICATIONS TO RARE NUCLEAR AND COSMOLOGICAL PROCESSES ARCHIVED IN GEOLOGICAL SAMPLES

Claudio TUNIZ

Dipartimento di Fisica, Universita' degli Studi, Trieste, Italy
Istituto Nazionale di Fisica Nucleare, Sezione di Trieste, Italy

Jeff KLEIN

Department of Physics, University of Pennsylvania, Philadelphia, PA 19104 USA

1. INTRODUCTION

Interactions among particle physicists, cosmologists and astrophysicists are creating a new discipline, *astroparticle physics*[1], of great scientific importance. This new discipline is encouraging a variety of experiments, not involving high-energy accelerators, whose main aim is to test the predictions of Grand Unified Theories and neutrino physics.

A theoretical overview of the present situation in particle physics, with a particular emphasis on these new aspects, is given in Ref. 2. A recent review on experimental particle physics without accelerators is given in Ref. 3.

Two principal detection methods are employed in these experiments: Direct detection using characteristic radiation or kinematic reconstruction for identification, and passive accumulation of reaction products followed by chemical separation and identification by highly sensitive analytic techniques. Significant progress in the direct measurement of neutrinos and rare decays is forseen with the development of cryogenic detectors like superheated superconducting granules, bolometers, and tunneling junctions. Ultra-sensitive analytic techniques, developed in the last few years, such as laser-based Resonance Ionization Spectrometry (RIS) and Accelerator-based Mass Spectrometry (AMS) allow new and as yet unforseen opportunities for the detection of rare processes using *radiochemical* and *geochemical* methods. We will give an overview of the use of ultra-sensitive mass spectrometry in detecting the products of very rare processes like proton decay and double-beta decay, and the detection of particles of special interest such as relics from the Big Bang (particles whose inclusion results in isotopes with anomalously high masses or with fractional charges) and superheavy elements. In particular, we will describe and examine the potential of several new possibilities afforded by AMS to search for the products of double-beta decay accumulated in natural samples over geological times.

2. EXPERIMENTAL CONSIDERATIONS

2.1. Integration of rare physical processes with geologic samples

Although several experiments have been proposed for the direct detection of the decay of the proton, inverse beta decay by solar neutrinos and double-beta decay, the small cross sections and decay constants hamper their measurement by direct counting techniques. An alternative exists, however, if the product nucleus or its presence is somehow distinctive (e.g. if the product nucleus is a rare isotope, or if its accumulation results in an isotopic anomaly). Radiochemical methods can be used to overcome the small cross sections by integrating the process over the lifetime of the daughter isotope (thereby capitalizing on the small decay constant); if the lifetime is long enough (or if the daughter is stable), integration can be extended to geologic time scales (the so called geochemical method). The geochemical method is most promising when suitable targets can be identified that are deeply buried (to reduce interference from competing reactions induced by cosmic rays) and chemically relatively pure (ores).

In evaluating possible geochemical experiments, the following parameters have to be considered:

a) Amount of sample necessary to perform the experiment,

b) Difficulty in the chemical separation of the product,

c) Purity of sample and ore location,

d) Difficulty of detection of product isotopes.

The amount of sample needed depends on the cross section (or decay constant in the case of p-decay), on the geologic age of the ore or the lifetime of the product (daughter), and on the abundance of the target (parent). An idea of the relative rarity of these processes is seen in these examples from Ref. 4:

i) proton decay: one decay per 1000 tons of target per year,

ii) solar neutrino detection: one reaction per ton of target per year,

iii) double-beta decay: one decay per 100 g of target per year.

Ultimately, the amount of material needed depends on the detection efficiency of the final analytic technique and the precision of the answer required. For example, if the detection efficiency is 10^{-3} and a precision of 10% is desired, then about 10^5 atoms must be made by the process in the target material. If the target is exposed for only a year, a sample of 10^8 tons would be necessary to measure proton decay, 10^5 tons for solar neutrinos, and 10 tons for double-beta decay. However, if a suitable geologic sample could be identified, the "duration" of the experiment might be extended to a time comparable to the age of the Earth (4.5×10^9 a). This would reduce the sample-size requirement by 10^9 and increase the concentration of the product nucleus proportionately. In the case of proton decay, instead of requiring the identification of 1 particle in 10^{31}, the problem would be reduced to 1 in 10^{22}. Although even this lower figure sounds formidable, it is within the range of the analytic techniques described in the next section.

The major limitations restricting the usefulness of geologic experiments come from blanks and backgrounds. If the resulting nucleus is stable, then at some level it already existed in the target material before the "experiment" began, and the requirement of sufficient detection sensitivity is replaced by the requirement to measure isotopic ratios with sufficient precision that the isotopic anomaly produced by the process is detectable. Using precision isotope-ratio mass spectrometers, isotopic anomalies of 10^{-5} can be measured. Typically, every element is present in every geologic sample at least at the ppb level. If the nucleus produced by the process is a rare isotope, another factor of 10^4 in sensitivity may be possible. Hence, in a pure ore consisting of the appropriate target, it might be possible (under the best of circumstances) to detect a process with a daughter/parent ratio of 10^{-18} ($10^{-9} \times 10^{-4} \times 10^{-5}$). If the integration period is the age of the Earth, then conceivably, the lowest detectable rate for a process that produces a stable (or very long-lived) nuclide is $10^{-18} \times 10^{-9}$ or 10^{-27} a^{-1}. An additional factor of 10^2 to 10^5 in sensitivity is possible when the final analysis is preceded by a stage of isotope enrichment, either as part of the experiment or the result of the sample collection process. Several examples of these will be given in the following.

For radioactive daughters, blanks pose less of a problem because primordial contamination is absent. However, backgrounds from alternative reactions induced by cosmic rays (mainly muons if the target is deeply buried) and natural radioactivity (uranium and thorium) interfere with experiments irrespective of whether the daughter is stable or unstable. Ores at a depth of more than 1500 m with uranium and thorium contents at the ppb level will reduce these backgrounds to an acceptable level for some of the particle-physics problems with geologic solutions discussed in the following sections.

2.2. Analytic capabilities of accelerator mass spectrometry

One of the most promising new analytic techniques is accelerator mass spectrometry. AMS is an analytic technique that uses an ion accelerator and its beam transport system as an ultra-sensitive mass spectrometer [Fig. 1] to provide several stages of mass analysis and, in most cases, element (Z) identification in the final detector. Detailed reviews of AMS and its applications can be found in Refs. 5 and 6. AMS has been successfully employed in the analysis of long-lived cosmogenic radioisotopes present in natural samples with isotopic ratios $10^{-9} - 10^{-15}$ (Refs. 7, 8 and 9). Usually

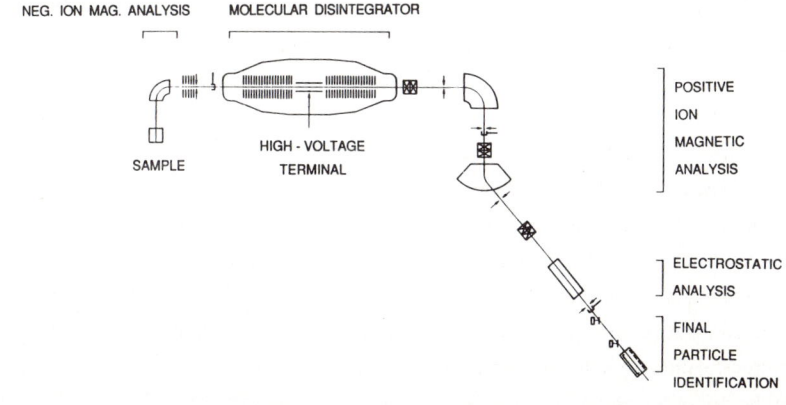

FIGURE 1

Schematic diagram of the XTU tandem accelerator mass spectrometry system (from Ref. 10)

a tandem accelerator, so called because negative ions are first accelerated to the center of a machine whose terminal is at a potential of several million electron volts (MV), converted there to positive-ions by removal of several electrons and subsequently accelerated back to ground potential (two sequential stages of acceleration), is used. The sensitivity of AMS is unaffected by the half-life of the isotope being measured since the atom itself is detected, and not the radiation or particles that result from its decay. And since definitive isotopic identification is normally possible, it is almost unaffected by the backgrounds that limit conventional mass spectrometry (elemental and molecular isobaric interferences, and "tails" of neighboring masses). AMS has higher sensitivity than decay counting for isotopes with lifetimes of > 600 years counted for periods of < a month, and is 5 to 9 orders of magnitude more sensitive than conventional mass spectrometry.

The technique does of course have limitations that affect its selectivity, precision and sensitivity.

2.2.1 Selectivity

The backgrounds that plague conventional mass spectrometers are reduced in several ways. Molecular interferences are eliminated by analyzing ions in the 3+ or higher charge state: without outer binding electrons, molecular bonds are so weakened that molecular lifetimes are less than 1 microsecond and molecules don't survive long enough to make it to the detector. Elemental isobars are more difficult to eliminate and several different approaches have been adopted:

i) At the ion source. Several elements do not form negative ions (or specific molecular negative ions may be stable for one element but not for neighboring elements) so that the competing isotopic isobar is eliminated at the ion source. ^{14}C dating of organic samples is a well known case, since the ubiquitous ^{14}N does not form negative ions but copious quantities of C$^-$ can be formed. We'll discuss a similar case for the study of ^{70}Zn($\beta\beta$)^{70}Ge , where a rejection of about 10^{12} was obtained at the ion source for the interefering isobar. More general elemental specificity in negative-ion formation can be obtained by combining sputtering with RIS. A laser produces selective multi-step ionization of the atomic species of interest. The use of sputter-initiated -RIS AMS was recently discussed in Ref. 11.

ii) Range separation. When the interfering isobar has a higher Z than the rare isotope, it loses energy faster when passing through matter and can be made to stop in a thick absorber placed in front of the detector, while the isotope of interest passes through. For example, ^{10}Be is separated from ^{10}B in this manner.

iii) Complete stripping followed by magnetic analysis. At sufficiently high energies, all electrons can be removed leaving the bare nucleus. Since beam transport systems select a given m/q ratio -- isobaric interferences can be eliminated if the interfering isobar has a lower Z than the rare isotope. The drawback of this approach is that very high energies are sometimes needed to efficiently fully strip the isotope of interest.

iv) Measurement of energy, specific energy loss, and range in the ion counter (Fig. 2). At sufficiently low count rates ($< 10^4$ s^{-1}) the particle detectors employed in AMS are able to identify the Z of the the incoming ion if it enters with an energy above the Bragg peak. This and the break up of molecules are the aspects of AMS that largely account for its high sensitivity (selectivity).

v) Gas filled magnet. Charge changing collisions of an ion travelling through a gas result in an average charge state that depends on the Z of the ion. Since the trajectory of an ion in a magnetic field depends on its charge, ions with different average charges (hence different Zs) will follow different trajectories resulting in isotope separation. Application of this technique will be discussed for a solar neutrino detector.

2.2.2 Precision

The precisions of the isotopic ratios measured by AMS are much lower than those obtainable with conventional mass spectrometers. In general, precision depends on counting statistics, isotopic fractionation and blank and standard corrections. A precision of better than 1 % is obtained at dedicated AMS facilities in the measurement of ^{14}C (see Ref. 12). More typically, isotopic ratios are measured with an uncertainty of 3-10%. This limitation comes from the complicated beam optics of AMS systems and/or from the small number of ions counted. We have already discussed the importance of high precision in isotope ratio measurements when looking for anomalies (section 2.1). Because of its lower precision, AMS has a sensitivity of about 10^4 times lower than conventional mass spectrometry in detecting isotopic anomalies -- but this is often compensated for by its higher selectivity. For isobaric processes that convert an isotope of one element into an isotope of another element (β decay, $\beta\beta$ decay ...) lack of sufficient isotopic selectivity may limit the sensitivity of conventional mass spectrometers to an extent that the sensitivity of AMS is actually greater.

2.2.3 Efficiency

The efficiency of an AMS system is the ratio of the number of the atoms (of interest) counted in the final detector to the number loaded in the ion source. This ratio is the product of (i) the negative-ion formation efficiency, (ii) the probability of forming the analyzed charge state, (iii) the efficiency of transporting the beam from the source to the ion counter, and (iv) the detection efficiency. The overall efficiencies of AMS systems range from 1% for ^{14}C detection to 0.02% for ^{129}I detection[9]. This is one of the major limitations in the application of AMS to solar neutrino detectors. AMS efficiency is much lower than that obtainable with other laser- or ion-based analytical techniques or with Neutron Activation Analysis.

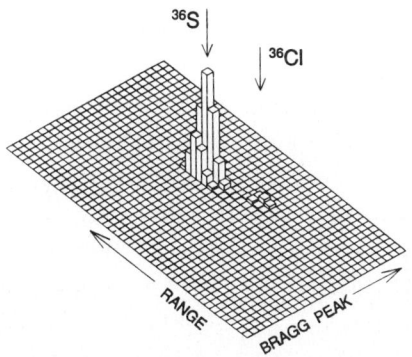

FIGURE 2
Isobaric separation with the Bragg detector (XTU term. volt. 7.75 MV, Q= 7+, $^{36}Cl/Cl = 10^{-11}$)

2.2.4 Background

Backgrounds include all the signals in the final detector that cannot be separated from those corresponding to the nuclides of interest contained in the samples. They include: (i) backgrounds deriving from the limited resolution of the ion detector in the measurement of energy, velocity and specific energy loss, (ii) ion cross contamination with other samples in the ion source, (iii) ion-source background and (iv) contamination before or during sample preparation. Factors (iii) and (iv) are the main limitations in the detection of stable isotopes (see section 8). Special ion sources made with ultra-pure materials have to be designed in order to detect traces of stable elements below the level of ng/g. For most long-lived cosmogenic isotopes, backgrounds corresponding to isotopic ratios below 10^{-15} are possible without any special ion source considerations.

3. SOLAR NEUTRINO EXPERIMENTS

Two reviews of the experimental and theoretical status of the solar neutrino puzzle have recently appeared[13, 14]. The experimentally measured average flux of ν_e from the sun is still ~ 3.8 times lower than the prediction of the standard solar model and this has been variously attributed to problems with the physics of the Sun and problems with the physics of neutrinos (most notably, neutrinos might have mass which would allow them to "oscillate": the missing ν_es would arrive as ν_μ or ν_τ). The so called "non-standard" models are constructed by changing something, physics or parameters, primarily to reduce the central temperature of the Sun to severely reduce the flux of ^{8}B neutrinos. Some of these models suggest that the Sun is not in steady-state. It clearly would be advantageous to have more experimental data, especially on the flux of neutrinos from the *p-p* reaction (all models give about the same flux for these neutrinos) or on the flux integrated over a few million years (to check for existence of variations in solar output).

3.1 Geochemical experiments

AMS allows the direct detection of rare radioisotopes, present in natural samples with abundances 10^{-15} of the matrix. New solar neutrino detectors have been proposed based on this new technique that do not require the prompt decay of the reaction product as in radiochemical experiments. Furthermore, geochemical experiments can integrate neutrino reactions over geologic times. In these experiments, natural detectors are employed, consisting of appropriate elements contained in minerals buried underground.

In a search for this kind of neutrino detector, the following factors have to be considered: a) equilibrium product-to-target ratio, b) isotopic abundance of the target, c) background due to competing reactions, d) difficulty in the detection of the product isotope.

In the following we summarize the present experimental status of several proposed neutrino detectors using long-lived product isotopes, updating, in many cases, the reaction rates based on recent revisions of the standard solar model and newly measured cross sections.

3.1.1 $\quad ^{41}\text{K} + \nu \rightarrow e^- + ^{41}\text{Ca}$ ($T_{1/2} \cong 1.0 \times 10^5$ a) (Ref. 15)

The effective threshold for this reaction is 2.36 MeV. Therefore, the process is sensitive only to ^{8}B neutrinos. A cross section of $(1.45 \pm 0.05) \times 10^{-42}$ cm^2 has been estimated. Assuming a ^{8}B neutrino flux of 5.8×10^{-6} cm^{-2}s^{-1}, a capture rate for ^{41}K of 8.4 ± 0.3 SNU (1 Solar Neutrino Unit

= 10^{-36} captures/ s . target atom) is obtained.

If the potassium ore is buried at a depth of 1500 m, cosmic-ray muons will contributo only 0.77 SNU to the ^{41}Ca. The contribution of ^{40}Ca(n,γ) and (α,p) followed by ^{41}K(p,n) will be below 10% if uranium and thorium contents of the potassium ore are less than 2.2 and 3.5 ppb, respectively.

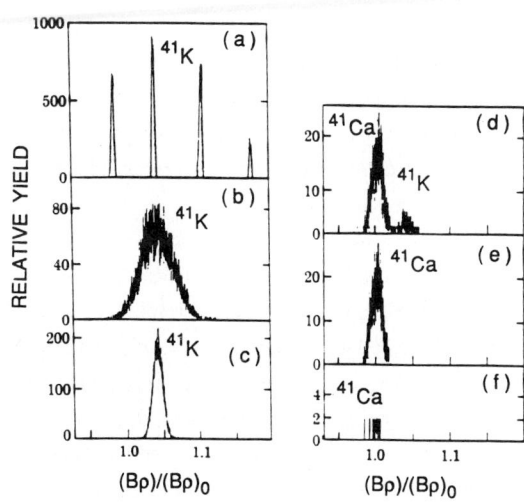

FIGURE 3
Spectra from a magnetic spectrograph focal-plane detector. Figures (a)-(c) show the collapse of the magnetically dispersed charge-state spectrum of ^{41}K into a single line as nitrogen gas is introduced into the magnetic field region. Spectra from standard (d) and (e) and natural (f) samples illustrate the identification of ^{41}Ca at a concentration level corresponding to ^{41}Ca/Ca = 10^{-12} (from Ref. 16). ^{41}Ca^{10+} ions accelerated to 200 MeV

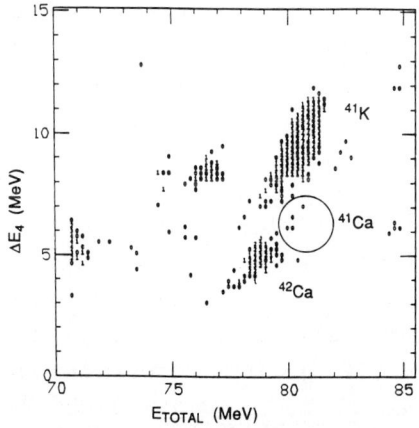

FIGURE 4
E-ΔE spectrum from a commercial CaH$_2$ (blank) in which 4 counts were detected in 90 minutes, corresponding to ^{41}Ca/^{40}Ca=6 x 10^{-16}. ^{41}Ca^{9+} ions accelerated to 84.42 MeV

There are many deposits of potassium satisfying the above criteria. The KCl deposit at Regina, Sasketchewan, Canada, was suggested in Ref. 15. In order to measure the neutrino flux with 10% accuracy we need to observe 100 ^{41}Ca events. With an overall efficiency of 1%, 10^4 atoms are needed for this analysis and they can be extracted from 0.5 tons of KCl (the natural abundance of ^{41}K is 6.77%). Since the Ca contamination is approximately 0.1%, the above extraction will produce 1 kg of natural calcium.

Because the sample size for AMS analysis is 10 -100 mg, a pre-concentration of ^{41}Ca relative to natural calcium by at least 10^5 is required. This results in a ^{41}Ca/Ca ~ 10^{-16}. Particle identification using a gas-filled spectrograph (Fig. 3) gives a ^{41}Ca/Ca sensitivity of 6×10^{-14}. Recent measurements at the University of Pennsylvania[17] have improved the detection level to ^{41}Ca/^{40}Ca < 6×10^{-16} using CaH$_3^-$ (Fig. 4).

3.1.2 $\quad \nu + {}^{205}\text{Tl} \rightarrow e^- + {}^{205}\text{Pb}^m \rightarrow {}^{205}\text{Pb}$ ($T_{1/2} = 1.6 \times 10^7$ a) (Ref. 18)

This neutrino reaction leads to the excited state of ^{205}Pb (excitation energy 2.3 KeV), followed by a fast isomeric decay to the ^{205}Pb ground state. The threshold is 0.048 MeV and therefore this detector is sensitive to most (85%) of the low-energy neutrinos produced by the *p-p* fusion in the Sun. Unfortunately, the cross sections for the neutrino absorption are unknown, and can not be calculated accurately. In a recent reappraisal of this experiment, based on (p,n) data of Krofcheck, Bahcall and Ulrich[14] estimate a capture rate of 263 SNU, with about half the rate (173 SNU) coming from neutrinos produced by the *p-p* reaction.

The main background comes from the reaction ^{205}Tl(p,n)^{205}Pb where the protons are produced by high-energy cosmic-ray muons. According to Ref. 18, a depth of 300 m for the ore would keep this background at an acceptable level. Uranium and thorium induced reactions leading to ^{205}Pb should also be kept below the 10% level.

Thallium minerals are very rare. A Lorandite (TlAsS$_2$) deposit of 10 Ma is located in Southern Macedonia, Yugoslavia, at a depth of about 120 m. Due to the erosion of the rocks, this

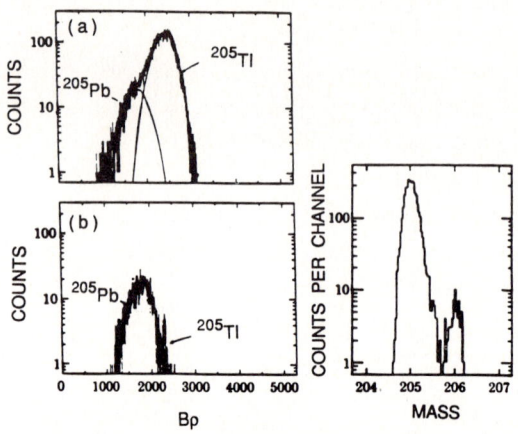

FIGURE 5
Charge (left) and mass (right) spectroscopy with an incident energy of 2.3 GeV (from Ref. 16)

deposit was much deeper during the last million years, providing sufficient shielding against the cosmic-ray muons. Estimates show that backgrounds from cosmic ray and natural radioactivity are acceptable.

The reaction $^{205}Tl(\nu,e^-)^{205}Pb$ gives the most favorable equilibrium product-to-target ratio: $^{205}Pb/^{205}Tl = 3 \times 10^{-19}$. From the measured Pb contamination in the ore, 3 ppm, a $^{205}Pb/Pb$ concentration of $10^{-13} - 10^{-14}$ is expected in the final sample material. To study the feasibility of this solar neutrino detector, a series of experiments have been performed at the GSI heavy ion facility UNILAC[16]. Pb and Tl ions were accelerated to 2.3 GeV and identified using a time-of-flight system and a magnetic spectrometer (Fig. 5). The present performance of the system is sufficient to suppress neighboring isotopes at the level of 10^{-17}. Isobaric separation between ^{205}Tl and ^{205}Pb is at best 10^{-3} using energy-loss measurements. Hence, chemical separation is required to reduce the Tl/Pb ratio below the level of 10^{-11}. A technique based on isobar separation with a gas-filled magnet will allow the reduction of requirements for the chemical purification. The major problem is the low efficiency of the source, $10^{-6} - 10^{-5}$, giving an overall efficiency of $10^{-9} - 10^{-8}$ for the AMS system. It has been estimated that 10^7 atoms of ^{205}Pb are contained in 100 kg of Lorandite which are obtained from 20 tons of mineral. A recent review on solar neutrino detection with ^{205}Tl is given in Ref. 19.

3.1.3 $\quad \nu + {}^{98}Mo \rightarrow e^- + {}^{98}Tc$ ($T_{1/2} = 4.2 \times 10^6$ a) (Ref. 20)

The effective threshold for this reaction is > 1.7 MeV and can be induced only by high-energy ^8B neutrinos. A capture rate of $17.4\,^{+18.5}_{-11}$ SNU is obtained using the standard model ^8B neutrino flux of Ref. 14 and the cross sections in Ref. 21. A mineral containing molybdenum is molybdenite (MoS_2). (α,p) reactions on Mo can produce ^{98}Tc but they are suppressed by high-energy thresholds. ^{98}Tc can be produced with higher rates by reactions $S(\alpha,p)Cl$ and $S(n,p)P$ followed by $^{98}Mo(p,n)^{98}Tc$ and by the (n,p) reaction induced by fission neutrons on ^{98}Ru. Another source of ^{98}Mo background is the spontaneous fission of uranium and reactions induced by cosmic-ray muons. According to Ref. 20, the latter backgrounds can be kept at the 10% level if the mineral is at a depth of 1400 m.

The Henderson ore body, Red Mountain, Colorado, was suggested as a source of Molybdenum for this experiment. 2600 tons of ore would produce 13 tons of molybdenite, yielding 10^7 atoms of ^{98}Tc. Because there are no stable Tc isotopes the ^{98}Tc atoms could be extracted using rhenium as a carrier since it is a chemical homologue of technetium. A sample of about 10^7 atoms (obtained from 20 boxcars of ore) is now essentially ready for mass spectrometric analysis. Results from this experiment can be expected in 1989.

The main problem in the AMS measurement is the separation of ^{98}Tc from the ^{98}Mo remaining after the chemical separation. A feasibility study was performed by the Rochester group[22]. It was shown that ^{98}Mo contamination must be at the 10 ppt level in the Re matrix. This means that the 10^7 atoms of ^{98}Tc should be contained in a mg of Re with no more than 10^8 atoms of ^{98}Mo. Less restrictive chemical purity would be possible using a gas-filled magnet for isobar separation.

An interesting variation of the ^{98}Mo experiment is given in Ref. 23. Haxton and Johnson suggest using the reaction

$$^{98}Mo(\nu_e,e^-)^{97}Tc + n$$

($T_{1/2} = 2.6 \times 10^6$ a, E(threshold) = 8.96 MeV)
to measure the neutrino flux from stellar collapse. This proposal is very attractive. They predict, at a stellar collapse rate of 0.09 a^{-1} with an average of 0.8×10^{53} erg in the electron-neutrino channel, that the average ν_e flux at the Earth due to supernovae will be 3.6×10^3 v/s. Although this is ~ 300 times smaller than the flux of ν_e from the ^{8}B reaction (using the low-metallicity non-standard solar model adjusted to fit the ^{37}Cl counting rate), the galactic neutrino contribution to ^{97}Tc is predicted to be 40% of the solar contribution. As just described, the ^{98}Mo to ^{98}Tc experiment is already underway. It costs nothing to look for ^{97}Tc at the same time, and integrated over the lifetime of ^{97}Tc this may provide a good estimate of the rate of stellar collapse in the gallaxy.

3.2. A radiochemical experiment (using AMS)

A new application of AMS and radiochemical methods to solve the "solar neutrino puzzle" has been recently suggested[24].

The following reaction is considered:

$$\nu + {}^7\text{Li} \rightarrow e^- + {}^7\text{Be} \quad (T_{1/2} = 54 \text{ d}).$$

According to the standard solar model, the capture rate in ^{7}Li corresponds to 51.8 ± 0.31 SNU. 100 tons of lithium would produce 3000 ^{7}Be atoms. Such a low number of atoms cannot be measured by counting techniques. In fact, 90% of ^{7}Be decays emitting 50 eV electrons and only 10% decays with 0.477 MeV γ-rays. Direct measurement of the atoms is possible with AMS. The use of 10 μg of ^{9}Be carrier would produce a ^{7}Be/^{9}Be of 2.2×10^{-15}. A major problem is the efficiency of the AMS system, as we mentioned in section 2. Fireman et al. suggest a multiple sputtering procedure to concentrate BeO on the source electrode and obtain at least 100 events in the final detector.

4. PROTON DECAY

According to the Grand Unified Theories, the proton is expected to decay with a lifetime of 10^{31} years (Refs. 25, 26 and 27). The geochemical method can be used in this case to search for the disappearance of the proton and is inclusive of all the decay channels.

One proposal is the decay chain

$$^{130}\text{Te} \rightarrow {}^{129}\text{I} \rightarrow {}^{129}\text{Xe} \quad (\text{Ref. 28})$$

where the proton decays in a tellurium mineral of geological age (10^9 a), producing an isotopic anomaly in the ^{129}Xe abundance. A limit for the proton decay half-life of 1.6×10^{25} a was obtained by Evans and Steinberg, analyzing the data obtained from geochemical $\beta\beta$ experiments[29].

It was later pointed out that there exists the possibility of detecting the ^{129}I, $T_{1/2} = 1.6 \times 10^7$ a, directly, using accelerator mass spectrometry[30]. ^{129}I does not have a stable isobar and the isotopic background can be separated by time-of-flight measurements

Assuming a half-life of 10^{30} a for the proton decay, there are 1325 atoms of ^{129}I per kg of tellurium at equilibrium. The amount of stable iodine per kg of Te must be kept below 2 mg in order that the isotopic ratio of ^{129}I/^{127}I $\geq 10^{-16}$. The present AMS background for ^{129}I detection corresponds to ^{129}I/^{127}I = 2×10^{-14} (see Ref. 9) but the above isotopic ratio should ultimately be measurable. Overall efficiency for an AMS system is 2.3×10^{-4} and 300 kg of tellurium would be necessary to obtain 100 counts of ^{129}I in the detector for a precision of 10% in the measurement of the isotopic ratio.

Several background sources in the tellurium mineral have to be considered: spontaneous fission, reactions induced by cosmic-ray muons and neutron capture in ^{128}Te. At depths greater than 4000 m, the most important background is due to the reaction.

$$^{128}Te(n,\gamma)^{129}Te \rightarrow ^{129}I$$

induced by neutrons produced by high-energy muons. It has been estimated that 3000 ^{129}I atom per kg of tellurium would be produced at a depth of 8000 m water equivalent, so that much greater depths would be necessary.

5. SEARCHES FOR STABLE HEAVY HADRONS

Many recent Big Bang cosmologies[31, 32] have predicted the formation of heavy (10 GeV to 100 TeV) stable or nearly stable (lifetimes comparable to the age of the universe) particles during the early moments of the universe at abundances with respect to ordinary hadronic matter of 10^{-10} to 10^{-12}. There are many candidates for this material: the lightest of the technicolor baryons[33], the fermion partners of bosonic particles (predicted by Supersymetric theorists)[34], the so called X$^-$ and X$^+$ particles of Cahn and Glashow that only have electromagnetic interactions with ordinary matter[33], or the lightest six-quark state[35], to name just a few. Regardless of their origin, if they exist and if they are charged, they should be observable in ordinary matter. If they are positively charged, they probably have similar chemical properties to hydrogen and would appear as an anomalously heavy isotope of hydrogen. If they are negatively charged, they would probably exist bound to an otherwise ordinary nucleus (with nuclear charge Z) and would behave as an anomalously heavy isotope of an element with Z-1. Neutral particles may or may not be bound to ordinary nuclei. If they are, then the searches that will be described shortly would detect them; but if they exist as heavy isolated neutral particles, they could have heretofore escaped notice. Recently[36], the possible

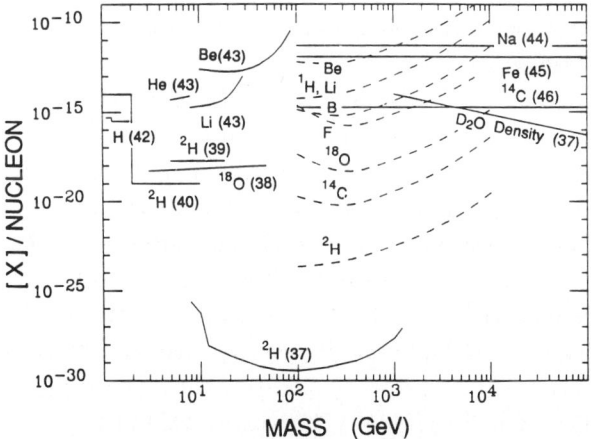

FIGURE 6

Concentration limits (90% confidence level) for the existance of heavy isotopes in matter[41]

existence of yet another class of particles has been proposed that would result in isotopes that are lighter than ordinary ones; so far there have not been searches to look for them. Before the advent of accelerator searches, the limits on anomalous nuclei of arbitrary mass were generally rather poor, with an upper limit for X/nucleon ~ 10^{-6} (where X is a hadron of anomalous mass).

Fig. 6 summarizes the recent searches for anomalous isotopes. More then half of these searches used AMS. The limits below the level of 10^{-15} include a "pre-enrichment" stage such as electrolysis[37] or isotope enrichment[38, 39, 40]. The measurement shown with dashed lines were made using an all electrostatic AMS analysis system[41] which meant that the system was largely mass independent, and anomalous isotopes over the entire mass range indicated could be looked for simultaneously. The other AMS searches required changing all the magnetic elements (bending magnets and magnetic quadrupole lenses) in the injector and high-energy analysis systems in overlapping steps to continuously scan the indicated mass regions. If one assumes that geologic fractionation has not dramatically reduced the concentration of heavy isotopes in ordinary materials, these searches appear to rule out the existence of charged (and possibly even neutral particles) with masses of 1 GeV to 10^5 GeV.

6 SEARCHES FOR SUPERHEAVY ELEMENTS

There is a certain fascination connected with the notion that there is an island of nuclear stability lying just beyond the known peninsula of nuclei forming our world. Several accelerator experiments have attempted to create these nuclei, but they have not been successful, perhaps because the island does not exist, but maybe because of the difficulties of creating such elements by the collision of two heavy nuclei. An alternative to the accelerator experiments is to search for superheavies in nature.

A few years ago, Nilsson, Thompson, and Tsang[47] predicted that the superheavy element with $Z = 110$ and $A = 294$ ($^{294}110$) might have a lifetime on the order of 10^8 years. Although there is some uncertainty whether superheavy elements can be produced by the r process during stellar collapse, the prediction of Shramm and Fowler[48] indicates that the ratio of $^{294}110$/Pt (Pt is its chemical homologue) should be about 1/100. This prompted a number of searches for superheavy elements in various platinum ores.

These searches have consisted of looking for the radiation from a characteristic decay (alpha emission, fission tracks or gammas from spontaneous and induced fission) and one accelerator search[49]. The AMS experiment consisted of a time-of-flight detector coupled to a gas cell to measure energy and a silicon surface barrier detector to provide one of the fast signals for timing. The result of the AMS measurement was to place an upper limit on the fractional number abundance of $^{294}110$/Pt $< 1 \times 10^{-11}$. While consistent with the other searches, it is inconsistent with a lifetime on the order of 10^8 years or a production fraction as high as 0.01. Nonetheless, this search demonstrated the ability of AMS to detect (or set limits on the detection of) heavy, high-Z nuclei.

7. SEARCHES FOR FRACTIONALLY CHARGED PARTICLES

AMS offers an attractive method of searching for fractionally charged particles (FCP) because of the variety of materials that can be studied, and, assuming a positive result, the additional information about the FCP that can be determined, e.g. mass. In general, the energy requirements for these searches is low, most of them have been performed with 3 MV machines or smaller. Tandem accelerators offer an additional advantage of even more positive identification because of analysis as both negatively charged and positively charged ions. Often, very clever schemes can be employed to extract fractionally charged materials from the ion source.

The first AMS search for FCPs was done by Elbert[50] with a single-ended 1 MV accelerator, an RF ion source, an electrostatic and magnetic analysis system and a silicon detector. He was able to set a limit of about 10^{-13} per nitrogen molecule. Subsequent searches by Boyd[51, 52], Schiffer[53], Kutschera[54], Chang[55] and McKeown[56] were inspired by the claims of FCPs detected by the Fairbanks group. The sensitivities (10^{-15} to 10^{-20} per atom over a mass range of about 0.2 to 300 GeV/c^2) of these studies are comparable to the levels at which Fairbanks claims to have seen FCPs.

The Rochester group constructed a special all electrostatic analysis system to search for FCPs and anomalous-mass isotopes[57]. In an experiment on fractionally distilled xenon enriched an assumed 10^6 times, they set a limit of 10^{-18} on the existence of FCPs.

Despite the fact that no AMS search has detected a single FCP, searches will undoubtably continue, especially as new source materials are suggested, e.g. meteorites, or as new detection schemes are developed. Without pre-enrichment of some sort, however, the limits set by AMS are not likely to be much better than 10^{-16} to 10^{-18}, because of the ion source limitations.

8. NEW OPPORTUNITIES TO DETECT ββ

The importance of Double Beta Decay (DBD) experiments as tests of lepton number conservation, mass and charge conjugation properties of the electron neutrino, and right-handed admixtures in the weak leptonic currents[58] is well known.

There are more than 30 DBD reactions that can be looked for among naturally occurring parent isotopes (Table 1). Direct detection is being attempted using solid state detectors, e.g. ^{100}Mo(ββ)^{100}Ru (Ref.59) and time projection chambers, e.g. ^{82}Se(ββ)^{82}Kr (Ref. 60) but the best detection efficiency is obtained when the detector is made from the parent material, e.g. ^{76}Ge(ββ)^{76}Se (Ref. 61) and ^{136}Xe(ββ)^{136}Ba (Ref. 62) .The latter method can be extended to ^{100}Mo ββ-decay and other cases with cryogenic detectors[63].

An alternative to direct detection is the geochemical method, where the signal is an isotopic anomaly in the daughter isotope present in an ancient , natural occurring sample of geological age. Maximization, and indeed detection, of the anomaly require a sample with the parent as the major element and the daughter as an ultratrace element. At best DBD will yield about 10^9 daughter atoms/g. As an illustration of the magnitude this limitation represents, consider a daughter with an isotopic abundance of 15% and an atomic weight of 50. if the the endogenous or primordial concentration of the daughter element is 50 ppt in the geological sample, the corresponding isotopic anomaly will be 1%. Such extraordinarily low concentrations lie well below the detection limits of conventional chemical analytical methods thus restricting the preliminary selection of promising systems to inferences based on well-established geochemical trends.

Considering the possible DBD cases of Table I, four of them produce daughter isotopes that

TABLE 1 - $\beta^-\beta^-$ decay transitions from naturally occurring parent isotopes[64].

Transition	T_0 (KeV)	Abundance (%)
^{46}Ca → ^{46}Ti	985	0.0035
^{48}Ca → ^{48}Ti	4272	0.187
^{70}Zn → ^{70}Ge	1001	0.62
^{76}Ge → ^{76}Se	2045	7.8
^{80}Se → ^{80}Kr	136	49.8
^{82}Se → ^{82}Kr	3005	9.2
^{86}Kr → ^{86}Sr	1249	17.3
^{94}Zr → ^{94}Mo	1148	17.4
^{96}Zr → ^{96}Mo	3350	2.8
^{98}Mr → ^{98}Ru	111	24.1
^{100}Mo → ^{100}Ru	3033	2.8
^{104}Ru → ^{104}Pd	1301	18.7
^{110}Pd → ^{110}Cd	2014	11.8
^{114}Cd → ^{114}Sn	540	28.7
^{116}Cd → ^{116}Sn	2808	7.5
^{122}Sn → ^{122}Te	358	4.56
^{124}Sn → ^{124}Te	2278	5.64
^{128}Te → ^{128}Xe	869	31.7
^{130}Te → ^{130}Xe	2533	34.5
^{134}Xe → ^{134}Ba	843	10.4
^{136}Xe → ^{136}Ba	2481	8.9
^{142}Ce → ^{142}Nd	1414	11.1
^{146}Nd → ^{146}Sm	61	17.2
^{148}Nd → ^{148}Sm	1928	5.7
^{150}Nd → ^{150}Sm	3367	5.6
^{154}Sm → ^{154}Gd	1250	22.6
^{160}Gd → ^{160}Dy	1731	21.8
^{170}Er → ^{170}Yb	655	14.9
^{176}Yb → ^{176}Hf	1077	12.6
^{186}W → ^{186}Os	489	28.6
^{192}Os → ^{192}Pt	408	41.0
^{198}Pt → ^{198}Hg	1043	7.2
^{204}Hg → ^{204}Pb	414	6.9
^{232}Th → ^{232}U	850	100.0
^{238}U → ^{238}Pu	1146	99.275

are noble gases; three of these cases (^{82}Kr, ^{128}Xe and ^{130}Xe) have already been studied with geochemical methods by conventional mass spectrometry[65]. Among the remaining cases, 3 have as parents either Kr or Xe which do not form solids on Earth. Six other parent elements, Ce(Nd), Nd(Sm), Sm(Gd), Gd(Dy), Ru(Pd) so resemble their daughters chemically that finding an ore of the parent not containing a significant amount of the daughter is unlikely. We are left then with the following elemental pairs ; Ca(Ti), Zn(Ge), Ge(Se), Zr(Mo), Mo(Ru), Pd(Cd), Cd(Sn), Sn(Te) and U(Pu). With the exception of U(Pu), most of these elemental pairs share one discouraging feature, namely a tendency to concentrate in sulphur-bearing phases.

Of the three most promising cases (see also Ref. 66), Ca(Ti), Zn(Ge) and Zr(Mo), in light of the AMS requirements, the decay ^{70}Zn($\beta\beta$)^{70}Ge is most attractive.

8.1 The ^{70}Zn($\beta\beta$)^{70}Ge case

The feasibility of this measurement must be evaluated with respect to the following parameters: backgrounds from the accelerator and the sample, AMS selectivity and precision in measuring the isotopic ratios and the overall efficiency of the AMS system.

The following formula relates germanium contamination (F, fraction by weight) to the anomaly (ΔR) in the natural isotopic ratio (R) induced by the DBD of ^{70}Ge (decay constant λ) in a mineral of geological age T:

$$\Delta R/R = 0.0302 \times \lambda \times T / F$$

As an example, if the germanium background is 10^{-12} (by weight), with a DBD half-life of 10^{21} a, an isotopic anomaly larger than 4% is produced in a sample with a geological age of 1.8×10^9 a. A background of 10^{-9} will produce the same effect with 10^{18} a half life.

8.1.1. Ion Source background

Our measurements at the XTU Tandem of the Laboratori Nazionali di Legnaro show that

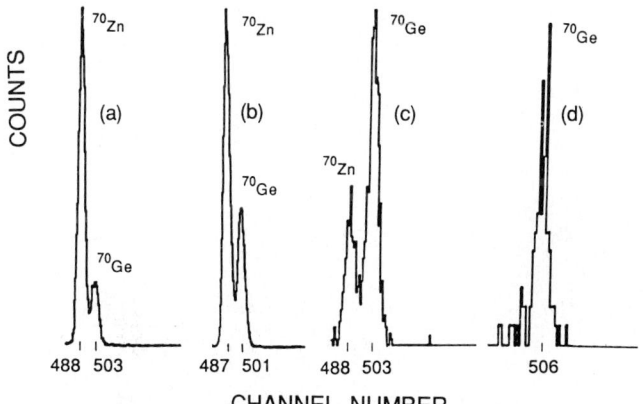

FIGURE 7
Bragg peak spectra, (a), (b) and (c), from a zinc sample and with the slits after the injection magnet progressively closed, (d) from a cathode of pure tantalum. XTU terminal vltage 8.3 MV, charge state 9$^+$.

with a cathode of very high purity tantalum, a ^{70}Ge counting rate of 5 cps is observed when injecting and transmitting mass 70 through the accelerator at 8.3 MV, equivalent to a germanium contamination of about 1 ppb (Fig. 7d). The Cu⁻ current was 0.2 µA. The same background was observed at the FN tandem at the University of Pennsylvania, where a ^{70}Ge counting rate of 65 cps was observed with a Cu current of 2.75 µA (Fig. 8-c).

J.M. Anthony and D. J. Donahue suggest that modification of the Cs ion source and sample holder can lower the source background for germanium to the ppt level[67].

8.1.2. Sample background

The principal zinc ores are sphalerite and wurtzite, two forms of ZnS, and a variety of others evidently formed from it under oxidizing conditions: ZnO, $ZnCO_3$, Zn_2SiO_4, etc. Sphalerite often contains appreciable amounts (100 ppm) of germanium. We could not find Ge analysis for the minor minerals listed above but suspect that some Ge accompanies the Zn, perhaps to form germanates. Some sphalerite, namely that formed at high temperatures, contains "little" germanium. Thus it may be, depending on the detailed conditions, that some of the secondary zinc minerals that originate as low-Ge sphalerite could wind up with a few ppt of germanium. Louis Cabri, from Energy, Mines and Resources (Canada), provided some pure sphalerite from different ores. They were analyzed for germanium contamination by M. Rivers, S. Sutton, A. Hansen and K. Jones using synchrotron radiation - based X-ray fluorescence at the Brookhaven National Light Source with the following results: i) Wiarton ore (430x10^6 a), 730 ppm germanium, ii) Gaina ore (900 x 10^6 a), < 50 ppm germanium, iii) Nygroven ore (1850 x 10^6 a), < 50 ppm germanium. Ge was detected using the Kβ peak at 11 KeV because of the Zn interference at the Kα energy. Neutron Activation Analysis of samples ii) and iii) will be performed before attempting AMS measurements.

8.1.3. AMS selectivity

The ^{70}Ge/^{70}Zn ratio in sphalerite samples will be 10^{-12} if DBD half-life is 10^{21} years.

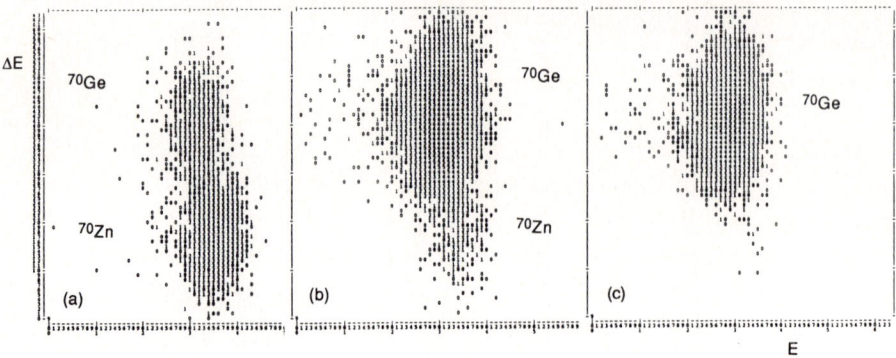

FIGURE 8

E-ΔE spectrum from a cathode of commercial zinc (a) with source slits open, (b) with source slits closed to 3.55 mm, and (c) from a cathode of pure tantalum obtained with the FN tandem at the University of Pennsylvania at an incident energy of 80 MeV (for a description of this AMS system, see Ref. 68)

Measurements at the XTU tandem of Legnaro and at the FN tandem of the University of Pennsylvania show that the ratio of the Zn⁻ ionization probability to Cu⁻ ionization probability is 10^{-11} - 10^{-12}. In fact, a ^{70}Zn counting rate of few cps was observed in both cases with a cathode containing commercial zinc (Figs. 7-a-b-c and 8-a-b). There is a strong evidence that this intereference does not come from the injection of ^{70}Zn⁻ but from the ^{70}ZnH⁻ tail. The alignment of the beam optics was obtained by transmitting ^{65}Cu^{9+} at the appropriate terminal voltage.

In conclusion, the sputter source and the final particle identification system are the key elements of the AMS system which make it possible to obtain the required separation between ^{70}Ge and ^{70}Zn during the ^{70}Ge analysis of sphalerite minerals. AMS selectivity is sufficient but the background (mainly from the original sample) appears to be the major factor limiting sensitivity.

The remaining cases will be briefly discussed in the following.

8.2. The ^{46,48}Ca($\beta\beta$)^{46,48}Ti case.

Extraterrestrial materials offer the advantage of old (4.5x10⁹ a) ages but the disadvantage of significant cosmic ray bombardment. The curve of cosmic abundances is such that of all the pairs in the list of Table 1, only the Ca and Zr form independent minerals in meteorites. In practice, only Ca compounds are abundant enough in meteorites to be worth further consideration. Most Ca-bearing minerals in meteorites contain significant quantities of Ti. The phosphates whitlockite and apatite may not. Unfortunately these minerals are enriched in U and Th and the decay products of extinct ^{244}Pu. To our knowledge, no Ti analyses are available for the phosphate minerals in meteorites.

Terrestrial materials are customarily classified as igneous, sedimentary or metamorphic. The lowest Ti content observed in feldspars is about 10 ppm. It is improbable that a sedimentary, Ca-bearing mineral would exclude Ti. It might be that some minor mineral such as fluorite (CaF$_2$) excludes Ti because of a mismatch in ionic radii.

8.3. The ^{92,94}Zr($\beta\beta$)^{92,94}Mo case.

This case seems promising from the geochemical point of view. Unlike Ca and Ti, the two elements have decidedly different chemical affinities. Zr is strongly lithophile while Mo may be siderophile or chalcophile. In addition, Zr forms a well-defined igneous mineral, zircon, which in a recent report of trace element analyses of zircon, gave no results for Mo.

9. CONCLUSION

The use of AMS to measure the products of rare nuclear and cosmological processes accumulated or archived for geologic times in natural samples would appear to offer a fantastic potential for determining very low rates and very small cross sections. To some extent, this potential has been realized. The limits set by AMS on the existence of relic particles leftover from the Big Bang are comparable to, or far exceed the limits that can be set by any other analytic technique. The limits on rare isotopes of many types: anomalously heavy, fractionally charged, or of superheavy elements, have in general been good enough to challenge the predictions made by theory. These successes have occurred when the full sensitivity of AMS could be exploited: when the isotope being sought was in itself distinctive.

On the other hand, major difficulties arise when attempting to use AMS to exploit the advantages of natural samples in accumulating products of rare processes over geologic time. The advantage of natural samples, compared with the detectors used in direct counting experiments whose duration is a year or so, are their comparatively small size and the 10^6 to 10^9 times larger concentrations of product. But because the nuclei that result from double-beta decay, inverse beta decay or proton decay are "ordinary" isotopes, high sensitivity is only possible when the resulting element has been geochemically excluded from the target material. This rarely occurs to a level of better than a part per billion and finding samples where levels are this low is difficult. In addition, backgrounds may occur from handling the sample, or from contamination in the ion source. Finally, as selective as AMS machines are, they do not have the precision of isotope-ratio machines, restricting the size of measurable isotopic anomalies to being larger than they might otherwise have to be.

Improvements in the efficiency, background and precision of AMS are constantly being made. Increased sensitivity is possible because of the development of high-intensity ion sources and techniques of universal usefulness in removing isobaric interferences. But in overcoming the geological and geochemical restrictions we must depend on the kindness of Nature (and on the collaborative attitude of geologists). The ultra-pure separations from enormous amounts of mineral material required in some of the geological experiments needs the contributions of large and well organized groups. This interdisciplinary effort is justified only as long as we think these methods offer the possibility of testing some of the fundamental predictions made by the new physical theories.

ACKNOWLEDGEMENT

We would like to thank G.F. Herzog for pointing out some of the geochemical problems involved in the $\beta\beta$ studies with AMS.

REFERENCES

1) A. Salam, Preprint ICTP Trieste 1988, IC/88/109

2) A. Salam, Preprint ICTP Trieste 1987, IC/87/402

3) J. Rich, D. Lloyd Owen and M. Spiro, Physics Reports 5&6 (1987)239

4) W. Kutschera, Nucl. Instr. and Meth. B5(1984)420

5) Proc. Third Int. Symp. on Accelerator Mass Spectrometry, Zurich, Switzerland (1984), eds. W. Woelfli, H.A. Polach and H.H. anderson,, Nucl. Instr. and Meth. B5 (1984) 91-448

6) Proc. Fourth Int. Symp. on Accelerator Mass spectrometry, Niagara-on-the-Lake, Ontario, Canada (1987),eds. H. E. Gove, A.E. Litherland and D. Elmore, Nucl. Instr.and Meth. B29(1987)1-445

7) A.E. Litherland, Annu. Rev. Nucl. and Part. Sci. 30 (1980) 437

8) L. Brown, Annu. Rev. Earth. Plan. Sci. 12(1984)39

9) D. Elmore and F.F. Phillips, Science 236(1987) 543

10) C. Tuniz et al. Nucl. Instr. and Meth. B29(1987)133.

11) N. Thonnard et al. Nucl. Instr. and Meth. B29(1987)398

12) G. Bonani et al. Nucl. Instr. and Meth. B29 (1987) 87

13) J.N. Bahcall, R. Davis J.r, and L. Wolfenstein, Nature 334 (1988) 487

14) J.N. Bahcall and R.K. Ulrich, Rev. Mod. Phys. 60 (1988) 297

15) W. C. Haxton and G.A. Cowan, Science 210(1980)897

16) W. Henning, Phil. Trans. R. Soc. Lond. A323(1987)87

17) J. Klein, R. Middleton, D. Fink, and P. Sharma, Submitted Radiocarbon

18) M. S. Freedman et al. Science 193(1976)1117

19) Proc. of the Int. Conf. on Solar Neutrino Detection with ^{205}Tl, and Related Topics, Dubrovnik, Yugoslavia (1986), eds. B.C. Maglich, J. Norwood Jr. and A. Newman, Nucl. Instr. and Meth., A271(1988)237.

20) G.A. Cowan and W.C. Haxton, Science 216(1982)51

21) J. Rapaport et al. Phys. Rev. Lett. 54(1985)2325

22) D. Elmore et al. Nucl. Instr. and Meth. B5(1984) 109

23) W.C. Haxton and C.W. Johnson, Nature 333 (1988) 325

24) E.L. Fireman, A.E. Litherland and J.K. Rowley, Nucl. Instr. and Meth. B29 (1987)387

25) J.M. Losecco, Comments Nucl. Part. Phys. 15 (1985)23

26) W. Lucha, Comments Nucl. Part. Phys. 16(1986)155

27 D.H. Perkins, Non-accelerator experiments, AIP Conference Proceedings, Lake Louise, Canada (1986), ed. Donald F. Geesman, American Institute of Physics, New York, 1986.

28) C. Evans and R.I. Steinberg, Science 197 (1977)989

29) L. W. Hennecke et al., Phys. Rev. C 11 (1975) 1378

30) K.W. Allen, Nucl. Instr. and Meth. 186 (1981) 479

31) S. Wolfram, Phys. Lett. 82B (1979) 65

32) C. B. Dover, T.K. Gaisser, and G. Steigman, Phys. Rev. Lett. 42 (1979) 1117

33) R. Cahn and S. Glashow, Science 213 (1981) 607.

34) J. Wess and B. Zumino, Nucl. Phys. B70 (1974) 39

35) H. Fritzsch, Phys. Lett. 78B (1978) 611

36) C.B. Dover and H. Feshbach, Phys. Rev. Lett. 59(1987)2539

37) P.F. Smith et al. Nucl. Phys. B206 (1982) 333

38) R. Middleton, R.W. Zurmuhle, J. Klein and R.V. Kollarits, Phys. Rev. Lett. 43 (1979) 4293

39) T. Alvager and R.A. Naumann, Phys. Lett. 24B (1967) 647

40) R. Muller, L.A. Alvarez, W.R. Holley, and E. J. Stephenson, Science, 196 (1977) 521

41) T.K. Hemmick et al. Nucl. Instr. and Meth. B29 (1987) 389

42) R.N. Boyd et al. Phys. Lett. 72B(1978) 484

43) J. Klein, R. Middleton, and W.E. Stephens, Symp. on Accelerator Mass Spectrometry, Argonne National Lab. (1981) 136

44) W. J. Dick, G.W. Greenlees, and S.L. Kaufman, Phys. Rev. Lett. 53 (1984) 431

45) E. B. Norman, S.B. Gazes, and D.A. Bennett, Phys. Rev. Lett. 58 (1987)1403

46) A. Turkevich, K. Weilgoz, and T.E. Economou, Phys. Rev. D30(1984)1876

47) S.G. Nilsson, S.G. Thompson, and C.F. Tsang, Phys. Lett. B28 (1969) 458

48) D.N. Schramm and W.A. Fowler, Nature 231 (1971) 103

49) W. Stephens, J. Klein, R. Zurmuhle, Phys. Rev. C21 (1980) 1664

50) J.W. Elbert et al. Nucl. Phys. B20(1970)217

51) R.N. Boyd, D. Elmore, A.C. Melissinos, E. Sugarbaker, Phys. Rev. Lett. 40(1978)216

52) R.N.Boyd et al. Phys. Rev.Lett. 43(1979)1288

53) J.P.Schiffer, T.R. Renner, D.S.Gemmell, F.R. Mooring, Phys. Rev.D17(1978)2241

54) W. Kutschera et al. Phys. Rev. D29(1984)791

55) K.H. Chang, Ph.D. thesis, University Toronto

56) R.D. McKeown, Phil. Trans. R. Soc. Lond. A323(1987)145

57) D. Elmore et al. Nucl. Instr. and Meth. B10(1985)738

58) S.P. Rosen, Comments Nucl. Part. Phys. 18(1988)31

59) M. Alston-Garnjost et al. Phys. Rev. Lett. 60(1988)1928.

60) S.R. Elliott, A.A. Hahn and M.K. Moe, Phys. Rev. Lett. 59(1987)1649

61) E. Bellotti et al. Il Nuovo Cimento 95A(1986)1

62) A. Alessandrello et al. Nucl. Physics A478(1988)453.

63) E. Fiorini and T.O. Niinikoski, Nucl. Instr. and Meth. 224(1984)83

64) W.C. Haxton, Progress in Particle and Nuclear Physics, 12(1984)409

65) T. Kirsten, H. Richter, and E. Jessberger, Phys. Rev. 50(1983)474

66) H.E. Gove, Nukleonika 25(1980)31

67) J. M. Antony and D. J. Donahue, Nucl. Instr. and Meth. B29 (1987) 77

68) R. Middleton, J. Klein, Phil. Trans. R. Soc. Lond. A 323(1987)121

V

EXPERIMENTAL TECHNIQUES AND PARTICLE ACCELERATORS

INTRODUCTION TO ACCELERATORS AND ACCELERATOR PHYSICS

Modesto PUSTERLA

Dipartimento di Fisica "G.Galilei" - Via Marzolo 8, 35131 Padova (Italy)
Istituto Nazionale di Fisica Nucleare - Sezione di Padova.

Description of the main types of accelerator.

If one wants to introduce an elementary course on accelerators and accelerator Physics one cannot avoid mentioning the electrostatic machines which initiated the long story of collisions between beams of particles and target and are still useful nowadays in several versions.

The desired voltage is normally obtained by the Van de Graaff electrostatic generator which, by means of a belt that carries charges from a generator ($\sim$ 10KV) induces electrostatic charges in the so-called "hot" terminal (usually spherical) so that one may reach values of the order of several M Volts.

TANDEM. The high-voltage generator is the same electrostatic charging belt as used in the Van de Graaff. The new idea, namely to utilize the high voltage for two-stage acceleration, is based on the possibility of creating a well defined beam of negative ions. Negative ions are formed at ground level and injected into the first stage when acceleration to the terminal with a potential + V_T takes place. In the first stripper canal (thin foil), negative ions lose at least two electrons and change charge to positive ions with the charge state+n. For hydrogen ions, n=+1, but for heavier ions, it is possible to obtain n>1. In the second stage, the positive ions gain the energy neV_T (in eVolts) and total energy at the exit is $e(n+1)V_T$. Today, the highest terminal voltage obtainable in a tandem is about 25MV, which is perhaps near saturation. Only singly charged, negative ions can be formed while it is possible for heavy ions to obtain multicharged positive ions by stripping. Now it is possible to form negative ion beams of almost all elements. At sufficiently high energies we can obtain fully stripped ions.

For instance the ion Au⁻ at a terminal voltage V_T=15 MV get stripped to n=+13 and an energy of 208 MeV can be obtained in the second stage of the tandem.

The tandem accelerator is still a very important tool in nuclear physics, but it can be envisaged that a combination of tandem accelerators and strippers will be used as injectors to relativistic heavy-ion colliders.

RADIO-Frequency Accelerator.

It is, in practice, difficult and expensive to use DC (direct corrent) accelerators for obtaining the highest energies. Acceleration over many gaps, i.e. radio-frequency acceleration is therefore here an attractive solution. Two types of RF accelerators will be considered, namely linear accelerators and cyclic accelerators.

Linear accelerators. The LINAC is a linear machine the development of which started before 1930. Its improvement went parallel with the improvement of the RF technology. Today the highest obtainable Linac energy is 15 GeV for electrons and a Linac for 50 GeV is almost ready for the single-pass collider at Stanford, CAL-USA.

Ion Linacs are present is several laboratories (UNILAC at G.S.I.) and RF and induction linacs are under study for future research in pure and applied physics (inertial fusion).

RFQ. Radio-frequency quadrupoles. A new type of low energy accelerator for very high currents has been developed during the last few years. Both acceleration and focusing are obtained for a specially shaped RF field. Superconducting cavities are introduced nowadays. It can be said that RFQ's are going to replace the huge Cockroft-Walton accelerators as injectors to high energy accelerators.

Cyclic Accelerators. The principle of these machines is based on a combination of RF acceleration and bending of charged particles in a magnetic field. In this way, it is possible to use acceleration gaps over and over. There are many types of cyclic accelerators and, since the early thirties, they have played a decisive role in accelerator Physics from the lowest to the highest energies.

The Cyclotron is the first example of a circular machine: a homogeneous magnetic field B_0, supplied by an H-shaped magnet, bends back the particles to the same RF gap (between the two D-shaped electrodes) twice each period of radiofrequency. The path of ions with the electric charge e is circular, with a radius given by $R = \frac{mV}{eB_0}$. Since R is proportional to the velocity V, the period of circulation T, and thereby also the frequency f, is constant for all radii. Once in resonance, the ion will receive an energy gain each time it passes the acceleration gap between the D's. We have then:

$$T = \frac{2\pi R}{V} = 2\pi \frac{m_0}{eB_0} \quad \text{and} \quad f = \frac{1}{T} = \frac{1}{2\pi} \frac{eB_0}{m_0}.$$

For each passage through an acceleration gap, the energy gain is $\Delta E_p = eV_m \cdot \sin\varphi$ where V_m and φ are the RF maximum voltage and phase respectively. The final kinetic energy (particle energy) is $E_p = \frac{1}{2} \frac{e^2}{m_0} B_0^2 R_{MAX}^2$.

It can be shown that the ions oscillate in a certain phase region around the synchronous phase φ_s without being lost. This stable-phase oscillation is vital for obtainable current intensity and the principle of phase stability is general for all RF accelerators.

Limitations. The synchronism breaks down because of firstly, the relativistic effect which for protons sets around $E_p \sim 20$ MeV and secondly the decrease of the magnetic field B for $R \sim R_{MAX}$. Mainly due to relativistic mass, the saturation for the fixed frequency cyclotron takes place for $E_p \sim 20$ MeV, corresponding to a magnet diameter of $\sim$ 1m.

Synchrocyclotron. The new idea came in 1945. A decreasing field is given by $B_z \cong B_0 \left(\frac{R_0}{R}\right)^n$ n>0 (z, vertical). One easily sees that the field component B_R (horizontal, radial) yields a restoring force for 0<n<1, the result being harmonic oscillations around the equilibrium orbit with frequencies $f_R = f_0 \sqrt{1-n}$ and $f_z = f_0 \sqrt{n}$ (0<n<1); the resulting effect is a focusing one (called weak focusing) and these oscillations are called betatron oscillations.

From the condition $B = B_0 - \Delta B$ and $m = m_0 \gamma$ we obtain

$$f = \frac{1}{2\pi} \frac{eB}{m} \cong \frac{e}{2\pi} \frac{B_0(1-\Delta B/B_0)}{m_0 \gamma}$$

From $E_0=m_0c^2$ and $E=m_0c^2\gamma=E_0+E_p$ we have

$$\gamma = \frac{E}{E_0} \text{ and } f=f_0 \frac{(1-\Delta B/B_0)}{1+E_p/E_0} \cong f\left(1-\frac{\Delta B}{B_0}\right)\left(1-\frac{E_p}{E_0}\right)$$

It is easily seen from this eq. that the lack in synchronism due to the relativistic effect and the decrease of B with R can be overcome by a decreasing frequency f of the RF system with the energy. The periodic change of frequency f is carried out with a fast rotating condenser: synchrocyclotrons are also called frequency modulated or, in short FM cyclotrons. In this way we can reach a maximum energy of 1 GeV for protons. The disadvantage stays in the reduction of the current since acceleration takes place only when f is decreasing. Other important drawbacks are the hugeness of the massive magnet and the cost (which increases with R^3).

<u>Isochronous cyclotron. (AVF cyclotron)</u>. The problem connected with the lack of resonance due to the increase óf the relativistic mass in a homogeneous magnet is solved with an <u>Azimuthally Varying</u> <u>Field (AVF) cyclotron</u>. Here, the edge focusing effect is used to overcome the defocusing effect due to an <u>increasing magnetic field B_z with increasing R</u>. In this way it is possible to operate the cyclotron with fixed frequency even to relativistic velocities. Modern superconducting magnets make this type of cyclotron very promising fro heavy ions.

<u>Synchrotrons</u>.

The principle is as follows: we have a Linac as injector and the massive magnet of the cyclotron is here replaced by a ring of bending magnets. An RF system is used for acceleration and the magnetic field increases with energy in such a way that the radius is kept constant during acceleration: the bunches of charged particles circulate within the vacuum chamber. This vacuum chamber has transversal dimensions considerably large in the weak-focusing synchrotrons (width more the 100cm and height ~ 30); hence only few weak-focusing synchrotrons have been built. The real breakthrough in synchrotrons came with the invention of "strong focusing" magnets.

The first two large synchrotrons (AGS at Brookhaven and PS at CERN) of this type were combined-function accelerators: in addition to bending the magnets have a strong lens effect. The newer synchrotrons are built as separate-function accelerators. Here simple dipole magnets are used for bending and strong quadrupoles take care of the focusing. The two largest synchrotrons of this type are SPS at CERN and FNAL in USA (both with diameters more than 2 Km long and an energy $E_p \sim 500$ GeV). FNAL is now equipped with superconducting magnets and the beam energy is raised to 1 TeV, the highest in the world. This machine is named TEVATRON . In a fixed target proton accelerator collisions take place between nucleons with the same rest mass m_0c^2; in high energy physics the goal is to produce new particles or to study new reactions; at high energies, the relativistic mass of the incident proton is very large compared to the target nucleus at rest; consequently the main part of the particle energy E_p is wasted as kinetic energy of colliding particles and their reaction products. Only a small fraction, the centre of mass energy E_{cm}, is available for new reactions and for production of new particles. More specifically for nucleon-nucleon collision, the available energy is $E_{cm}=2m_0c^2 \left[\sqrt{1+\frac{E_p}{2m_0c^2}} -1 \right]$ and $E_p \gg m_0c^2 (\sim 1$ GeV) gives $E_{cm} \cong \sqrt{2 \; E_p}$ GeV.

Colliding Beam Accelerator (colliders). To overcome the above form of insufficiency at high energies physicists thought of making the c.m. system of collisions coincide with the laboratory system by head-on collisions between bunches of particles. In these cases (for equal rest mass particles) $E_{cm}=E_p+E_p=$ $=2E_p$. This argument holds both for electrons, protons and heavy ions.

In order to acquire a good density in the bunches you need to store the particles and therefore the physics of the colliding beams passes through the storage ring technique. (We can remind that colliders and storage rings were first conceived for electrons in the pioneering work at Frascati by a team directed by B. Touschek).

Double ring hadron colliders. We mention the old ISR (30 GeV + 30 GeV) and the present projects LHC and SSC (still under RCD and not found yet).

Single ring colliders: we here quote the SPS p$\bar{\text{p}}$ and the Tevatron p$\bar{\text{p}}$. They produced a lot of Physics and particularly the fundamental results in the field of electroweak field theory giving experimental detailed confirmation of the standard-model of Weinberg-Salam, the latter begins its runs during this year and physicists expect further confirmations of the SPS-UA1, UA2 experiments plus something more because of the higher energy (more than 1 TeV in the c.m.).

LEP, HERA, TRISTAN, SLC and SLD which involve e^+e^- collisions and e-p collisions, are on the point of entering the adventure of high-energy physics and one expects further discoveries plus a better understanding of what has already been seen.

Applications in other fields of pure and applied Physics. We like here to recall the special accelerators for medical and industrial uses.

Synchrotron Radiation. The synchrotron radiation, from electron synchrotrons, has become so important in solid state and biophysics that dedicated machines, named light-sources are now under construction in URSS, Europe and USA.

Implanters. They are used by several industries interested in the modification of the surface properties of metals, an implanter is a small accelerator in the energy range of 50-500 KeV equipped with a universal ion source. Since isotopically pure beams are often required, implanters are furnished with rather powerful magnets for isotope separation.

Inertial Confinement Fusion. It is the great future interest of obtaining energy in the controlled light nuclei fusion, as an alternative approach to the complicated magnetically confined plasmas: the possibility of depositing energy on a D-T pellet in a fusion reactor from the impact with heavy ion beams that come from a system of accelerating machines and storage rings.

I. Motion of the beam particles (Linear theory of perfect machines)

The equilibrium orbit is defined as the periodic solution of the equation of motion

$$\frac{d}{dt}(m\gamma\vec{V}) = e\ \vec{V} \times \vec{B} \tag{1}$$

In machines with a median plane it lies in that plane and in the hard-edge approximation consists of straight lines in the straight sections and quadrupoles, and arcs of circles in the bending magnets. We here study small deviations from the equilibrium orbit and introduce a curvilinear coordinate system: the s-coordinate runs along the equilibrium orbit and the x and y are perpendicular to it: x is radial, in the median plane and y is perpendicular to it.

Eq. (1), once we introduce the system xys above defined, gives in the linear approximation limit, in the x,y (transverse motion) planes the following equations

$$\frac{d^2x}{ds^2} + \left[\frac{1}{\rho^2(s)} - K(s)\right] \cdot x = \frac{1}{\rho(s)} \frac{\Delta p}{p} \tag{2}$$

$$\frac{d^2y}{ds^2} + K(s)\ddot{y} = 0 \tag{2'}$$

$\rho(s)$ is the radius of curvature, $K(s)$ defines the focusing strength of an element and $\Delta p/p$ is the momentum error of the particle;

$$K(s) = -\frac{1}{B\rho} \frac{\partial B_y}{\partial x} \tag{3}$$

and $B\rho = p/e$ is the magnetic rigidity of the particle, a constant of the motion in purely magnetic fields.

Eqs. (2), (2') neglect correctly, because they are smaller values, the linear coupling terms which link the two equations; terms which are at least quaratic in x and/or y give rise to nonlinear phenomena that may be very relevant and are under investigation in the theoretical and simulated dynamics in the lattices nowadays.

We may start discussing the transversal motion under the limiting case $(\Delta p/p)=0$. The two eqs. (2), (2') are obviously similar (betatron oscillations):

$$\frac{d^2u}{ds^2} + n(s) u(s) = 0 \tag{4}$$

where the exact meaning of $n(s)$ depends of the axis we consider (x or y). Since the equilibrium orbit is closed, $n(s)$ is a periodic function with period, the circumference (we may also have superperiods): $n(s+L)=u(s)$.

The solution of any second order differential equation (4) can be written as follows

$$\begin{aligned} u(s) &= C(s,s_0) u(s_0) + S(s,s_0) u'(s_0) \\ u'(s) &= C'(s,s_0) u(s_0) + S'(s,s_0) u'(s_0) \end{aligned} \tag{5}$$

where the cosine - like and the sine-like functions $C(s,s_0), S(s,s_0)$ depend on

s_0 and s and s are normalized $C(s_0,s_0)=S'(s_0,s_0)=1$ $S(s_0,s_0)=C'(s_0,s_0)=0$.

Eqs. (5) can be written in matrix form:

$$\begin{vmatrix} u(s) \\ u'(s) \end{vmatrix} = M(s|so) \begin{vmatrix} u(s_0) \\ u'(s_0) \end{vmatrix} = \begin{vmatrix} C(s,s_0) & S(s,s_0) \\ C'(s,s_0) & S'(s,s_0) \end{vmatrix} \begin{vmatrix} u(s_0) \\ u'(s_0) \end{vmatrix} \quad (6)$$

which separates the properties of the machine lattice, M, from the initial conditions. The determinant of M is equal to the Wronskian W and C and S. W is a constant of the motion as it can be shown directly and can be normalized to 1. Successive application of (6) for positions $s_1, s_2 \ldots$ shows $u(s_i), u'(s_i)$ related to $u(s_0)$, $u'(s_0)$ by a product of matrices (6). This fact is fundamental and is used in the transfer matrix for lattices and insertions.

The matrices become particularly simple when $n(s)$ is piece-wise constant

$$M(s|so) = \begin{vmatrix} \cos\varphi & n^{-\frac{1}{2}}\sin\varphi \\ -n^{\frac{1}{2}}\sin\varphi & \cos\varphi \end{vmatrix} \quad (6')$$

where $\varphi = n^{\frac{1}{2}} \cdot (s-s_0)$. For $n<0$ a more convenient way of writing $M(s|s_0)$ is

$$M(s,so) = \begin{vmatrix} \cosh\psi & (-n)^{\frac{1}{2}}\sinh\psi \\ (-n)^{\frac{1}{2}}\sinh\psi & \cosh\psi \end{vmatrix} \quad (6'')$$

$\psi = (-n)^{\frac{1}{2}}(s-s_0)$

To ensure stability of the betatron oscillation it is sufficient that the function C and S remain bounded for all s. In order to find this condition we study the matrix for a full period (or superperiod) $L: M = M(s_0+L|s_0)$. We recall that

$$M \begin{vmatrix} u_0 \\ u'_0 \end{vmatrix} = \lambda \begin{vmatrix} u \\ u' \end{vmatrix}_0$$

and therefore it can be written as follows

$$M = \begin{vmatrix} \cos\mu + \alpha\sin\mu & \beta\sin\mu \\ -\gamma\sin\mu & \cos\mu - \alpha\sin\mu \end{vmatrix} \quad (7)$$

with the constraint $\beta\gamma - \alpha^2 = 1$.

The eigenvalues of M are $\lambda = \exp\{\pm i\mu\}$ and $\cos\mu = \frac{1}{2} \text{Tr}|M|$

M can be alternatively represented by:

$$M = I \cos\mu + J \sin\mu \tag{8}$$

when I is the unit matrix and J has the form

$$J = \begin{vmatrix} \alpha & \beta \\ -\gamma & -\alpha \end{vmatrix} \text{ with } J^2 = -I \tag{9}$$

and

$$(I \cos\mu_1 + J \sin\mu_2)(I \cos\mu_2 + J \sin\mu_2) = I \cos(\mu_1 + \mu_2) + J \sin(\mu_1 + \mu_2) \tag{10}$$

and, consequently,

$$M^k = I \cos k\mu + J \sin k\mu \tag{11}$$

Since the elements of M^k are bounded for all k the betatron oscillations are stable, if μ is real, which implies the stability criterion

$$|\text{Tr}(M)| \leq 2 \tag{12}$$

Analytical developments

The Hill equation:

$$\frac{d^2u}{ds^2} + n(s)u = 0 \tag{13}$$

gives by making use of the Floquet theorem, the quasi-periodic solution:

$$u(s) = u_A(s) + u_B(s) \tag{14}$$

with $u_A(s) = A w(s) \cos\psi(s)$, $u_B = B w(s) \sin\psi(s)$

A,B are arbitrary constants determined by the initial conditions, u_0 and u'_0
W(s) and $\psi(s)$ are periodic functions of the cell-length L.

By imposing that u_A and u_B verify eq. (13) we obtain:

$$W'' - W \psi'^2 + nW = 0 \tag{15}$$

$$2 W' \psi' + W \psi'' = 0 \tag{16}$$

The latter equation yields $\psi' = \frac{\text{const}}{W^2}$ the constant can be chosen 1 without any lacking of generality; the former becomes

$$W'' + n(s) W - \frac{1}{W^3} = 0 \tag{17}$$

The final form of the matrix $M(s|s_0)$ is then:

$$M(s|s_.) = \begin{vmatrix} a & b \\ c & d \end{vmatrix} \quad (18)$$

where $a = \dfrac{W(s)}{W_0} \cos(\psi(s)-\psi_0) - W(s) W_0' \sin(\psi(s)-\psi_0)$ (18')

$b = W(s) W_0 \sin(\psi(s)-\psi_0)$ (18")

$c = -\dfrac{1+W(s)W_0 \, W'(s)W'_0}{W(s) W_0} \sin(\psi(s)-\psi_0) - \left[\dfrac{W_0}{W(s)} - \dfrac{W'(s)}{W_0}\right] \cos(\psi-\psi_0)$ (18''')

$d = \dfrac{W_0}{W} \cos(\psi-\psi_0) + W_0 \cdot W' \cdot \sin[\psi-\psi_0]$ (18^{iv})

if we define $\beta(s) = W^2(s), \alpha(s) = -\tfrac{1}{2}\beta' = WW'$, $\mu(s) = \psi-\psi_0$

the matrix $M(s|s_0)$ can finally be written as follows

$$\begin{vmatrix} \sqrt{\dfrac{\beta(s)}{\beta_0}} [\cos \mu(s) + \alpha_0 \sin \mu(s)] & \sqrt{\beta_0 \beta(s)} \sin \mu(s) \\ -\dfrac{(\alpha(s)-\alpha_0)\cos \mu(s) + [1+\alpha_0\alpha(s)]\sin \mu(s)}{\sqrt{\beta_0 \beta(s)}} & \sqrt{\dfrac{\beta_0}{\beta(s)}} [\cos \mu(s) - \alpha(s) \sin \mu(s)] \end{vmatrix}$$

Since for $s=s+L$ one has $\mu(s)=\mu$ $u(s)=u_0$, $u'=u'_0$ $W=W_0$ $W'=W'_0$ the Matrix (18) coincides with eq. (7). From the above definitions one writes:

$$\mu = \int_s^{s+L} \psi' ds = \int_s^{s+L} \dfrac{ds}{W^2} = \int_s^{s+L} \dfrac{ds}{\beta(s)} = Q \dfrac{2\mu}{m} \quad \text{(phase advance per cell)}$$

$$Q_H = \dfrac{N}{2\pi} \int_s^{s+L} \dfrac{ds}{\beta_{H,V}} \quad N = \text{number of cells.}$$

Emittances. From the use of the matrix (18) we obtain:

$$u(s) = \sqrt{\dfrac{\beta}{\beta_0}} (\cos \mu + \alpha_0 \sin \mu) \cdot u_0 + \sqrt{\beta_0 \beta} \cdot \sin \mu \cdot u'_0 \quad (19)$$

$$u'(s) = -\dfrac{(\alpha-\alpha_0)\cos \mu + (1+\alpha_0 \alpha)\sin \mu}{\sqrt{\beta_0 \beta}} u_0 + \sqrt{\dfrac{\beta_0}{\beta}} (\cos \mu - \alpha \sin \mu) u'_0 \quad (20)$$

and after setting $\dfrac{u_0}{\sqrt{\beta_0}} = \sqrt{\varepsilon} \cos \delta_0$ and $-\left(\dfrac{u_0}{\sqrt{\beta_0}} \alpha_0 + \sqrt{\beta_0} u'_0\right) = \sqrt{\varepsilon} \sin \delta$

one has $u(s) = \sqrt{\epsilon\beta(s)} \cos(\mu(s)+\delta_0)$

$$u'(s) = -\sqrt{\frac{\epsilon}{\beta}} \{\alpha(s) \cos(\mu+\delta_0) + \sin(\mu+\delta_0)\}$$

$$\epsilon = \gamma_0^2 u_0^2 + 2\alpha_0 u_0 u_0' + \beta_0 u_0'^2 = \gamma^2 u^2 + 2\alpha uu' + \beta u'^2 \qquad (21)$$

Then ϵ is a <u>constant of the motion and is invariant</u>: it is called emittance (horizontal and vertical emittance).

Eq. (21) describes an ellipse in the (u,u') plane. This ellipse may refer either to the capability of the lattice to "accept" the beam or the beam considered a family of trajectories. In the first case one looks at a lattice property and defines its acceptance, in the second case we refer to the beam properties and consider it as the emittance.

Adiabatic invariants

Particles are accelerated; the question arises whether an invariant exists which can describe the acceleration cycle. Since the energy increase is extremely slower than the betatron oscillations, the so-called adiabatic approximation applies letting the use of the Boltzman-Ehrenfest theorem:

$$I = \text{const.} = \oint pdq \qquad (22)$$

where p and q are canonical-conjugate variables and the integral is meant to be taken over a whole oscillation period.

Let us consider the horizontal motion, then $q=x$, $p=\gamma mx=px =$

$= \gamma m \dfrac{ds}{dt} \dfrac{dx}{ds} = \beta m\gamma cx' = px'$:

$$I_H = p \oint x' dx = p\pi \cdot \epsilon_H = \pi\, mc\, \beta\gamma\, \epsilon_H = \pi mc\, \epsilon_H^* \qquad (23)$$

Similarly

$$I_V = \pi\, mc\, \epsilon_V^* \qquad (24)$$

having defined the normalized emittances $\epsilon_H^* = \beta\gamma\epsilon_H$ and

$$\epsilon_V^* = \beta\gamma\epsilon_V \qquad (25)$$

<u>Dispersion</u>. A momentum spread in the beam yields a modification in the horizon-

tal motion equation namely:

$$\frac{d^2x}{ds^2} + n(s)x = \frac{1}{\rho}\frac{\Delta p}{p} \qquad (26)$$

In analogy to what has been done in the homogeneous equation the general matrix M(s) transforms into:

$$M(s) = \begin{vmatrix} C(s) & S(s) & D(s) \\ C'(s) & S'(s) & D'(s) \\ 0 & 0 & 1 \end{vmatrix} \qquad (27)$$

where D(s) is defined as <u>dispersion</u> on the initial vector $(x_0, x'_0, \frac{\Delta p}{p})$ we obtain:

$$x(s) = C(s) x_0 + S(s)x'_0 + D(s)\frac{\Delta p}{p} \qquad (28)$$

$$x'(s) = C'(s) x_0 + S'(s) x'_0 + D'(s)\frac{\Delta p}{p} \qquad (28')$$

$$\frac{\Delta p}{p} = \frac{\Delta p}{p}$$

and if we take into account the splitting $x(s)=x_\beta(s)+x_p(s)$ where $x_\beta(s)$ refers to the homogeneous equation solution and $X_p(s)$ is a particular solution of the inhomogeneous equation:

$$x_\beta(s) = C(s)x_0 + S(s) x'_0 \qquad (29)$$

$$x'_\beta(s) = C'(s) x_0 + S'(s)x'_0 \qquad (29')$$

$$x_p(s) = D(s)\frac{\Delta p}{p} \qquad (30)$$

$$x'_p(s) = D'(s)\frac{\Delta p}{p} \qquad (30')$$

<u>Momentum compaction</u>

We define the momentum compaction $\alpha = \left[\frac{\Delta C}{C}\Big/\frac{\Delta p}{p}\right]$ with $C = \oint ds$ and $C+\Delta C = \oint d\sigma$ where $d\sigma$ is an infinitesimal arch of trajectory referred to an off-momentum particle, while ds is referring to a particle of given momentum p:

$$d\theta = \frac{d\sigma}{\rho + x_p} = \frac{ds}{\rho} \qquad (31) \qquad d\sigma = ds\left(1+\frac{x_p}{\rho}\right) \qquad (32)$$

and, by integration $C + \Delta C = C + \oint \frac{Xpds}{\rho}$ (33)

$$\Delta C = \oint \frac{X_p ds}{\rho} = \frac{\Delta p}{p} \oint \frac{D(s)}{\rho} ds$$ (34)

and finally

$$\alpha = \frac{1}{C} \oint \frac{D(s)ds}{\rho}$$ (35)

II. Longitudinal dynamics (synchrotron oscillations)

Transition energy. The angular frequency of a particle revolving in a synchrotron is

$$\omega = \frac{2\pi}{T} = \frac{2\pi\beta c}{C}$$ (36)

By differentiation of eq. (1) we obtain:

$$\frac{d\omega}{\omega} = -\frac{dT}{T} = \frac{d\beta}{\beta} - \frac{dC}{C} = \left(\frac{1}{\gamma^2} - \alpha\right)\frac{dp}{p}$$ (37)

where $\eta = \frac{1}{\gamma^2} - \alpha$ can be rewritten $\eta = \frac{1}{\gamma^2} - \frac{1}{\gamma_{tr}^2}$

$\eta = 0$ means $\gamma = \frac{1}{\sqrt{\alpha}} = \gamma_{tr}$ (38)

We then define transition the energy value (for the monoenergetic beam particle) such that its corresponding Lorentz γ function verifies eq. (38).
One sees immediately that:

$\eta < 0$ for weak focusing machines and for the strong focusing case, so far the beam energy is bigger than $E_{tr} = \gamma_{tr} mc^2$

$\eta > 0$ for strong focusing synchrotrons when $E < E_{tr}$ (injection) and for linear accelerators where the average radius is infinite and consequently $\alpha = 0$.

Phase stability principle. The Electric field of the RF in the accelerating cavities ideally has to be synchrotron with the particles revolving in the circular machine: perfect synchronism means equal phase φ_s between the crossing particles and the oscillating field. This practically happens for a very little amount of the particles present in the beam. All other particles will cross the RF cavity (or cavities) at different times (or phases) either anticipating or retarding with respect to the synchronous particle.

Thus, below transition eq. (37) gives $\frac{dT}{T} = -|\eta| \frac{dp}{p}$ meaning that more energetic particles go faster than less energetic ones, exhibiting quite a natural behaviour. Hence the accelerating voltage must be chosen on the positive slope (ϕ_s between 0 and 90°) so that faster (slower) particles receive lower (higher) energy kick at every next crossing in the cavity, thus undergoing an energy oscillation around the synchronous particle (phase stability).

Opposite situation above transition where $\frac{dT}{T} = |\eta| \frac{dp}{p}$: this implies that particles with higher energy tend to go slower, contrary to any physical intuition acting as if their masses were negative. In this case the phase-stability requirement imposes ϕ_s between 90° and 180°.

During the accelerating cycle, if $E_{injection} < E_{transition}$, as has been occurring so far in proton synchrotrons, there will be a stage where transition-crossing must take place. A fast phase-jump has to be performed from ϕ_s to $\pi-\phi_s$, to continue keeping the synchronism.

<u>Energy oscillation equation</u>. The physical quantities related to a generic particle and to the synchronous one (labelled by the letter s) are connected by the following relations:

total energy $E = E_s + \Delta E$ (39)

momentum $p = p_s + \Delta p$ (40)

angular frequency $\omega = \omega_s + \Delta\omega$ (41)

revolution period $T = T_s + \Delta T$ (42)

and again $\omega_{RF} = h\omega_s$ (43)   h≡harmonic number.

Defining as energy gain per turn $\delta E = eV \sin\phi$, $\delta E_s = eV \sin\phi_s$ with V=total crest voltage.

From the approximation $\dot{E} = \frac{dE}{dt} \simeq \frac{\delta E}{T}, \frac{\dot{E}}{\omega} \simeq \frac{\delta E}{2\pi}$

But $\delta E \ll E$ and we obtain:

$$\frac{\dot{E}}{\omega} - \frac{\dot{E}}{\omega_s} = \frac{eV}{2\pi}(\sin\phi - \sin\phi_s) \qquad (44)$$

and after cumbersome easy steps we have:

$$\dot{w} = eV(\sin\phi - \sin\phi_s) \qquad (45)$$

introducing the canonical conjugate variable of ϕ

$$w = \frac{2\pi}{\omega_a} \Delta E = C\Delta p \quad (C\text{-circumference}) \tag{46}$$

as it will be clarified later.

After a revolution of the synchronous particle, the phase of a generic particle is changed by $T_s \cdot \frac{d\phi}{dt}$; on the other hand the same phase-variation is seen by the RF cavities as $\omega_{RF}\Delta T$:

$$\Delta\phi \cong T_s \frac{d\phi}{dt} = \omega_{RF} \Delta T \tag{47}$$

We therefore from the previous formulae derive the following equation:

$$\frac{d\phi}{dt} = - \frac{h\eta \; \omega_s^2}{2\pi\beta_s^2 \; \gamma_s \; mc^2} \; w \tag{48}$$

Combining all previous eqs. and particularly eqs. (45), and (46) we arrive at:

$$\frac{d^2\phi}{dt^2} + \frac{\Omega_s^2}{\cos\phi_s} (\sin\phi - \sin\phi_0) = 0 \tag{49}$$

where

$$\Omega_s = \omega_s \sqrt{\frac{h\eta\cos\phi_s}{2\pi\beta_s^2 \; \gamma_s} \; \frac{V}{V_0}} = \omega_s \; Q_s \tag{50}$$

Notice that Ω_s is the synchrotron-oscillation frequency and Q_s is the so--called synchrotron tune, (normally $Q_s \ll 1$).

<u>Small Oscillations</u>. If $\phi=\phi_s+\varphi$ with $\varphi\ll\phi,\phi_s$ one has $\sin\phi \cong \sin\phi_s + \varphi\cos\phi_s$ and eq. (50) reduces to harmonic oscillations:

$$\ddot{\varphi} + \Omega_s^2 \varphi = 0 \tag{51}$$

<u>Large Oscillations</u>. The RF system is usually capable to accept phase excursions much bigger than ; therefore the approximation of small oscillations is no longer valid and the energy (or phase) equation remains the non-linear one (49) which is integrable in terms of non-elementary functions (elliptic functions). Nevertheless a first integration of eq. (49) is possible and gives:

$$\frac{1}{\Omega_s} \dot{\phi} = \pm \sqrt{\frac{2}{\cos\phi_s} (\cos\phi - \cos\phi_s) + 2(\phi - \phi_s) \, \text{tg}\phi_s + \frac{1}{\Omega_s^2} \dot{\phi}_0^2} \qquad (52)$$

where ϕ_0, $\dot{\phi}_0$ are initial conditions, that exhibit either stable solutions (closed curves) or unstable ones (open curves) in the phase plane $\left[\frac{\dot{\phi}}{\Omega}; \phi\right]$.

It is trivial to observe that there is a curve (same situation of the phase-space analysis of the pendulum) that separates the stable from the unstable trajectories. Dynamists call such a curve "SEPARATRIX". The phase-space area enclosed is <u>called</u> bucket and represents the whole capability of the RF system to contain particles. Stable trajectories within the bucket containing the actual beam make up the <u>bunch</u>. A bunch may completely fill the bucket but normally buckets are less than half filled.

One easily obtains the separatrix equation:

$$\dot{\phi}^2 = \frac{2\Omega_s^2}{\cos\phi_s} \left[\cos\phi + \cos\phi_s + (\phi - \pi + \phi_s) \sin\phi_s\right] \qquad (53)$$

Hamiltonian formalism. Eqs; (45) and (48) are hamiltonian type

$$\dot{\phi} = \frac{\partial H}{\partial w}, \quad \dot{w} = -\frac{\partial H}{\partial \phi} \qquad (54)$$

where

$$H = -\frac{h\eta\omega_s^2}{4\pi\beta_s^2 \, \gamma_s \, mc^2} w^2 + eV \left[\cos\phi - \cos\phi_s + (\phi - \phi_s)\sin\phi_s\right] \qquad (55)$$

As we already noticed, w and ϕ are conjugate variables. The interest of the hamiltonian formulation stays in the possibility of adding easily contributions due to the <u>space-charge potential</u> which have been ignored up to now and might become important for intense dense beams.

<u>Storage rings and stacking problems</u>. In storage rings when particles are stored but not accelerated or in capturing injected particles via RF, varying the lenght of a bunch, bunching an unbunched beam (for instance RF capture of a multiturns - injected beam) debunching a bunched beam etc. ..., stationary buckets are required. The main aspect of such buckets is that the synchronous particle is not accelerated:

$\sin\phi_s = 0$ $\phi_s = 0$,   , according to the fact that the machine work below or above transition.

We obtain simplified formulae for the equations of the phase oscillation and the separatrix; the latter becomes:

$$\frac{1}{\Omega_s} \frac{d\phi}{dt} = \pm \sqrt{2(1+\cos\phi)} = \pm \cos\frac{\phi}{2} \tag{56}$$

In the stationary-bucket case the bucket area (in phase space) is easily evaluated: it comes out after trivial steps

$$\text{Bucket Area} \equiv 16\tau_\infty \cdot \sqrt{\frac{\gamma s c^2 V_0 V}{2\pi h \eta}} \tag{57}$$

$$\tau_\infty = \frac{C}{2\pi c}$$

References

1) M. Conte, W.W. MacKay "Principles of particle accelerator Physics" Lectures notes (to be published)

2) CERN - 85-19 (CERN ACCEL. SCHOOL); In particular K.O. Nielsen.

3) CERN - 87-03 (C.A.S. advanced school)

4) E. Ferrari, E. Persico, S.E. Segre' "Principles of Particle Accelerators) W.A. Benjamin, Inc. N.Y. 1968 (chapter on RF by M. Puglisi).

THE PHYSICS CASE FOR EHF

Franco BRADAMANTE

Dipartimento di Fisica dell'Università degli Studi di Trieste
and Sezione di Trieste dell'INFN

1. INTRODUCTION

The description of the sub-atomic world in terms of leptons and quarks, interacting via gluons and electroweak bosons in the framework of the Standard Model, represents a superbe intellectual achievement of the last few decades. Still in spite of its many successes, the Standard Model is not regarded as a complete theory, and there is a general feeling that there is new physics to be discovered, and that to this end both new theory and new experiments are needed.

Three different arenas are being explored and will be explored in the next two decades in a monumental world-wide effort:
- the Energy Frontier, now ranging between .2 and 20 TeV; fundamental issues are the Higgs, the W/Z physics, the supersymmetric quarks and leptons, the new Families;
- the Passive Physics; addressing at monopoles, proton decay, solar neutrinos, neutrino oscillations, fifth force;
- the Precision Frontier; looking at rare and forbidden processes, CP violation, hadron dynamics and spectroscopy, neutrino physics.

It is impossible to-day to predict in which of the three arenas the breakthrough to new physics will occur. The three arenas are clearly complementary, and are being investigated in parallel. In Europe two major facilities, LEP and HERA, are about entering into operation to address the high-energy frontier, while initiatives are already taken towards the next step, a Large Hadron Collider in the LEP tunnel. The Gran Sasso Underground Laboratory is also close to completion, and will give Europe a unique tool to address the questions of the second arena. A modest effort is going on at present in Europe in the third arena, most of the work being done at Brookhaven and at KEK. The European Hadron Facility (EHF) is a proposed new research facility[1] at intermediate energy optimized for the problems of the third arena, and in this talk I will review the physics potential of this project.

2. THE EUROPEAN HADRON FACILITY

The EHF is conceived as an international project of European Countries active in Nuclear and Particle Physics at intermediate energy, and as such it has been proposed to the Bundesministerium für Forschung und Technologie (BMFT), Federal Republic of Germany, and to the Istituto Nazionale di Fisica Nucleare (INFN), Italy. For details on the proponents and on the physics, which has been discussed for about three years, a vast literature exist. I will just mention the most relevant contributions, namely the proceedings of the Workshop in Freiburg (in 1984),[2] of the Winter School in Folgaria (1986),[3] of the International Conference in Mainz (1986),[4] and, of course, the EHF proposal (1987). In particular, I would like to recall here that the first edition of this school was <u>dedicated</u> to the physics at EHF, so that the proceedings of that course can be regarded as a textbook for the field, and everything I will say can be traced back and spelled out in detail there.

EHF is a complex of accelerators intended to produce a high intensity (100 µA) proton beam of 30 GeV kinetic energy, and is meant to provide a broad range of intense, high quality secondary beams of neutrinos, muons, pions, kaons and antiprotons. An essential characteristic of this complex is that it has been designed to accelerate polarized proton beams to full energy.

The design intensity of 100 µA primary protons, which is the prerequisite for the envisaged research program, can be realized and safely handled with present-day technology. The design energy of 30 GeV is determined mainly by the desire to produce copious beams of antiprotons.

The three main components of EHF are a high energy Linac, accelerating on H^- beam to 1.2 GeV, and two fast cycling synchrotrons, a 9 GeV Booster Ring and a 30 GeV Main Ring, with radii and repetition rates in the ratios 1:2 and 2:1 respectively. The repetition rates of the Linac and of the Booster are the same, 25 Hz. Two more rings complement the system: a 9 GeV Holding Ring, with the same radius as the Booster, where the Booster pulses are stored before being transferred to the Main Ring, and a Stretcher Ring, having the same circumference as the Main Ring synchrotron, where the fast extracted 30 GeV beam from the Main Ring is stored and then slowly extracted to produce 100% duty factor secondary beams.

Two experimental Areas are foreseen: a Fast Extraction Hall, for neutrino and pulsed muon physics, and a slow Extraction Hall for counter experiments, with nine secondary beam lines from a sophisticated 3-targets system. A schematic lay-out of the facility is given in Fig. 1.

If realized as a new laboratory in a virgin site, the facility would require an investment cost of about 870 MDM (millions of Deutsch Marks), which includes also laboratory buildings, common services and utilities. As far as I can tell, the availability of this sum of money is still an open problem!

EUROPEAN HADRON FACILITY

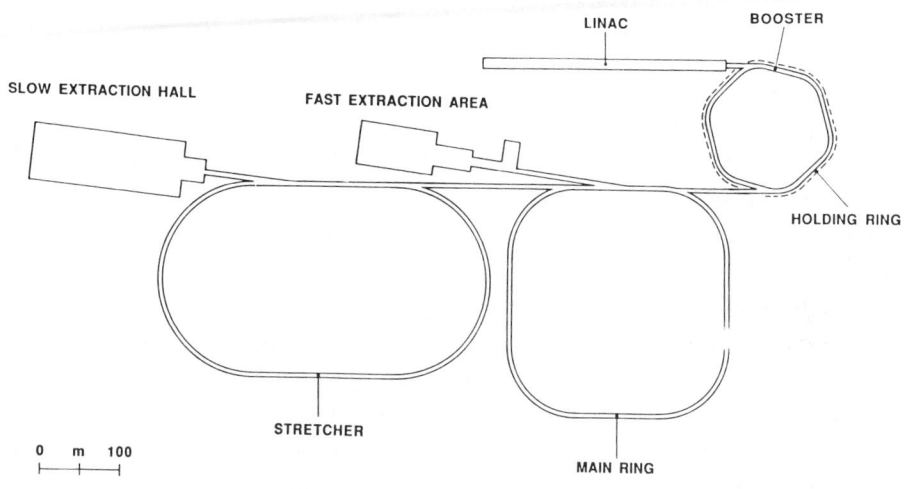

FIGURE 1
Possible schematic lay-out of the accelerator complex of the European Hadron Facility.

3. GENERAL REMARKS

A facility of the size and of the scope of EHF opens up a rich and novel spectrum of experimental possibilities in Nuclear Physics and in specific domains of Elementary Particle Physics, and the list of topics whose study will become possible by means of dedicated experiments is impressive. Nevertheless, there appears one central theme of investigations for which the proposed accelerator complex provides the unique facility which cannot be replaced by any other facilities. This theme may be summarized by Hadronic Interactions in the Confinement Regime. It concerns, generally speaking, the properties of hadrons embedded in the nuclear medium, scattering properties of hadronic systems at non-asymptotic momentum transfers, the domain of nonperturbative Quantum Chromodynamics (QCD), as well as the physics of elementary hadronic systems and the properties of simple hadronic matrix elements such as they occur in radiative and weak (charged and neutral) interactions.

This theme has become to-day a common research goal to nuclear physicists and to particle physicists, who speak very much more the same language today than they did, say ten years ago. EHF as a facility for low energy particle physics and advanced nuclear physics will provide a bridge between these fields and will strengthen the common research goals, to the benefit of both. At the same time, given the size of a collaboration and the time scale for a typical experiment at EHF, EHF will be rather close to the universities and to their role as training institutions for the younger generation.

Thus, EHF not only opens up a rich novel spectrum of experimental possibilities in nuclear physics of the future and in specific domains of elementary particle physics, it also provides the meeting ground of the two disciplines and constitutes the complement to the very large accelerator projects necessary for the advancement of subnuclear physics as a whole.

In the Proposal, we divided the physics for which EHF is the optimal tool into three major themes, each of which will cover a long-range research programme of considerable depth, i.e..

- Electroweak Interactions and Signals of New Physics
- Quantum Chromodynamics in the Quark Confinement Regime
- Nuclear Substructure as Revealed by Nonnucleonic Probes.

Here I will follow essentially the same subdivisions, with the warning, however, that there are many cross-relations and overlaps between these topics. For instance, the investigation of a rare, but existing, semi-leptonic decay of a hadron is relevant for testing QCD in the regime of confinement and chiral symmetry. Conversely, probing hadronic weak current may reflect the limits of the Standard Model, i.e. of flavour physics.

4. ELECTROWEAK INTERACTIONS AND BEYOND

The point of reference in the discussio in of electroweak interactions, at all energies, is the standard $SU(3) \times SU(2) \times U(1)$ model of unified local gauge interactions. Among the basic questions to which EHF can provide partial answers, important hints or clues are the following:
- The nature and the level of parity violation.
- Signals of unification beyond the minimal $SU(3) \times SU(2) \times U(1)$ gauge structure.
- Radiative corrections, i.e. quantum effects which test and establish the standard model as a renormalization quantum field theory through the finite quantum effects it predict.
- Nature of leptonic quantum numbers and structure of the leptonic mass sector
- Quark masses and quark state mixing.
- Lepton-quark symmetry.
- Signals of "horizontal" symmetry, technicolor/or technicolor symmetries.
- Signals of compositeness.

There are many reasons why experiments at EHF energies and intensities can provide important information, complementary to the pionering investigations at higher energies:

(i) Experiments at low energies with high quality beams can be made selective and dedicated in the sense that complete information on final states can be obtained and selection rules can be utilized as "filters" for specific aspects of interest.

(ii) Many of these experiments can reach high precision. This is particularly important, for instance, in precision tests of the standard model, and in the search for signals of new physics in the TeV range. For example, the recent series of precision experiments on muon decay have shown that charged current weak interactions are sensitive to virtual mass scales in the range of 0.5 to 1 TeV. Similarly, muon number changing decays, at the level of 10^{-11} to 10^{-12} in branching ratio, test mass scales up to 100 TeV.

(iii) Rare and ultrarare decay processes are, and will continue to be, sources of information and inspiration for a variety of fundamental questions. These decays can only be studied, mainly for reasons of backgrounds, with the dedicated, high intensity beams of EHF.

The possibilities and options that EHF will open up, include the following:

4.1 NEUTRINO PHYSICS

4.1.1 Neutrino Oscillations

The discovery of transitions between neutrinos of different flavours would provide a fundamental step towards the Grand Unification schemes. Such transitions would be possible if the neutrinos were massive, hence the importance of the issue, also for cosmology. In the simple case of only two neutrino species, ν_e and ν_μ, it is well known that there would be $\nu_e \leftrightarrow \nu_\mu$ oscillations with probability.

$$P = \sin^2 2\vartheta \, \sin^2 \pi \frac{\ell}{L}$$

where ϑ is the mixing angle, ℓ the distance between the neutrino source and the detector, and L the oscillation length, given by

$$L = \frac{2.5 \, E}{\Delta m^2}$$

E in GeV
Δm^2 in eV^2
L in km.

The quantity $\Delta m^2 = m_1^2 - m_2^2$ refers to the linear combinations of ν_μ and ν_e which are mass eingenstates.

At present no experimental evidence exists of neutrino oscillations. The very high intensity of the neutrino beam which can be constructed at EHF makes it possible a most challenging neutrino oscillation experiment, which could improve on the present limits by more than three orders of magnitude. As shown schematically in Fig. 2, the experiment could be performed detecting in the ICARUS detector, in the Gran Sasso Underground Laboratory, $\nu_\mu \leftrightarrow \nu_e$ oscillations in the neutrino beam produced at EHF.

The possibility of using an underground detector placed in the Gran Sasso Laboratory to detect neutrino beams created at CERN has been discussed for a long time[5]. In the EHF proposal the limits on the ν-oscillation parameters Δm^2 and $\sin^2 2\vartheta$ are worked out for the configuration of Fig. 2, where a crucial role is played by the ICARUS detector, a liquid argon chamber with large mass

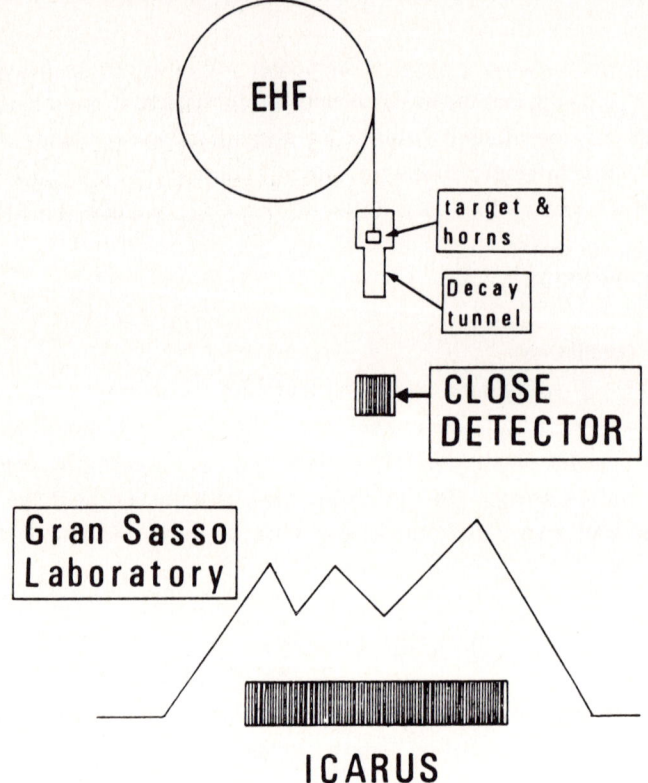

FIGURE 2

Possible experimental arrangement to measure neutrino oscillations between EHF and the Gran Sasso Underground Laboratory.

(5×10^3 tons) and excellent electron identification. These limits are shown in Fig. 3, together with the limits obtained in previous experiments[6]. The distance ℓ between the source and the detector has been taken equal to 700 km, which results in 10^3 interactions in 60 days of 100 μA primary proton beam (one should note that the sensitivity of this experiment has a broad maximum for $\ell \simeq 1500$ km). The background due to atmospheric neutrinos is $\sim 10^3$/year and has a negligible effect on the measurement.

Fron Fig. 3 the gain in sensitivity over the existing experiments is impressive. One should also note that the proposed experiment will reach a domain, $\Delta m^2\ 10^{-4}$ eV², where it is possible to distinguish between the ν-oscillation hypothesis and a possible breakdown of the solar model, the two hypothesis currently put forward to explain the discrepancy between the measured solar neutrino flux[7] and the flux estimated from the Standard Solar Model[8].

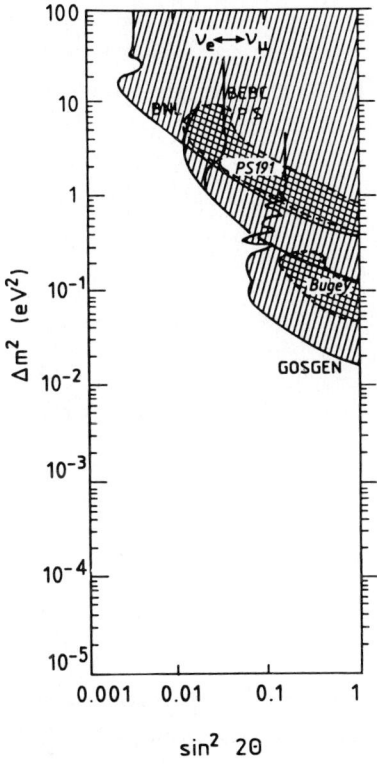

FIGURE 3

Present limits (90% C.L.) for the neutrino oscillation parameters[6] and predicted limits in the EHF + ICARUS experiment, assuming 10^3 (a), 2×10^3 (b), and 10^4 (c) interactions.

4.1.2 Measurement of ϑ_W.

The neutrino oscillation experiment described in the previous paragraph will compare the ν_e flux measured in ICARUS and the same flux measured in a similar detector, close to the neutrino source. Clearly this detector will collect millions of events per year, and will provide extremely interesting physics information in a domain in which statistics has always been the big problem.

One of the processes which could conveniently be studied is the elastic neutrino-electron scattering at low energy, which provides a precise measurement of $\sin^2 \vartheta_W$. Once all radiative corrections are taken into account, a comparison of the $\sin^2 \vartheta_W$ values given at small Q^2 and at the collider constitutes a very important test of the Standard Model. In 200 days a "typical" experiment at EHF could collect 10^5 $\nu_\mu e \to \nu_\mu e$ events, with an average neutrino energy of 0.8 GeV; Particularly interesting is also the possibility of studying the process $\nu_e e \to \nu_e e$, described by the diagrams of Fig. 4. It is a particularly "clean" process, with contributions from charged currents, neutral currents and their interference, so it provides an excellent test for the existence of scalar, pseudo scalar or tensor weak currents. In the wide band EHF neutrino beam in 100 days a "typical" 500 tons detectors could collect $\sim 10^3$ $\nu_e e \to \nu_e e$ events.

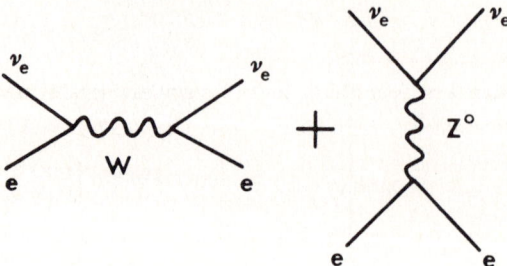

FIGURE 4

Feynmann Diagrams for $\nu_e e \to \nu_e e$ elastic scattering.

4.2 RARE KAON DECAYS

There is overall agreement in the scientific community on the very special role played by the kaon and on the importance of studying its rare decay modes. Indeed, the projects for hadron facilities originated in the late seventies as project for kaon factories, and even recently a low energy proton synchrotron, ASTOR[9], been proposed at SIN specifically as a kaon factory.

The study of the rare k decay is somewhat similar to the study of the rare μ decay at the Meson Factories, and it is worthwhile to recall that the best present limit on the existence of right-handed weak currents is given by a precision experiment on the $\mu^+ \to e^+ \nu \bar{\nu}$ decay at TRIUMF[10]. On the other hand the kaon carries strangeness and therefore belongs to the second generation, so the variety of decay channels accessible to the kaons is much larger than in the case of the muons, and the physics potential is much richer. Moreover, CP violation manifests itself in the kaon system. Although very accurate experiments are at present being performed (NA31 at the CERN SPS and PS195 at LEAR), the study of CP violation in the kaon system will surely be pursued in the future hadron facilities, although it will be a real challange to design an experiment much more accurate than the ones mentioned above.

The rare and ultrarare processes, when seen in the light of the standard model, can be divided into three classes:
- processes which are allowed in the standard model, but in which the basis electro-weak vertices are affected by the strong interactions. These processes give access to hadronic matrix elements of currents which couple to the various gauge bosons, and therefore provide an important "laboratory" and testing ground for chiral dynamics, QCD in the confinement regime, and nonperturbative analysis (such as QCD sum rules and lattice calculations).
- processes which are not forbidden in the standard model, but which depend on unknown parameters of the model, such as fermionic mass scales and mixing matrix elements.
- processes which are forbidden in the standard model and whose existence would provide unambiguous hints "new physics" beyond the model.

As an example of forbidden decays, in the EHF proposal the decay $K^o_L \to \mu e$ has been considered. This decay violates lepton number conservation, hence strictly forbidden in the Standard Model. If observed, it is a clear signal of new physics, and its branching ratio allows to set a new mass scale. In analogy with the $K^+ \to \mu^+ \nu$ decay, mediated by the weak boson W, one can introduce a new heavy boson Y^o to mediate the decay $K^o_L \to \mu e$, as in Fig. 5.

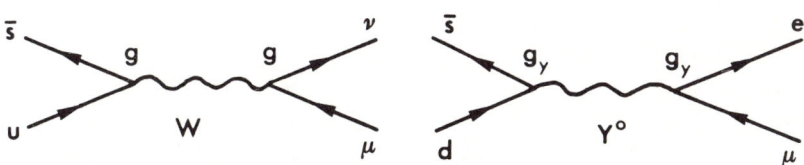

FIGURE 5

Feynmann Diagrams for the decays $K^+ \to \mu^+ \nu$ and $K^o_L \to \mu e$.

Assuming for the coupling constant g_Y a value similar to the weak coupling constant g, one can write for the branching ratio BR

$$BR\ (K_L^o \to \mu e) \sim (M_W/M_\varphi)^4$$

In so far this decay has never been observed, and the present limit to the BR is 10^{-8}, which corresponds to a family boson mass $M_Y > 20$ TeV. At Brookhaven a presently running experiment (E791) is trying to push this limit down to 10^{-12}, which would give $M_y > 200$ TeV, a mass scale unaccessible even to the future generation of colliders! In the EHF proposal it is shown how one hopes to improve on this branching ratio by two more orders of magnitude, mainly by improving on the beam intensity and quality. On top of this, present experience with muon decays shows that improvement of the beam is then followed by improvements in the experimental techniques, so that in the long run the sensitivity is improved by many orders of magnitude.

Among the kaon decays that are not completely forbidden in the Standard Model, but are so strongly suppressed that any observation at a rate above the expected level would be a signature for a failure of the model, particular attention has been given to $K^+ \to \pi^+ \nu \bar\nu$. This decay is suppressed by the GIM mechanism, and is described at the lowest order by the graphs of Fig. 6.

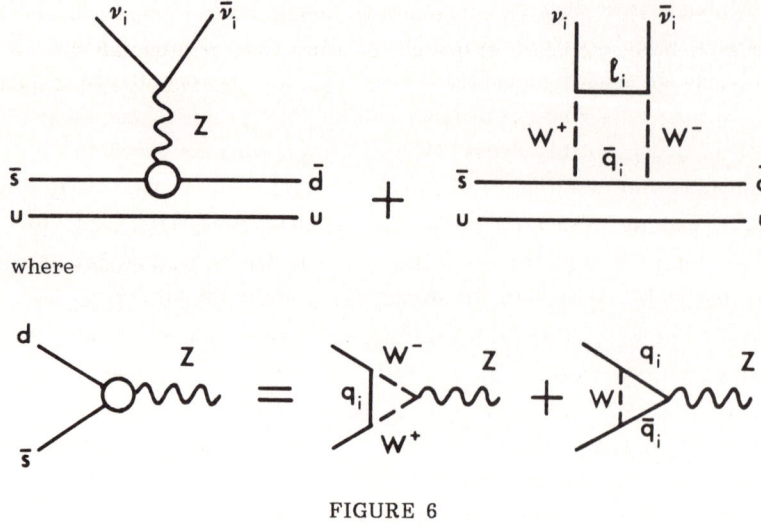

FIGURE 6
Feynmann Diagrams for the decay $K^+ \to \pi^+ \nu \bar\nu$

The branching ratio for this decay has been estimated already in 1974[11] to be BR $\simeq$ N $\cdot$ 3 $\cdot$ 10^{-11}, with N the number of generations with light neutrinos.

More recent calculations give values $2 \cdot 10^{-11} < BR < 9 \cdot 10^{-11}$, for N = 3, where the uncertainty is due to the uncertainties in the top mass, in the B-meson lifetime, and in the semileptonic B-meson decay data. The importance of measuring the branching ratio of this decay is self-evident. At present, it has not yet been observed, and $BR < 10^{-7}$. At Brookhaven experiment E787 aims at a 10^{-10} sensitivity, which is the level predicted by the Standard Model. In the EHF proposal it is shown how one could reach a sensitivity of 10^{-12} and provide a good measurement of this decay.

4.3 MATTER-ANTIMATTER SYMMETRY, GRAVITATIONAL EXPERIMENTS

Low-energy antiproton physics has become a major research field since LEAR, the CERN Low Energy Antiproton Ring, entered in to operation in 1983. The possibility of decelerating to Van der Graaf energies intense (10^{10}) bunches of $\bar{p}$'s stored and cooled in the ACOL complex has opened up extremely interesting physics programs, which, as it usually happens, go beyond the original motivations which pushed for the constructions of LEAR.

Present technology allows to trap bunches of antiprotons and to cool them to very low temperature ($10°K$)[12]. Further progress can be expected in the near future in the trapping efficiency, the cooling rate and in the maximum numbers of $\bar{p}$s that can be trapped at a time.

These facts open up the possibility of comparing the fundamental properties of p and $\bar{p}$ to a high degree of accuracy, using modern versions of classical experiments on both particles, either at the same time or alternatively.

The physics of these measurements is mainly related to two fundamental problems:

- CTP invariance,
- gravity.

The inertial masses (or actually the e/m ratios) can be compared with a relative accuracy of $10^{-9} \div 10^{-10}$). A measurement of the matter-antimatter gravitation interaction ($\bar{p}$s against the earth's gravitational field) will be very selective, in the framework of gravitation theories (supergravity), that deduce from general principles a vector and a scalar potential of the Yukawa type. This measurement is in fact the only one to see the sum and not the difference of these two Yukawa terms. Both measurements (inertial mass and gravitational acceleration) have been approved at CERN[13].

A beautiful test of CTP in electromagnetic interactions can be performed with $\bar{H}_o$[14]. The proposed experiments need development of the cooling techniques in the ring and in the trap, and a substantial improvement in the $\bar{p}$ production rate, but the key problem is how to confine the $\bar{H}_o$s. Overcoming this last problem also needs high production rates of $\bar{H}_o$s, consequently the long term future of this type of physics also will probably depend on the availability of the intense p̄s beams at EHF.

5. QUANTUM CROMODYNAMICS IN THE QUARK-GLUON CONFINEMENT REGIME:

5.1 PARTICLE PHYSICS AND NUCLEAR PHYSICS ASPECTS.

With reference to the standard model, one may divide the basic issues in our present understanding of elementary particle physics into questions of flavour dynamics and, as far as strong interactions are concerned, problems of colour dynamics. Althouhg this subdivision is not sharp, one may say that the highest energy machine address primarily flavour physics, whereas intermediate energy facilities like EHF allow the study of colour physics, i.e. strong interaction in a regime where it exhibits its most interesting and most characteristic features. This applies equally well to experiments with elementary particle targets and to experiments with nuclear targets. As already remarked, the distinction between these is somewhat artificial for the purposes of QCD in the confinement regime, insofar as an experiment in nuclear medium can be as informative as, or complementary to, experiments on elementary targets. In the next section, we discuss first a few general aspects of QCD in the confinenement regime. We then illustrate our remarks with a couple of specific topics in strong interaction physics for which EHF provides the optimal tool of investigation.

5.2 COLOUR PHYSICS: GENERAL ASPECTS AND FUNDAMENTAL QUESTIONS

Local colour symmetry, formulated quantitatively by QCD, appears today as the key to understanding the filigree of strong interaction physics revealed by the accelerators of the sixties: AGS, PS, SLAC, etc.. This "revelation", however, has not become a full and deep understanding for a number of reasons, both experimental and theoretical. On the theoretical side, QCD has not really developed into a complete and unfailing calculational tool, as compared to QED for instance, due to the considerable problems posed by a full-fledged quantum field theory with a complex ground state structure and essential nonperturbative features[15]. Some of the basic issues in QCD have not been solved in a satisfactory manner, such as the closeness of the hadronic world to the chiral limit, or the correlations in the QCD ground state as manifested by the appearance of quark and gluon condensates. Although these problems are difficult, and although nonperturbative methods to solve them are still in their infancy, important progress is expected towards understanding a quantum field theory such as QCD in its nonperturbative regime (the physics of confinement) by testing the basic ideas of the theory with subtle and diversified sets of data.

On the experimental side, the initial phase of enthusiasm throughout the sixties was followed by a period of disenchantment with hadronic physics, and a concentration of efforts on "asymptotic" physics at the highest energies. It would be unwise, however, to put aside this whole domain of physics as a kind of "chemistry" of strong interactions, too complex and too difficult to be tractable. On the contrary, I believe that real progress is possible in the future, due to very much improved experimental techniques (as compared to the sixties), and thanks to the interplay between theory and experiment which helps in understanding and spelling out the subtleties of the hadronic world at low energies.

In the assessment of the standard model of electroweak and colour interactions, the hard problems of QCD are in fact thought to be best approached through large-scale computer calculations, i.e. by means of Monte Carlo simulations of local gauge theories on lattices. Personally, I do not consider satisfactory this solution to the problem of strong interactions, and I believe our goal should still be to calculate in a <u>simple</u> way the pion mass, as it was done for the Rydberg at the beginning of this century. To this end, hints from Nature are essential, and to dispose in the nineties of a tool capable of producing a quantity of hadronic data is seen as a crucial step towards finally understanding the key problems of hadronic matter.

5.3   SPECTROSCOPY

The quark model has achieved considerable success. The aim now is to refine the study of the quark dynamics and search for new states with constituent gluons.

<u>Ordinary Hadrons</u>   One should complete a detailed scanning of radial and orbital excitations of mesons to see whether there is room left for new states like multiquarks, hybrids or gluonia. For baryons too there is the possibility of multiquarks or hybrids, and also some open questions concerning the three-quark states. Most observed baryon resonances, for instance, are compatible with a quark-diquark picture where one relative distance is frozen. States where both oscillators (corresponding to the two Jacobi relative coordinates) are excited should be identified to confirm the three-body nature of baryons, i.e. to count the number of degrees of freedom in hadronic matter.

Gluonia Gluonia have been studied mostly in J/Ψ decay. Other entrance channels should be considered: $N\bar{N}$ annihilation, production in meson-baryon or baryon-baryon scattering, etc. with a variety of final orbits. The couplings to strange and non-strange particles are necessary to unravel the gluonium nature of the state.

Hybrids As soon as the gluon is acknowledged as a coloured constituent, as in gluonium, one expect hybrid mesons or baryons: $q\bar{q}g$, $qqqg$. These states are likely to decay into orbitally excited mesons or baryons. This means a rather high total multiplicity, requiring sophisticated detectors. Hybrids might have escaped detection in previous spectroscopy experiments.

Heavy Flavours (hidden) Quarkonia have been studied with e^+e^- machines, with a great deal of success. However, it allows only for the formation of JPC = 1^{--} states and the observation of some C = +1 states via γ emission. The $q\bar{q}$-spectrum is thus incomplete and one cannot determine the spin structure of the $q\bar{q}$ potential. In $p\bar{p}$ collisions, all $q\bar{q}$ states are, in principle, accessible. The feasibility of $p\bar{p} \to c\bar{c}$ has been demonstrated by the R 704 experiment at ISR. Apart from the mass spectrum, $p\bar{p}$ experiments will allow for a systematic study of the decay of the χ-states (similar to the very rich information obtained from the study of J/Ψ decay).

Heavy Flavour (open) The spectroscopy of open charm and beauty is rather poor so far. We know only some ground states, like D, D*, F, F*, Λ_c, Σ_c, B, etc., with one exception, a candidate for an orbital excitation of the D. These states are however extremely interesting, since the light quark, when associated with heavy quarks, experience more relativistic corrections than in ordinary hadrons (remember that pe^- is more relativistic than e^+e^-).

5.4 THE CASE FOR SPIN PHYSICS IN THE LIGHT OF QCD

At the "confinement frontier", spin observables provide very powerful tools to penetrate the intricacies of hadronic phenomena at intermediate energies. At the hadron level, spin emerges as a combination of spins and angular momenta of quarks. Thus the observation of spin effects will shed light on the dynamics that confines spin-1/2 quarks within well-defined regions of space-time (bags). In particular, as already documented, the presence of sizeable spin effects at large momentum transfer runs counter to the expectations of perturbative QCD, although according to general belief, QCD should be directly applicable in this region. Thus, in the experimental programme of EHF, the investigation of polarization phenomena at large angles and transverse momenta is of primary importance. The results will be crucial for obtaining information on how quark confinement works at short distances in space-time.

6. QUARK CONFINEMENT AND NUCLEAR PHYSICS

Much of what we said above applies also to the investigation of nuclear structure at the scale of quark and gluon degrees of freedom. For instance, one of the oldest, and yet to a large extent unresolved, basic problems of nuclear physics is the nature of the hard core observed in nucleon-nucleon scattering. In the light of the quark model, the hard core seems to be linked to the inhibiting effects of the Pauli principle between constituent quarks. If the simple explanations that were put forward have some truth in them, what would one find in Λ - N or Σ - N scattering?

QCD as a non-Abelian local gauge theory has much more structure than one would expect on the basis of perturbative quantum field theory. Just as in the case of electrodynamics, it may well be that QCD in an extended medium like nuclear matter exhibit new phenomena of many-body collective type. Such effects may show up in processes like

- Production and propagation of resonances in nuclei.
- Antiproton annihilation in nuclei.

As very little is know to date, this field is in a stage of exploration and speculation. Quite obviously, many experiments are needed in order to clarify the basic mechanisms.

Even at a more modest and less speculative level there are basic issues and questions in nuclear physics that must be clarified, such as: Where and how does the quark-gluon substructure of nuclei shows up? How can one reconcile the description of nuclear dynamics in terms of meson and baryon degrees of freedom with the QCD description of nuclei in terms of quarks and gluons?

6.1 PRODUCTION AND PROPAGATION OF RESONANCES IN NUCLEI

Some examples of experimental projects in this area are the following:

(i) Production and scattering of baryonic resonances in nuclei. We already have some information on $\underline{\Delta}$, Λ(1116) and Σ interactions with nuclei, from which a very simple pattern seems to emerge, in agreement with naive model picture but still a long way from real understanding. Very little is known, however, about the formation of higher resonances in nuclei and their interactions in the nuclear medium. For instance, the states Λ(1405) and Λ(1520), which would be produced via the (K-p) channel, are of particular interest. In the constituent quark model, both are simple orbital excitations of the Λ(1116), and their interaction with nuclei should be accurately calculable. Investigating their spatial structure and their interaction properties in K^--nucleus interactions would give clues for additional structures if the data does not follow the naive expectations.

(ii) The spin-orbit couplings of $N, \Delta, \Lambda, \Sigma$ and Ξ show simple regularities and may help to discriminate between models typical of QCD and conventional one-boson-exchange pictures.

(iii) K^+-nucleus interactions are of great interest because they should be calculable with high accuracy, and yet the comparison of experimental results with calculations shows serious discrepancies[16].

(iv) Production in nuclei of $(q\bar{q})$ meson states such as $\eta'(958)$ which can have a glueball admixture may turn out to be the best way to reveal these gluonic components, typical for QCD[16].

(v) Interactions of antiprotons with nuclei, under various experimental conditions, will continue to provide an important source of information on hadron dynamics in the nuclear medium.

6.2 ANTINUCLEON-NUCLEUS INTERACTION

Some rare antiproton annihilations on light nuclei could give very important hints on the existence of deconfined quarks inside the nuclear medium. This is the case, for instance, of the so-called Pontecorvo reactions:

i) $\bar{p}d \rightarrow p + \pi^-$ $\qquad\qquad$ $\bar{n}+d \rightarrow n + \pi^+$
$\qquad\quad\searrow n + \pi^0$ $\qquad\qquad\qquad\quad\searrow p + \pi^0$

ii) $\bar{p} + {}^3He \rightarrow n + p$ $\qquad\qquad$ $\bar{n} + {}^3He \rightarrow p + p$

iii) $\bar{p} + d \rightarrow \Lambda + K^0$ $\qquad\qquad$ $\bar{n} + d \rightarrow \Lambda + K^+$
$\qquad\quad\searrow \Sigma^0 + K^0$ $\qquad\qquad\qquad\searrow \Sigma + K^+$

These reactions cannot proceed through the antinucleon annihilation on a single nucleon, but rather on quarks from different nucleons. As an example Fig. 7 shows the quarks diagrams describing the second reactions.

For all these reactions one can write a set of recurrence relations, in which the main contribution comes from the (possible) quark degrees of freedom in the nuclei.

In this field even more than the others mentioned above, close collaboration between theory and experiment is vital. The quark-gluon structure of nuclear matter raises challenging problems of non-linear dynamics which require new concepts. New theoretical ideas must be checked against experiment before one can proceed. This requires a high-quality programme of dedicated experiments which shed light on specific facets of hadron dynamics. Thus, close coordination of experimental and theoretical research will enable decisive progress in this field. Again, this ambitious goal calls for a dedicated facility such as EHF.

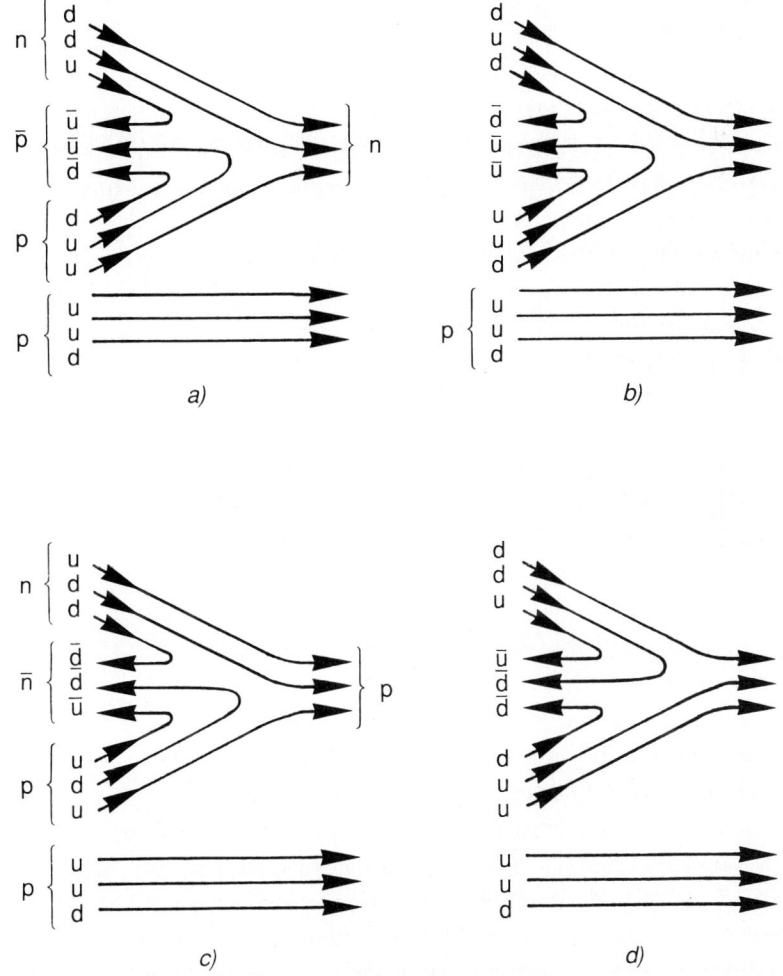

FIGURE 7

Quark Diagrams for the reactions $\bar{p} + {}^3He \rightarrow p+n$ a), b) and $\bar{n} + {}^3He \rightarrow p+p$ c) and d).

7. AKNOWLEDGEMENTS

Most of the material I have presented can be found in the EHF Proposal, and I am indebted to all the members of the EHF Study Group for permission to use it.

For more information on the role of INFN in the project and on the significance of an Italian option for the site of the new laboratory, I'd like to recommend the interested reader the article T. Bressani, P. Dalpiaz and myself recently wrote for "Il Nuovo Saggiatore"[17].

REFERENCES

(1) Proposal for a European Hadron Facility, Edited by J.F. Crawford, EHF-87-18, 18 May 1987.

(2) Proceedings of the "Workshop on the Future of Intermediate Energy Physics in Europe", Freiburg im Breisgau, Germany, 10-13 April 1984, Edited by S. Galster, KfK Karlruhe, Germany.

(3) "Hadronic Physics at Intermediate Energy", Proceedings of the Winter School, Folgaria, Italy, 17-22 Feb. 1986, Edited by T. Bressani and R.A. Ricci, North Holland (1986) (in the following referred to as Folgaria I).

(4) Proceedings of the "International Conference on a Europea Hadron Facility", Mainz, Germany, March 10-14, 1986, Edited by Th. Walcher, North Holland, reprinted from Nucl. Phys. B279 (1987) nos.1,2.

(5) A. Zichichi, The Gran Sasso Project, GUD Workshop, Rome, 1981.

(6) The compilation is taken from F. Vannucci, Les Rencontres de Physique de la Vallée d'Aoste, La Thuile, Valle d'Aosta, 1-7 March 1987.

(7) R. Davis et al., Bulletin of the American Physical Society, April 1984, p. 331.

(8) J. Bahcall et al., Rev. Mod. Phys., 54 767 (1982).

(9) ASTOR Proposal, SIN, Dec. 1986.

(10) J. Carr et al., Phys. Rev. Lett. 51, 627(1983).

(11) M.K. Gaillard and B.W. Lee, Phys. Rev. D, 10, 897(1974).

(12) G. Gabrielse et al., Phys. Rev. Lett. 57, 2504(1986).

(13) G. Gabrielse et al., "Precision Comparison of Antiproton and Proton Masses in a Penning Trap", CERN proposal PSCC/85-21/P83-/22, March 1985, N. Beverini et al., "A Measurement of the Gravitational Acceleration of the Antiproton", CERN proposal PSCC/86-2/P94/16, January 1986.

(14) N. Beverini and G. Torelli "Antiproton Traps and Related Experiments, Workshop on Intermediate Energy Physics, Trieste 1.-3.4. 1985, "Nuclear and Particle Physics at Intermediate Energy with Hadrons", Eds. T. Bressani and G. Pauli, Conf. Proc. SIF Vol. 3, p. 111.

(15) G. Preparata, in Folgaria I, p. 3.
(16) F. Lenz "Exotic Hadronic States in Nuclei" in: Proceedings of International Conference on a European Hadron Facility, Mainz 1986, Nucl. Phys. B279 (1987)Nos.1.2, p. 119.
(17) F. Bradamante, T. Bressani and P. Dalpiaz, "L'European Hadron Facility", Il Nuovo Saggiatore, Bollettino della Società Italiana di Fisica, Nuova Serie Anno 3, N. 4, luglio-agosto 1987.

HADRON CALORIMETRY AND THE MECHANISM OF COMPENSATION

Livio PIEMONTESE

INFN, Sez. di Ferrara - Italy

1. INTRODUCTION

In the large array of detectors[1] used in the experimental research on high-energy particle physics, calorimeters are those instruments which measure the total energy of a primary particle (or of a collimated jet of particles). They do this by degrading the energy of the primary particle, via a multiple cascade of interactions in a dense material, into the creation of a large number of secondary charged particles which are stopped in the calorimeter through the usual mechanisms of energy loss (ionization and excitation of atoms).

The classical method to measure energy (or momentum, which is nearly the same at high energy) of charged particles relies on the measurement of curvature or deflection in a magnetic field. Calorimetry provides an interesting alternative, owing to the fact that its merits and limitations complement those of magnetic spectrometers.

The principle of operation of calorimeters lies in the interaction of primary particles with matter: two elementary forces are at play in calorimetry, according to the particles whose energy has to be measured. High-energy photons and electrons interact with matter via the electromagnetic force in the strong Coulomb field of a high-Z nucleus. Two processes dominate: electrons lose energy by radiation of high-energy photons (bremsstrahlung), and photons convert into electron-positron pairs. The average energy loss of an electron traversing a layer dx of material being proportional to the electron energy E, the constant of proportionality is the inverse of a length X_0

$$\frac{dE}{dx} = \frac{E}{X_0} . \qquad (1)$$

The parameter X_0 is called the radiation length, and its value depends on the atomic number Z of its nuclei, $X_0 \simeq 180\ A/Z^2 (g/cm^2)$. The second process which contributes is the pair creation, i.e. the interaction of a high-energy photon with the Coulomb field of a nucleus in which the photon energy is transferred to an electron-positron pair. The probability of a pair creation to take place in a layer dx of material is $dP/dx = (7/9\ X_0)$, i.e. nearly the same constant we have seen in bremsstrahlung.

A photon (or electron, or positron), incident on a block of absorber many radiation lengths long undergoes a series of interactions (pair creation, followed by bremsstrahlung of the generated electrons, etc.) which eventually leads to creation of a large number of low-energy charged particles. On average the number of particles behind each layer doubles until a shower maximum where the energy has been sufficiently degraded and the multiplication process stops, all particles losing their (low) energy via ionization and excitation of atoms. This happens when the electrons and positrons reach the so-called critical energy, the energy where the average energy loss for radiation equals that due to ionization; the critical energy ε depends on the absorber material via the approximate formula $\varepsilon \simeq 550$ MeV/Z and takes values of the order of 5 to 10 MeV for the heavier materials. This fact, combined with the low value of the electron mass, has the consequence that, already at the level of a few GeV, an em shower consists of thousand(s) secondaries. From the definitions of X_0 and ε it follows also that the ionization energy loss of a minimum ionizing particle mip is $dE/dx_{mip} \simeq \varepsilon/X_0$.

Under the assumption of constant ionization energy loss the total length T of all the tracks of the charged particles is proportional to the initial energy,

$$E = \frac{dE}{dx}_{mip} \times T \qquad (2)$$

and the energy E can be measured by measuring T.

Electromagnetic calorimeters have been built since a long time, and their performance is very well known and understood. By comparison, hadron calorimeters - which measure the energy of hadrons, i.e. protons, pions, neutrons etc. via their strong interaction with the nuclei - are relatively new instruments, in which the physics of the elementary processes is much more complicated. Hadrons are stopped with a relatively few inelastic interactions in which most of their energy is distributed between a large number of secondaries created in the hadron-nucleus interaction. The characteristic length in hadron interactions is the absorption length λ, with

$$\lambda = \frac{A}{(N_{AV} \times \sigma)} \simeq 35 \, A^{1/3} \, g/cm^2 \qquad (3)$$

where N_{AV} is the Avogadro number and σ, is the absorption cross section.

Table 1 reports the values of X_0 and of ε for the most commonly used materials: note how, as the Z of the material increases, the radiation length becomes much shorter than its absorption length. While the energy deposition in an em shower is a very regular and continuous process (Fig. 1) in hadron calorimetry although the distribution of average energy release is also a smooth function of depth, individual showers (Fig. 2) can vary by a much larger extent than em showers. The direct consequence of this diversity in the basic processes involved is that the energy resolution of hadron calorimetry is much poorer than that of em calorimeters.

Table 1

Radiation and absorption lengths of materials normally used in calorimeters

MATERIAL	Z	X_0 (cm)	λ (cm)
Scintillator		42.4	80
Aluminium	13	8.9	39.4
Liq. Argon	18	14.0	83.7
Iron	26	1.76	16.8
Lead	82	0.56	17.1
Uranium 238	92	0.32	10.5

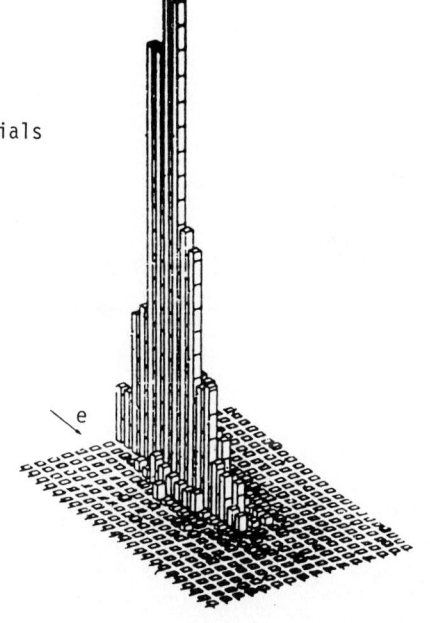

Fig. 1 Energy deposition^2 of a 27 GeV electron on a Lead/Liquid Argon calorimeter. One cell corresponds to a $1X_0 \times 1X_0$ segmentation.

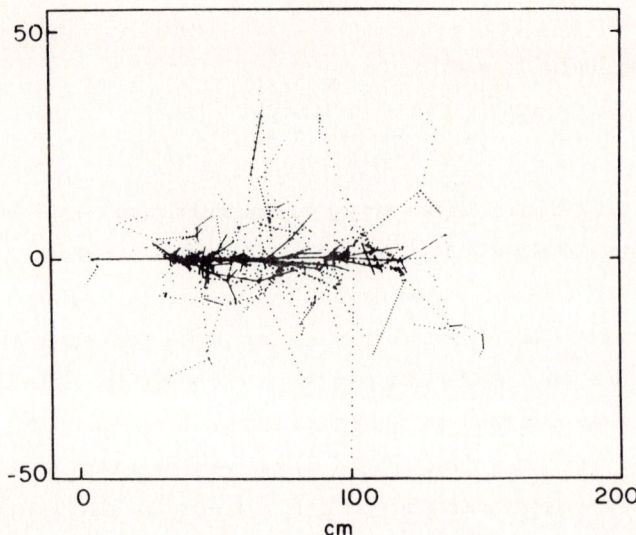

Fig. 2: Monte Carlo simulation[3] of a 100 GeV π^- shower in iron, dotted lines represent neutrons.

In the following, after a brief description of the main techniques of calorimeter construction (Section 2), in Section 3 the energy resolution will be discussed and compared for em and hadron calorimeters. Section 4 will discuss and present ways to improve the resolution by the mechanism of compensation.

2. MAIN CONSTRUCTION TECHIQUES

The hadron calorimeters, although already used in cosmic ray research have entered particle physics only around 1970. The physics case was the need of neutron detection, for neutron-proton elastic scattering. Figure 3 shows the hadron calorimeter[4] made: it consisted of a sandwich of absorbers with 40 plates of scintillator, each 40 × 40 cm^2, connected to a single photomultiplier. Iron plates inserted between the scintillators provided the amount of matter necessary for absorbing the incoming hadron. The calorimeter was tested with various thickness of absorber, with a maximum value of 80 cm, corresponding to about 5λ. The hadron energy was measured via pulse height analysis of the photomultiplier (PM) signal: the resolution dE/E varied with E, with a 1/$\sqrt{E}$ dependence: at 23.5 GeV/c it was 11% which, assuming a dE/E = k/$\sqrt{E}$ parametrization, implies for k the value 54%.

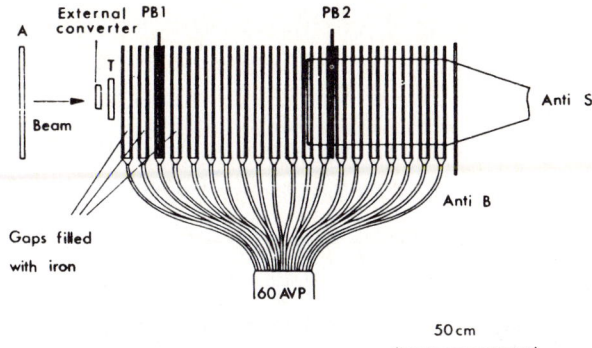

Fig. 3: The iron-scintillator/calorimeter used for neutron detection[4].

Such counter is called a sampling counter, meaning by this that the energy released in the absorber is not fully measured, but only sampled at regular depths of absorber.

Since then some tens of hadron calorimeters have been constructed for experiments at the higher energy accelerators. The reasons for this increased use of the calorimetry as an alternative to magnetic spectrometers are:

a) capability of detection/measurement of neutral hadrons;

b) measurement of total energy of jets;

c) operation as a large mass target/detector, p.ex. in neutrino physics;

d) energy resolution dE/E improving with increasing energy (it deteriorates for magnetic spectrometers);

e) by comparison, of the measured total energy with the initial energy, they allow a measurement of "missing energy" and therefore a detection of escaping neutrinos.

Also the construction techniques have evolved from the first counter of Ref. 4 made with the cumbersome method of glued light guides.

Em calorimeters can be - and have been - made with homogenous materials[5], i.e. with a material which acts at the same time as absorber and detector. Such materials are detectors (scintillator or glass, in which Čerenkov light is produced) containing a sizeable proportion of a high-Z element, which accounts for a reasonably short radiation length: examples are lead-glass (Čerenkov light) and crystals of BGO, NaI or CsI, all three scintillators. These counters provide the best energy resolutions attainable for em calorimetry, since for the scintillator crystals the number of photoelectrons collected is so high that the main contribution to resolution is the statistics of the primary showering mechanism, yielding resolutions of the order of $\sim 1\%/\sqrt{E}$, while for the lead glass the resolution is limited by the photoelectron statistics to a low $\sim 3\%/\sqrt{E}$.

Homogeneous calorimeters are not employed for hadron calorimetry since the interaction lengths of materials are much longer than their radiation lengths, and homogeneous calorimeters would become very large, and expensive tools, not to mention the impossibility to grow crystals of the size required.

In fact, hadron calorimeters are most often made with iron slabs (iron being a cheap material with a short interaction length) interspersed with detectors.

New types of detectors are employed, which avoid the main problems of the geometry of Fig. 3, i.e. a large dead space around the counter (taken by the adiabatic light-guides) and a complicated construction. A new system of light collection from scintillator layers is depicted in Fig. 4: the light from the scintillators is collected in another scintillator, running along one side of the calorimeter, called wavelength shifter wls. The wls absorbs the light from primary scintillators and re-emits it isotropically at a longer wavelength. The re-emitted light is then guided, via internal reflection, to a PM located behind the calorimeter. This technique allows, at the expense of a reduction of the total number of detected photons, the construction of compact modular calorimeters, which can be installed side to side with minimal dead space. An extra feature is the possibility of splitting the calorimeter longitudinally in two sections with separate readouts (Fig. 4): a front part where the em showers are fully contained, and is therefore the em calorimeter, with optimized choice of absorber sheets; and a back part, which together with the first one is the hadron calorimeter.

Besides detectors relying on the generation of light, detectors sensitive to ionization are also employed. In a first category, primary ionization in liquids[7] can be measured directly. This technique allows the construction of rugged, self contained calorimeters in which the density of the liquid provides an ionization large enough to be measured, and the energy calibration problems are greatly simplified by the absence of a detector amplification mechanism (like would be p.ex. the amplification of a gas MWPC). The first substance employed for such purpose was liquid argon LA: only recently organic liquids[8] are used which,

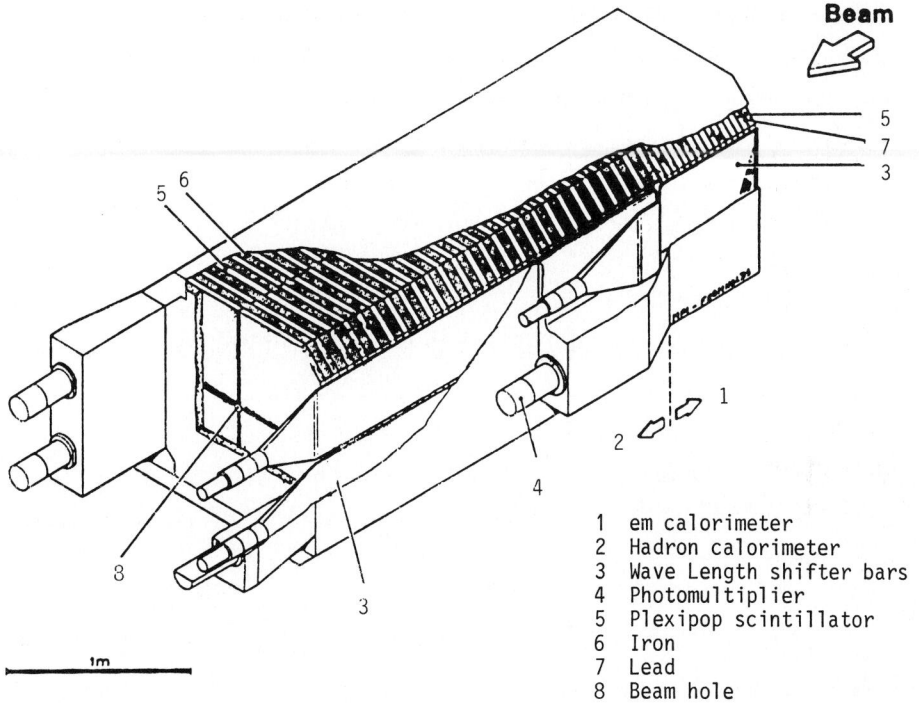

Fig. 4: A sampling calorimeter[6] with a front em section (Pb absorber sheets) and a back hadron section (Fe). The two sections are read independently via separate wls bars.

although at the expense of added complications in purity requirements, operate at STP and therefore do not need a cryogenic equipment.

The last category of detectors employed are gas-wire detectors, be they operated in proportional mode (MWPC) or in saturated mode (limited streamer tubes). Again, due to the limited space needed for support and readout they lend themselves very well to the construction of modular calorimeters.

Another important concept in calorimetry is segmentation, both longitudinal and especially lateral. Lateral segmentation is related both to the measurement of original particle/jet position and direction and to the separation of energy measurements for different, but near particles, while longitudinal segmentation allows to measure the development of the shower process inside the absorber material, and helps in the distinction between em and hadronic showers.

Different types of detectors have different characteristics and lend themselves more or less well to segmentation, the easiest to segment being gas and liquid ionization detectors.

3. THE ENERGY RESOLUTION

Both for em and hadron calorimeters the relative energy resolution has usually an inverse square root dependence on energy,

$$\frac{dE}{E} = a + \frac{k}{\sqrt{E}}. \qquad (4)$$

The reason for this dependence is that, whatever the detection mechanism, the type of shower and the absorber material and thickness, the number of detected signals, be they photoelectrons or primary ionization electrons, is proportional to the particle energy, and the error on this number is determined by Poisson statistics to be $dN/N = 1/\sqrt{N}$. The constant term a accounts for basic limitations of the particular detector made when, at high energy, the resolution would tend to be very small.

3.1 Electromagnetic calorimeters

In em calorimetry the resolutions are in general very well understood. In homogeneous calorimeters, two factors are at play: photoelectron statistics, the main contribution for lead-glass counters; and detailed statistics of the shower development at the low energy shower end. This because the low-energy electrons at a certain "cut-off energy", which depends on the detection, can no longer transfer energy to the medium to be detected. This energy, multiplied by the number of charged particles in the final state - variable from shower to shower - is lost to measurement, and its fluctuations are reflected in the so-called intrinsic energy resolution: $0.7\%/\sqrt{E}$ for a NaI crystal.

For sampling calorimeters, the main contribution to the resolution comes from the fact that the total energy loss of the particle is not measured, but only sampled at regular depths of absorber. This implies that one infers the total energy loss by measuring, for each layer, the *number of particles* which cross the detector, and then multiplies this number by the energy loss of a mip traversing one calorimeter cell (i.e. absorber + detector). The total energy is thus measured by applying

$$E = \frac{dE}{dx}\bigg|_{mip} \times \sum_i \delta x_i \, N_i \qquad (5)$$

where the sum is applied to all the detector layers of the calorimeter.

Now, the point is that, dE/dx_{mip} being a constant, same as the layer thickness δx (absorber + detector) of a given calorimeter the only variable parameter is the total number N of particles detected. This being again a case of Poisson statistics, its variance is given by $\sqrt{N}$ and with a little arithmetics we can compute the energy resolution

$$\frac{dE}{E} = \sqrt{\frac{1}{N}} = \sqrt{\frac{\delta x}{T}} = \sqrt{\delta x \times \frac{dE}{dx_{mip}} \times \frac{1}{E}} = 3.2\% \sqrt{\frac{\delta E}{E}} \quad (6)$$

where δE is the energy loss of a charged particle of energy E (E in GeV and δE in MeV: this is where the 3.2% comes from) across one calorimeter layer.

Equation 6 gives the dependence of the energy resolution of a sampling em calorimeter (sampling fluctuations) on the primary energy *and* on the absorber thickness. It applies when the number of particle crossings in each detector layer is measured precisely; this case is usually referred to in the literature as "digital readout". In fact what usually happens is that in each detector layer what is actually measured is the total detected charge, and the number of crossings is inferred from this quantity assuming a constant charge measured per single particle crossing.

This assumption is more or less true, according to the type of detector and also on the absorber employed. The charge Q released by a shower particle is distributed according to a distribution which depends on various detector parameters, but which is well understood[5], and contributes an additional term to the Eq. 6.

$$\left(\frac{dE}{E}\right)^2 = \left(\frac{dE}{E}\right)^2_{sampling} + \frac{1}{N}\left(\frac{dQ}{Q}\right)^2 \quad (7)$$

Figure 5 shows a compilation[2] of measured resolutions of various calorimeters which have been built together with the resolution computed according to Eq. 6: the generally good agreement shows how well em calorimetry is understood.

3.2 Hadron calorimeters

At variance with what happens in em calorimeters, the energy resolution of hadron calorimeters can not be accounted for by simple statistical considerations. The situation is depicted in Fig. 6, which is the analog of Fig. 5 for hadron calorimeters. The lines for the various energies have been computed with the formula

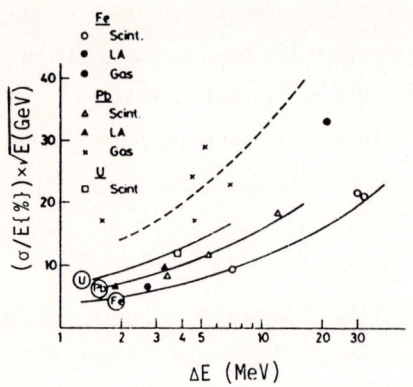 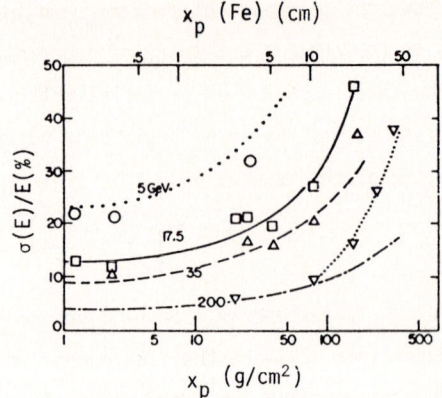

Fig. 5: Energy resolution for various em calorimeters, together with estimates from (Eq. 7).

Fig: 6: Energy resolution for various hadron calorimeters. Curves from Eq. 8.

$$\left(\frac{dE}{E}\right) = (50\%/E)^2 + \left(46\% \frac{X}{X_0} \frac{1}{E}\right)^2 \tag{8}$$

where the second term represents the contribution to the error from the already seen "sampling fluctuations", and the first is an empirically determined contribution which is ascribed to the fluctuations inherent in the hadron shower process, and is conventionally called "shower fluctuations".

From the numbers it is easily seen that, while the contribution of the sampling can be controlled by acting on the thickness of the absorber layer, the shower fluctuation puts a physical limit of $\sim 50\%/\sqrt{E}$ on the attainable resolution. Moreover, the same sampling fluctuation term is not $15\% \times /(X_0\sqrt{E})$, which would be expected for iron, but a factor 3 higher.

The reason for the poor performance of hadron calorimeters, compared to em ones, became clear from both Monte Carlo simulations of hadronic showers and from the results of test calorimeters.

Monte Carlo programs have been made for practically every large calorimeter which has been built. These programs are very complicated objects, since they have to take into account all the details of the hadronic interaction of high-energy particles with nuclei, reproduce the fission of nuclei and the subsequent path of fission products, and also the production of π^0's, which originate em showers. From these programs came a qualitative understanding of the "shower

Table 2

Average fractional energy deposition by particle type for 10 GeV proton interactions in an iron-argon calorimeter.

Type of energy deposition	Percent of total
Primary proton ionization	2.3
Secondary ionization	39.8
Electromagnetic cascade	21.0
Z > 1 ionization	2.4
Residual nuclear excitation energy	3.7
Neutrons	8.8
Nuclear binding energy plus neutrino energy	20.6

fluctuations": From Ref. 10, a 10 GeV proton loses energy in different channels (Table 2); the first four channels of Table 2 correspond to detectable energy, i.e. to energy which eventually turns out under the form of ionization or of atomic excitation of the medium, while the last three channels correspond to energy which is not detected by the calorimeter. The contribution of these channels is not negligible since it amounts, in this particular case, to more than 30% of the total energy. Fig. 7 shows, according to the calculation of Ref. 11, the fraction of the energy of an incoming π^+ eventually degraded into em shower as a function of energy, together with its fluctuations.

According to the early Monte Carlo programs, the explanation for the hadronic energy resolution lies in the fact that the hadronic shower proceeds via two distinct channels, i.e. electromagnetic, which ultimately leads to all the energy being detected and hadronic, in which a large part of the energy goes undetected; the sharing of energy between the two components is dominated

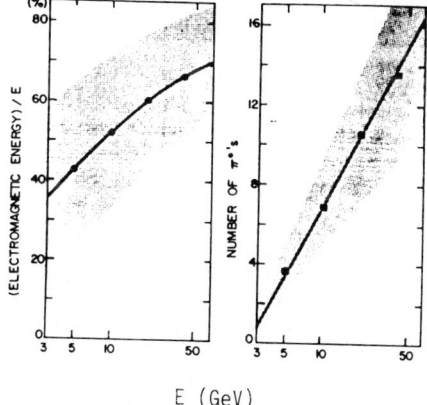

Fig. 7: Fraction of total energy converted in π^0 in a hadronic shower and number of generated π^0's. Shaded areas indicate 1σ.

by the statistics of a few processes in which large amounts of energy are exchanged.

This interpretation was confirmed by the results obtained in the operation of test calorimeters exposed to monochromatic hadron beams and of large experiment calorimeters. In particular, a large test calorimeter[12], which introduced the technique of liquid argon as a sensitive medium, was built in the early seventies, with very thin iron sheets as absorber, in order to minimize the influence of sampling fluctuations. The calorimeter was exposed to hadron and electron beams of known energy, and the resolution was measured as a function of beam energy: the results are plotted in Fig. 8, together with the contribution of sampling fluctuations, which have been measured with a clever method: the charges from alternate collecting electrodes were ganged together in two separate measuring chains, easily indicated as E(ven) and O(dd); the energy was measured by O + E, and the standard deviation of the distribution of O + E was the measurement error; the average value of the distribution of O - E was of course zero, but the rms of its distribution was indicative of the contribution of the sampling fluctuations to the measurement error. This is confirmed by the experimental results: for em showers, the energy resolution, which is known to be dominated by sampling fluctuations, is identical to the rms of O - E (Fig. 8).

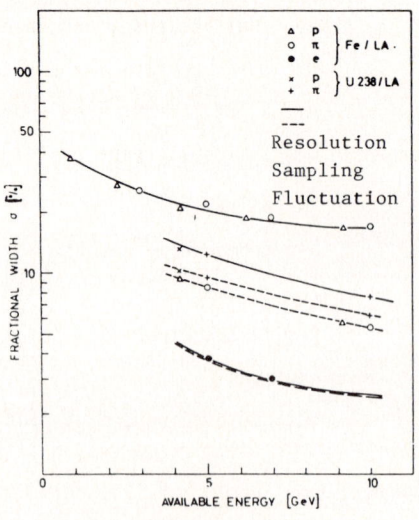

Fig. 8: Energy resolutions for a LA calorimeter, for electron and pion beams, and with Fe and U_{238} absorbers. The contribution of sampling fluctuations is also indicated.

For the case of hadronic showers instead, not only the sampling fluctuations are much larger than the em case, but the energy resolution itself is larger than the sampling fluctuations, thus indicating that another source of fluctuation is at work, with a contribution dominant wrt that of sampling.

That the mechanism at work is the different response of the calorimeter to em and to hadronic showers was confirmed by the measurement that between the total charges collected for electron and pion beams of the same energy there was a factor 1.5. This situation is customarily referred to as the "e/π" ratio, which is dependent on energy and, at least for iron, takes the value of ∼ 1.5.

The group which performed these tests had the idea that, after finding an explanation for the effect, they could also try and do something about it. This something was the replacement of iron as absorber material with U_{238}; the idea behind it being that the fraction of undetected hadronic energy in hadron showers is connected also with low energy neutrons produced in the fission or subsequent evaporation of struck nuclei; and these neutrons could start nuclear reactions with the uranium, resulting eventually in an amplification of the energy detected in the neutron channel. A calorimeter was built with uranium slabs, the data analysed, and the results were stunning: the total charge measured for hadrons and electrons were very nearly the same, and the hadronic resolutions improved by a factor ∼ 2 (Fig. 8). The mechanism at work for the optimization of the resolution, that is the equalization of the calorimeter response to hadrons and electrons, was then called "compensation".

4. UNDERSTANDING THE COMPENSATION

The group who introduced uranium compensation was convinced that it worked because of amplification of hadronic energy loss due to fission of uranium; they used uranium/scintillator calorimeters in experiments and, after a few years' experience they confirmed their original results, although with other types of absorber/detector combinations: in Fig. 9 are shown the results quoted (Ref. 13) for the e/π ratio as a function of energy, for various types of sampling calorimeters.

Precise hadron calorimetry was starting to be an essential ingredient in many of the experiments proposed for the new generation of hadronic and e^+e^- colliders, and for Hera as well; the collaborations formed around these accelerators proposed various types of compensated calorimeters, and engaged themselves in a thorough program of construction and test of prototypes, especially of liquid argon/uranium calorimeters with, as an alternative, lead/liquid argon.

The results[14] confirmed only partially the original results of Ref. 11, although they showed that U_{238}/LA does perform better than Iron/LA: the hadronic energy resolution was poorer than expected and at the same time the e/π ratio was nearer to 1.2 than it was to 1.0.

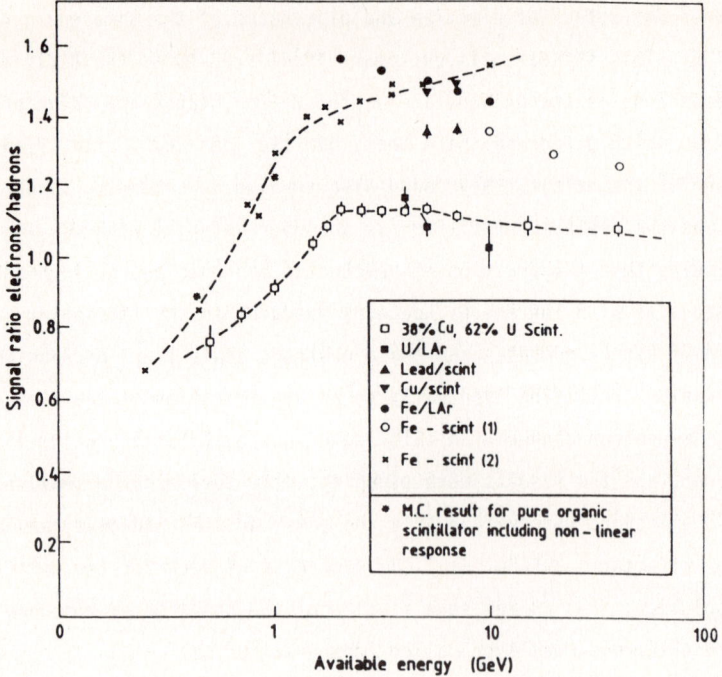

Fig. 9: Compilation of experimental values for the e/π ratio.

What was more interesting was that the partial compensation obtained with uranium was very nearly obtained also with the use of lead. The comparison of signals (charges) measured with pion, electron and muon beams impinging on the test calorimeter allowed to intercalibrate the response of the calorimeter to different types of energy (muons lose, in the traversal of the calorimeter, a fixed amount of energy, determined mainly by the energy loss at minimum dE/dx_{mip} particle mip). It turned out that in uranium as well as in lead the compensation is due not to an amplification of the hadron signal, but to a reduction of the electron signal compared to the mip signal: there seems to be a large loss of detected energy also for em showers, at least for high-Z materials (lead and uranium). Once discovered, the effect was quickly explained[15,16] again by the use of more and more sophisticated Monte Carlo programs. Two effects are at play to reduce the signal of em showers in high-Z materials: the first effect is due to the preferential absorption of low-energy bremsstrahlung photons, produced especially near the end of the cascade, in the high-Z absorber than in the low-Z detector, thus effectively subtracting from detection a large component of the total energy. The second is related to both the shorter radiation length and to the smaller critical length of high-Z materials: the bulk of the electron signal being given by electrons (and positrons) of energy equal to the critical energy or lower, they suffer severe multiple Coulomb scattering and, being the absorber much thicker than the detector in units of radiation length low-energy electrons travel very long paths inside the absorber layers, thereby effectively increasing the ratio of absorber/detector layer thicknesses.

The existence of compensation was therefore confirmed in LA calorimeters, although at a smaller level than previously believed and was explained correctly as due to a mechanism of suppression of em energy measurement rather than to a mechanism of amplification of the neutron signal; this latter mechanism being not very effective in increasing the signal in liquid argon, since low-energy ($\sim$ 2 MeV) neutrons can not release their energy to Ar atoms, given the large mass difference between the two bodies.

Still, experimental data (Ref. 17) confirmed the production of a large number of evaporation neutrons in high-energy hadron-uranium (and hadron-lead) collisions. Typically, a 10 GeV proton releases $\sim$ 320 neutrons in U_{238} and $\sim$ 220 neutrons in lead. If these neutrons can not be detected in LA calorimeters, they can in scintillator calorimeters, owing to the presence in plastic scintillator of a large number of hydrogen atoms. The low-energy neutrons release a large fraction of their original kinetic energy in elastic collisions with these free protons, which in turn ionize and excite the neighbouring atoms. A very detailed Monte Carlo calculation[16] stressed the role these neutrons play in increasing the hadronic signal in scintillator/uranium calorimeters (Fig. 10).

Again, tests have been made with U/scint. calorimeters, in particular with a calorimeter[18] built for the experiment WA 78 at CERN, which turned out to be very easy to modify and adapt to various scintillator/absorber configurations.

The test setup consisted in a set of 100 scintillator sheets viewed by 26 PM's, each with its own ADC channel, thus allowing a very detailed measurement of the longitudinal shower development. Between the sheets were inserted the absorber slabs, either 25 mm thick iron or 5, 10 or 15 mm uranium.

Figures 11 (a) and (b) are scatterplots - for Fe and U data respectively - with, in abscissa, the energy measured by the PM with the largest signal and in

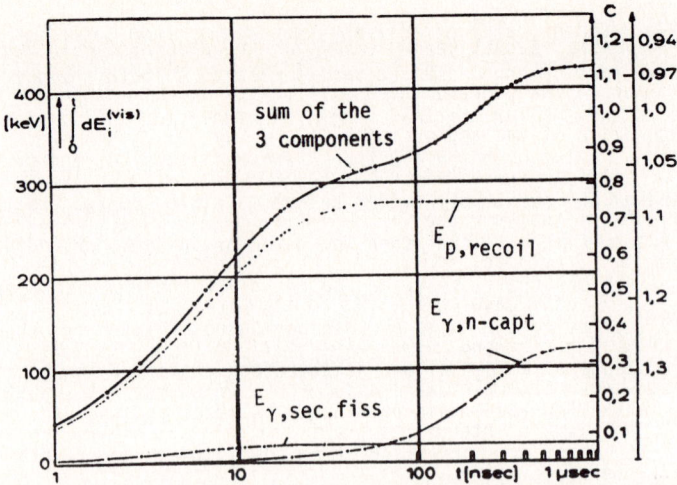

Fig. 10: Contribution of various channels of neutron energy detection to the total signal in uranium/scintillator calorimeters as a function of time.

ordinate the total energy. Figure 11a indicates a positive correlation between the two quantities for hadrons, as already reported in Ref. 19: this means that, when a shower has an especially large energy release concentrated in a thin layer, the total energy is higher. This effect is explained as being due to events in which a large-than-normal fraction of the energy goes into π^0's,

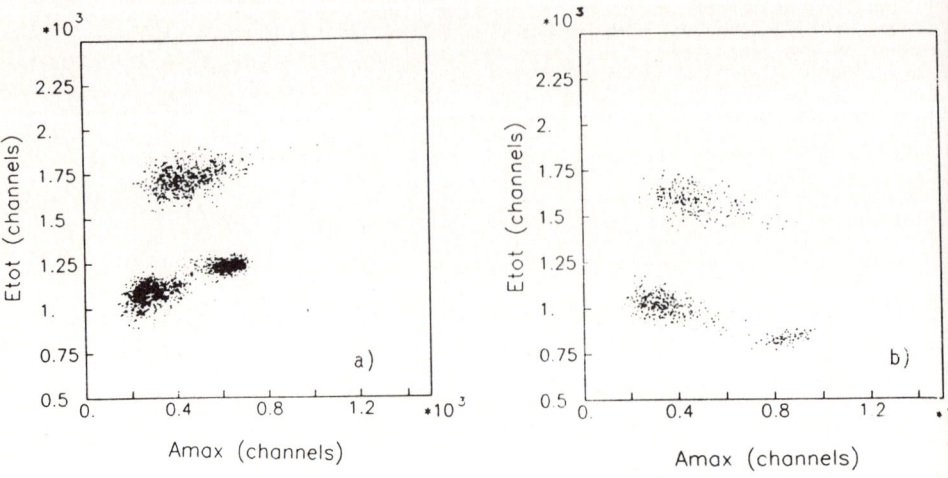

Fig. 11: Correlation between maximum energy release in one single layer and total energy, for hadrons (low A_{max}) and electrons (high A_{max}), 135 and 210 GeV data, from (11a) and Uranium (11b) absorbers.

which give a higher signal (e/π ≈ 1.51) and are absorbed in a short distance. As a confirmation the electron events (the data at high values of A_{max}) sit on the extrapolation of the hadron correlation line.

The most striking result is shown in Fig. 11b which refers to Uranium: the difference is that now the result is reversed. Instead of having a positive correlation between large, localized energy releases and larger than average total energy, in this case to a large localized energy deposit, indicative of a large em shower, correspond events with total energy lower than average. Events corresponding to electrons impinging on the calorimeter are also shown and confirm nicely the interpretation of large, localized energy releases being due to high-energy em showers.

The explanation of this phenomenon is that at high energy the hadronic energy is larger than the em, i.e: the ratio e:π, which is already a slowly decreasing function of energy, becomes smaller than 1. Table 3 reports the experimental values of this ratio determined by WA 78. In fact, it was expected, and alo measured that in scintillator sampling calorimeters the ratio e/π depend on the ratio of absorber/detector thickness (Fig. 12).

The mechanism responsible for this puzzling behaviour is connected with the detection of the energy carried bw neutrons: these take their kinetic energy away from the absorber - where they are mostly created - via elastic collisions with the much heavier atoms in which they lose only a tiny fraction of their energy, and then lose it in only a very few elastic collisions with the free protons contained in the scintillator. The neutrons deliver thus a fixed

Table 3

Measured values of e/π ratio in a scintillator (5 mm thick) sampling calorimeter. (Ref. 18)

Energy (GeV)	Fe (25 mm)	U (5mm)	U (10 mm)
135	1.11	0.88	0.80
210	1.07		0.79

amount of energy in the scintillator or, more precisely, a fraction of the total energy which depends, of course, on the distribution of the initial energy into the em and hadronic channels, but does not depend on the ratio between absorber and detector thickness. This is a fixed signal, which is added to the hadronic part of the signal. Now, being e/π larger than 1, the missing signal $(e - \pi)/e$ can be tuned, by varying the absorber/detector thickness ratio, to be exactly equal to the neutron signal, thereby achieving perfect compensation, and this independently of the choice of absorber material.

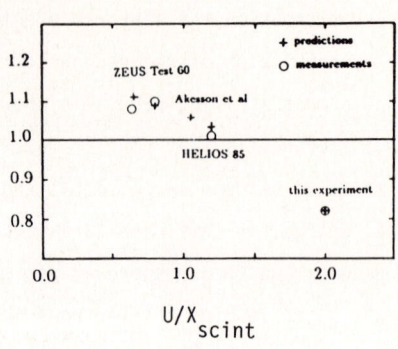

Fig. 12 e/π experimental ratio (Ref. 17) vs. the ratio of absorber/scintillator thickness.

As an additional proof, a lead/scintillator prototype has been built and tested[2] by the group which was developing calorimeters for the Zeus Collaboration at Hera, with the thickness ratio optimized for perfect compensation. They obtained a nearly perfect compensation, $e/\pi = 1.05$. Still, since unfortunately perfection compensation for lead requires a thickness ratio of Pb/Scin ~ 4, they used a sampling thickness of ~ 1 cm Pb, leading to a resolution $\sim 44\%/\sqrt{E}$. It is hoped that, by reducing the contribution of the sampling fluctuations[22], one may achieve a resolution $\sigma(E) \simeq 33\%/\sqrt{E}$.

REFERENCES

1) K. Kleinknecht: "Detectors for particle radiation", University Press, Cambridge, 1986.

2) Courtesy P. Sergiampietri, Pisa.

3) A. Grant, Nucl. Instrum. Methods 131 (1975), 167.

4) J. Engler et al., Nucl. Instrum. Methods 106 (1973), 189.

5) S. Iwata, DNPU 13-80.
   U. Amaldi, Physics Scripta 23 (1981), 409.

6) C. DeMarzo et al., Nucl. Instrum. Methods 217 (1983), 405.

7) W. Willis and V. Radeka, Nucl. Instrum. Methods 120 (1974), 221
    J. Engler et al., Nucl. Instrum. Methods 120 (1974), 157.

8) M.G. Albrow et al., Nucl. Instrum. Methods A265 (1988), 303.

9) C. Fabjan and T. Ludlam, Ann. Rev. Nucl. Part. Sci. 32 (1982), 335.

10) T.A. Gabriel and W. Schmidt, Nucl. Instrum. Methods 134 (1976), 271.

11) A. Baroncelli, Nucl. Instrum. Methods 118 (1974), 445.

12) C. Fabjan et al., Nucl. Instrum. Methods 141 (1977), 61.

13) T. Akesson et al., Nucl. Instrum. Methods A241 (1985), 17.

14) R. Dubois et al., IEEE Trans. NS 33 (1986), 194.
    D0 Collaboration, Nucl. Instrum. Methods 263 (1988), 78.

15) J. Brau, Proc. Workshop on Copmensated Calorimetry, Pasadena (1985),
    CALT-68-1305.

16) H. Bruckmann et al., Nucl. Instrum. Methods A263 (1988), 136.

17) W.A. Gibson et al., Oak Ridge Nat. Lab. Report ORNL-3940 (1965).
    L.R. Veeser et al., Nucl. Instrum. Methods 117 (1974), 509.

18) M. De Vincenzi et al., Nucl. Instrum. Methods A243 (1986), 348.

19) H. Abramowicz et al., Nucl. Instrum. Methods 180 (1981), 429.

20) A. Nigro, Nucl. Instrum. Methods A263 (1988), 102.

21) E. Bernardi et al., Nucl. Instrum. Methods A262 (1987), 229.

22) R. Wigmans et al., LAA Proposal in "The LAA Project", CERN 1987.

SEMICONDUCTOR DETECTORS FOR COLLIDERS PHYSICS

Guido TONELLI

Universita' di Pisa, INFN Sezione di Pisa, Via Vecchia Livornese 582/a
56010 S.Piero a Grado, Pisa, Italy.

ABSTRACT

The use of Silicon detectors for heavy flavours tagging in high energy Colliders and the physics motivations of high precision vertex detectors are discussed. Properties and limitations of microstrip detectors in this field are presented together with some experimental applications. The possibility of using solid state tracking devices in the new generation of high energy, high luminosity Colliders is finally considered.

1. INTRODUCTION

Since a few years, semiconductor detectors have been extensively used in high energy physics. After the pioneering experience of NA1 at Cern [1], Silicon and Germanium detectors have been successfully used as active target, dE/dx and tracking systems in many fixed target experiments. The high spatial resolution obtainable with solid state devices has allowed the study of interesting properties of short living particles, in particular the lifetimes of many charmed particles ($D^{\pm}$ D^o, F, Λ_c) have been precisely measured [2,3,4,5]. At present several experiments are under way to study, with similar techniques, the properties of beauty flavoured particles. While Silicon detectors have become a widely used tool in high energy physics, the whole solid state detector field is rapidly moving ahead. Already existing devices have found new interesting applications: charge-coupled devices (CCD's) employed as particle detectors [6], large silicon diodes used to build very compact calorimeters [7]. New devices, like semiconductor drift chambers, have been fabricated and tested [8]. New ideas such as fully depleted CCD's, pixel devices, radiation hardened detectors are being developed [9,10]. This fast growth of interest has involved also colliding beam experiments to such an extent that almost all the experiments at SLC, Fermilab and CERN Colliders, LEP and HERA are building tracking systems which include also solid state detectors.

The purpose of this paper is to discuss briefly the physics motivation of this choice and the properties and limitations of silicon detectors in this field. I wish also to present some of these experimental applications, showing design concepts and first results obtained on test beam. In doing so I shall limit myself only to the experiments in which our laboratory in Pisa is involved. Finally I shall give a brief look to the future discussing the possibility of using solid state tracking devices in the new generation of high luminosity, high energy Colliders.

2. PHYSICS MOTIVATIONS

2.1. Tagging Short-living Particles

The interest in heavy quarks starts with the prediction by Kobayashi and Maskawa in 1973, of the third quark family, t and b, and by the discovery of charm, the massive partner of the quark s, in 1974. Since then a full set of new particles (D, F, Λ_c, B) was discovered and an extensive study of their properties began : their lifetimes, branching ratios, decay modes, masses and other observable quantities have been widely investigated to determine or confirm the different theoretical predictions.

In particular the lifetimes of these massive particles were measured to be $10^{-12} < \tau < 10^{-13}$ s, the range of the so-called short-living particles to distinguish them both from stable and long-living particles ($\tau < 10^{-10}$ s) and from the resonances subject to prompt decays ($10^{-18} < \tau < 10^{-23}$ s). Since then decay-lenghts in the range between 10 µm and 100 µm have become an important signature of the presence of heavy quarks, particularly in experiments where high multiplicity inelastic processes are involved. Although the production cross sections are usually large, the combinatorial background increases with increasing the energy of the reaction and it becomes practically impossible to observe these new states as invariant mass peaks. This is particularly true for high energy p$\bar{\text{p}}$ colliders and an example can be seen in fig 1a) which is an ISAJET simulation of the process p$\bar{\text{p}} \to $ b$\bar{\text{b}}$ + X at $\sqrt{s}$ = 2TeV . The horizontal scale of the picture is 126 cm.

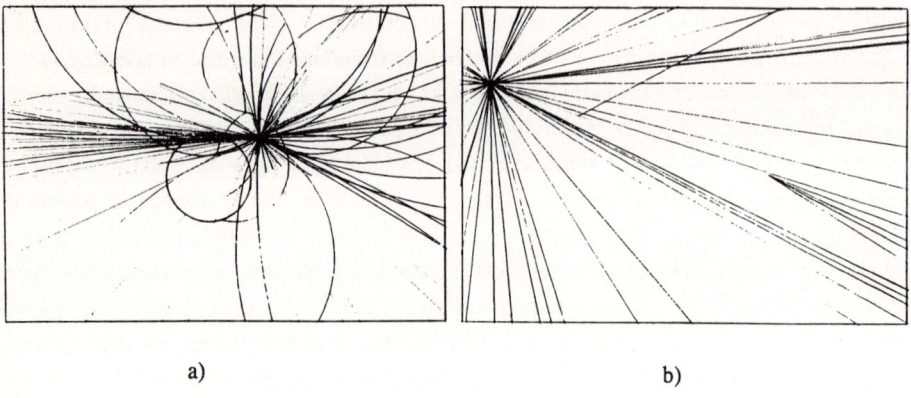

a)   b)

Fig.1 : p$\bar{p} \to b\bar{b}$+X at $\sqrt{s}$=2 TeV. a) Hor. scale=126 cm b) Hor. scale=0.3 cm.

In events like this one, there is no hope to disentangle the few tracks belonging to decay products of the b-quark by using combinatorial methods. But looking at the same event on a smaller scale (0.3 cm in fig.1b) it becomes very simple to identify the two secondary vertices produced by the sequence b $\to$ c $\to$ s and to distinguish the few tracks of the decay products of beauty and charm from the many tracks coming out from the primary vertex.

The traditional tools used to identify this class of events, high p_t leptons, missing p_t or M_t, Kaons in the final state, are strongly affected by the complexity of the events which increases with energy and by the presence of a full sequence of decays which produce some high p_t leptons and undetectable neutrinos and without a clean vertices reconstruction it is very difficult the identification of the process.

What happens in the first few millimeters just close to the primary vertex is a crucial point for this kind of physics. The information on secondary vertices, once available, can be coupled with the requests of high p_t leptons, missing E_t and Kaons in the final state. Apart from measuring lifetimes, such information can provide orders of magnitude suppression of combinatorial background which would otherwise obliterate signals from heavy quark states. From here stems he importance of high precision vertex detectors for the observation of short-lived states in clean conditions.

2.2. Heavy Flavours Physics

The possibility of measuring with high precision the lifetime or the different lifetimes of beauty-particles can improve our knowledge of the Cabibbo-Kobayashi-Maskawa matrix. By collecting big samples of fully reconstructed beauty events it becomes possible not only to measure branching ratios and to study rare decays but also to measure the mixing parameter B^0_s-$\overline{B}^0_s$ which is expected to be large and deeply connected with the number of flavours. The measurement of an anomalous mixing could disprove the 6-quark model even before the direct observation of a higher flavour.

Inside the $b\bar{b}$ sample one can look for top signature if the reaction energy is high enough to produce an object which we expect to be as massive as 80-100 GeV. The lifetime of top is expected to be very short but we know from the past experience that not necessarily larger mass means shorter lifetime (I like to remind that, until 1982, the general prediction for the beauty lifetime was $\tau < 10^{-13}$ s). Therefore one must look carefully to all high p_t leptons not belonging to a beauty or charm vertex.

Similar methods can be used to search for new flavours but, obviously, the topology of these events become even more difficult. If a fourth generation exists we could have in the same event something as 8 vertices plus products of the primary interaction and the event reconstruction becomes difficult even for the most sophisticated vertex detector at present conceivable (Fig.2).

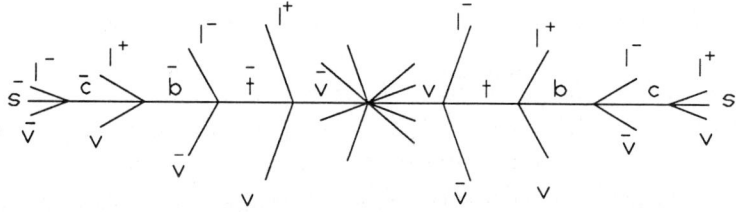

Fig. 2 : Semileptonic decay chain of v quark. Any lepton pair can be replaced by $q\bar{q}$.

Heavy flavours physics is also a powerful tool for the identification of other high mass particles which decay in the standard b, c quarks. For example, if the scalar Higgs has a mass in the range 40 GeV< M_H < 50 GeV, it can be seen at LEP in the reaction $e^+ e^- \to Z^0 \to H^0 e^+ e^- \to b\bar{b} e^+e^-$ but it is very difficult to identify the process without a good vertex detector, since the small amount of good events are actually masked by a large background of standard heavy flavour decays.

In general the experimental environment of e^+e^- Colliders is considerably cleaner for this kind of physics. Sitting on the Z^0 peak a large amount of heavy flavours are produced due to the democratic coupling of the Z^0 to many $q\bar{q}$ states and no other products are presents. But the available energy limits in this case the possibility of discovering new particles, as is the case of the top which seems actually to be outside the limits of LEP phase 1. On the other hand, in hadron colliders the production of heavy flavours is ~10^{-3} of the total cross section and therefore the identification of the few interesting high mass events decaying in heavy flavours depends critically on the background rejection.In this case the possibility of triggering on secondary vertices should be helpful to collect large samples of c and b jets and in fact several studies are actually under way on this particular area [11].

3. VERTEX DETECTORS

The range of lifetimes we are considering $10^{-13} < \tau < 10^{-12}$ s corresponds to decay lengths $l = \gamma \beta c \tau = (p/m) \cdot c\tau$ with $c\tau$ ranging between 30 and 300 μm. Despite the short lifetimes a high γ factor makes the mean flight path easily measurable for detectors with spatial resolution of the order of 10 μm. Fig.3 shows the fraction of 10 GeV/c particles with decay length grater than a given l as a function of l. The plot refers to a simulation of heavy quarks produced at the LEP energies in $e^+e^- \to Z^0 \to q\bar{q}$ and the effect of stretching on the decay length is clearly visible. Nevertheless the region below 1 mm, in which the most part of the interesting events are concentrated, is now not accessible by using conventional tracking devices ; this region can be explored only with a high resolution vertex detector.

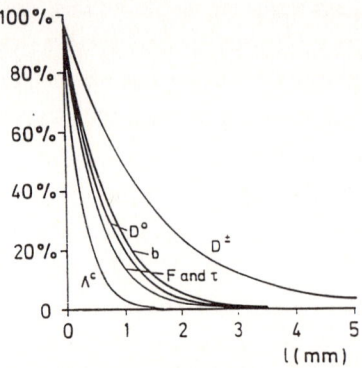

Fig. 3 : Fraction of 10 GeV/c particles with decay length >l at LEP energy.

Hadron colliders, which produce particles with higher longitudinal momenta, from this point of view, seem to offer some advantage with respect to electron-positron colliders but, unfortunately, the opening angle of the decay products follows roughly the simple relationship $\theta \sim 1/\gamma$ and the impact parameter, b, of the tracks coming out from the secondary vertex becomes again fairly independent from the momentum of the decaying particle: $b \sim l \theta \sim c \tau$.

On the other hand longitudinal momentum reduces the multiple scattering which degrades the space resolution of any tracking system and must be taken into account when measuring flight paths of the order of a fraction of a millimeter.

Another critical parameter is the distance from the beam, mainly determined by the radius of the vacuum pipe. Since we are looking for secondary vertices we measure very precisely each track and then we extrapolate them back to the interaction region to check if their impact parameters are compatible or not with the primary vertex. In doing so we introduce an error, which depends obviously on the precision by which we measure each track, but is also linearly dependent on the distance from the interaction point. The ideal situation would be to have detectors placed only few millimeters far away from the interaction region, but this is partially possible only at SLC where the beam pipe radius is actually 2.6 cm and there are perspectives to reduce it down to 1 cm. Neither at the Tevatron, nor at CERN Collider this ideal situation is achievable, the vacuum pipes being wide, the vertex detectors must be installed at several centimeters distance from the beam line. The situation is even worse at LEP where the beam pipe radius is about 8 cm.

All in all the good spatial resolution, although necessary, is not the unique parameter to consider when evaluating the performance of vertex detectors. Several kinds of detectors (i.e. pressurized drift chambers, nuclear emulsions, bubble chambers etc.) can provide high spatial resolutions, but only few can fulfill the peculiar requirements related with their use as vertex detectors in the core of big multi-purpose Collider experiments.

Often the number of channels increases rapidly with the dimensions of the detector (usual numbers are 50.000- 100.000 channels) and huge problems are connected with the read-out electronics (density, power dissipation, costs, read-out speed).The region around the beam pipe is the most critical one from the point of view of the amount of material that can be placed just in front of all the rest of the detector; multiple scattering is a limiting factor in the position measurement resolution for low p_t tracks and the vertex detector material becomes itself a source of background for the rest of the apparatus due to interactions and conversions which occur in it. The overall material cannot exceed 1-3% radiation lengths and the total weight must be limited to few kilograms. At the same time everything must be extremely rigid and stiff in order to have reproducible measurements with time. Finally alignment and survey must be performed with accuracy comparable with the space resolution of the detectors (5-15 µm).

4. SILICON DETECTORS

Silicon detectors can fulfill many of the basic requirements for vertex tracking systems.The high spatial resolution obtainable with microstrip tracking devices (3-30 µm) makes these detectors particularly suitable for heavy flavours tagging. Being solid devices they are mechanically stable and can be very compact. Very precise electrodes can be produced on the surfaces by using integrated circuit technology. The accuracy in manufacturing each individual read-out strip is very high (~2 μm) and the alignment itself is greatly simplified.

The multiple scattering and the photon conversions can be limited by using thin devices and this is possible the ionization in Silicon being very high : 3.6 eV are needed to produce an electron-hole pair in Silicon, compared to 30 eV/pair for gas and 300 eV needed to generate one photoelectron at the photocathode when using plastic scintillators [12]. Finally Silicon detectors can operate properly inside magnetic fields or in vacuum itself and, in principle, can be used even inside the accelerator pipe[13].

4.1. Silicon Diode Basic Structure

Fig.4 shows the basic structure of a silicon diode detector. The principle is based on reverse biased n-p semiconductor junction (metal-semiconductor). The bulk material is n-doped high resistivity silicon, usually $\rho > 4$ KΩ·cm. High resistivity means low depletion voltage: for a given applied voltage V, the depleted thickness is $d = (2\varepsilon\mu\rho V)^{1/2}$ where ε, μ are constants and ρ the resistivity of the material. To avoid large leakage currents or even internal breakdown it is preferable to have full depletion voltage, V_d, in the range 50-100 Volts. The n^+ layer is typically doped with As or P (~10^{12} ions/cm^2), while the p^+ layer is obtained by implantation of Boron ~10^{14} ions/cm^2. When an external reverse bias is applied, the depletion region extends from the junction side (p^+) toward the ohmic side (n^+) and the charge produced by ionization in the depleted region can be detected by using low noise charge sensitive preamplifiers. For bias greater than the full depletion voltage, the whole volume becomes sensitive to ionizing radiation.

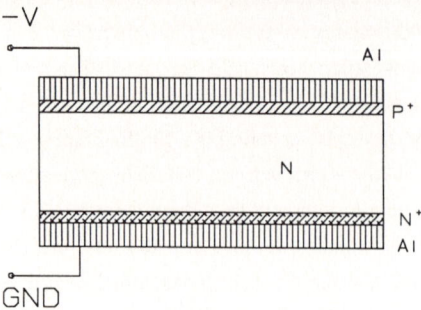

Fig. 4 : Reverse biased semiconductor diode.

To reduce the amount of material it would be convenient to use very thin detectors but many practical reasons limit this possibility. First of all the signal for minimum ionizing particles decreases almost linearly when reducing the thickness: as a rough estimate one can consider that, on average, ~80 e-h pairs are produced per micron of silicon. Unfortunately the capacitance of the detector increases, when reducing the thickness, as 1/d and this capacitance, at the input of the amplifier, is a source of noise. Thinner detectors produce not only smaller signals but also a higher level of noise; the signal to noise ratio comes out to be proportional to d^2 and therefore degrades very quickly with reducing the thickness.

For very special applications there are preamplifiers with levels of noise so low that few thousands of electrons are clearly visible, but they are of no practical use to amplify thousands of channels, densely packed together in a small volume at room temperature. At the present level of development of low noise analog electronics, a good compromise between reducing at the most the thickness and still having a good signal to noise ratio is around a thickness of 300 μm which corresponds to 0.3% radiation length and gives a signal of about 24.000 electrons.

The signal collection time in silicon is very fast and, in fully depleted detectors, is practically independent from the thickness : a useful formula for electrons is t (nsec)= 2ρ (KΩ·cm) which gives a collection time of 8 nsec for material with resistivity 4.000 Ohm·cm. The exact value can be calculated considering the drift speed v=-μ·E where μ (mobility at 300° K) is 1350 $cm^2 \cdot V \cdot$ sec for electrons and 480 $cm^2 \cdot V \cdot sec$ for holes. This fast response allows to stand very high rates of incoming particles : 1- 10 MHz.

4.2. Microstrip Detectors

Fig 5 shows the basic structure of a microstrip Silicon detector. It is a simple diode fabricated using techniques similar to those developed for Large Scale Integrated electronics ; one surface in particular, usually the junction side, is highly segmented in read-out electrodes. For reference purpose we consider here a readout pitch of 50 μm and assume that each electrode be connected to charge sensitive preamplifiers. A ionizing particle crossing the detector close to a particular strip releases a charge which is detected by the corresponding preamplifier : from the position of the fired strip we get the information on the impact point of the incoming particle.

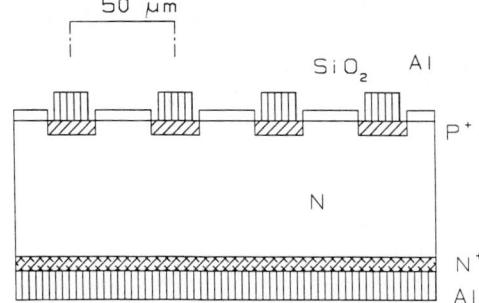

Fig. 5 : Basic design of a microstrip Silicon detector.

The space resolution obtainable with this kind of device depends on the density of the read-out electrodes : single track accuracy of 10 μm and double track resolution of 50 μm are easily achievable. At this level of resolution magnetic field effects must be taken into account, in particular if these detectors must operate inside the big Collider experiments where usually strong magnetic fields are present to help tracks separation and to analyze momenta of particles. Since the mobility is different for electrons and holes, also the corresponding Lorentz angles θ_L are different: $\tan\theta_L = 0.15$ B for electrons and 0.0035 B for holes when B is measured in Tesla. For the reference detector 300 µm thick and B=1T, the deflection of the collected holes is ~ 10 µm.

Lateral diffusion of the ionization cloud can be another source of degradation of the spatial resolution. The lateral spread of the cloud after a transit time t is given by $s=\sqrt{2Dt}$ where D is again different for electrons and holes. At 300° K, D=35 cm^2/sec for electrons and 12 cm^2/sec for holes. Considering again the same 300 µm thick detector, the charge blow up by diffusion is ~ 5 µm for holes collected at the junction side.

Production of delta rays during ionization may be another source of error, but the most part of these energetic electrons have energy below 10 KeV corresponding to a maximum range in silicon of ~1 µm. For 300 µm silicon we have 10% probability that energetic δ rays produce a displacement of the centroid ~ 5 µm [14].

4.3. Charge Partition Read-out

If the read-out electrodes are produced on the junction side, the charge released inside the crystal along a perpendicular trajectory is collected by a single strip at the surface. Since the depletion starts from the junction side, in the region between two electrodes there is lack of free carriers and the interstrip resistance is very high (R >100 MΩ). Holes are collected by one electrode or by the adjacent one and the expected position resolution in this coordinate is simply σ = $p/\sqrt{12}$ where p is the read-out pitch. The only way to improve the resolution in this case is to decrease the pitch, but there are practical limits : it is inconvenient to handle large area counters with many thousands of channels each one and, in particular, the connections to preamplifiers are very difficult.

Increasing the resolution without increasing the number of channels is possible by subdividing the total charge produced among different electrodes, that means operating the crystal in charge partition mode. An usual scheme is shown in fig. 6.

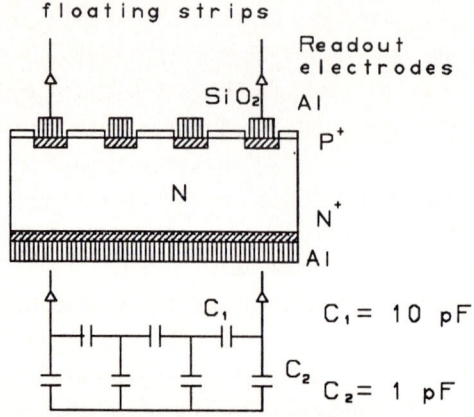

Fig. 6 : Capacitive partition read-out

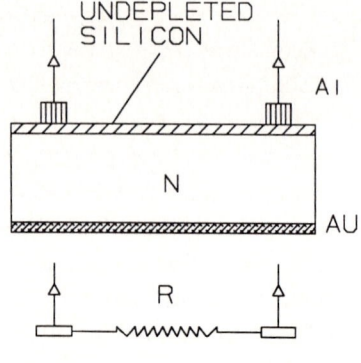

Fig. 7 : Resistive partition scheme.

The junction surface is subdivided into many collecting electrodes with a typical pitch of 20 μm, but preamplifiers are connected only to one strip out of n, while the remaining n-1 are left floating. In this configuration each electrode is capacitively coupled with the adjacent ones (the capacitance to ground is usually one order of magnitude smaller) and the charge collected on the floating strips induces a signal on the two read-out strips. For a pitch 50-100 μm this signal comes out to be linearly proportional to the distance of the read-out strips from the impact point which can be reconstructed through an interpolation algorithm with a precision which is much better than the read-out pitch itself [15].

Obviously by increasing the pitch the resolution worsens but there is also a loss of linearity as the interstrip capacitance becomes comparable with the capacitance to ground. Deviations from linearity come out also from the very high impedance between the floating strips that can produce significant voltage drops at the junction side and inefficiencies in charge collection for particles crossing in the middle of two read-out electrodes.

Finally the high capacitive coupling between strips is a source of noise since the interstrip capacitance becomes the dominant capacitance at the input of the amplifier (a typical value for interstrip capacitance is 1pF per cm of length of read-out electrode).

Other schemes of charge partition are possible . Fig 7 shows the mechanism of resistive charge partition in a surface barrier silicon detector with electrodes produced on the ohmic side [16]. In this case no floating electrodes are left between two readout strips and therefore the capacitive coupling is very weak. If the detector is biased at a voltage slightly under the full depletion (i.e.10% less than the nominal V_d) the read-out channels become resistively coupled .In these conditions, in fact, a thin layer of undepleted silicon is left underneath the strips on the ohmic side and, if we measure the interstrip resistance at this voltage , we get values around 50 -100 KΩ. This region acts as a linear resistive chain which divides the charge proportionally to the impact point of the electron cloud. The method can be sensitive to temperature and impurity density variation , furthermore the interstrip resistance is a source of noise at the input of the amplifier and must be as high as possible, but, in general , this method is less sensitive to large read-out pitch , can be easily applied to inter-electrode gaps as large as 500 μm or 1 mm without significant loss of linearity and is particularly useful when strong limitations on the total number of channels are required.

In both charge partition method, for a perpendicular track, the charge is basically divided between the strips according to the relative distance; by reconstructing the center of gravity of the charge it is possible to get the position of the incident particles with a precision which depends mainly on the read-out pitch, the released charge and the noise of read-out electronics according to simple relationships. Figg.8 and 9 shows results obtained with the two methods: in the first plot is shown the extreme resolution obtained with 20 μm read-out pitch detectors with capacitive coupling [17].The other histogram shows a resolution of 22 μm obtained with resistive charge partition with electrodes spaced 200 μm apart [16] .

Although the interpolation approach offer many advantages, one basic limitation is the unavoidable degradation of the two-track resolution which must be taken into account when dealing with the high granularity needed for distinguish tracks belonging to secondary vertices superimposed to tracks coming out from the primary interaction.

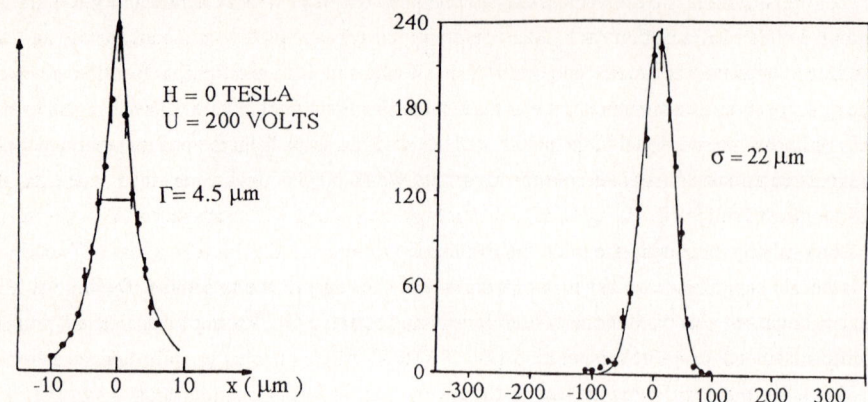

Fig. 8 : Charge distribution collected by the diode strips (Ref.17)

Fig.9 : Residuals distribution of impact position of particles (Ref.16)

4.4. Read-out Electronics

Even if charge partition can reduce the overall number of read-out channels, a full size Silicon vertex detector will anyway contain something of the order of 50.000 individual read-out channels within a small volume around the interaction region. This extremely high channel density makes conventional electronics and standard schemes prohibitively difficult . The solution was found using custom-designed microelectronics devices with a read-out pitch matching the electrode pitch of the detector. This approach was pioneered at Stanford where a custom VLSI circuit named Microplex was developed using n-MOS technology [18]. This device contains 128 charge sensitive preamplifiers on a 50 μm pitch and a multiplexed serial output. Thus many channels can be reduced to a single read-out line and even the problem of cables to carry out signals is simplified. The generic read-out scheme for a silicon strip detector using this circuit is shown in fig 10.

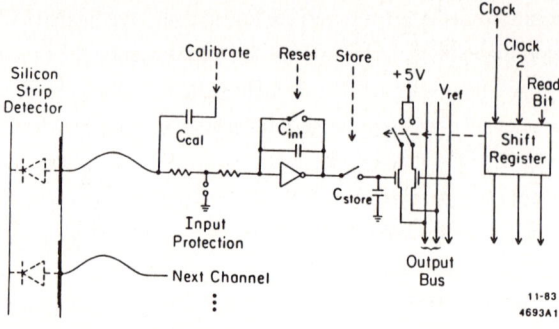

Fig.10 : Block diagram of read-out electronics on the Microplex chip (Ref.18)

The electrodes are directly connected to the inputs of the amplifiers by using ultrasonic bonding. The signals are fed through a charge sensitive amplifier with a 500 gain onto a storage capacitor. In Collider operation the storage capacitor would be reset between bunch crossing. The stored information is read-out at ~ 1MHz (~150 μs to read-out a complete chip). The analog signal from each capacitor in turn is connected to a line driver for remote digitization. The first generation of this chip has shown that the principle can work but still had high level of noise (~2500 electrons) and a strong dependence on the input capacitance. Since then other similar approaches have been successfully tried using CMOS technology and also J-FET transistor at the input stage in order to get better performance.

So far the main limitations to a further increase in granularity of the detectors come from the far from simple problem of making the connections from the detector to the read-out chip. Ultrasonic bonding under 50 μm pitch is very difficult and possible alternative solutions are being investigated: among them are various schemes of indium bump bonding and, hopefully, the integration of read-out electronics on the same substrate used for the detector which, if successful, will eliminate drastically the problem.

5. EXPERIMENTAL APPLICATIONS

5.1. CDF Silicon Vertex Detector

Fig.11 shows a schematic view of CDF which detects $p\bar{p}$ collisions at $\sqrt{s}=1800$ GeV at the Tevatron Collider. Inside the big hadronic and electromagnetic calorimeters a large drift chamber (CTC) provides tracking in the central region where a superconducting solenoid produces a magnetic field of 1.5 Tesla. Eight modules of small time projection chambers (VTPC) are installed inside the CTC to select real beam-beam interactions, giving a rough reconstruction of the primary vertex. To show the limitations of this sophisticated system in tagging secondary vertices one can look at a typical jet event as shown in fig 12.

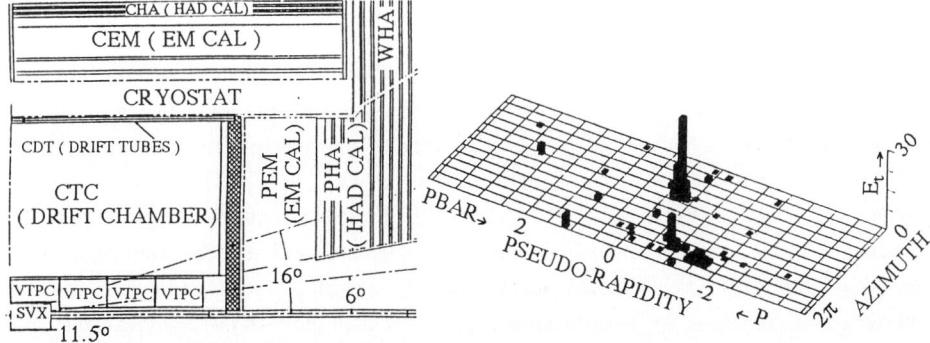

Fig.11 : Quarter view of CDF (Ref.19)   *Fig.12 : Typical jet event in CDF.*

Here the transverse energy released in the calorimetry is plotted versus pseudorapidity η and the azimuthal angle ϕ. The two jets structure is clearly visible in energy and the event seems extraordinary clean. Now we take a look to the tracks belonging to the two jets; fig 13 shows a cross view of the CTC tracks in the same event and the high density of tracks, when the interaction region is approached, is clearly visible. The situation is even worse if we look at the same event in the VTPC, still closer to the beam pipe (fig 14). In extrapolating back the reconstructed tracks we get so large errors in the impact parameters that, even if they would be present in this event, there is no hope to distinguish secondary vertices fraction of millimeter far away from the primary interaction.

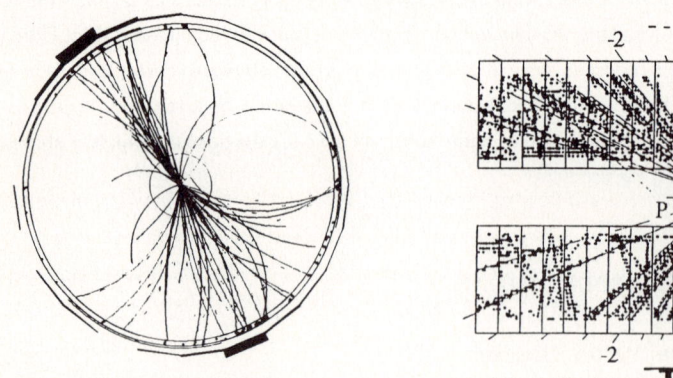

Fig.13 : CTC tracks of the event of fig.12   Fig.14 : The same event seen by the VTPC.

These limitations brought the CDF collaboration to plan the construction of a Silicon Vertex Detector (SVX) to be installed inside the VTPC in 1990 in the small radial region which is left between these chambers and the beam pipe. The responsibility for the construction of this new device inside CDF was given to physicists from Pisa, Berkeley and Fermilab[19].

The interaction region in CDF is extended longitudinally with a sigma of 35 cm ; in order to guarantee a good geometrical acceptance the detector must be long enough to cover a substantial part of the interaction spot.

Although, in principle, only two points are necessary, in order to have a certain amount of redundancy in the track position it has been decided to have four radial layers of silicon strip detectors surrounding the 2 -inch diameter beam pipe(fig 15).The SVX overall layout is shown in fig. 16. It consists of two cylindrical modules placed end to end with their axes coincident with the beam axis. It is meant to track primarily in the r-ϕ plane as the CTC does. The intention is to link tracks in CTC with segments in the SVX. In each module the detectors are arranged as a twelve sided barrel at each radial position. The strip detectors are 280 μm thick. The inner three layers have strips on a 60 μm pitch while the outer one has a 110 μm pitch. The detectors are electrically connected together along the beam direction in groups of three and each group is read-out at both ends. Each individual detector is 8.5 cm long that means that an individual channel sees a 25.5 cm long strip. The source capacitance of each strip is about 30 pF.

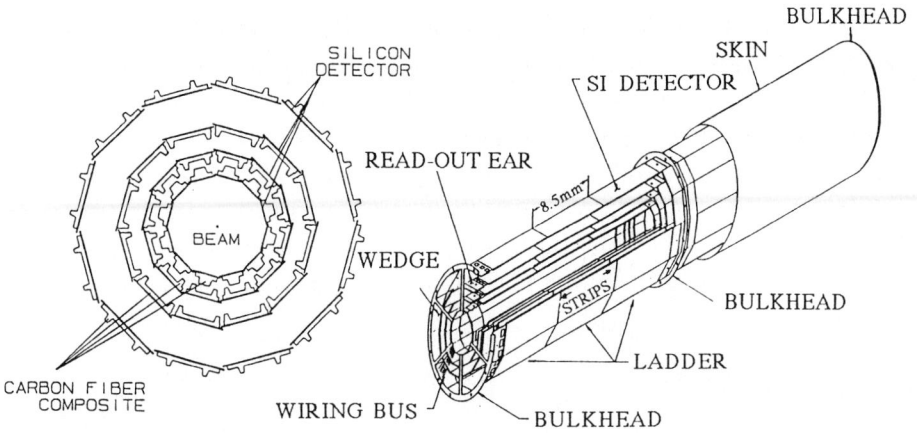

Fig.15 : Cross sectional view of the CDF SVX.   Fig16 : Overall layout of the SVX (Ref.19)

The mechanical structure is very light and stiff. Groups of three detectors are mounted on a "ladder" made-out of a light weight foam reinforced with thin carbon fiber strips. At the end of each ladder is mounted a small circuit board which contains the read-out chips as well as auxiliary components needed for the operation of the chips. The total radiation length traversed by particles crossing the SVX is, on average, 3% x_0. The assembly of each ladder is very demanding in term of the mechanical tolerances required; since the strips of three detectors will be electrically connected, it is mandatory that they will be aligned to better than their intrinsic spatial resolution (~5 µm). First tests show that this careful alignment is possible.

The CDF SVX will contain approximately 40.000 individual read-out channels within a cylindrical region 8 cm in radius and 51 cm long. To deal with this a custom VLSI circuit has been developed at LBL to amplify, sample and multiplex data from the SVX. The device has been fabricated using CMOS technology and analog and digital functions are integrated onto the same chip. It contains 128 high gain charge sensitive preamplifiers, each followed by a sample and hold stage, a threshold storage stage, a comparator and latch, and digital circuitry to control a serial multiplexed read-out. Bench tests have indicated an open loop gain of about 2000 for the integrator which has an effective capacitance of 600 pF (feedback capacitance 0.3 pF), far in excess of the strip capacitance, as it is needed to have a good charge collection. The dependence of noise on detector capacitance was studied by varying a test capacitor on the input. The results are indicated in fig.17 for two different settings of sample and hold circuitry. In both cases the level of noise corresponding to 30 pF at the input is compatible with having a signal to noise ratio greater than 10.

The total power dissipation of the device is ~1.5 mW per channel that means less than 100 Watts for the entire detector. This amount of power dissipation is quite small, nevertheless it must be removed and some sort of cooling will be necessary.

The expected performance of the SVX are a position resolution of ~10 µm per each layer which will result in ~30 µm resolution on impact parameter of 1 GeV/c p_t tracks. The asymptotic impact parameter resolution is ~10 µm (for tracks with $p_t > 10$ GeV) obtained combining the information from CTC and from SVX. For tracks at lower momenta the effects of multiple scattering dominate and the resolution is degraded to about 60 µm at a p_t of 500 MeV/c.

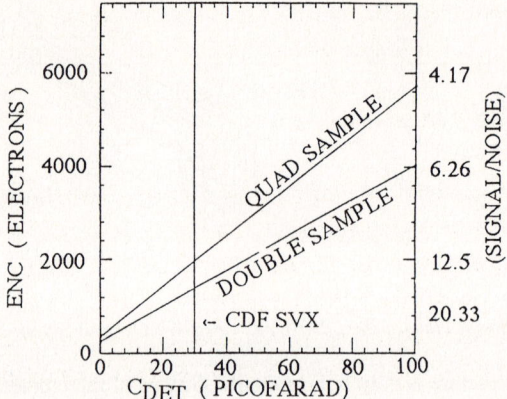

Fig 17 : Noise and S/N for 300 µm Silicon vs detector capacitance (Ref.19).

An extended simulation work has shown that the calculated efficiency for reconstructing secondary vertices from heavy quarks, with the strong signature of three charged tracks, is between 20% and 40%. Some of the performance have been already tested, in particular the position resolution of the SVX layers has been measured. A four layers silicon telescope has been assembled in the CDF test-beam. For each layer only one detector was mounted onto a ladder and connected to integrated preamplifiers. Several thousands 230 GeV/c pion events were collected and analyzed. One of the four layers failed to work properly. In fig.18 a typical event is shown: for each one of the three working layers the pulse height of the 128 strips is shown versus the strip position.

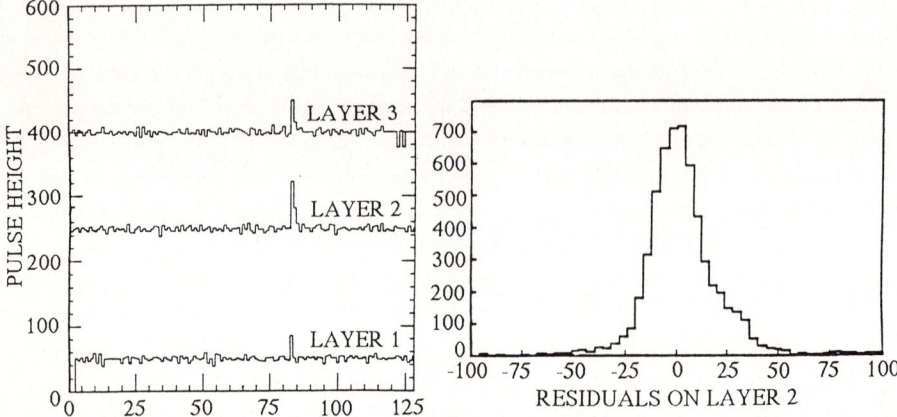

Fig.18 : Typical event from the SVX beam test (Ref.19)   Fig.19 : Single layer resolution (Ref.19)

The charge released by the pion passing through the three detectors is clearly visible. The signal to noise ratio was measured to be 12 and the efficiency for minimum ionizing particles larger than 99% per layer.

Fig. 19 shows the the distribution of the residuals between the extrapolated track position as measured by the two outer detectors and the position reconstructed by the inner one. The width of this distribution corresponds to a space resolution $\sigma = 8$ μm.

5.2. ALEPH Mini Vertex Detector

The central tracking system of the Aleph detector is made out of a large, very sophisticated TPC which reconstructs tracks in r-ϕ and r-z coordinates, performs momentum analysis and provides also some dE/dx particle identification. Mainly devoted to trigger purposes, a cylindrical drift chamber (ITC) surrounds the beam pipe providing other useful informations for tracking, mainly in the r-ϕ coordinate. The residual space available between the 8 cm radius beam pipe and the inner wall of the ITC (r = 13.2 cm) will be used to install the final element of the tracking system : a mini vertex detector mainly used to identify event topology with multistep decays [20].

The overall view of the system is shown in fig 20. Two layers of silicon strip detectors are arranged on two concentric surfaces around the beam pipe. The average radius of the inner barrel is 90 mm, of the outer 121 mm. Four Silicon detectors are mounted on each face of the barrels for a total active length of about 20 cm. Each individual detector has a size 48x50 mm2 and a thickness of about 300μm.

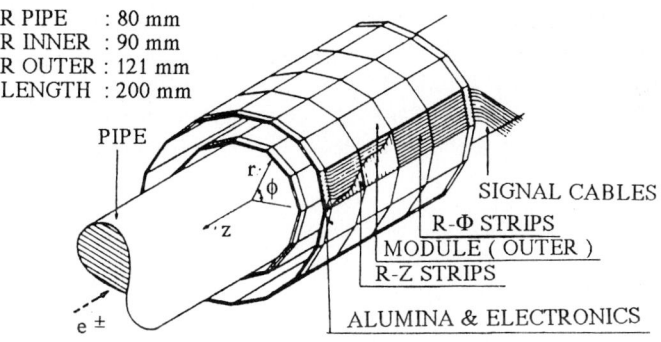

Fig.20 : Schematic view of ALEPH Mini Vertex Detector.

The original feature of the ALEPH MVD is that all detectors are double side read-out microstrip wafers. The detectors have strips on both sides: on one side the electrodes run parallel to the beam and measure the azimuthal angle ϕ, on the other side they are perpendicular to the beam and measure the z coordinate parallel to the beam. With the radius r given by the detector position, the cylindrical coordinates r-ϕ-z of the particles passing through the detectors are measured. The read-out pitch is 100 μm in r-ϕ and 200 μm in r-z while the strip pitch is 25 μm and 50 μm respectively. Three additional strips are left between read-out strips to improve the position resolution by capacitive charge division. The expected resolution is 10 μm in r-φ and 20 μm in r-z with a two-track resolution of 200 and 400 μm respectively. A total number of 53.000 strips will be read-out by monolithic multiplexing electronics which is located at the edges of detectors.

The simultaneous measurement of the two orthogonal projections of the particle impact point offers some advantage over the more conventional single sided detector. First of all the impact parameter of one track can be measured as a three dimensional vector by using only two layers of Silicon. The measurement of space points helps in matching with the rest of the tracking system and the two-dimensional information is obtained without any additional scattering material on the trajectory of the particles. In addition , when just two particles cross the same detector , the Landau fluctuations can help to solve ambiguities. The charge collected on each side is the same and the correlation can be exploited in coupling projections to space points. This feature is particularly useful for LEP experiments where the number of tracks hitting a single wafer is expected to be 1.2 in average.

Unfortunately, the production of double-side read-out silicon detectors it is not the simple extension to the ohmic side of the same technique used for the junction side, and an intensive work of development has been necessary in Pisa to fabricate this kind of device [21].

The problem is illustrated in fig.21a. If one simply produces n^+ strips on the ohmic side (by implanting Phosphorus) finds out that the interstrip resistance on this side is too low to avoid short circuits between the n^+ strips so preventing any position information. The reason of this low resistance is well understood: in the Silicon-Silicon dioxide interface the fixed positive charge builds up an accumulation layer of electrons in the n-type substrate. This n-type channel cannot be depleted by any reasonable bias voltage applied across the wafer and dramatically lowers the resistance between n^+ strips on the ohmic side, making the signal charge to spread over many electrodes. To solve the problem a p^+ blocking strip has been produced between the n^+ strips in order to interrupt the conduction channel and thus to increase the interstrip resistance (fig.21 b).

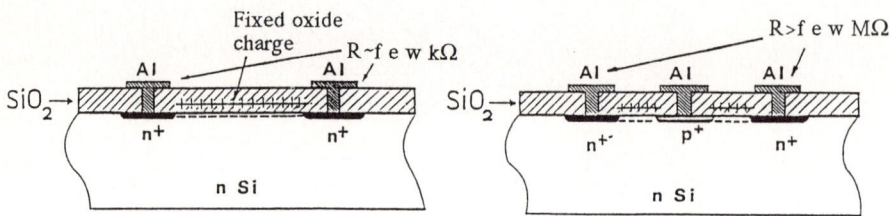

Fig. 21 : a) the problem of double side read-out   b) the proposed solution (Ref. 21)

The principle seems very simple but the realization has requested a long term development effort for setting-up the right processing of the two surfaces. The basic fabrication procedure is shown in a simplified scheme in fig.22.

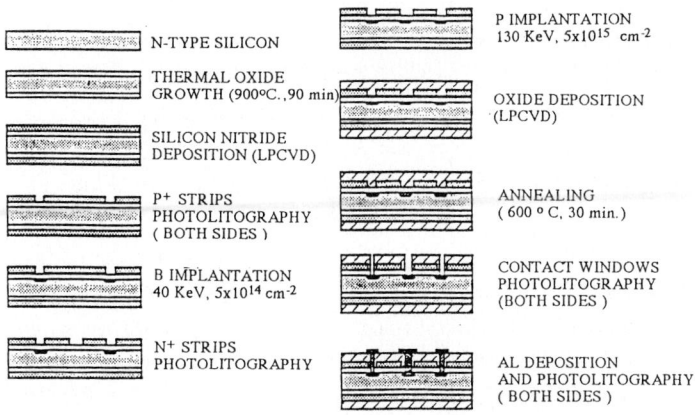

Fig. 22 : The fabrication process (Ref.21).

Many prototype detectors have been produced so far and, recently, a set of full size 5x5 cm^2 detectors have been fabricated and tested with a radioactive source [22]. The action of the blocking strip is clearly shown in fig.23 where the interstrip resistance on the ohmic side is plotted versus the bias voltage. When the applied reverse bias approaches the full depletion V_d, the interstrip resistance rises of about 3 orders of magnitude. The same resistance, measured in reference detectors with no blocking strips, shows no significant rise with the voltage and sits around values as low as few KΩ.

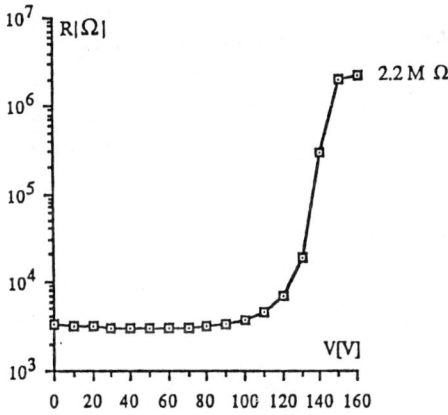

Fig. 23 : Interstrip resistance vs bias voltage (Ref.21).

The detectors have been tested using conventional hybrid electronics and highly penetrating electrons selected from a β source. At the full depletion voltage, signals from the the two sides are clearly visible and the spectrum of charge collected on two faces shows a typical Landau distribution (fig. 24 a) and b)).

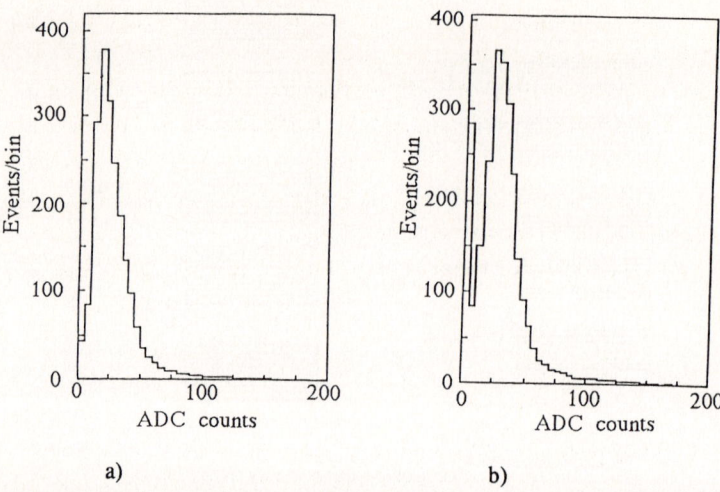

Fig. 24 : Spectrum of charge collected a) on the junction side b) on the ohmic side (Ref. 21)

The scatter plot of fig.25 shows the correlation between charge collected on the ohmic side and charge at the junction side. Further informations on the performance of this kind of device are expected from test beam measurements now under way. In particular the signal charges at the ohmic side are electrons and we know that they have higher mobility in E and B fields compared to holes. Charge collection is faster but sensitivity to magnetic field is increased. Furthermore the electric field strength decreases toward the n^+ side of the detector and the electrons passing through the low field region near the ohmic contact suffer more diffusion than holes collected by the p^+ strips. All these effects can affect the ultimate space resolution obtainable with this new device and must be tested carefully.

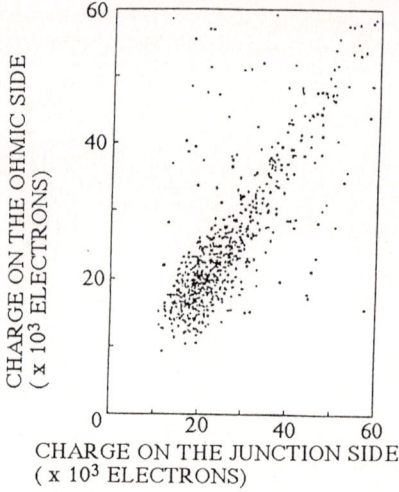

Fig. 25 : Correlation plot between charge collected on the two faces (Ref. 22)

The monolithic multiplexing electronics has been developed at MPI Munich and integrated using CMOS technology. CMOS techniques allow the realization of single-stage high gain preamplifiers, stable against unwanted oscillations; fast power switching is also possible to reduce the heat dissipation. These preamplifiers will perform a 60:1 multiplexing and reduce the total number of output channels to 1008. Equivalent noise charge with the detectors connected at the input is expected to be less than 1000 electrons. The operating power during read-out is 2mW/channel while the power switching reduces the average power consumption to 0.4 mW/channel. The total power consumption will be enough low (20 mW) to allow a soft cooling with gas.

6. A LOOK ON THE FUTURE AS FINAL REMARK

At the actual stage of development, microstrip Silicon detectors accomplish quite well the requirements for improved tracking and heavy flavour tagging at the actual Colliders. But what about the next generation of particle accelerators ? In the multi-TeV energy domain we shall probably face events like that one shown in fig. 26. The event is an ISAJET simulation of a $b\bar{b}$ di-jet produced in pp collisions at $\sqrt{s}$ = 40 TeV; the tracks are bent by a strong magnetic field of 1.5 Tesla but even a closer look to the same event as shown in fig.27 does not simplify the topology. Tracking and hunting for secondary vertices in this case comes out to be really difficult even for the most sophisticated Silicon detector so far available.

The situation seems to be hopeless if we consider the basic parameters of Colliders like SSC or LHC. Center of mass energy of 20-40 TeV, luminosity 10^{32}-10^{34} cm$^2 \cdot$ s^{-1}, time interval between bunch crossings between 25 and 35 nsec. The real question is : what tracking detector can: a) survive in this environment b) provide some useful measurements for physics ?

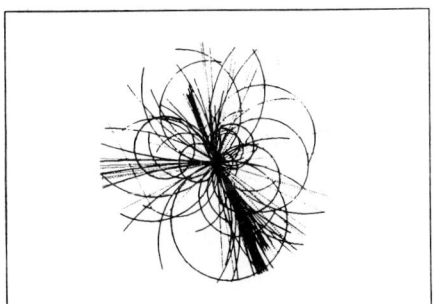 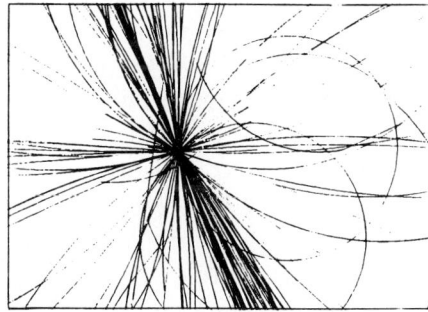

Fig. 26 : pp $\rightarrow b\bar{b}$+X at $\sqrt{s}$=40 TeV   *Fig. 27 : Enlarged view of the same event.*

In principle Silicon detectors are fast enough to stand such high rates but their major limitation is radiation damage. Fig.28 shows the radiation dose, expressed in minimum ionizing particles/cm^2, for a 10 cm long detector, after one year of operation, for different luminosities and different distances from the beam [23]. Silicon detectors can actually stand, without macroscopic effects, radiation dose corresponding to 10^{12} mip/cm^2 while at 10^{14} mip/cm^2 severe degradation of the performance are reported. Furthermore a high flux of neutrons is expected from interactions in the beam pipe and from a sort of neutron gas, with energy around 1 MeV, which is scattered backward by the massive calorimeters themselves. It is well known that sensitivity to neutrons is orders of magnitude higher in Silicon detectors but the environment is even more critical for front-end electronics which usually stops working properly for about 10^{11} mip/cm^2.

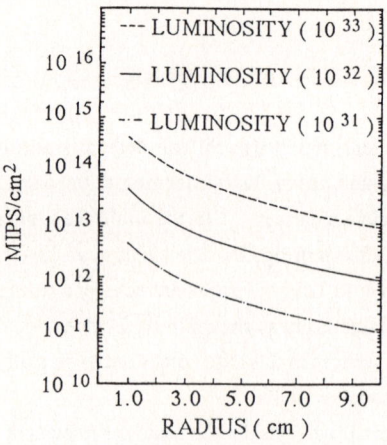

Fig. 28 : Radiation dose in m.i.p../cm^2 for a 10 cm long Silicon detector in one year (Ref.23)

Meanwhile many promising ideas are under development : multiplexing analog preamplifiers using J-FET at the input stage can stand a flux of particles in the range of 10^{14} mip/cm^2; microstrip detectors made-out of GaAs could take profit of the fast response of this semiconductor (1.5 nsec collection time in 300 μm detector) and there are also clues of a lower sensitivity to radiation damage ; amorphous silicon can be studied to build large area detectors at low cost ; pixel devices, which give directly space points with resolution of few μm in both coordinates, seem to be a very promising approach, providing that a sort of local intelligence be integrated on the same substrate .

The conclusion is that an extraordinary effort in development of detectors and electronics is required in order to build reliable tracking systems for the future generation of Colliders.

REFERENCES

[1] S.R.Amendolia et al. Nucl. Instr. Meth. 176 (1980) 449.

[2] E.Albini et al. Phys. Lett. 110 B (1982) 339.

[3] S.R.Amendolia et al. Europhys. Lett. 5 (1988) 407.

[4] R.Bailey et al. Phys. Lett. 139 B (1984) 320.

[5] S.R.Amendolia et al. Z. Phys. C 36 (1987) 513.

[6] R.Bailey et al. Nucl. Instr. Meth. 213 (1983) 201.

[7] S.Penzotti et al. A 257 (1987) 538.

[8] E.Gatti and P. Rehak, Nucl. Instr.Meth. A 225 (1984) 608.

[9] J.Kemmer and G.Lutz, Nucl. Instr. Meth. A 253 (1987) 365.

[10] L.Struder et al. Nucl. Instr. Meth. A 253 (1987) 386

[11] M.Dell'Orso and L.Ristori, Proceedings of the International Conference on " Impact of Microelectronics and Microprocessors on High Energy Physics", Trieste, Italy, 28-30 March 1988 (in print).

[12] A.H.Walenta, Nucl. Instr. Meth. A 253 (1987) 558.

[13] G.Apollinari et al. Nucl. Instr. Meth. A 252 (1986) 467.

[14] C.J.S.Damerell RAL-84-123 December 1984.

[15] J.B.A.England et al. Nucl. Instr. Meth. 185 (1981) 43.

[16] S.R.Amendolia et al. Nucl. Instr. Meth. 226 (1984) 82.

[17] E.Belau et al. Nucl. Instr. Meth. 214 (1983) 253.

[18] J.T.Walker et al. Nucl. Instr. Meth. 226 (1984) 200.

[19] CDF Collaboration, Proposal P-775, Fermilab, June 1988.

[20] L.Bosisio et al. ALEPH-Note 84/130, Cern, November 1984.

[21] G.Batignani et al., Procccedings of the "London Conference on Position Sensitive Detectors", London, 7-11 September 1987 (in print).

[22] G.Batignani et al., Proceedings of the "3rd Topical Seminar on Perspectives for Experimental Apparatus at Future High Energy Machines and Underground Laboratories",S.Miniato, Italy, 7-11 March 1988 (in print).

[23] M.G.D.Gilchriese, Proceedings of "1984 Summer Study on the Design and Utilization of the Superconducting Super Collider " Eds. R.Donaldson and J.G.Morfin (Division Particles and Fields of the APS), 1984.

LIST OF PARTICIPANTS

M. AGNELLO
Politecnico di Torino
Corso Duca degli Abruzzi, 24
10125 Torino
(Italy)

E. AMALDI
Dipartimento di Fisica
Università "La Sapienza"
Piazzale A. Moro, 2
00185 Roma
(Italy)

G. AURIEMMA
Dipartimento di Fisica
Piazzale A. Moro, 2
00185 Roma
(Italy)

M. BALDO CEOLIN
Dipartimento di Fisica
Via F. Marzolo, 8
35100 Padova
(Italy)

G. BELLETTINI
Sezione INFN
Via Livornse, 582/A
56010 S. Pietro a Grado (Pisa)
(Italy)

P. BERNARDINI
Dipartimento di Fisica
Via Arnesano
73100 Lecce
(Italy)

E. BERSANI
Dipartimento di Fisica
Universita' degli Studi
"La Sapienza"
Piazzale A. Moro, 2
00185 Roma
(Italy)

A. BETTINI
Dipartimento di Fisica
Via F. Marzolo, 8
35131 Padova
(Italy)

V. BOLOGNESI
Dipartimento di Fisica
Via Paradiso, 12
44100 Ferrara
(Italy)

G. BONAZZOLA
Dipartimento di Fisica
Via P. Giuria, 1
10125 Torino
(Italy)

S. BOSSOLASCO
Dipartimento di Fisica
Via P. Giuria, 1
10125 Torino
(Italy)

F. BRADAMANTE
Dipartimento di Fisica
Via A. Valerio, 2
34127 Trieste
(Italy)

M. BREGOLA
Dipartimento di Fisica
Via Paradiso, 12
44100 Ferrara
(Italy)

T. BRESSANI
Dipartimento di Fisica
Via P. Giuria, 1
44100 Torino
(Italy)

R. BROGLIA
Dipartimento di Fisica
Via Celoria, 16
20133 Milano
(Italy)

R. BRUGNERA
Sezione di INFN di Padova
Via F. Marzolo, 8
35100 Padova
(Italy)

M.P. BUSSA
Dipartimento di Fisica
Via P. Giuria, 1
10125 Torino
(Italy)

D. CALVO
Dipartimento di Fisica
Via P. Giuria, 1
10125 Torino
(Italy)

L. CANESCHI
Sezione INFN di Firenze
Largo E. Fermi, 2
50125 Firenze
(Italy)

G. CASINI
Dipartimento di Fisica
Largo E. Fermi, 2
50125 Firenze
(Italy)

R. CASTALDI
Sezione INFN
Via Livornese, 582/a
56010 S. Pietro a Grado (Pisa)
(Italy)

F.R. CAVALLO
Sezione INFN di Bologna
Via Irnerio, 46
40126 Bologna
(Italy)

F. CAVANNA
Sezione INFN di Roma
Piazzale A. Moro, 2
00185 Roma
(Italy)

R. CHERUBINI
Laboratori Nazionali di Legnaro
Via Romea, 4
35020 Legnaro (Padova)
(Italy)

S. CITTOLIN
CERN
Division DD
1211 Geneve 23
Switzerland

P. DALPIAZ
Dipartimento di Fisica
Via Paradiso, 12
44100 Ferrara
(Italy)

R. D'ALESSANDRO
Dipartimento di Fisica
Largo E. Fermi, 2
50125 Firenze
(Italy)

S. DALLA TORRE
Sezione INFN di Trieste
Via A. Valerio, 2
34127 Trieste
(Italy)

G. DE CATALDO
Sezione INFN di Bari
Via Amendola, 173
70126 Bari
(Italy)

S. DELL'UOMO
Sezione INFN di Roma
Piazzale A. Moro, 2
00185 Roma
(Italy)

U. DOSSELLI
Dipartimento di Fisica
Via F. Marzolo, 8
35131 Padova
(Italy)

A. FACCO
Laboratori Nazionali di Legnaro
Via Romea, 4
35020 Legnaro (Padova)
Italy

R.M. FINI
Dipartimento di Fisica
Via Amendola, 173
70126 Bari
(Italy)

E. FIORAMONTI
Sezione INFN di Torino
Via P. Giuria, 1
10125 Torino
(Italy)

P. FORTINI
Dipartimento di Fisica
Via Paradiso, 12
44100 Ferrara
(Italy)

G. FORTUNA
Laboratori Nazionali di Legnaro
Via Romea, 4
35020 Legnaro (Padova)
(Italy)

F. FRONTERA
Dipartimento di Fisica
Via Paradiso, 12
44100 Ferrara
(Italy)

P. GALEOTTI
Dipartimento di Fisica
Via P. Giuria, 1
10125 Torino
(Italy)

R. GARFAGNINI
Dipartimento di Fisica
Via P. Giuria, 1
10125 Torino
(Italy)

U. GASTALDI
CERN
Division EP
1211 Geneve 23
Switzerland

M. GIORGI
Dipartimento di Fisica
Via A. Valerio, 2
34127 Trieste
(Italy)

E. GORINI
Dipartimento di Fisica
Pad. 19 Mostra d'Oltremare
80125 Napoli
(Italy)

G. INGROSSO
Dipartimento di Fisica
Via Arnesano
73100 Lecce
(Italy

R. IASEVOLI
Dipartimento di Fisica
Mostra d'Oltremare Pad. 19
80125 Napoli
(Italy)

F. IAZZI
Politecnico di Torino
C.so Duca degli Abruzzi, 4
10100 Torino
(Italy)

R. LEONARDI
Universita' degli Studi di Trento
Facolta' di Fisica
38050 Povo (Trento)
(Italy)

E. LODI RIZZINI
Universita' di Brescia
Facolta' di Ingegneria
Viale Europa, 39
20149 Brescia
(Italy)

G. MANCARELLA
Dipartimento di Fisica
Via Arnesano
73100 Lecce
(Italy)

G. V. MARGAGLIOTTI
Sezione INFN di Trieste
Via A. Valerio, 2
34127 Trieste
(Italy)

G. MARON
Laboratori Nazionali di Legnaro
Via Romea, 4
35020 Legnaro (Padova)
(Italy)

A. .MARTIN
Dipartimento di Fisica
Via A. Valerio, 2
34127 Trieste
(Italy)

M. MARTINI
Dipartimento di Fisica
Via Paradiso, 12
44100 Ferrara
(Italy)

M. MASALA
Dipartimento di Fisica
Via P. Giuria, 1
10125 Torino
(Italy)

A. MASONI
Dipartimento di Fisica
Via Ospedale, 72
09100 Cagliari
(Italy)

G. MIELE
Dipartimento di Fisica
Mostra d'Oltremare Pad. 19
80125 Napoli
(Italy)

B. MINETTI
Politecnico di Torino
Via Duca degli Abruzzi, 24
10129 Torino
(Italy)

M. MORETTI
Dipartimento di Fisica
Via Paradiso, 12
44100 Ferrara
(Italy)

F. NAVARRIA
Dipartimento di Fisica
Via Irnerio, 46
40126 Bologna
(Italy)

A. NANNINI
Dipartimento di Fisica
Largo E. Fermi, 2
50125 Firenze
(Italy)

R. ONOFRIO
Dipartimento di Fisica
Università "La Sapienza"
P.le A. Moro, 2
00185 Roma
(Italy)

O. PALAMARA
Dipartimento di Fisica
P.le A. Moro, 2
00185 Roma
(Italy)

N. PAVER
Università di Trieste
Dipartimento di Fisica Teorica
Strada Costiera, 11
34014 Miramare - Grignano (Trieste)
(Italy

C. PERONI
Dipartimento di Fisica
Via P. Giuria, 1
10125 Torino
(Italy)

L. PESANDO
Dipartimento di Fisica
Via P. Giuria, 1
10125 Torino
(Italy)

S. PETRERA
Dipartimento di Fisica
Piazzale A. Moro, 2
00185 Roma
(Italy)

L. PIEMONTESE
Dipartimento di Fisica
Via Paradiso, 12
44100 Ferrara
(Italy)

P. PISTILLI
Dipartimento di Fisica
Via Arnesano
73100 Lecce
(Italy)

E. PREDAZZI
Dipartimento di Fisica
Via P. Giuria, 1
10125 Torino
(Italy)

G. PREPARATA
Dipartimento di Fisica
Via Celoria, 16
20133 Milano
(Italy)

G. PUDDU
Dipartimento di Fisica
Via Ospedale, 72
09100 Cagliari
(Italy)

M. PUSTERLA
Dipartimento di Fisica
Via F. Marzolo, 8
35131 Padova
(Italy)

E. REMIDDI
Dipartimento di Fisica
Via Irnerio, 46
40126 Bologna
(Italy)

P. SALVINI
Dipartimento di Fisica
Via Bassi, 6
27100 Pavia
(Italy)

R. SANTONICO
II Università di Roma
Via O. Raimondo
00178 Roma
(Italy)

M. SAVRIE'
Dipartimento di Fisica
Via Paradiso, 13
44100 Ferrara
(Italy)

S. SERCI
Dipartimento di Fisica
Via Ospedale, 72
09100 Cagliari
(Italy)

L. SILVESTRIS
Dipartimento di Fisica
Via Amendola, 173
70125 Bari
(Italy)

A. SURDO
Dipartimento di Fisica
Via Arnesano
73100 Lecce
(Italy)

G. TONELLI
Sezione INFN
Via Livornese, 582/a
56010 S. Pietro a Grado (Pisa)
(Italy)

A. TORNAMBE'
CNR
Istituto Astronomico Spaziale
Casella Postale 67
00044 Frascati (Roma)
(Italy)

C. TUNIZ
Dipartimento di Fisica
Via A. Valerio, 2
34127 Trieste
(Italy)

C. VANNINI
Sezione INFN
Via Livornese, 582/a
56010 S. Pietro a Grado (Pisa)
(Italy)

L. VANNUCCI
Laboratori Nazionali. di Legnaro
Via Romea, 4
35020 Legnaro (Padova)
(Italy)

G. VEDOVATO
Laboratori Nazionali di Legnaro
Via Romea, 4
35020 Legnaro (Padova)
(Italy)

C. VOCI
Dipartimento di Fisica
Via F. Marzolo, 8
35131 Padova
(Italy)

L. VOTANO
Laboratori Nazionali di Frascati
Casella Postale 13
00044 Frascati (Roma)
(Italy)

AUTHOR INDEX

AGLIETTA, C., 389
ALPAT, B., 389
ALYEA, D., 389
ANZIVINO, G., 389
ARTEMI, F., 389
ARYAL, M., 389
AURIEMMA, G., 295
BADINO, G., 389
BARBAGLI, G., 389
BARI, G., 389
BARRANCO, F., 103
BASILE, M., 389
BEREZINSKY, V.S., 389
BERGAMASCO, L., 389
BETTINI, A., 169
BIANCO, S., 389
BRACCO, A., 111
BRADAMANTE, F., 465
BREGOLA, M., 61
BRESSANI, T., 151
BROGLIA, R.A., 103, 111
BRUNI, G., 389
CAPPELLETTI, C., 389
CAPUTI, L., 389
CARA ROMEO, G., 389
CASACCIA, R., 389
CASTAGNOLI, C., 389
CASTALDI, R., 211
CASTELLINA, A., 389
CASTELVETRI, A., 389
CHINCELLATO, J.A., 389
CIFARELLI, L., 389
CINDOLO, F., 389
CINI, G., 389
CONFORTO, G., 389
CONTIN, A., 389
D'ALI, G., 389
DADYKIN, V.L., 389

DAI, Y., 389
DARDO, M., 389
DE PASQUALE, S., 389
DE, K., 389
DEFELICE, M., 389
DEL PAPA, C., 389
DI SCIASCIO, G., 389
DIN, L., 389
DIODATI, P., 389
DOBRIGKEIT CHINCELLATO, C., 389
DONG, Y., 389
DOSSELLI, U., 257
ENORINI, M., 389
FABBRI, F.L., 389
FAUTH, A.C., 389
FORTINI, P., 303
FRONTERA, F., 361
FULGIONE, W., 389
GALEOTTI, P., 319, 389
GHIA, P., 389
GIUSTI, P., 389
GRIANTI, F., 389
HAFEN, E.S., 389
HARA, T., 389
HARIDAS, P., 389
HUANG, H.H., 389
IACOBUCCI, G., 389
IAZZI, F., 123, 151
INOUE, N., 389
JECKELMANN, B., 389
JI, G., 389
JING, C., 389
JING, G., 389
KHAICHUKOV, F.F., 389
KITAMURA, T., 389
KLEIN, J., 425
KOCHAROV, G.E., 389
KOROLKOVA, E.V., 389

KORTCHAGUIN, P.V., 389
KORTCHAGUIN, V.B., 389
KUDRYAVTSEV, V.A., 389
LAAKSO, I., 389
LANDI, G., 389
LAU, K., 389
LEONARDI, R., 79
LIPPS, F., 389
LODI RIZZINI, E., 163
LU, Z., 389
MACCARRONE, G., 389
MAO, C.S., 389
MARKOV, A.S., 389
MASSAM, T., 389
MAYES, B., 389
MINETTI, B., 123
MINORIKAWA, Y., 389
MISAKI, A., 389
MO, G.H., 389
MORELLO, C., 389
MOROMISATO, J., 389
NANIA, R., 389
NAVARRA, G., 389
NAVARRIA, F.L., 229
O'SHEA, V., 389
PALMONARI, F., 389
PAVER, N., 39
PELFER, P., 389
PERIALE, L., 389
PERONI, C., 283
PEROTTO, E., 389
PETRERA, S., 409
PICCHI, P., 389
PIEMONTESE, L., 485
PINSKY, L., 389
PISTILLI, P., 339
PITAS, A., 389
PLESS, I.A., 389
PREDAZZI, E., 27
PREPARATA, G., 3
PUDDU, G., 123
PUSTERLA, M., 447
PYRLIK, J., 389
QIAN, S., 389
REMIDDI, E., 117
RINDI, A., 389
ROHRBACH, F., 389

ROTELLI, P., 389
RYASSNY, V.G., 389
RYAZHSKAYA, O.G., 389
SAAVEDRA, O., 389
SALETAN, E., 389
SALVADORI, P., 389
SANDERS, D., 389
SANTONICO, R., 377
SARTORELLI, G., 389
SCRIMAGLIO, R., 389
SHAMBROOM, D., 389
SHAPIRO, A.M., 389
SHELDON, W.R., 389
SHEN, P., 389
SHI, Z., 389
SPALLONE, A., 389
SUN, Y., 389
SUSINNO, G., 389
TAKAHASHI, N., 389
TALOCHKIN, V.P., 389
TONELLI, G., 505
TORNAMBE', A., 327
TRINCHERO, G.C., 389
TUNIZ, C., 425
TURTELLI, A., 389
VALLANIA, P., 389
VASILEYEV, V., 389
DEUTSCH, M., 389
VERNETTO, S., 389
VETRANO, F., 389
VIGEZZI, E., 103
VON GOELER, E., 389
VOTANO, L., 389
WADA, T., 389
WANG, S.W., 389
WEINSTEIN, R., 389
WIDGOFF, M., 389
WILLUTZKY, M., 389
WU, Y.R., 389
YAKUSHEV, V.F., 389
YAMAMOTO, I., 389
YUAN, Y.R., 389
ZALLO, A., 389
ZATSEPIN, G.T., 389
ZHAO, C.Z., 389
ZHU, Q., 389
ZICHICHI, A., 389

SUBJECT INDEX

Accelerator based Mass Spectrometry	427
– physics	447
ACD (Anisotropic ChromoDynamics)	7, 8
– analysis of e^+e^- annihilation	13
Adiabatic invariants in accelerator	457
Adler sum rule	16
ALEPH mini-vertex detector	519
Ambiguities in Nucleon-nucleus OMP	132
Antinucleon-nucleon potential	164
– – interaction	480
– –nucleus OMP	142
Antiproton annihilation	163
ARES accelerator	151, 157
Astroparticle physics	425
Asymptotic Freedom	3, 7
B-meson lifetimes	35
BCDMS (NA4) experiment	266
– experiment results	278
Betatron oscillations	453
Bjorken limit	4, 15
– scaling	4, 5, 16
– sum rule	17
Bosonic string theory	62
– – – , quantization procedure in	65
CDF silicon vertex detector	515
Centroid energies	80
CERN UA4 experiment	222
CERN-SPS muon beam-line	259
Chandrasekar mass limit	298

Changes in nuclear shape	103
Charge of quark	238
– partition read-out in silicon detectors	512
Charm decay	27
– –, exclusive	34
– –, hadronization scheme in	29
– –, inclusive	33
– –, non-spectator diagram in	28
– –, spectator diagram in	27
Chew-Frautschi plot	8
CHLORINE experiment	378
Colliding beam accelerators (Colliders)	451
Colourless sea	229
Comparison between scintillator and Cerenkov detectors	323
Construction techniques of hadron calorimeters	488
Cosmic rays at ultra high energy	351
Cosmological principle	304
Covariant reformulation of the hadronization scheme	31
Cyclic accelerator	448
Cyclotron	449
Damping mechanism in hot finite many body systems	113
Darmstadt effect	117
Darmstadton	117
Data analysis in Neutrino observations	321
Deep inelastic scattering	283
	4, 14
Deeply inelastic reactions	39
Diameter of quarks	206
Double beta decay	437
Drell effect	412
e^+e^- annihilation	12
ECM effect	283
–, theoretical ideas in	290
_ experiment	266
_ _ analysis	267
_ _ results	274
Effective beam height	218
– mass approximation	128

Subject Index

— nucleon number	157
Electromagnetic calorimeters	485
— —, energy resolution	492
Electroweak interactions	468
Ellis-Jaffe sum rule	17
Emittance in accelerator	456
Energy released in Gravitational collapse	299
Energy resolution of hadron calorimeters	493
Equivalent potential	129
European Hadron Facility (E.H.F.)	466
Evidence for quark	230
Evolutionary behaviour of Supernova 1987A	333
Exotic decay	108
Explosion mechanism in supernovae	329
Extraction of the structure function in the ECM exp.	273
Fire-strings	10
Fractionally charged particles	436
Froissart bound in hadron total cross-sections	211
GALLIUM experiment	379
Gamma and neutrino astronomy in MACRO experiment	413
Gamma astronomy	344
Λ_b^+	152
Λ_c^+	152
Geometrical scaling in hadron total cross-sections	211
Giant dipole resonance	111
— — —, frequencies of	111
— — —, widths in	115
Gottfried sum rule	17
Gravitational collapse	295, 297
	389
Gravitational experiments	475
Greenlees theory	137
Gross-Llewellyn Smith sum rule	16
Hadron calorimeters	485
— —, compensation in	497
— —, wavelenght shifters in	490
— era in Radiation dominated Universe.	315
— total cross-sections at CERN-ISR	215

– – – – measurement at the CERN pp-Collider 222
– – – – – at a fixed target accelerator 213
– – – – – at collider machine 214
Hadronic Interactions in the Confinement Regime 467
– jets 40
– physics 3
Heavy flavours physics 507
High p_T Physics 21
High-energy hadron total cross-sections 211
Hilbert variational principle 303
Hill equation 455
Hubble's law 304
Hydrogen cycle 377
Impulse approximation 140
Instability of the stellar core 297
Interaction of magnetic monopoles with matter 412
Intermediate vector boson 170
– – – , detection (in the UA1 experiment) 181
– – – , mass measurement 176
– – – , production of 171
– – – , production properties of 194
Isobar Analog State (IAS) 91
Isospin dependence in Nucleon-nucleus OMP 131
Isotensor excitations 99
Isovector excitations 81
– – , monopole mode 91
– – , quadrupole mode 94
– – , dipole mode 87
– resonances 80
Jet angular distributions 49
– cross-sections 46, 52
Kinematics of lepton-nucleon scattering 257
Lepton families 317
Linear accelerator 448
Local OMP 125
Low energy antiproton annihilation 165
Luminosity measurements at the CERN pp-Collider 220
– – at the CERN-ISR 218

Subject Index

LVD detector 392
– experiment 391
MACRO experiment 409
– –, cosmic rays detection in 415
– –, detection of magnetic monopoles in 410, 412
– –, detector of 415
– –, neutrino bursts from stellar collapse in 414
– –, supermassive particle detection 419
Magic momentum in supernuclei production ... 155
Magnetic monopoles 410
– –, bounds in flux and velocity of 411
Matter-Antimatter symmetry 475
Microscopic antinucleon-nucleus OMP 146
– isospin dependent force 84
– nuclear potential 84
– nucleon-nucleus OMP 137
– spectroscopy 80
Microstrip silicon detectors 511
Mini-jets 53
Momentum compaction 458
– sum rule 242
Motion of the beam particle in accelerator 452
Motional narrowing 113
Muon scattering 283
Neutrino astronomy 348
–, ^{51}Cr source 379, 383, 384
– emission 378
– from the Gravitational collapse 300
– observations 320
– –, remarks on 325
– – interpretation 325
– oscillations 379, 385
– physics 469
Neutrinos and gamma rays astronomy telescope ... 340
– – – – – –, DAQ system for 353
– – – – – – EAS detector 342
– – – – – – muon detector 342

– – – – – – , off-line rejection power 353
– – – – – – , μ -tracking and discrim. system in 342
ν -emission, "non-standard" model predictions on 324
– – , "standard" model predictions on 323
Non-local OMP 128
Nuclear cosmology 295
_ shape 103
Nucleon form factors 231
_ structure function 232, 234, 238, 244
............257
............284
– – – , scaling violations 246
– – – , scaling violations analysis 246
Nucleon-nucleus OMP 129
Observational information in supernovae events 328
Octupole tunneling 108
Optical model potential (OMP) 123
Origin of cosmic rays 405
Parity violation in weak interactions 187
Parton model (PM) 229
Partons densities 238
Penning effect 412
Phenomenological Antinucleon-nucleus OMP 143
Point sources 340, 344
Pontecorvo reactions 480
p-p cycle 377
PQCD (Perturbative QCD) 7
– analysis of deep inelastic scattering 14
– – – e^+e^- annihilation. 13
Precocity 5
Production of elements 316
Proton decay 434
QCD in the quark-gluon confinement regime 476
– – – – – – , Colour Physics 476
– – – – – – , Particle and Nuclear Physics 476
– – – – – – , spectroscopy 477
_ parton model 43
QGD (Quark GeometroDynamics) 7, 8

Subject Index

– analysis of deep ineleastic scattering	19
Quark confinement and nuclear physics	479
– – in Radiation dominated. Universe.	314
Quark-parton model (QPM)	229, 230
Radiation dominated Universe	310
– – – , decupling time and temperature	314
– – – , quantum gravity in	313
Radiochemical techniques	378
– – , solar neutrino flux measurement with	378
Random Phase Approximation (R.P.A.)	85
Rare kaon decays	472
Read-out electronics in silicon vertex detectors	513
Robertson-Walker line	305
Role of gluons	242
SAX (Satellite Astronomia X) mission	361, 370
– – , satellite characteristics in	371
– – , scientific objectives in	374
Scaling law in deep inelastic.scattering	4
Scattering in OMP	125
Semiconductor detectors	505
Shadowing	285
Short distance of Hadrónic Physics	3
Silicon detectors	509
– diode structure	510
Solar neutrino experiments	430
– – puzzle	430
Spin Physics in the light of QCD	478
Stacking problems	462
Standard cosmological model	296
	303, 304
Standard model, Probing the	169
– – , extensions and modifications of the	203
– – , comparison of UA1 experimental. results with the	185
– solar model	377
Static quark model (QM)	229
Storage rings	462
String theory	61
Sum rules	83

Superheavy elements 436
Supernova 295
_ 1987A 330
Supernovae 327
Supernuclei 151
_ , binding energy of 153
_ , methods of production 153
Superstring theory 62
- - , compactification in 69
- - , conformal symmetry in 73
Supersymmetric boson string theory 68
Syncrocyclotron 449
Syncrotron 450
_ oscillations 459
Tagging short-living particles 506
TANDEM accelerator 447
Thermodinamic of the Universe 308
Time scale of the Gravitational collapse 299
TOP, search for (in the UA1 experiment) 200
Total strength 80
Transmission experiment in a fixed target accelerator 213
Trigger system for an extensive air shower 351
Type of supernovae 329
_ Ib supernovae 336
- II supernovae 396
UA1 detector 177
_ experimental results 169
Valence quarks 242
Van de Graaff accelerator 447
Vertex detectors 508
Wilson model 402
X-ray astronomy 361
_ _ _ missions 363
_ _ _ observations 364
_ _ extra-galatic sources 368
_ _ galatic sources 366

JAN 0 4 1990